CHEMICAL ENGINEERING METHODS AND TECHNOLOGY SERIES

BULK MATERIALS: RESEARCH, TECHNOLOGY AND APPLICATIONS

Chemical Engineering Methods and Technology Series

Treatment of Tannery Effluents by Membrane Separation Technology
Sirshendu De, Chandan Das, and Sunando DasGupta
2009. ISBN: 978-1-60741-836-8

Handbook of Hydrogels: Properties, Preparation & Applications
David B. Stein (Editor)
2009. ISBN: 978-1-60741-702-6

Handbook of Hydrogels: Properties, Preparation & Applications
David B. Stein (Editor)
2009. ISBN: 978-1-61668-167-8 (Online Book)

Non-Ionic Surfactants
Pierce L. Wendt and Demario S. Hoysted (Editors)
2009. ISBN: 978-1-60741-434-6

Handbook of Membrane Research: Properties, Performance and Applications
Stephan V. Gorley (Editor)
2010. ISBN: 978-1-60741-638-8

Chromatography: Types, Techniques and Methods
Toma J. Quintin (Editor)
2010. ISBN: 978-1-60876-316-0

Green Composites: Properties, Design and Life Cycle Assessment
François Willems and Pieter Moens (Editors)
2010. ISBN: 978-1-60741-301-1

Environmentally Harmonious Chemistry for the 21st Century
Masakazu Anpo and K. Mizuno (Editors)
2010. ISBN: 978-1-60876-428-0

Bulk Materials: Research, Technology and Applications
Teodor Frías and Ventura Maestas (Editors)
2010. ISBN: 978-1-60692-963-6

CHEMICAL ENGINEERING METHODS AND TECHNOLOGY SERIES

BULK MATERIALS: RESEARCH, TECHNOLOGY AND APPLICATIONS

TEODOR FRÍAS
AND
VENTURA MAESTAS
EDITORS

Nova Science Publishers, Inc.
New York

For permission to use material from this book please contact us:
Telephone 631-231-7269; Fax 631-231-8175
Web Site: http://www.novapublishers.com

LIBRARY OF CONGRESS CATALOGING-IN-PUBLICATION DATA

Bulk materials : research, technology, and applications / editors, Teodor Frías and Ventura Maestas.
p. cm.
Includes bibliographical references and index.
ISBN 978-1-60692-963-6 (hardcover : alk. paper)
1. Bulk solids. 2. Granular materials. I. Frías, Teodor. II. Maestas, Ventura.
TA418.78.B85 2009
620.1'1--dc22
2009046139

Published by Nova Science Publishers, Inc. ✝ New York

CONTENTS

PREFACE

Due to the effect of "colossal" magnetoresistance, bulk lanthanum manganites have become one of the most fashionable objects of investigations in solid state physics. This book illustrates that at radio frequencies, the magnetic permeability exceeds unity far above the temperature of the magnetic phase transition. Furthermore, it is demonstrated that radio-frequency properties of bulk lanthanum manganites can be useful for magnetoelectronic applications: sensors of static and high-frequency electromagnetic fields, nonlinear devices with magnetic field modulation. In addition, lightening structural components is becoming more and more important for a lot of applications, particularly in the automotive and aerospace fields. In this book, a review of several recent studies about the effect of surface treatments on the fatigue and contact and rolling fatigue behavior of different bulk materials is reported. Moreover, the chemical routes to elemental selenium and a number of metal chalcogenide materials are reviewed. Some of the synthesized materials characterized by their structural aspect and also their opto-electrical properties and sensing potentials were investigated. Other chapters investigate the advantageous applications of mercury cadmium tellurium in the military engineering and space technology, the electrical and mechanical properties for bulk polycrystalline materials, and the impact of the effective-mass theory on nanocrystal structures.

Chapter 1 - Due to the effect of "colossal" magnetoresistance lanthanum manganites have become one of the most fashionable objects of investigations in solid state physics. Most publications devoted to investigations of physical properties of doped lanthanum manganites report about the results obtained by static (under the *dc* excitation current or at frequencies below 1 kHz) methods. Investigations of properties of doped manganites in the radio-frequency range (from 10 kHz to 100 MHz) received much less attention. Meanwhile in the case of manganites the range of radio frequencies can be of particular interest since the magnitude of effects observed can be much higher than under *dc* current excitation. The value of skin depth can be tailored by variation of frequency and can be comparable with the size of a sample. In such case different mechanisms of interaction between an electromagnetic wave and spin subsystem of a sample under investigation can be analyzed in detail. When the thickness of a sample exceeds the value of skin depth variations of the penetration coefficient in the *dc* magnetic field are mainly determined by variation of impedance of a sample. Variations of impedance are caused by resistivity and dynamic magnetic permeability. Therefore, important information about dynamic magnetic properties of doped manganites can be obtained from investigations of penetration of electromagnetic field, especially near the temperature of magnetic phase transition.

In manganites the magnetic susceptibility measured at low frequencies in the *dc* magnetic field abruptly decreases above the temperature of magnetic phase transition. We show that at radio frequencies the magnetic permeability exceeds unity far above the temperature of the magnetic phase transition. This result can be interpreted within the phase separation scenario as preservation of local short-range magnetic interaction in doped manganites in the paramagnetic range of temperatures.

It is demonstrated that radio-frequency properties of bulk lanthanum manganites can be useful for magnetoelectronic applications: sensors of static and high-frequency electromagnetic fields, nonlinear devices with magnetic field modulation.

Chapter 2 - Manganese is the cheapest alloying element with high hardening effect commonly used in wrought steels. Nevertheless, manganese was as an alloying element studied in powder metallurgy since ~1948 mainly under laboratory conditions. The hard reducibility of Mn oxides due to high affinity of manganese for oxygen, especially of MnO, requiring the purity of the sintering atmospheres not attainable in practice, is regarded for a technical problem. In this chapter are given the data (oxygen partial pressure, dew points of the atmosphere) required for the thermodynamic equilibrium conditions for Mn – O system. Presented mechanical and other properties of manganese steels attained prove successful sintering. The used sintering atmospheres for these steels did not fulfill the thermodynamic requirements for the purity. Material and processing factors applied are summarized. Some representative maximal tensile strength values attained are presented. High vapor pressure characterizes manganese, and its values in dependence on temperature are presented demonstrating in a spontaneous sublimation. Effect of manganese vapor on sintering process and alloying of iron powder matrix in the compacts is explained. The reaction of gaseous manganese with the oxygen in the atmosphere is characterized. The result of this is the reduction ("self-cleaning/protection" effect) of manganese for sintering atmosphere to equilibrium conditions for Mn – O system. This process, given by physical property of manganese, is only possible to ensure the effective sintering of manganese containing steels without other special measures.

Chapter 3 - Lightening structural components is becoming more and more important for a lot of applications, particularly in the automotive and aerospace fields. From this point of view, the choice of the most effective surface treatments can play a fundamental role in the enhancement of the fatigue resistance, and hence in the mass reduction, of mechanical components. In this chapter, a review of recent studies on the effects of surface treatments on the fatigue and contact fatigue behaviour of different bulk materials is reported. Fatigue and contact fatigue tests on treated specimens and components made it possible to collect and analyze a large amount of data. Theoretical-numerical models were also developed to verify and justify the experimental results and to predict the fatigue behaviour of surface treated components. Materials ranging from structural steels to light alloys, such as titanium and aluminum ones, are considered. With regard to the surface treatments, particular attention is paid to physically deposited thin hard coatings - i.e. with PVD (Physical Vapor Deposition) and CVD (Chemical Vapor Deposition) techniques. CrN, TiN, WC/C, DLC are some of the coatings which have been studied. Furthermore, the more traditional shot peening treatment is also discussed at the end of the chapter.

Thin hard coatings deposited with PVD and CVD techniques are, for the time being, used over an increasing number of structural applications. They were first introduced in the field of cutting tools and their use has been extended also to the oil, aerospace, transport and biomedical industries. The improvement of the tribological and corrosion-resistant properties of thin hard-coated components has been demonstrated by a number of studies and good results for the fatigue behaviour have been collected too. The surface residual stress field induced by the deposition process is supposed to be one of the main parameters affecting the fatigue behaviour. Both experimental and numerical studies on thin hard-coated components have been carried out. The presence of the cited residual stresses was simulated with finite element models which allowed to foresee the fatigue life of coated components. The features of the fracture surfaces of coated samples, including the possible modifications introduced by the surface treatments, can be highlighted with microscope observations. A collection of scanning electron microscope (SEM) micrographs is shown in the present chapter. The shot peening treatment has been studied with numerical models and by applying the statistical method Design of Experiments (DoE) to optimise the treatment parameters. The proposed procedure could be used to select the best shot peening parameters for each application and bulk material.

Chapter 4 - The chemical routes to elemental selenium (both the amorphous, *i.e.* red and hexagonal, *i.e.* gray allotropic modifications) and a number of metal chalcogenide materials (such as Bi_2Se_3, CuSe, Cu_3Se_2, NiO, Co_3O_4, Fe_2O_3) which have recently been developed by our group are reviewed. Some of the materials, in particular the elemental Se and Cu_3Se_2, were synthesized for the first time by the chemical deposition method implementing these routes. The developed chemical methods are controllable in the sense that they enable synthesis of the mentioned materials with predefined (*i.e.* designed) properties, as both bulk precipitates and also in the form of thin films on various substrates. The synthesized materials were characterized from structural aspect and also their opto-electrical properties and sensing potentials were investigated. All synthetic approaches were based on controlled homogeneous precipitation reactions in aqueous solutions. Elemental selenium and metal selenides were synthesized using aqueous solution of selenosulfate ions as a precursor of the selenide ionic species, while the synthesis of metal oxides was essentially based on thermally-induced hydrolysis of urea (carbamide) in aqueous solutions. Measurements of the optical and thermal band gap energies of the synthesized Bi_2Se_3 enabled fundamental conclusions concerning the material's microstructure to be derived and also a higher bound to the Bohr's excitonic radius for bulk bismuth(III) selenide to be estimated for the first time in the literature. Optoelectrical investigations of the two copper selenides (CuSe and Cu_3Se_2) indicated the fundamental differences in the bulk band structures of these semiconductors. The synthesized elemental selenium, aside from its substantial engineering importance in xerographic technologies, was shown to exhibit a highly pronounced tendency towards conversion to other metal selenides by very simple chemical treatments. Chemically deposited α-phase of Fe_2O_3, especially when this material is synthesized in thin film form, was shown to exhibit a pronounced sensitivity of its electrical resistivity to moisture. Calibration of the dependence of the α-Fe_2O_3 thin films' resistivity on the moisture content in the atmosphere indicated potential applicability of this material in moisture sensing systems. Important information concerning the reaction pathways to nickel and cobalt hydroxide and oxide phases from the "urea system" has been derived by structural analyses of as-deposited and thermally treated materials ($Ni(OH)_2$, $Co(OH)_2$, NiO and Co_3O_4).

Chapter 5 - Combinatorial friction materials are novel brake friction composites defined as the raw materials used were optionally selected according to scientific criteria using a combinatorial approach and the multi-component friction material formulations were optimized by a optimization technique including design of friction material formulations by a Golden Section approach, relational grade analysis of the relationships between friction performance and compositions, and comprehensive evaluation of friction performance. The combinatorial friction materials research (CFMR) was proposed and performed to solve the critical problems related to automotive friction materials to meet the requirements of brake industry including mainly (1) to screen raw materials with good friction performance and find out the interaction effects among ingredients by a combinatorial approach; (2) to design friction material formulation by a Golden Section approach; (3) to optimize friction material formulations by statistical analysis and multi-objective optimization method; (4) to characterize the structural development of friction layer formed on the friction surface and explain the friction mechanisms of friction materials during the brake process; and (5) to minimize the environment impact and development of eco-friendly friction materials. The objective of the CFMR is to shift the friction materials research from experience to science.

Chapter 6 - For the bulk polycrystalline materials, the size of microstructures such as the grain size shrinking to the nanometer scale will result in an apparent change of the physical and mechanical performance of materials, which are distinguished remarkably with the ones of coarse-grain polycrystalline materials. In order to understand the influence of grain size on the properties of polycrystalline materials completely, the theoretical models with the point of view of the microstructures are presented to analyze grain size effects on the phonon thermal conductivity, electrical conductivity and the elastic modulus. For the phonon thermal conductivity, the phonon transport in both the grain interior and grain-boundary are discussed by using the phonon radiative transfer equation, and the analytical expression of conductivity is addressed to describe the influence of grain size on the thermal properties of polycrystalline materials. The electron grain-boundary scattering in the polycrystalline materials is taken into account as the boundary conditions of electron transport in the grain interior, to investigate the grain size effect on electrical conductivity. The predictions of proposed models for the phonon thermal conductivity and the electrical conductivity can agree well with experimental results. For elastic properties of nanocrystalline materials, a framework based on the continuum theory of mechanics is developed by considering the interface energy effects and grain-boundary effects to describe grain-size effects on elastic modulus. Base on the numerical results, it is illustrated that the effects of interface energy and grain-boundary can not be considered for large grain size. When the grain size is under several tens nanometers, the grain-size effects are significant and the effects of interface energy and grain-boundary must be taken into account.

Chapter 7 - As a most important material for the infrared photodetection, mercury cadmium tellurium (HgCdTe) has advantageous application in the military engineering and the space technology. With the development of the infrared detection technology, the knowledge about the properties of HgCdTe is not just confined to the empirical parameters, but extends to the physical mechanism of the crystal growth and even device performance based on the quantum theories. Some important issues for the HgCdTe, such as the evaluation of mercury vacancy (V_{Hg}) and *p*-type doping technical *et al.*, have been investigated from the first-principles calculations.

In the mercury vacancy of $Hg_{1-x}Cd_xTe$, the effects of the mercury vacancy on the electronic structure have been shown by the structural relaxation, local charge density, and densities of states. It is indicated that the nearest neighbor Te around the Hg vacancy defect contract toward to the defect site. It is attributed to the concentration of the electrons around the defect accompanying the dangling bond reconstruction, which induces extra Coulomb interactions between the defect site and the nearest neighbor Te. The Jahn-Teller distortion has been also found in some mole fraction and some disordered configurations of HgCdTe with mercury vacancy. The double acceptor levels introduced by the vacancy are determined in agreement well with the experimental results.

There are the amphoteric behaviors of arsenic *in situ* impurities in $Hg_{1-x}Cd_xTe$. It is found that the Jahn-Teller distortion gives rise to the change of the degenerate situation of the impurity levels. The amphoteric behaviors of arsenic *in situ* impurities in HgCdTe have been proved based on basic theoretical methods. At the same time, we have found that the impurity has the covalent bonding characteristics with host atoms. Based on the calculated formation energy, the chemical trends of the *in situ* arsenic impurity between As_{Te} and As_{Hg} in $Hg_{1-x}Cd_xTe$ have been studied systematically.

The understanding of the microstructures of the arsenic complex in $Hg_{1-x}Cd_xTe$ is important to explain the high electrical compensation of molecular beam epitaxy (MBE) samples and the conversion to *p*-type behavior. According to the defect formation energy and the binding energy, we predicted the most probable configurations for the absorption of As_4 and As_2 cluster defects on the grown surface. By the comparison with the experimental results, the arsenic dimer combined with the mercury vacancy is believed to account for the highly electrically compensated *n*-type characteristics in arsenic-doped $Hg_{1-x}Cd_xTe$ by MBE.

Chapter 8 - Investigation of crystal structure, dielectric, magnetic and local ferroelectric properties of the diamagnetically- substituted $Bi_{1-x}A_xFeO_{3-x/2}$ (A= Ca, Sr, Pb, Ba; x= 0.2, 0.3) polycrystalline samples has been carried out. It has been shown that the heterovalent A^{2+} substitution result in the formation of oxygen vacancies in the host lattice. The solid solutions have been found to possess a rhombohedrally distorted perovskite structure described by the space group *R3c*. Piezoresponse force microscopy have revealed signs of existence of the ferroelectric polarization in the samples at room temperature. Magnetization measurements have shown that the magnetic state of these compounds is determined by the ionic radius of the substituting elements. A- site substitution with the biggest ionic radius ions has been found to suppress the spiral spin structure of $BiFeO_3$ giving rise to the appearance of room-temperature weak ferromagnetism. The result indicates possible way for improving multiferroic properties of $BiFeO_3$- based magnetic ferroelectrics.

Chapter 9 - In 2005, with a group of colleagues, we reported original results obtained in experimenting with laser-processed high-quality Sr_2RuO_4 crystals. These results have been treated as evidence of high-temperature superconductivity at temperatures as high as 200-250K (*cond-mat/0509313*). Both four-probe resistive transition and diamagnetic response have been reported. At the next stage (*arXiv:0705.0641*) resistive transition was documented simultaneously via four- and two-probe techniques; in the same publication it was reported that the composition of samples is more complex than it originally had been conceived: in addition to *Sr*, *Ru* and *O* it contained *Fe*, *Ni*, and *Cr*. This fact, at the time of its finding (2006) was very disappointing since the ferromagnetic compositional elements were regarded not in favor of superconductivity. The discovery of iron-based superconducting materials and the raising of their critical temperatures up to *55K* within a year or so have been quite encouraging. Moreover, in one of the iron-based compositions (*arXiv:0904.0772*) three of the four components are *Sr*, *Ru* and *Fe*! In consideration of the cross-analysis between these results it is suggestive to pursue the ways of reaching the room-temperature limit of T_c. To facilitate it, we summarize here our findings.

Chapter 10 - The k·p method is particularly appropriate to study many physical properties of semiconductors which depend on a small region of **k** space. Since it was proposed, the **k·p** effective-mass envelope function theory has been extensively used in all kinds of semiconductor structures, such as bulk materials, heterojunctions, superlattices and even nanostructures. In this section the energy levels in the conduction band and valence bands of narrow-gap semiconductor is studied in the framework of the eight-band effective-mass method. Especially the eight-band effective mass Hamiltonian is written in the spherical and cylindrical coordinates in order to study the electronic structures of quantum dots and wires. The interaction between conduction and valence bands has been treated exactly, and a finite spin-orbital coupling effect is taken into account. This method is particularly adapted to such problems as arise in connection with the electric and magnetic properties in the external fields. The eight-band Hamiltonian of an electron or hole in a diluted magnetic semiconductor, in the presence of an external homogeneous magnetic field, is given using a mean-field approximation. We applied the effective-mass theory to the electronic structures of semiconductor quantum dots and quantum wires, and discussed some physical properties in these nanocrystal structures. Exciton transition behaviors, polarization properties of emission, Landé electron g factors, and giant Zeeman splitting effect were discussed, and our theoretical results are compared with the available experimental data.

In: Bulk Materials: Research, Technology and Applications
Editors: T. Frias, V. Maestas, pp. 1-74
ISBN: 978-1-60692-963-6

Chapter 1

RADIOFREQUENCY ELECTRODYNAMIC PROPERTIES OF BULK LANTHANUM MANGANITES

***A. Rinkevich*[*] *and A. Nosov*[**]**
Institute of Metal Physics, Ural Division of the Russian Academy of Sciences,
18 Sofia Kovalevskaya St., GSP-170, Ekaterinburg 620041, Russia.

ABSTRACT

Due to the effect of "colossal" magnetoresistance lanthanum manganites have become one of the most fashionable objects of investigations in solid state physics. Most publications devoted to investigations of physical properties of doped lanthanum manganites report about the results obtained by static (under the *dc* excitation current or at frequencies below 1 kHz) methods. Investigations of properties of doped manganites in the radio-frequency range (from 10 kHz to 100 MHz) received much less attention. Meanwhile in the case of manganites the range of radio frequencies can be of particular interest since the magnitude of effects observed can be much higher than under *dc* current excitation. The value of skin depth can be tailored by variation of frequency and can be comparable with the size of a sample. In such case different mechanisms of interaction between an electromagnetic wave and spin subsystem of a sample under investigation can be analyzed in detail. When the thickness of a sample exceeds the value of skin depth variations of the penetration coefficient in the *dc* magnetic field are mainly determined by variation of impedance of a sample. Variations of impedance are caused by resistivity and dynamic magnetic permeability. Therefore, important information about dynamic magnetic properties of doped manganites can be obtained from investigations of penetration of electromagnetic field, especially near the temperature of magnetic phase transition.

In manganites the magnetic susceptibility measured at low frequencies in the *dc* magnetic field abruptly decreases above the temperature of magnetic phase transition. We show that at radio frequencies the magnetic permeability exceeds unity far above the temperature of the magnetic phase transition. This result can be interpreted within the phase separation scenario as preservation of local short-range magnetic interaction in doped manganites in the paramagnetic range of temperatures.

---

[*] Corresponding authors: E-mail: rin@imp.uran.ru
[**] E-mail: nossov@imp.uran.ru

It is demonstrated that radio-frequency properties of bulk lanthanum manganites can be useful for magnetoelectronic applications: sensors of static and high-frequency electromagnetic fields, nonlinear devices with magnetic field modulation.

INTRODUCTION

Starting from 1990 doped lanthanum manganites attract considerable interest of scientific community. This class of oxide materials has perovskite structure and their general chemical formula can be expressed as $R_{1-X}M_XMnO_3$, where, for example, R = La, Pr, Nd, and M = Ca, Sr, Ba, Pb. These oxides were known from 1950 when the general regularities of variations of their physical properties were described [1-4]. The burst of novel investigations of these materials was stimulated in 1988 by discovery of the effect of giant magnetoresistance (GMR) in metallic multilayers with antiferromagnetic interlayer coupling [5]. Subsequent intense researches of fundamental physical processes, which define magnetic and transport properties of 3*d* ferromagnetic-based low-dimensional metallic nanostructures, resulted in discovery of thickness oscillation of noncollinear magnetic coupling [6], spin valves [7], spin transistors [8], and magnetic tunnel structures [9]. Additional attention was paid to all materials in which application of magnetic field resulted in variations of resistivity. This effect was also found in oxide ferromagnetic. Strictly speaking these properties were reported in the first publications devoted to investigations of physical properties of doped manganites [2]. However for some doped lanthanum manganites the magnetoresistance was found to be so strong (in comparison to the values observed in metallic materials based on 3*d* ferromagnetic elements) that this phenomenon was called "colossal" magnetoresistance (CMR) [10]. It is worth noting that substantial values of magnetoresistance were observed in simple oxides even earlier. For example, in nonstoichiometric EuO variations of resistivity under application of *dc* magnetic field can reach nine orders of magnitude in the vicinity of the Curie temperature T_C [11]. However, this effect is observed at low (≈100 K) temperatures. Similar temperature dependence of CMR was found for manganites. Their principal advantage is that by proper choice of doping cation it is possible to obtain compositions which have the values of T_C by ≈100 K exceeding the room temperature and substantial magnetoresistance in a wide temperature range [12].

For the GMR sensors based on 3*d* metallic nanoheterostructures it was shown [13] that the maximum value of magnetoresistance is defined by the value of spin polarization of carriers at the Fermi level of the ferromagnetic material which does not exceed 40% for iron, cobalt, or nickel [14]. One can overcome this fundamental limitation by the choice of materials with higher value of spin polarization. This fact initiated search for new materials of such type, which are known as semimetallic ferromagnets [15]. It was found that in doped manganites the value of transport spin polarization of carriers can reach about 80% [14, 16]. Therefore, doped manganites are very attractive as materials for injection of spin polarized carriers in spintronics [17].

After burst of interest to doped manganites in the middle of 1990-th numerous publications were devoted to investigations of their physical properties (see, for example the reviews [12, 18-22]). However, up till now the fundamental questions about physical mechanism of colossal magnetoresistance remain without answer. One can state that doped manganites belong to the class of strongly correlated materials with rich phase diagram and very complicated interactions between the charge, spin, and orbital subsystems. That is why such new phenomena as magnetic polaron and charge ordering [20] were found in manganites. As a rule only general tendencies of variations of static magnetic and transport properties can be explained within the existing theoretical models. Historically the double exchange model was proposed as the first one [23]. Later the polaron [0.20] and phase separation [18] theories were suggested.

In recent years several theoretical works, which permit to understand the nature of carriers and kinetics of transport phenomena in doped manganites more clearly, were published [24, 25]. The results of *ab-initio* calculations of band structure and Fermi surface of the $La_{2/3}A_{1/3}MnO_3$ (A=Ca, Ba, Sr) manganites showed [25] that it has both hole and electron sheets, which was experimentally confirmed in [26]. Within the framework of relatively simple phenomenological model the mechanism of CMR phenomenon was explained in [27]. This approach is based on the concepts from physics of disordered systems with an account for movement of a mobility edge which result in strong variations of resistivity in the vicinity of the magnetic phase transition temperature.

Among the numerous methods of investigations of physical properties of doped manganites the dynamic electromagnetic technique is of particular interest. It permits to monitor the response of the spin subsystem to the external perturbations in a wide frequency range. For doped manganites the techniques of ferromagnetic resonance and antiresonance at microwave frequencies were most widely used [28-30]. At the same time investigations of doped manganites at radiofrequencies received much less attention. This frequency range is of particular interest since at these frequencies the effects observed have are much more pronounced than at low frequencies. Owing to variation of frequency the value of skin depth can be comparable to the size of a sample. This permits to monitor the smooth transitions between different mechanisms of interaction of electromagnetic wave with spin subsystem of a material under investigation. When we started our investigations, only one publication [31] was known in which the properties of doped manganites at radiofrequencies (335 kHz and 10 MHz) were described. Strong influence of the static magnetic field on magnetic permeability was found and a simple *dc* magnetometer was described.

This chapter contains the results of investigations of the dynamical electromagnetic properties of doped lanthanum manganites. This permit to get important information about the mechanism of metal-dielectric transition and the nature of interaction of high-frequency electromagnetic field with the spin subsystem of manganites. The experimental investigations were carried out in a wide frequency range. The specific peculiarities of skin effect in the oxide half-metallic ferromagnets were revealed and their theoretical description was given. Bulk polycrystalline manganites with compositions $R_{1-X}M_XMnO_3$, where R = La, Y, M = Ca, Sr, Ba, Pb, and $0{,}25 \leq X \leq 0{,}60$ were studied.

In general, doped lanthanum manganites are of great interest from the fundamental point of view as the objects in which the effects of doping on formation of the principal magnetic state in the oxide half-metallic ferromagnets can be investigated. Moreover, they can be considered as the advanced materials for injection of carriers with high degree of spin polarization for spintronic applications.

1. Dynamical Properties of Bulk Doped Manganites at Radiofrequencies

Most investigations of physical properties of doped manganites were carried out using static (*dc* current or *ac* current with frequencies below 1 kHz) methods. Investigations of manganites in the radiofrequency range (from tens of kilohertz to hundreds of megahertz) received much less attention. When we started our investigations only one work was published [31]. Meanwhile, the range of radiofrequencies is of particular interest since for this frequency range the effects observed have much greater values than under static excitation conditions. Owing to variations of frequency the skin depth can become comparable with the dimensions of a sample. Under conditions when the thickness of a plate-shaped manganite sample is less than a skin depth, variations of penetration coefficient in a magnetic field are defined by variations of impedance of a plate. Variations of impedance arise because of changing in resistivity and dynamic magnetic permeability. Therefore, important information about the magnetic subsystem of material can be obtained from investigations of penetration of radiofrequency electromagnetic field in the temperature range of magnetic phase transition.

This section contains the results of investigation of interaction of radiofrequency electromagnetic field with bulk polycrystalline doped manganites of various compositions. The experiments were carried out in the frequency range from 20 kHz to 900 MHz and the temperature interval which included the region of the ferromagnetic-to-paramagnetic transition.

1.1. Solution of the Model Problem of Penetration of Electromagnetic Field through a Plate of a Conductive Ferromagnetic Material under Conditions of Normal Skin-Effect

Experiments on investigations of interaction of doped manganites with electromagnetic fields and waves of a radio-frequency range have been carried out in the penetration geometry, see the Figure 1. A manganite sample is placed between two identical coils. One is the excitation one since it is connected to the generator of the *ac* voltage, the other is the receiving one. The voltage in this coil is measured by a selective voltmeter. Actually, a manganite sample is used as an element screening the receiving coil from the *ac* electromagnetic field created by the primary coil. In our experimental conditions doped manganites under consideration are conductive and can be either in a ferromagnetic or in a paramagnetic state. Therefore, the value of the signal in the receiving coil will be defined by the frequency, conductivity, and the value of the *dc* magnetic field H since its application can vary the magnetic characteristics of a manganite. Before consideration of experimental results it is interesting to analyze the model problem about penetration of an electromagnetic field through a plate of a conducting ferromagnetic material under the conditions of normal skin-effect.

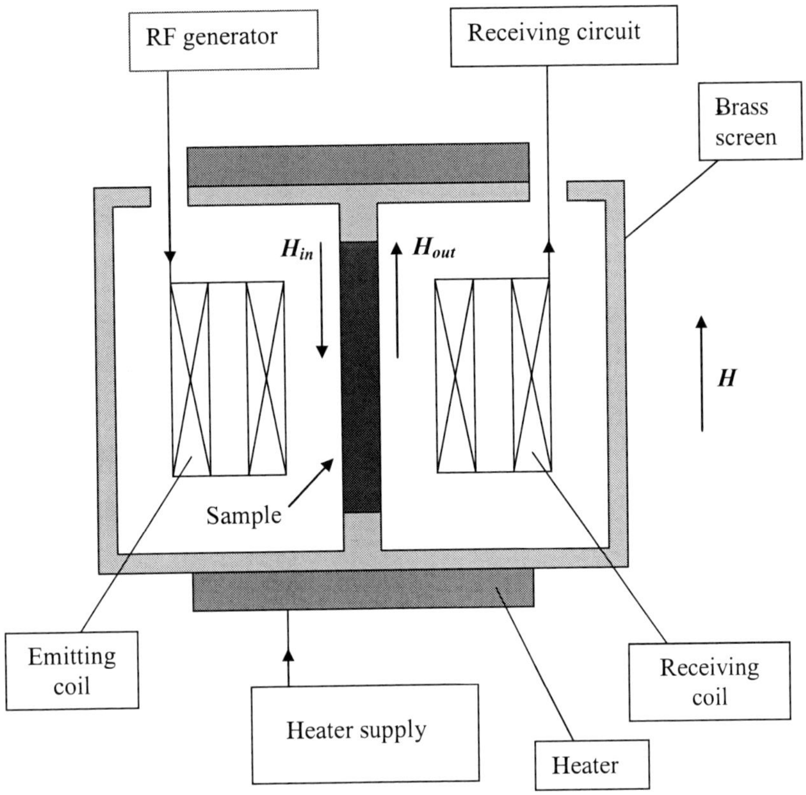

Figure 1. The scheme of experimental setup for measurements by the method of penetration of an electromagnetic field.

At higher frequencies of the microwave range (of the order of units and tens of gigahertz) a ferromagnetic resonance and an antiresonance [32, 33] essentialy influence on the penetration coefficient. We shall consider only much lower frequencies of a radio range. The problem about penetration of a radio-frequency electromagnetic field through a conducting material was extensively considered in the literature. For the theoretical description of the phenomena observed we will use the approximation within which the phenomena of a ferromagnetic resonance and an antiresonance will be neglected, but variations in wide range of the such parameters as frequency, skin- depth, and magnetic permeability will be supposed. We can use the formula for the transmission coefficient through a conducting plate at normal skin-effect which was given in [34]:

$$D = \frac{2Z_m}{2Z_m \,\mathrm{ch} k_m d + Z \,\mathrm{sh} k_m d}, \tag{1.1}$$

where k_m is the wave number a conductive media, $k_m = (1+i)/\delta$, $\delta = (2\rho/\omega\mu\mu_0)^{1/2}$, is the skin depth, μ is the relative dynamic differential magnetic permeability, $\mu = 1 + \chi$, $\omega = 2\pi f$. The transmission coefficient D depends on the ratio between the impedances of a plate Z_m and surrounding space $Z = (\mu_0/\varepsilon_0)^{1/2}$, and also on relation between the thickness of a plate d and skin- depth. The impedance of a plate can be calculated using the formula

$$Z_m = [(1+i) / \delta]\rho . \tag{1.2}$$

For a plate with high conductivity the impedance Z_m is always much less than Z, i.e. the condition $| Z_m | << Z$ is satisfied. Two cases are possible when in a denominator of the expression (1.1) one of the items prevails. The condition $2Z_m \mathrm{ch}\, k_m d >> Z\, \mathrm{sh}\, k_m d$ corresponds to the interval of low frequencies and a case of a thin plate. This case occurs when the following condition is satisfied

$$k_m d << 1, \quad d << \frac{2\rho}{Z}. \tag{1.3}$$

The transmission coefficient can be expressed by the formula

$$D = \frac{1}{\mathrm{ch}\, k_m d},$$

which, under condition $d << \delta$, takes the form

$$|D| \approx 1 - [\ 1/3\ (d / \delta)^4\ ]. \tag{1.4}$$

Relative variation of the module of penetration coefficient

$$r_m = \frac{(D(H) - D(0))}{D(0)} = \frac{\Delta D}{D(0)}$$

can be calculated by the equation

$$r_m = -\frac{1}{12} d^4 \omega^2 \mu_0^2 \left(\frac{\mu^2(H)}{\rho^2(H)} - \frac{\mu^2(0)}{\rho^2(0)} \right) \tag{1.5}$$

As $d << \delta$, one can assume that the value of r_m is small. If magnetoresistance is insignificant, that is $\rho(H) \approx \rho(0)$, then in the case when the initial permeability is substantial $\mu(0) >> 1$ but in saturation fields $\mu(H) \approx 1 << \mu(0)$□□□, variation of the module of the penetration coefficient is defined by the formula

$$r_m \approx \frac{d^4 \mu^2(0) \mu_0^2}{12 \rho^2(0)} \omega^2 \quad \text{for } H \rightarrow \infty. \tag{1.6}$$

Thus, variations of the penetration coefficient through a very thin plate at low frequencies and $\mu(0) >> 1$ are caused mainly by changes of magnetic permeability. They are small, $r_m << 1$, have a positive sign and, in the case of absence of a frequency dispersion of material constants, are proportional to ω^2.

If in the denominator of expression (1.1) $2Z_m$ ch $k_m d$ << Z sh $k_m d$, the transmission coefficient can be expressed the following formula

$$D = \frac{2Z_m}{Z \mathrm{sh} k_m d} = \frac{2(1+i)\rho}{Z\delta} \frac{1}{\mathrm{sh}\left(d/\delta\right)\cos\left(d/\delta\right) + i\,\mathrm{ch}\left(d/\delta\right)\sin\left(d/\delta\right)}. \quad (1.7)$$

For the thin screen under the condition of $d << \delta$ one can obtain from (1.7)

$$D = \frac{2\rho}{Z\mu d}. \quad (1.8)$$

In this limiting case the transmission coefficient is real, if the material constants are also real. From (1.8) the formula for relative variation of the module of the penetration coefficient can be written as

$$r_m = \frac{\Delta D(H)}{D} = \frac{\frac{\Delta\rho(H)}{\rho} - \frac{\Delta\mu(H)}{\mu}}{1 + \frac{\Delta\mu(H)}{\mu}}, \quad (1.9)$$

where $\frac{\Delta\rho\,(H)}{\rho}$ is the relative variation of resistivity in a magnetic field H, $\frac{\Delta\mu(H)}{\mu}$ is the relative variation of magnetic permeability in a magnetic field H. When $\left|\frac{\Delta\rho}{\rho}\right| << 1$ and in high applied fields where $\mu(H \to \infty) \to 1$ one obtains:

$$\frac{\Delta D(H \to \infty)}{D} \approx \mu(0) - 1. \quad (1.10)$$

The formula (1.10) gives the possibility to estimate the initial dynamic magnetic permeability from relative variation of the penetration coefficient of a radio-frequency field. In the same limiting case from (1.8) it is possible to obtain the following formula, which takes into account magnetoresistance of a material of a plate:

$$\left|\frac{D(H \to \infty)}{D(0)}\right| = \mu(0)\left(1 + \frac{\Delta\rho}{\rho}\right). \quad (1.11)$$

We can also analyze the case of a thick conductive plate, when $d >> \delta$. Under such condition from the formula (1.7) one can obtain

$$|D| = \frac{2\rho}{Z\delta\,\mathrm{sh}^2\left(d/\delta\right)} = \frac{8\rho}{Z\delta}e^{-\frac{d}{\delta}}. \tag{1.12}$$

It is obvious, that in this case the penetration coefficient mostly depends on skin-effect. For the external magnetic field exceeding a saturation field it is possible to obtain from the expression (1.12)

$$\frac{D(H\to\infty)}{D(0)} = \frac{\rho(H)\delta(0)}{\rho(0)\delta(H)}e^{-d\left(\frac{1}{\delta(H)}-\frac{1}{\delta(0)}\right)}. \tag{1.13}$$

From this expression, substituting an explicit form of the expression for the depth of a skin-layer under the condition of a normal skin-effect $\delta = (2\rho/\omega\mu\mu_0)^{1/2}$ and assuming $\mu(0) \gg 1$, one obtain

$$\left|\frac{D(H\to\infty)}{D(0)}\right| = \sqrt{\frac{\rho(H)}{\rho(0)\mu(0)}}\,e^{d/\delta(0)}. \tag{1.14}$$

The analysis of penetration of an electromagnetic field through a plate of a ferromagnetic conductor given above besides the restrictions, which were noted above, has also the following limitations. It was supposed, that a plate has the unlimited sizes within a plane. But the real experiments are carried out using a plate-shaped sample of the finite sizes. Besides this, in the theoretical analysis it is supposed, that the radio-frequency field, which falls on a plate, is homogeneous. In the experiment it is created by a non-uniform stray field of a coil. These circumstances affect the absolute value of transmission coefficient. However, when analyzing the field and temperature dependences of the transmission coefficient, the influence of these sources of errors on the value of r_m is much less. Therefore utilization of the simplified analysis presented above is justified for the analysis of experimental results.

1.2. Penetration of Radiofrequency Electromagnetic Field through Bulk Doped Lanthanum Manganites. Investigations at Room Temperature

High-frequency electromagnetic methods are effective in the research of spin subsystem of magnetic materials. The observed variations of high-frequency physical properties, as a rule, have much higher values than that observed at low frequencies [35, 36]. In a case of doped manganites the values of resistivity measured under *dc* and high frequency excitation can differ by two orders of magnitude [35, 37]. It is well-known, that structural and magnetic inhomogeneities of bulk samples considerably affect their kinetic properties. It is considered, that the grain boundaries give essential contribution in conductivity of polycrystalline samples. At high enough frequencies the conductivity is mostly defined by electric properties of the grain material. Besides, as it has been noted above, in the case of high-frequency investigations the relation between the thickness of a manganite sample and a skin depth is important. If a thickness of a sample *d* is greater than the skin depth δ then the surface resistance R_s is proportional to $\rho^{1/2}$ where ρ is resistivity. However, if $d < \delta$ unusual dependences can be realized: $R_s \sim \rho$ or even $R_s \sim \sigma$ ($\sigma = 1/\rho$) [37].

At the first stage it was necessary to investigate the general laws of penetration of an electromagnetic field through the bulk polycrystalline lanthanum manganites. Since skin depth is the fundamental parameter defining occurring processes it was necessary to reveal those physical characteristics which define variations of skin depth changes in doped manganites in a radio-frequency range. With this purpose penetration of an electromagnetic field in the range of frequencies from 20 kHz to 30 MHz through bulk polycrystalline $La_{0.60}Pb_{0.40}MnO_3$ and $La_{0.70}Y_{0.07}Ba_{0.23}MnO_3$ manganites has been experimentally investigated.

Figure 2 shows the *dc* magnetic field dependences of relative magnetoresistance *MR* and the differential magnetic permeability μ_{dif} for the bulk $La_{0.6}Pb_{0.4}MnO_3$ polycrystalline manganite. Magnetoresistance is defined here as $MR(H) = (\rho(H) - \rho(0))/\rho(0) \times 100\%$ where $\rho(H)$ and $\rho(0)$ are the values of resistivity in the *dc* magnetic field *H* and zero magnetic field, respectively. The *MR(H)* and $\mu_{dif}(H)$ dependences were measured with the *dc* magnetic field *H* applied parallel to the plane of samples. In Figure 3 the dependence of a signal in the receiving coil on the intensity of a *dc* magnetic field for the $La_{0.60}Pb_{0.40}MnO_3$ sample is shown. Data were recorded at frequency $f = 150$ kHz. The value of the penetration coefficient increases with *H*. Mutual orientation between the radio-frequency $H_{\sim}$ and *dc* magnetic field *H* strongly influences the shape of the dependences.

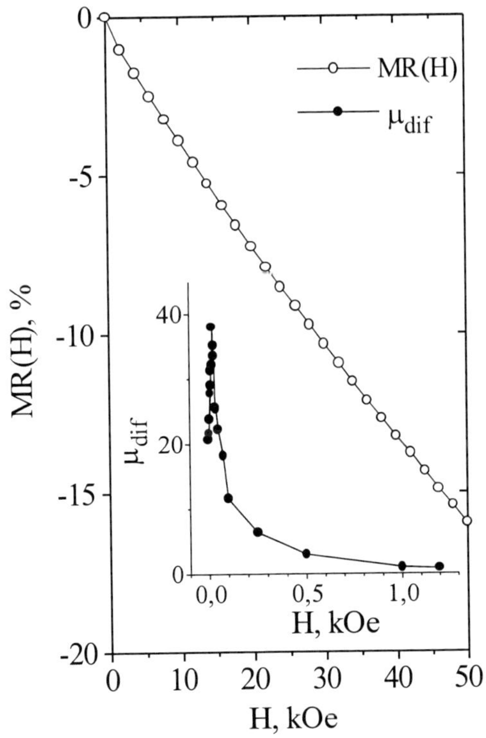

Figure 2. The *dc* magnetic field dependences of magnetoresistance and differential magnetic permeability for the $La_{0.60}Pb_{0.40}MnO_3$ sample.

Measurements were carried out under conditions

$$L<0.5\ \delta\ \frac{1}{(\sigma\delta Z)} << \frac{L}{\delta} < \left|\sinh\left[\frac{(1+i)L}{\delta}\right]\right|.$$

In this case the transmission coefficient D is purely real and is defined by the formula (1.8).

According to (1.8), the frequency dependence of D can be caused only by dispersion of conductivity or magnetic permeability. Comparison of data on magnetoresistance and magnetic permeability, presented in Figure 2, shows, that variations of $\Delta\varepsilon(H)/\varepsilon$ in an external *dc* magnetic field basically are mostly determined by with changes of magnetic permeability. Really, from (1.8) one can expect, that with decrease of ρ under the influence of the *dc* magnetic field the coefficient D should decrease. However, as is seen from Figure 3, an increase of D in a magnetic field is observed. Moreover, under the *dc* magnetic field of 10 kOe resistivity of the $La_{0.60}Pb_{0.40}MnO_3$ composition changes no more than by 3 %. Such changes are too small to explain observable $\Delta\varepsilon(H)/\varepsilon$ dependence.

The differences between the graphs shown in Figure 3, are caused by the dependence of magnetic permeability on mutual orientation between the *dc* and radio-frequency *ac* magnetic fields. When $\boldsymbol{H_{\sim}} \perp \boldsymbol{H}$ penetration of an electromagnetic field is related to the reversible dynamic transverse magnetic permeability $\mu_{\perp rev}$. The value of the recorded signal ε is proportional to $\dfrac{\omega}{\sigma(\omega)\mu_{rev}(\omega)}$. At low frequencies, where the frequency dependence of material constants σ and μ_{rev} is weak, the value of $\Delta\varepsilon/\varepsilon$ should increase with increase of frequency ω. For the $La_{0.68}Y_{0.07}Ba_{0.25}MnO_3$ composition this is illustrated by the data from the range of frequencies 1 kHz - 1 MHz presented in Figure 4.

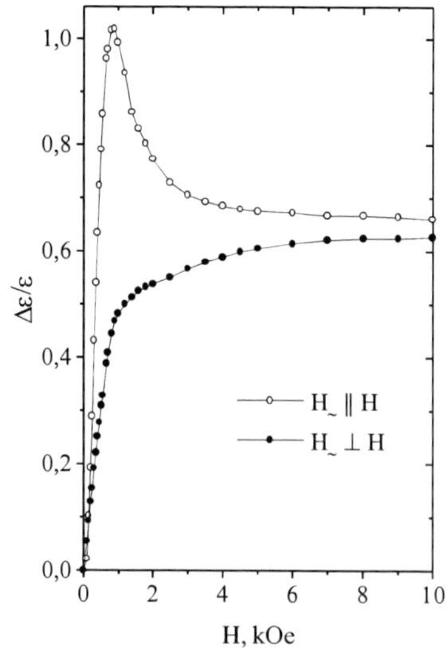

Figure 3. Dependence of a signal in the receiving coil on the strength of *dc* magnetic field for the $La_{0.60}Pb_{0.40}MnO_3$ sample at the frequency of $f = 150$ kHz for various orientations between *dc* and high-frequency *ac* magnetic fields.

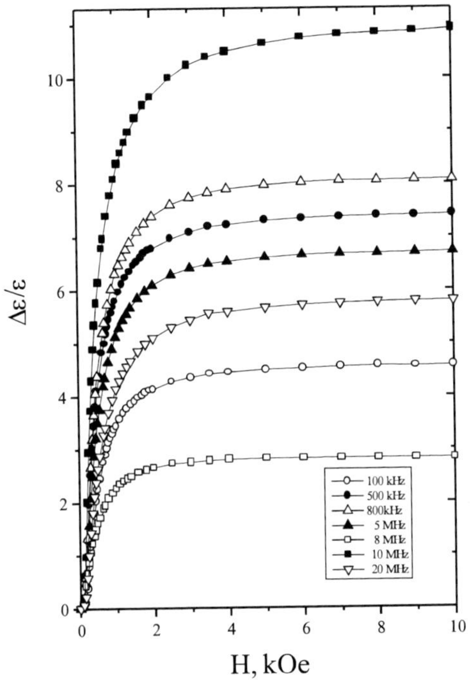

Figure 4. Dependence of the signal in the receiving coil on the strength of *dc* magnetic field for the $La_{0.68}Y_{0.07}Ba_{0.25}MnO_3$ (***H***~⊥***H***).

The dispersion of magnetic permeability becomes essential on higher frequencies. Thus, after the local minimum at f = 8 MHz and the maximum at f = 10 MHz variations of the $\Delta\varepsilon/\varepsilon$ gradually decrease with frequency. Such nonmonotonic variations are probably related to polycrystalline structure of the samples investigated and the response time of the material of intergranular boundaries [35].

The shape of the dependence in weak magnetic fields is also interesting. As it follows from Figure 4, the typical values of sensitivity to a *dc* magnetic field $\partial(\Delta\varepsilon/\varepsilon)\partial H$ in this range are from 0.7 to 2 % per Oersted. This circumstance makes doped manganites attractive materials for potential applications as *dc* magnetic field sensors.

Thus, in doped manganites variations of the penetration coefficient upon application of a *dc* magnetic field can be very strong (up to 10 times) and are much higher than the static magnetoresistance. Thus changes of skin-depth are mostly caused by variations of the dynamic reversible magnetic permeability.

1.3. Frequency Dependence of the Penetration Coefficient

As it was shown in section 1.2, upon penetration of a radio-frequency electromagnetic field through bulk polycrystalline doped lanthanum manganites strong variations of signal, which depend on mutual orientation between *dc* and high-frequency *dc* magnetic fields, are observed. It was interesting to investigate the frequency dependence of these variations. In the case of bulk samples conductivity of a material inside the grains essentially differs from that of a material of intergranular boundaries and one can expect occurrence of some specific features of the phenomenon of the penetration, which originate from the polycrystalline structure of the samples. With an account for the influence of percolation phenomenon on the process of penetration of an electromagnetic field [38], occurrence of anomalies in frequency dependence of penetration depth and strong variations of amplitude of the transmitted wave depending on the value of *dc* magnetic field is possible. By present time some results of investigations of magnetoimpedance in doped manganites [39] are available. It was necessary to reveal possible similarity or distinction of physical mechanisms of changes of both magnetoimpedance and penetration of an electromagnetic field, by carrying out measurements in the same range of frequencies using the manganites of identical composition. As shown in [39], in doped manganites occurrence of anomalies in the frequency dependences of impedance are possible.

Bulk polycrystalline $La_{0.67}Pb_{0.33}MnO_3$ and $La_{0.60}Y_{0.07}Ba_{0.33}MnO_3$ manganites were used to solve this problem. Typical dependences of the module of penetration coefficient on *dc* magnetic field and frequency are shown in Figure 5. After sharp increase in weak (≈500 Oe) fields, in strong fields (above 5 kOe) for all frequencies saturation is observed as for parallel ($\vec{H} \parallel \vec{H}_{\sim}$), and for perpendicular ($\vec{H} \perp \vec{H}_{\sim}$) orientations between *dc* and radio-frequency *ac* magnetic fields. In a limit of the highest frequencies the value of *emf* in strong magnetic fields is less than value in a zero magnetic field, i.e. $\varepsilon(0) > \varepsilon(10\ \text{kOe})$ (see the dependences for $f = 30$ MHz in Figure 5a and 5b). For the fixed frequency growth of *emf* in weak fields is less pronounced for parallel orientation, than for the perpendicular one. In the limit of low frequencies the $\varepsilon(H)$ dependence is monotonous and saturates at $H > 5$ kOe. A maximum appears in the $\varepsilon(H)$ dependences as the frequency increases. In strong fields there saturation occurs. The essential point is that in the fields close to saturation, the vale of $\varepsilon(H)$ does not depend on mutual orientation between *dc* and radio-frequency *ac* magnetic fields, that is $\varepsilon(H \to \infty)|_{(\vec{H} \parallel \vec{H}_{\sim})} = \varepsilon(H \to \infty)|_{(\vec{H} \perp \vec{H}_{\sim})}$.

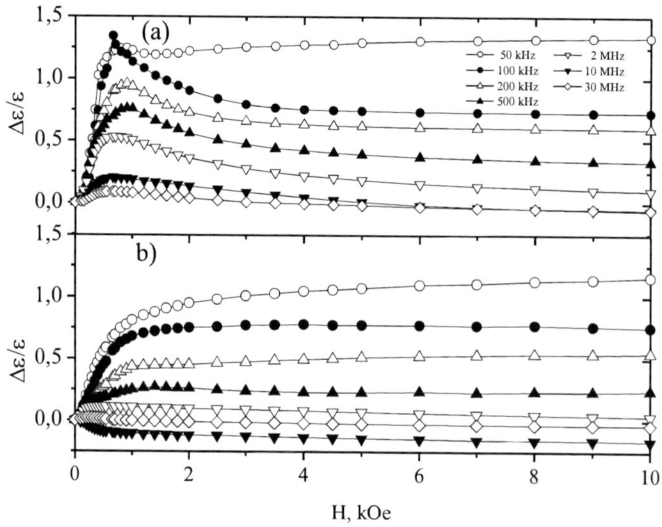

Figure 5. Dependence of *emf* in the receiving on an external *dc* magnetic field and frequency for the $La_{0.67}Pb_{0.33}MnO_3$ manganite at different orientations of *dc* and high-frequency *ac* magnetic fields: a) ($H_{\sim} \parallel H$), b) ($H_{\sim} \perp H$)

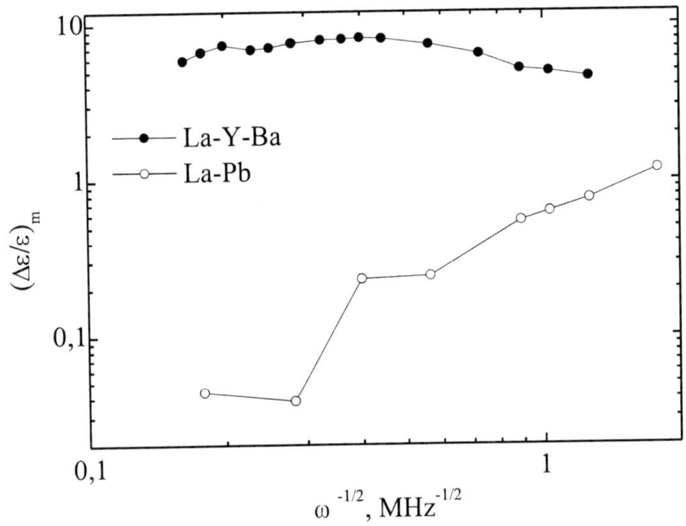

Figure 6. The dependences of $(\Delta\varepsilon/\varepsilon)_m$ on $\omega^{-1/2}$ (see definitions in the text) for the $La_{0.67}Pb_{0.33}MnO_3$ and $La_{0.60}Y_{0.07}Ba_{0.33}MnO_3$ manganites.

A granular system is spatially nonuniform. According to experimental data of [35], the *dc* conductivity in such system is by two orders of magnitude exceeds conductivity in the microwave frequency range. The magnetic permeability, which, alongside with conductivity, defines penetration of a high-frequency electromagnetic field through a manganite, is also frequency dependent. Therefore, variation of *emf* in a *dc* magnetic field $(\Delta\varepsilon/\varepsilon)$ also depends on frequency. In Figure 6 the dependence of the maximum variation of $(\Delta\varepsilon/\varepsilon)_m$ on $\omega^{-1/2}$ is presented for the samples of bulk polycrystalline manganites investigated. Measurements were carried out spent at frequencies where the skin-depth δ essentially exceeded the thickness of the sample d ($\delta>d$). In our case this is true at least for low frequencies. In this case the penetration coefficient of a high-frequency electromagnetic field through a plate from an investigated material is expressed by the formula (1.8).

Taking into account that $\delta=(2\rho/\omega\mu_{rev})^{1/2}$, one can obtain that $\Delta\varepsilon/\varepsilon=2\Delta\delta/\delta$, where $\Delta\delta/\delta$ is the relative variation of penetration depth through a plate of a manganite under investigation. This means, that relative variation of *emf* in a *dc* magnetic field is directly related to the change of penetration depth of a radio-frequency electromagnetic field. For a homogeneous media without frequency dispersion the skin depth is proportional to $\omega^{-1/2}$. For this reason in a horizontal axis of Figure 6 the scale proportional to $\omega^{-1/2}$ is chosen. For the $La_{0.60}Y_{0.07}Ba_{0.33}MnO_3$ manganite variations of $(\Delta\varepsilon/\varepsilon)$ in the considered frequency range are small. However for the $La_{0.67}Pb_{0.33}MnO_3$ manganite the value of skin-depth sharply increases with frequency. Thus can be considered as the proof of appearance of frequency dependence of material constants.

Figure 7 shows relative variations of an impedance $\Delta Z/Z$ of manganites in a *dc* magnetic field, measured at different frequencies. The *dc* magnetic field was applied in a plane of samples. In fields up to 10 kOe relative changes of $\Delta Z/Z$ do not exceed 5×10^{-2}. In the same figure the *dc* magnetic field dependence of magnetoresistance is also shown for comparison. The shape of the $\Delta Z/Z$ dependences at relatively low frequencies (10 and 15 MHz) is similar to that of magnetoresistance measured under application of the *dc* magnetic field. However for the high frequencies of 20 - 30 MHz the shape of the *ΔZ/Z* dependence changes: under application of the *dc* magnetic field the value of impedance first increases, reaches a maximum, and then decreases. The maximum in the *ΔZ/Z* dependences is mostly pronounced at the frequency of $f = 25$ MHz. No saturation in strong fields is observed for all curves. Comparison of Figs. 5 and 7 shows, that the value of magnetoimpedance effect in manganites is defined by other physical mechanisms, than effects of penetration.

The maximum in the $\Delta\varepsilon/\varepsilon$, dependences presented in Figure 5 for the crossed ($\vec{H} \perp \vec{H}_{\sim}$) orientations of fields, is related to the fast that according to [40], the transverse magnetic permeability $\mu_{rev}(\omega,H)$ can be presented as: $\mu_{rev}(\omega,H)=\mu_{DW}(\omega,H)+\mu_{rot}(\omega,H)$. The first item is related to the domain wall movement and monotonously decreases with increasing an external *dc* magnetic field. The second item $\mu_{rot}(\omega,H)$ is related to rotation of magnetization vector. It has a maximum in the fields of the order of anisotropy field and also decreases with increasing an external *dc* magnetic field. In strong magnetic fields should be $\varepsilon(H\to\infty)|_{(\vec{H} \parallel \vec{H}_{\sim})} = \varepsilon(H\to\infty)|_{(\vec{H} \perp \vec{H}_{\sim})}$, which is confirmed by Figure 5b.

From the data presented in Figure 6 it follows, that in the case of La-Pb manganites the strong frequency dispersion of penetration of a radio-frequency electromagnetic field takes place. This effect can be caused by two reasons. The first one is related to the polycrystalline structure of the investigated materials and strong variations of conductivity of crystallite material and intergranular boundaries [35]. According to [38], the frequency dependence of skin depth for such inhomogeneous systems should deviate from the standard $\delta \sim \omega^{-1/2}$ dependence. The second reason is related to the frequency dependence of magnetic permeability. Considering our experimental data it is impossible to select unequivocally one of these reasons as the prevailing one though, in our opinion, the second reason seems to be more essential.

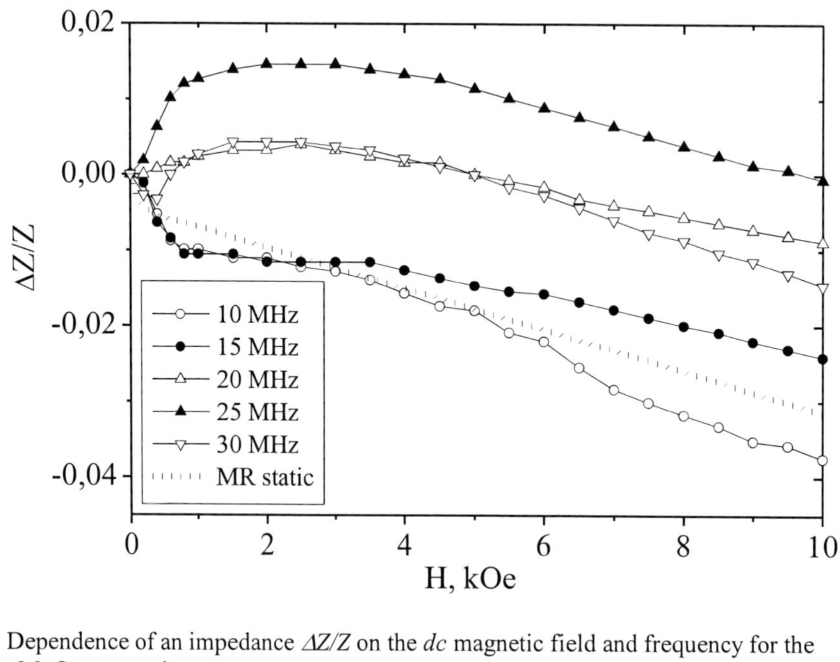

Figure 7. Dependence of an impedance *ΔZ/Z* on the *dc* magnetic field and frequency for the $La_{0.67}Pb_{0.33}MnO_3$ manganite.

Thus, as a result of experimental investigations of bulk polycrystalline Pb and Y-Ba manganites strong frequency dependence of penetration coefficient of radio-frequency electromagnetic radiation through them is established. It is shown, that the reason of this dependence are variations of magnetic permeability, but not of magnetoresistance. In a range of high frequencies it is possible to explain the frequency dependence of penetration coefficient by a frequency dispersion of magnetic permeability. It is shown experimentally, that the value of magnetoimpedance effect in manganites is defined by other physical mechanisms, than the penetration effects.

1.4. Rotation of Polarization Plane and Ellipticity of Radiofrequency Electromagnetic Waves upon Penetration through Bulk Doped Lanthanum Manganites in the Geometry of the Cotton-Mouton Effect

The phenomena of rotation of a polarization plane and ellipticity of electromagnetic waves in optics became the effective tool of research of magnetically ordered media. In principle, these phenomena can be used for investigations of a spectrum of spin or magnetostatic waves, restoration of dynamic tensors of dielectric and magnetic permeabilities and generally to establish relation polarizability and magnetization of media [41]. As a rule, in such researches, which are carried out in an optical spectral range, the media combining a transparency with large values of magnetooptic effects are used. The situation, when the skin depth is substantially larger than the thickness of a sample, is typically realized.

In conductive oxide ferromagnets – doped lanthanum manganites – it is possible to vary conductivity and the Curie temperature by proper doping. Thus in the radiofrequency range an arbitrary relation between skin depth and thickness of a sample can be realized. Investigations of the phenomenon of rotation of polarization plane and ellipticity of electromagnetic waves upon their penetration through doped lanthanum manganites are of special interest and have the specificity. In contrast to ordinary conditions of observation of the Cotton-Mouton effect in magnetically polarized media [41], in the case of a conductive manganite the real and imaginary components of wave number have the same order of magnitude. In principle manifestation of both magnetic birefringence and dichroism phenomena is possible.

In our experiments at lower frequencies of the frequency range investigated the skin depth was found to be much more, and on frequencies in tens megahertz – of the order and less than the thickness of the samples. The frequencies, at which the researches were carried out, have been chosen considerably below the frequencies of a ferromagnetic resonance and an antiresonance. In this case the peculiarities of penetration of radio-frequency electromagnetic field are caused by the processes of domain wall displacement and rotation of magnetization vector.

The experimental setup for radio-frequency measurements is similar to that which is shown in Figure 1. The axis of the receiving coil could be turned around its axis to determine the polarization of a field. The axes of exciting and receiving coils, and, accordingly the vector of *ac* magnetic field $\boldsymbol{H}_{\sim}$, were oriented in the plane of a sample. The *dc* magnetic field $\boldsymbol{H}$ was applied in the same plane. It was always oriented perpendicular to the wave vector $\boldsymbol{k}$. Thus, the geometry of the Cotton-Mouton effect was realized. The value of the *dc* magnetic field was always kept large enough ($|\boldsymbol{H}| >> |\boldsymbol{H}_{\sim}|$), so the magnetic state of a sample was defined only by the static magnetic field $\boldsymbol{H}$. Polarization of the radio-frequency *ac* magnetic field, which passed through the investigated sample, was determined from rotation of an axis of the receiving coil.

Bulk polycrystalline $La_{0.63}Y_{0.07}Ba_{0.30}MnO_3$ manganite samples were used as the objects of investigations. The plate-shaped samples with the thickness of $d = 0.54$ mm were used. For frequency of $f = 1$ MHz and magnetic permeability of $\mu = 1$ the calculated value of skin depth $\delta_l = (2\rho/\omega)^{1/2}$ was found to be $\delta_l = 0{,}8$ mm. Measurements were carried out in the frequency range of $f = 20$ kHz $\div$ 60 MHz.

In general, application of substantially strong *dc* magnetic field resulted in an essential, up to ten times, increase in the amplitude of the transmitted electromagnetic field. The result of measurement of magnetic field dependence of relative variation of *emf* in the receiving coil $\Delta E/E$ is shown in Figure 8 for $\boldsymbol{H}_{\sim} \perp \boldsymbol{H}$ and $\boldsymbol{H}_{\sim} // \boldsymbol{H}$ orientations. $\Delta E/E = (E(H)-E(0))/E(0)$, where $E(H)$ and $E(0)$ are the values of *emf* in the receiving coil in *dc* magnetic field H and zero magnetic field, respectively. The monotonous increase of *emf* E with growth of H up to saturation fields is observed for the perpendicular orientation $\boldsymbol{H}_{\sim} \perp \boldsymbol{H}$. For the parallel orientation $\boldsymbol{H}_{\sim} // \boldsymbol{H}$ the $\Delta E/E$ dependence measured at the frequency of $f = 30$ MHz has a maximum in the field of an order of anisotropy field, then the value of *emf* smoothly decreases. This dependence has the parity property relative inversion of the direction of the external *dc* magnetic field with an insignificant hysteresis.

The experimental dependences of the angle of rotation of polarization plane φ and ellipticity ε on the strength of an external *dc* magnetic field are shown in Figure 9. Ellipticity is defined here as the ratio of smaller to larger semiaxes of polarization ellipse of the transmitted electromagnetic field. Figure 9 presents the results of measurements recorded at the frequency of $f = 20$ MHz and orientation of the exciting coil at the angle of $\theta = -45^0$ relative to the *dc* magnetic field $\boldsymbol{H}$. It is seen, that the largest variations of polarization occur in fields less than 1 kOe, and the values $|\varepsilon| \sim 0.2$ and $\varphi \approx -24^0$ are reached. Close to saturation in stronger fields, the angle of rotation of polarization plane approaches to zero and ellipticity essentially decreases.

Let us introduce the transmission coefficient through a plate-shaped sample as $D = \dot{H}_{\sim}^{out} / \dot{H}_{\sim}^{in}$, where $\dot{H}_{\sim}^{in}$ and $\dot{H}_{\sim}^{out}$ are the complex amplitudes of the fields entering into a plate and leaving it. The *emf* of the coils E is proportional to $E \sim \omega\, \dot{H}_{\sim}^{in}\, |D|\, wS$,, where S and w are the area of cross-section and number of turns of the coil, $\omega = 2\pi f$. Under the condition $\boldsymbol{H} \perp \boldsymbol{k}$ the eigenmodes will be two waves with the distribution of field defined by skin effect. The vector of *ac* magnetic field for one mode is parallel $\boldsymbol{H}_{\sim} // \boldsymbol{H}$, while for another is perpendicular $\vec{H}_{\sim} \perp \vec{H}$ to the vector of an external *dc* magnetic field. For the transmission coefficients of these waves $D_{\parallel, \perp}$ we obtain

$$D_{\parallel,\perp} = \left|D_{\parallel,\perp}\right| e^{i\phi_{\parallel,\perp}} = \frac{2(1+i)}{Z\sigma_{\parallel,\perp}\delta_{\parallel,\perp}} \frac{1}{sht_{\parallel,\perp}\cos t_{\parallel,\perp} + i\sin t_{\parallel,\perp} cht_{\parallel,\perp}}, \quad (1.15)$$

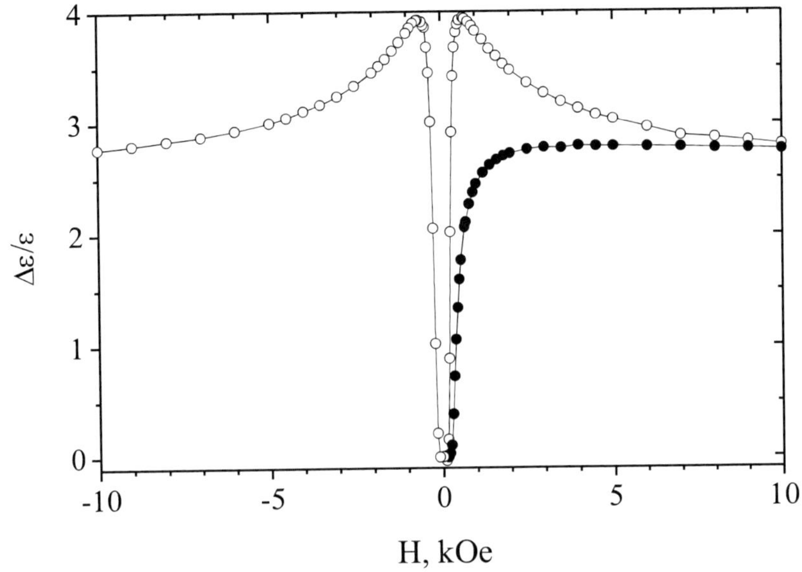

Figure 8. Penetration of an electromagnetic field through a plate from $La_{0.63}Y_{0.07}Ba_{0.30}MnO_3$ manganite. f = 30 MHz. Open circles – data for $\boldsymbol{H} \| \boldsymbol{H}_{\sim}$. Filled circles –data for $\boldsymbol{H} \perp \boldsymbol{H}_{\sim}$.

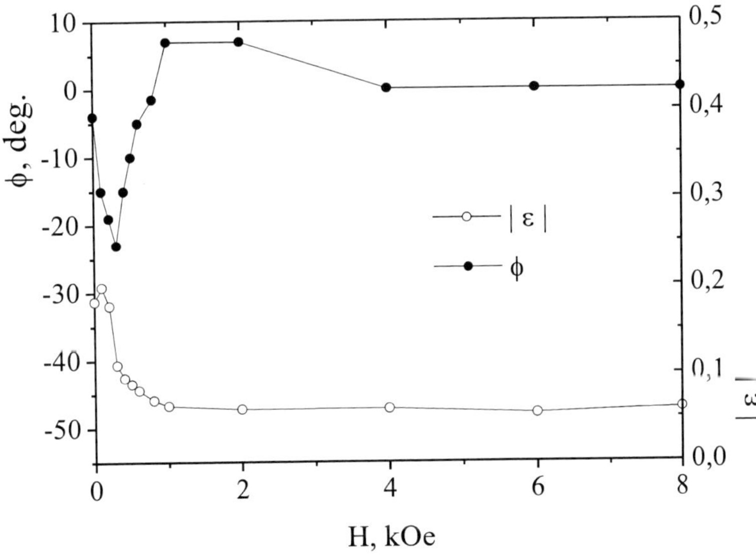

Figure 9. Ellipticity ε and an angle of rotation of a polarization plane φ upon excitation of the *ac* electromagnetic field $\boldsymbol{H}_{\sim}$ under an angle of θ=-45^0 relative to the *dc* magnetic field $\boldsymbol{H}$. f= 20 MHz.

where $\sigma_{\|,\perp}$ and $\delta_{\|,\perp}$ are the conductivity and skin depth for the $\boldsymbol{H}_{\sim}//\boldsymbol{H}$ and $\vec{H}_{\sim} \perp \vec{H}$ orientations, $\delta_{\|,\perp} = (2/\omega\mu_0\mu_{rev(\|,\perp)}\sigma_{\|,\perp})^{1/2}$, $\mu_{rev\ (\|,\perp)}$ is the relative reversible magnetic permeability, , $t_{\|,\perp} = d / \delta_{\|,\perp}$, $Z = 120\ \pi$ □is an impedance of a free space. Expression (1.15) is valid, if $d >> 2/(Z\sigma_{\|,\perp})$. This condition is fulfilled in our experiments. The difference between the transmission coefficients $D_{\|}$ and D □, arising because of anisotropy of material constants in a magnetic field, leads to variation of polarization of an electromagnetic field as a result of an interference of eigenwaves, which pass through a plate. One can show, that the angle of rotation of polarization plane and the angle of ellipticity $\gamma = tg^{-1}\,\varepsilon$ can be calculated using the formulas

$$\varphi = \frac{1}{2}\,tg^{-1}\,\frac{1-\xi^2 b^2 - b(1-\xi^2)\cos\phi}{2\xi b\cos\phi}, \qquad (1.16)$$

$$\gamma = \frac{1}{2}\,\sin^{-1}\frac{2\xi b\sin\phi}{1+\xi^2 b^2}, \qquad (1.17)$$

where $b=(|D_{||}|/|D_{\perp}|)$, $\phi=\phi_{||}-\phi_{\perp}$, $\xi=tg\theta$, θ - is the angle between the axis of the exciting coil and the $\vec{H}$ vector. The parameters b and ϕ from (1.16, 1.17) define the *dc* magnetic field dependence of polarization of the transmitted field.

In the limiting case of low frequencies $t_{||,\perp}$<<1, the relative variation of the transmission coefficient in a magnetic field is expressed by the formula (1.9). Since in manganites the relative variation of magnetoresistance in the fields $H\leq 10$ kOe does not exceed -5,7 %, variations of the transmission coefficient are mostly defined by variations of magnetic permeability in a field. In strong enough field the relation $\mu_{rev}(0) / \mu_{rev}(H) > 1$ is valid. So one can expect an increase of the amplitude of the transmitted signal owing to variation of magnetic permeability. The difference of the shaped of the curves for the cases $\boldsymbol{H_{\sim}//H}$ and $\vec{H}_{\sim}\perp\vec{H}$ is related to the difference between the types of the dynamic magnetic permeability which influence the process of penetration of an electromagnetic field. According to classification of Prof. V.K.Arkadiev [42], in the first case it is the transverse reversible magnetic permeability $\mu_{rev\,\perp}$, and in the second case – the longitudinal reversible magnetic permeability $\mu_{rev\,||}$. These permeabilities differ because of the various contribution of the processes of rotation of magnetization. Under the $\boldsymbol{H_{\sim}//H}$ condition this contribution to magnetic permeability increases from zero at $H=0$ up to a field of an order of an anisotropy field. Then it monotonously decreases up to $\mu_{rev} \to 1$ in saturation. Therefore for the $\boldsymbol{H_{\sim}//H}$ orientation the maximum in the fields of about 500÷700 Oe is observed. For these fields the differences between penetration for two orientations are the greatest, the ratio between the amplitudes of the transmitted fields is maximum. It is possible to make an estimation, that $\mu_{rev\,||} / \mu_{rev\,\perp} \sim 2$. On the basis of the formula (1.10) and the experimental results of Figure 8 it is possible to estimate the initial reversible magnetic permeability μ_{rev} (0). For the $La_{0.63}Y_{0.07}Ba_{0.30}MnO_3$ manganite $\mu_{rev}(0) \approx 3{,}8$ for the frequency of $f = 30$ MHz.

In the experiment, the results of which are presented in Figure 9, the *ac* electromagnetic field, with the polarization oriented under the angle of $\theta=-45^0$ relative to the direction of an external *dc* magnetic field, is excited in the transversely magnetized media. The structure of excitation field differs from that of the fields of normal waves. This will result in transformation of polarization similar to the Cotton-Mouton effect. It is necessary to note the differences in the conditions of observation of the polarization phenomena in a conductive manganite from typical conditions of observation of the Cotton-Mouton effect in the microwave and optical ranges of frequencies. First, transformation of polarization is observed here not for travelling electromagnetic waves, but for the fields which have the spatial distribution defined by skin-effect. The real and imaginary components of a wave number have one order of magnitude. Secondly, since at low frequencies of the interval investigated the condition $\delta_l >> d$ is realized, the amplitude of the transmitted field first of all depends on the ratio between the impedance of a ferromagnet and free space. As the impedance of normal waves in a ferromagnet is proportional to $\sim\sqrt{\mu_{\perp,||}}$ a contribution to polarization transformation arises from the differences between the longitudinal $\mu_{||}$ and transverse $\mu_{\perp}$ permeabilities. Since the impedance of a conductive media preserves both the real and imaginary parts, the result of an interference will be both in rotation of a polarization plane and in ellipticity of the transmitted electromagnetic field.

Thus, in a plate-shaped samples of conductive ferromagnetic lanthanum manganites in the geometry of Cotton-Mouton effect the phenomena of magnetic birefringence and dichroism were observed. They manifested themselves in rotation of polarization plane and ellipticity of a radio-frequency electromagnetic field. Variations of polarization are caused, basically, by anisotropy of dynamic magnetic permeability in an external polarizing magnetic field.

1.5. Penetration of a Radio-Frequency Electromagnetic Field through the Bulk Doped La-Pb and La-Y-Ba Manganites in a Wide Range of Frequencies

In case of high-frequency researches the relation between thickness of a manganite sample and skin depth is important. In turn, the value of skin depth is defined by frequency and material constants of a material under study. Therefore, investigations of the penetration effects in wide range of frequencies were of interest within which for the same series of samples it was possible to realize both inequalities: $d < \delta$ and $d > \delta$. For this purpose the *dc* magnetic field dependences of penetration of electromagnetic field in a frequency interval from 20 kHz to 300 MHz were investigated for manganites, doped by lead, and also barium and yttrium. The choice of these compositions is caused by that their dynamic properties at frequencies of a radio range differ strongly while static electric and magnetic properties are similar. All investigated compositions had the Curie temperatures above the room temperature.

The bulk polycrystalline $La_{0.60}Pb_{0.40}MnO_3$ and $La_{0.85}Pb_{0.15}MnO_3$ manganites were investigated. Some physical properties of these samples are presented in Table 1. The samples were pressed at room temperature. Also the hot pressed $La_{0.68}Y_{0.07}Ba_{0.25}MnO_3$ sample was studied. The procedure of hot pressing permits to increase the density and lower the porosity. It was carried out by heating the samples under the pressure of 3 kBar to the temperature of 900^0C with subsequent exposure at this temperature for 30 minutes. Thus obtained sample had the density of 76 % relative to the theoretical X-ray value. For comparison, the density of the sample of identical composition obtained by pressing at a room temperature was only 60 % relative to the theoretical X-ray value.

The plate-shaped samples of manganites with the thickness of $d = 0.6$ mm were used in experiments. Values of skin depth δ_l, calculated for the investigated compositions for the frequency of $f = 1$ MHz and assuming magnetic permeability $\mu = 1$ are presented in Table 1. From the resulted data it is seen, that in a range of frequencies $f = 20$ kHz ÷ 300 MHz different conditions are realized: $d \ll \delta$ (tens of kilohertz), $d \approx \delta$ (frequencies of an order of units of megahertz), and $d \gg \delta$ (frequencies of the order of tens and hundreds of megahertz).

Doping of lanthanum manganite by barium or lead permits to obtain compositions with the Curie temperature T_C above the room temperature. Additional doping of lanthanum-barium manganites with yttrium permits to investigate the influence of nonmagnetic disorder on physical properties since because of the differences in ionic radiuses structural distortions appear and the degree of nonmagnetic disorder increases. Thus results in slight differences in the static physical properties of the investigated compositions, see Table 1.

Table 1. Physical properties of the La-Pb and La-Y-Ba manganites

Composition	ρ(0, 300K), Ω cm	$MR(H$ = 10kOe, 300K), %	T_C, K	M_S, at T= 293K emu/g	δ_l, mm
$La_{0.85}Pb_{0.15}MnO_3$	0,640	-5,02	322,1	26,0	0,51
$La_{0.60}Pb_{0.40}MnO_3$	0,172	-3,91	340,5	38,0	0,27
$La_{0.68}Y_{0.07}Ba_{0.25}MnO_3$ annealed in oxygen for 6	1,560	-5,70	342,0	38,7	0,80
$La_{0.68}Y_{0.07}Ba_{0.25}MnO_3$ hot pressed	1,650	-6,50	340,0	39,6	0,82

ρ(0, 300K) is the resistivity at T = 300 K; $MR(H$ = 10kOe, 300K) is magnetoresistance in the field of H = 10kOe at T = 300 K; M_S, is the saturation magnetization at T = 293K; δ_l is the skin depth for f = 1 MHz, assuming μ = 1.

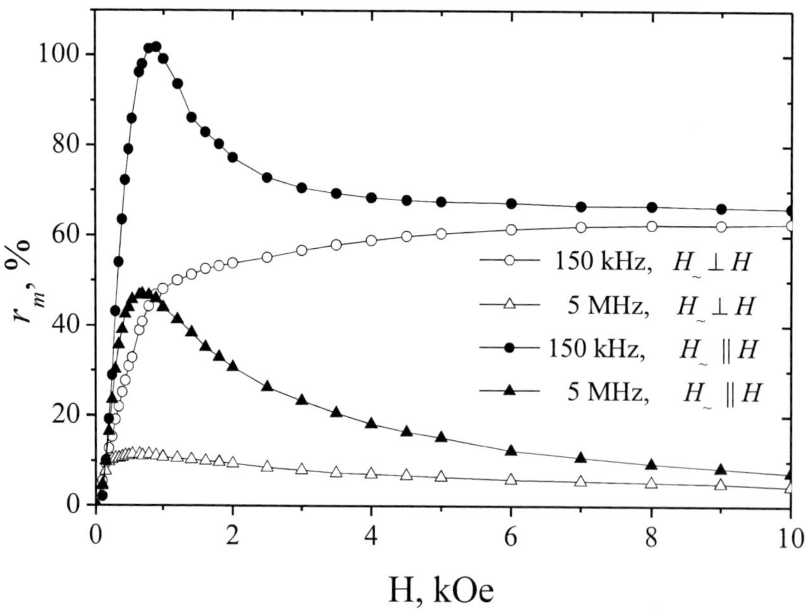

Figure 10. Relative variation of the module of the penetration coefficient of an electromagnetic field through a plate of $La_{0.60}Pb_{0.40}MnO_3$ manganite.

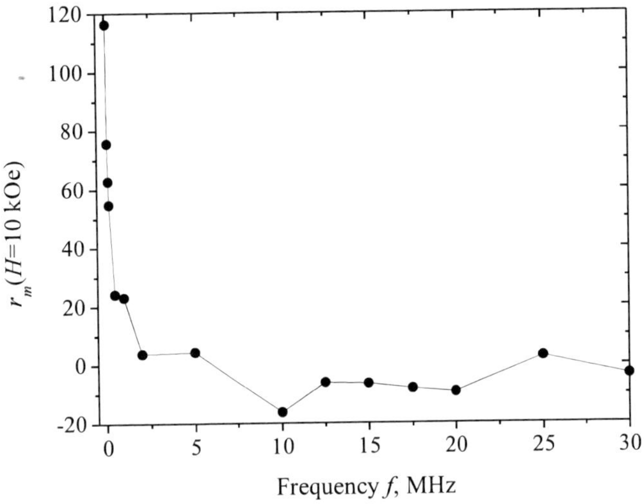

Figure 11. Frequency dependence of the r_m coefficient in the field of H = 10 kOe for the $La_{0.60}Pb_{0.40}MnO_3$ manganite. $\boldsymbol{H}\perp\boldsymbol{H}_\sim$.

Doped manganites can be considered as soft ferromagnetic materials since at room temperature their coercive forces has the typical values of ~10 Oe, saturation fields of ~5 kOe, and the initial magnetic permeability of ~ 100. If measurements are carried out below the Curie temperature, very large changes of penetration coefficient in the applied *dc* magnetic field are observed for a manganite sample because of high values of the initial magnetic permeability. In the experiments executed at a room temperature, up to 12 times increase in amplitude of the transmitted signal upon magnetization was observed (see section 1.2). Such large variations can not be caused by magnetoresistance since under these conditions its values does not surpass -10 %.

The values of r_m for manganites, doped with lead, were measured for two orientations between the static *dc* $\boldsymbol{H}$ and electromagnetic *ac* $\boldsymbol{H}_\sim$ fields: $\boldsymbol{H}//\boldsymbol{H}_\sim$ and $\boldsymbol{H}\perp\boldsymbol{H}_\sim$. The results are presented in Figure 10. For the $La_{0.60}Pb_{0.40}MnO_3$ manganite measurements were carried out at two frequencies: f = 150 kHz and 5 MHz. As is seen from Figure 10, the dependences measured for $\boldsymbol{H}//\boldsymbol{H}_\sim$, have a maximum. In a saturated state when $H\to\infty$ the values of r_m for both orientations are close to each other. As it has been shown earlier in section 1.1, relative variations of the penetration coefficient for $d \ll \delta$ are expressed by the formula (1.9). If $\left|\frac{\Delta\rho}{\rho}\right| \ll 1$ and $\mu(H\to\infty)\to 1$, then in the limit of strong fields (for $H\to\infty$) the formula (1.10) is valid. On the basis of this formula it is possible to assert, that variations of the penetration coefficient between the magnetized to saturation and non-magnetized states are defined by the initial magnetic permeability, and the magnitude of these variations does not depend on direction of an external *dc* magnetic field. Using the formula (1.10) for the data of Figure 10 it is possible to make the following estimations: $\mu(0) \approx 8$ for the frequency of f = 150 kHz and $\mu(0) \approx 1.7$ for the frequency of f = 5 MHz. The frequency dependence of r_m, measured in the field H = 10 kOe for the manganite of the same composition, is shown in Figure 11. It is seen, that for frequencies above 2 MHz the values of r_m sharply decreases with increasing frequency. This circumstance is possibly the result of a frequency dispersion of magnetic permeability.

The influence of doping with yttrium on properties of bulk polycrystalline lanthanum manganites was investigated in [35]. The radius of the Y^{3+} ion is less, than that of the La^{3+} ion. Therefore doping with yttrium changes the angle of the Mn-O-Mn chemical bond in the perovskite lattice and influences the degree of overlapping between the e_g-p-e_g orbitals. With increasing yttrium content the mobility of charge carriers is suppressed, and the magnetic state becomes "frustrated" [35]. As a result, the temperature of phase transition lowers and magnetoresistance increases.

Let us consider the influence of doping with yttrium and barium on dynamic magnetic properties. Measurements of penetration of an electromagnetic field showed that in such manganites considerable variations of the transmission coefficient are observed upon magnetization. The results of measurement of frequency dependence of the r_m coefficient in the field of H=10 kOe are shown in Figure 12. This dependence illustrates the differences of dynamic properties of yttrium and barium doped manganites from the compositions doped with lead (compare to Figure 11). Even for the frequencies of tens megahertz large values of r_m are observed. In particular, for the frequency of f = 40 MHz (where $d \ll \delta_l$) the module of the transmission coefficient through a plate changes by more than 8 times upon magnetization. The shape of the r_m dependence on the applied *dc* magnetic field for manganites doped with yttrium and barium is similar to the corresponding dependence for the composition doped with lead. This dependence for the frequency of f = 40 MHz is shown for the $La_{0.68}Y_{0.07}Ba_{0.25}MnO_3$ manganite is shown in Figure 13. Measurements were carried out at different values of high-frequency power. With decreasing power the amplitude of variation of r_m decreases. Strong manifestation of nonlinearity is unusual in this case taking into consideration the fact that in this case the amplitude of high-frequency fields does not exceed units of Oersted.

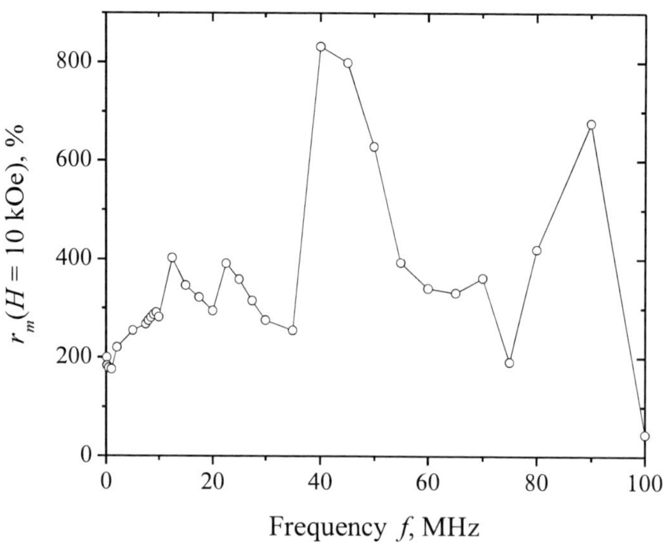

Figure 12. Frequency dependence of the r_m coefficient measured for the $La_{0.68}Y_{0.07}Ba_{0.25}MnO_3$ manganite at $\boldsymbol{H} \perp \boldsymbol{H}_{\sim}$ in the field of H = 10 kOe.

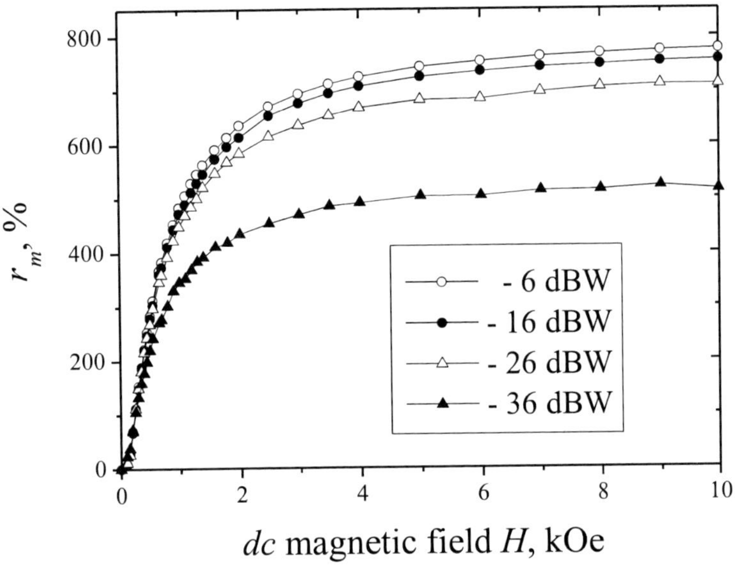

Figure 13. The *dc* magnetic field dependence of the r_m coefficient measured for $La_{0.68}Y_{0.07}Ba_{0.25}MnO_3$ manganiteon at the frequency of $f = 30$ *MHz* at $\boldsymbol{H}//\boldsymbol{H}_{\sim}$ and various values of the input power.

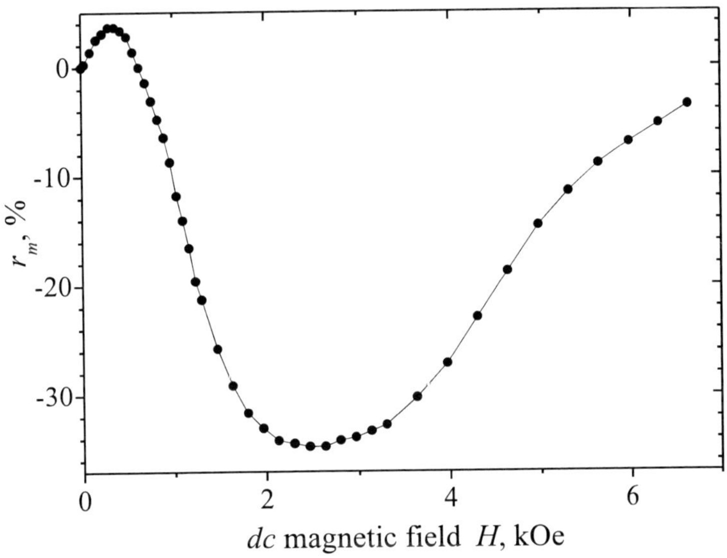

Figure 14. The ferromagnetic antiresonance and a resonance measured at the frequency $f=$ of 7412 *MHz* for the $La_{0.68}Y_{0.07}Ba_{0.25}MnO_3$ manganite.

As is seen from Figure 12, reduction of r_m is observed at the highest (about 100 MHz) frequencies. It can be supposed that this reduction is the result of frequency dispersion of magnetic permeability. With further substantial increase of frequency, when the $d>>\delta_l$ condition is fulfilled more exactly, the processes related to magnetic resonance are possible. We carried out experiments on measurement of absorption of elcctromagnetic waves at frequencies from 5,5 to 11 GHz. The sample of the $La_{0.68}Y_{0.07}Ba_{0.25}MnO_3$ manganite was placed in a hollow cylindrical resonator and the amplitude of an electromagnetic wave passed through a resonator with the sample was measured. Figure 14 shows the results of measurement at the frequency of $f = 7412$ *MHz*. With increasing the strength of *dc* magnetic field, first, the amplitude of the transmitted signal also increased. This is caused by the phenomenon of a ferromagnetic antiresonance. For higher values of the *dc* magnetic field the amplitude of the transmitted signal substantially decreases. This is caused by absorption of electromagnetic waves under conditions of ferromagnetic resonance.

The difference of conditions of this experiment from the previous results considered in this section, consists in the fact that at frequencies of thousands megahertz the skin depth is less than the sizes of a sample and at ferromagnetic resonance absorption of electromagnetic energy plays the basic role. An increase in skin depth and related decrease of absorption is important for antiresonance. For the experiments carried out in a radio-frequency range when $d << \delta_l$, the basic role plays variation of the surface impedance of a plate.

Figure 15 shows the *dc* magnetic field dependences of the r_m coefficient measured for the $La_{0.68}Y_{0.07}Ba_{0.25}MnO_3$ manganite at two frequencies: 30 and 175 MHz from which the following conclusions can be made. The shape of the r_m(H) dependences for the hot pressed manganite doped with yttrium and barium, is similar to that were considered above for manganites of other compositions. At the high frequency of $f = 175$ MHz relative variation of the penetration coefficient at the conditions of magnetic saturation is about 50%. In the maximum for $\boldsymbol{H} \perp \boldsymbol{H}_{\sim}$ its value exceeds 80 %. The distinctive feature of hot pressed manganite doped with yttrium and barium consists in essential influence of a magnetic field on the penetration coefficient even in the range of very high radio frequencies.

Thus, it is shown, that for doped lanthanum manganites of various compositions which have similar static properties, the dynamic properties are essentially different. At temperatures below the Curie temperature, the considerable, up to several times, variations of the penetration coefficient are observed upon magnetization. The shapes of the *dc* magnetic field dependences of penetration coefficient for manganites, doped with lead, are similar to those, which are observed for manganites, doped with barium and yttrium. In manganites doped with lead reduction of the dynamic magnetic permeability, which is observed at frequencies of units of megahertz, is caused by frequency dispersion. In manganites, doped with barium and yttrium, the influence of frequency dispersion of magnetic permeability manifests itself at frequencies of tens of megahertz. The hot pressed manganites, doped with barium and yttrium, preserve appreciable variations of penetration coefficient in a *dc* magnetic field up to the frequencies about 200 MHz.

1.6. Investigation of Magnetic Phase Transition in Bulk Doped La-Pb and La-Y-Ba Manganites by Penetration of Radiofrequency Electromagnetic Field

The properties doped manganites are most interesting in the range of temperatures, including the temperature of magnetic phase transition, since it is accompanied by abrupt variations of physical properties. Therefore the special attention has been given to detailed investigations of penetration of radio-frequency electromagnetic field through doped lanthanum manganites in the temperatures range including the Curie temperature T_C. At the first stage of these investigations the manganite samples doped by lead were used since they have the value of the Curie temperature above the room temperature. As the objects of research bulk polycrystalline manganites with compositions $La_{0.85}Pb_{0.15}MnO_3$ and $La_{0.75}Pb_{0.25}MnO_3$. were used. The temperature dependence of magnetic susceptibility χ_{ac}, measured for the $La_{0.75}Pb_{0.25}MnO_3$, manganite at the frequency of f = 40 Hz is shown in Figure 16a. The Curie temperature, determined from this dependence, was found to be 351 *K*. A is seen from Figure 16a, the magnetic susceptibility sharply decreases near T_C and it is rather small above this temperature. Magnetoresistance was measured by four-probe technique in the range of temperatures from 5 to 300 *K* and *dc* magnetic fields *H* up to 50 kOe at orientation of a current parallel to magnetic field. The temperature dependence of magnetoresistance *MR* in the *dc* magnetic field of 10 kOe is shown in Figure 16b. Magnetoresistance is defined as $MR(H,T) = (\rho(H,T)-\rho(0,T)/\rho(0,T)*100\%$ where $\rho(H,T)$ and $\rho(0,T)$ are the values of resistivity in a *dc* magnetic field *H* and zero magnetic field at temperature *T*, respectively. Near the temperature of phase transition a maximum of magnetoresistance is observed. Above T_C magnetoresistance considerably decreases. The value of magnetoresistance in the maximum is equal to -5.6 %. The value of resistivity for the $La_{0.75}Pb_{0.25}MnO_3$ manganite at room temperature and zero *dc* magnetic field is equal to 0.185 Ω*cm. The estimation of skin depth δ_l for the frequency of 1 MHz for normal skin-effect assuming the magnetic permeability μ =1 gives δ = 21.4 mm. For all frequencies used by us in experiments on penetration of an electromagnetic field, the condition $\delta > d$, where *d* is a thickness of a plate-shaped sample, was satisfied since in our case *d* = 1.2 mm. Radio-frequency measurements were carried out in the range of temperatures from 273 K to 365 K.

As shown earlier, in the case of a ferromagnetic state of a manganite the relative variations of penetration coefficient of a radio-frequency electromagnetic radiation upon magnetization, as a rule, are great. They have the positive sign, saturate in strong fields and can exceed magnetoresistance by several times. These specified circumstances force allow to link variations of penetration characteristics mainly with changes of the dynamic differential magnetic permeability. Then it is possible to assume, that the value of variations above the Curie temperature can becomes substantially smaller. However the experiments showed much more complicated picture of the phenomena. Comparison of the *dc* magnetic field dependences of penetration coefficient, measured both above, and below T_C at the frequency of f = 200 kHz is shown in Figure 17. The shape of the dependence changed substantially at higher temperature and a maximum appeared. However the size of changes of r_m in saturation fields in both cases differs not too much.

The results of measurements of *dc* magnetic field dependence of penetration coefficient at different frequencies for T = 365 K are shown in Figure 18. The dependence measured at the frequency of 20 kHz monotonously increases with H. At the higher frequency of f = 200 kHz a maximum appears in the $r_m(H)$ dependence. The values of r_m in strong magnetic fields $r_m\ (H \to \infty)$ are positive and their absolute values are by several times greater than the *dc* magnetoresistance measured in the same fields. The further increase of frequency leads to negative values of penetration coefficient in strong magnetic fields. The $r_m(H)$ dependence has no saturation in all interval of fields up to 10 *kOe* within which measurements were carried out. It is necessary to note, that in all measurements of penetration coefficient at temperatures below T_C only positive values of the $r_m\ (H \to \infty)$ parameter were found for all frequencies.

The temperature dependence of r_m, measured in external field of H = 9 kOe at the frequency of 8 MHz, shows sharp decrease of above T_C (see Figure 19). Moreover, at temperatures above the temperature of phase transition the value of r_m became negative.

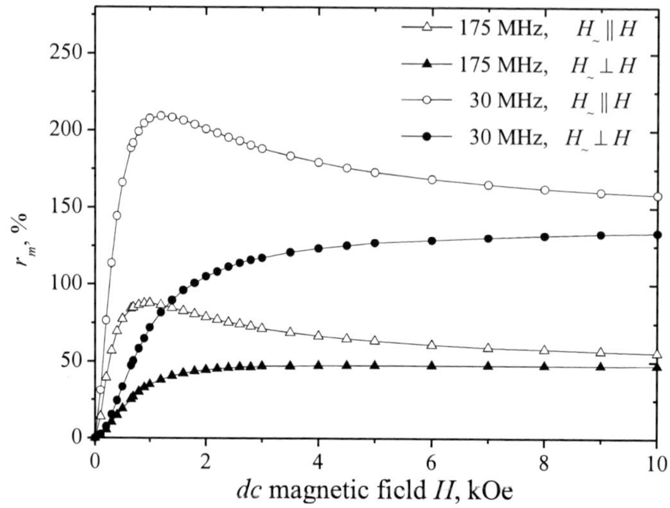

Figure 15. The *dc* magnetic field dependences of the r_m coefficient measured for the hot pressed $La_{0.68}Y_{0.07}Ba_{0.25}MnO_3$ manganite.

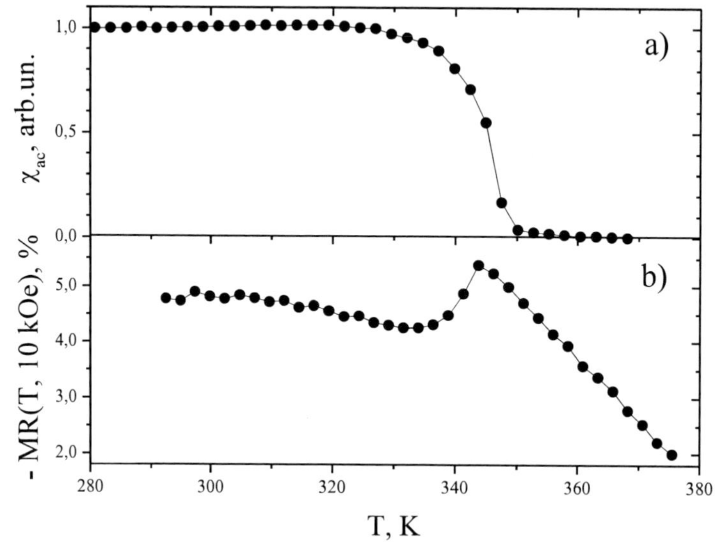

Figure 16. Temperature dependences of the magnetic susceptibility χ_{ac}, measured at the frequency of 40 Hz (a), and magnetoresistance *MR* (b) for the $La_{0.75}Pb_{0.25}MnO_3$ composition.

The temperature dependences of r_m, measured at the frequency of $f = 200$ kHz in different *dc* magnetic fields are interesting. Results of these measurements for the values of 1 and 9 kOe are shown in Figure 20. On both dependences the increase of r_m in a ferromagnetic state with increase of temperature is first observed which is followed by sharp decrease of r_m close to T_C.

At temperatures $T > T_C$ the sign of r_m remains positive. These two dependences intersect at the temperature of $T \approx 327$ *K*. In the temperature interval, where r_m *(H=9kOe)*> r_m *(H=1kOe)* no maximum is observed on the field dependence of r_m, while a maximum exists in the interval where reverse inequality is valid. Similar results were obtained for the

$La_{0.85}Pb_{0.15}MnO_3$ manganite. Near the Curie temperature the value of r_m decreases while above T_C substantial variations of r_m are preserved. The sign of these variations is positive.

We can analyze the experimental results taking into account formulas (1.10) and (1.11). First of all it is doubtless, that the data of measurements of the magnetic susceptibility shown in Figure 16 confirm that above T_C the low-frequency magnetic permeability tends to unity. A long-range magnetic order is absent in a system. However the data obtained on penetration of a radio-frequency *ac* electromagnetic field prove that the dynamic high-frequency magnetic permeability even at $T > T_C$ essentially differs from unity. Indeed, from the formula (1.11) follows, that for negative magnetoresistance $\Delta\rho/\rho < 0$, the positive sign of r_m is possible, only if the initial magnetic permeability $\mu(0) > 1$. In particular, for the conditions of Figure 17 at the temperature of $T = 362$ *K* which is larger than T_C, an estimate according to the formula (1.10) gives $\mu(0) \approx 1.4$. The fact of existence of the dynamic magnetic permeability distinct from unity at $T > T_C$ is confirmed by the data of Figure 20 and partly of Figure 18. This fact points to the principal possibility of existence in the doped manganites under consideration of a short-range magnetic ordering at temperatures considerably above the Curie temperature. Such assumption is in accordance with the available published data [43, 44].

Let us consider the results presented in Figure 18, in more details. The dependences measured at lower frequencies of 20 and 200 kHz, show saturation in strong fields, and $D(H \to\infty) > D(0)$. At these frequencies variations of penetration are defined by variations of magnetic permeability. However, different results were obtained for the frequencies of 1 and 3 MHz. In strong fields the values of r_m become negative. According to (1.11), it is possible only if the value of the initial magnetic permeability $\mu(0)$ is close to unity, and variations of D are mostly determined by magnetoresistance. Such point of view is confirmed by the data of Figure 18 by absence of saturation for the dependences having a negative sign in strong fields. We believe, that the decrease of magnetic permeability at frequencies of units of megahertz is related to its frequency dispersion.

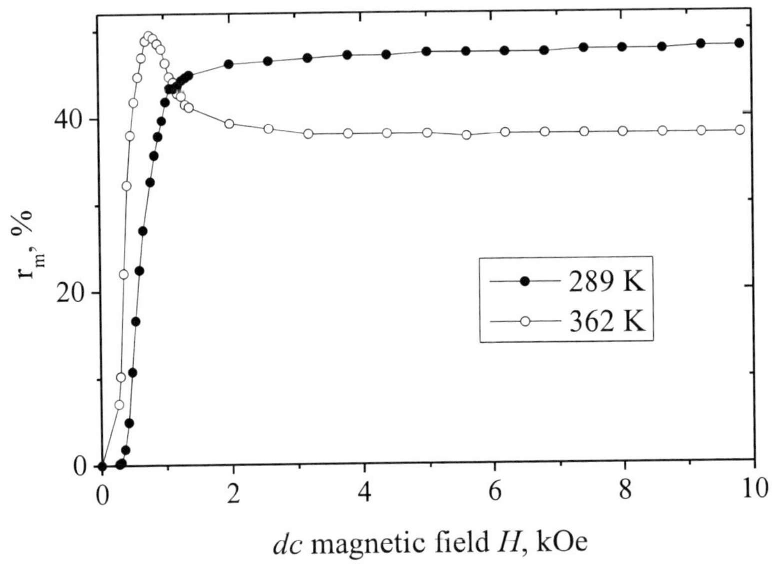

Figure 17. The *dc* magnetic field dependences of relative variations of penetration coefficient at the frequency of 200 kHz for the $La_{0.75}Pb_{0.25}MnO_3$ composition.

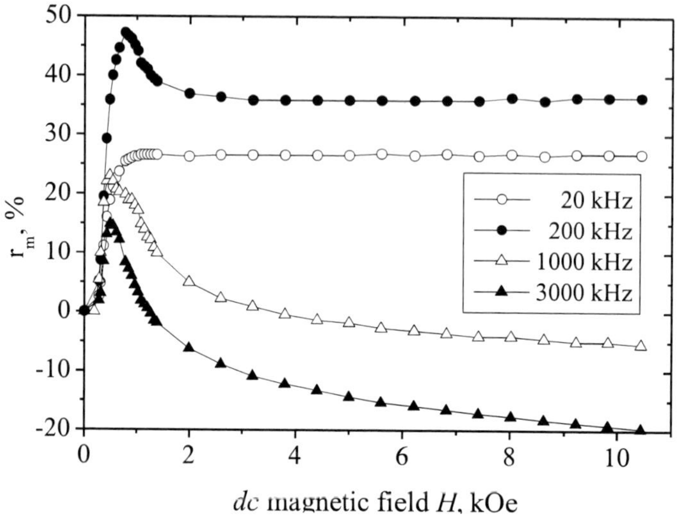

Figure 18. The *dc* magnetic field dependences of relative variations of penetration coefficient measured at several frequencies above T_C for the $La_{0.75}Pb_{0.25}MnO_3$ composition.

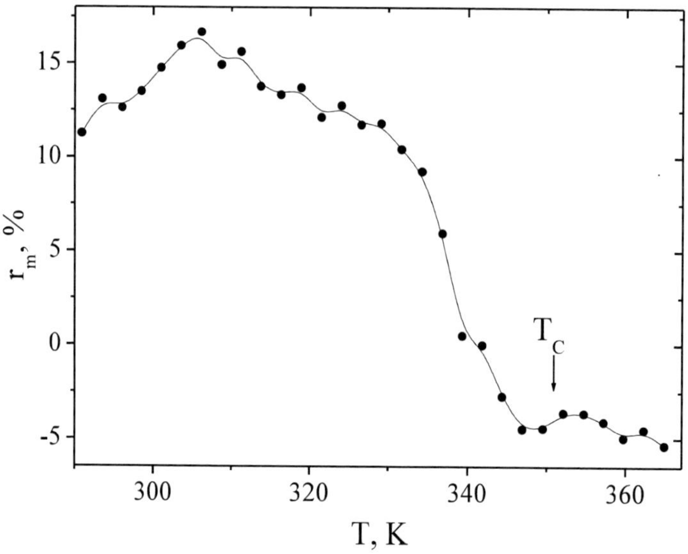

Figure 19. Temperature dependence of variations of penetration coefficient measured at the frequency of 8 MHz in the *dc* magnetic field of $H = 9$ kOe for the $La_{0.75}Pb_{0.25}MnO_3$ composition.

Thus, as a result of an experimental investigation of penetration of an electromagnetic field through a plate of the $La_{0.75}Pb_{0.25}MnO_3$ manganite in the vicinity of the magnetic phase transition temperature it is shown, that at radiofrequencies the dynamic magnetic permeability can essentially exceed unity even if the temperature by 10-14 degrees exceeds the Curie point. This fact can be interpreted as the experimental proof of existence of local magnetic ordering above the Curie temperature. It is established, that at frequencies of units of megahertz variations of penetration coefficient in a strong external magnetic field can become negative. In this case they are defined, mainly, by magnetoresistance of manganite.

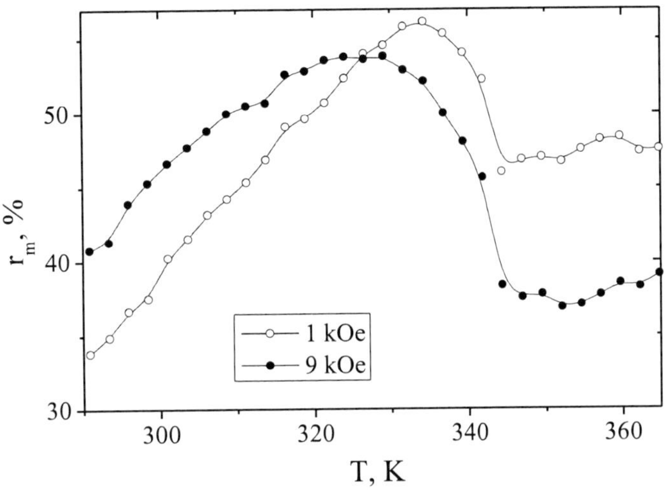

Figure 20. Temperature dependence of variations of penetration coefficient measured at the frequency of 200 kHz in the *dc* magnetic fields of $H = 1$ and 9 kOe for the $La_{0.75}Pb_{0.25}MnO_3$ composition.

2.3. Penetration of a Radio-Frequency Electromagnetic Field through Bulk Doped La-Er-Ba Lanthanum Manganites in Paramagnetic Range of Temperatures

As it was shown in section 1.6, at radiofrequencies the dynamic magnetic permeability can essentially exceed unity even if the temperature are by ~60 K higher than the Curie point. This fact can be interpreted as the proof of existence of local magnetic ordering in some range of the temperatures exceeding the Curie temperature. However the question of experimental establishment of the upper limit of this range remains opened.

For the answer to this question investigations of penetration coefficient were carried out for the $La_{0.60}Er_{0.07}Ba_{0.33}MnO_3$ composition. This manganite has T_C above the room temperature. The plate-shaped samples with diameter of 10 mm and thickness $d = 1.5$ mm were investigated. Experiments were carried out for the frequency range of 0.05-30 MHz. An estimation of skin depth under the assumption μ □= 1 gives the values of $\delta_l = 98$ mm and $\delta_l = 4$ mm for the bottom and top limits of the investigated range of frequencies, respectively. This means, that the condition $d \ll \delta$ is satisfied for all interval of frequencies investigated. Figure 21a shows the results of measurements of isotherms of *dc* magnetization near the Curie temperature. It is necessary to note an absence of saturation even in fields of an order of 60 kOe. Figure 21b shows the temperature dependence of resistivity. It is typical of bulk polycrystalline manganites having vacancies in oxygen sublattice: the wide peak near 240 K [45], caused by the presence of vacancies, the poorly expressed peak near T_c, and negative value of the $\partial\rho/\partial T$ derivative in paramagnetic range of temperatures. The results of measurements of *dc* magnetic field dependences of penetration coefficient r_m at different temperatures are presented in Figure 22. The data shown in Figure 22a were obtained for $T = 293$ K, that is for a ferromagnetic state of the sample. Measurements were made at several frequencies. At the lowest frequency of $f = 0.05$ MHz r_m monotonously increases with increasing field. A maximum appears in the $r_m(H)$ dependences as the frequency increases. Its position approximately corresponds to the value of an anisotropy field. The value of r_m

increases with increasing frequency. The presented dependences indicate that on frequencies below 30 MHz the frequency dispersion of magnetic permeability is insignificant. This circumstance is characteristic only of the $La_{0.60}Er_{0.07}Ba_{0.33}MnO_3$ composition. As it has been shown in section 1.5 in the case of lanthanum manganites, doped with lead, the frequency dispersion of magnetic permeability starts to manifest itself at frequencies above 1 MHz. In a ferromagnetic state variations of the penetration coefficient in the *dc* magnetic field are basically caused by changes of magnetic permeability.

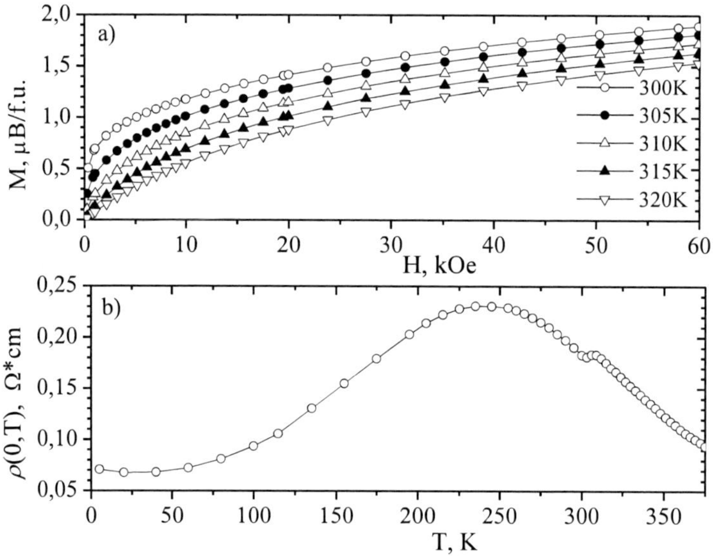

Figure 21. The *dc* magnetization curves for the $La_{0.60}Er_{0.07}Ba_{0.33}MnO3$ manganite near the Curie temperature (a). Temperature dependence of resistivity (b).

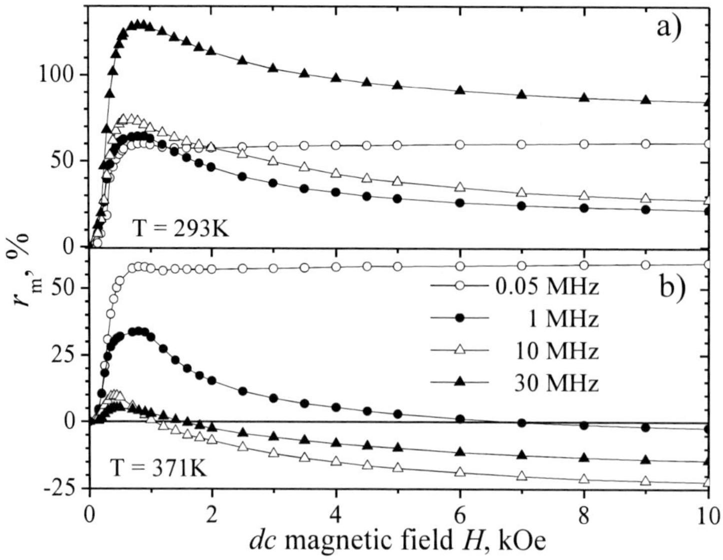

Figure 22. The *dc* magnetic field dependence of penetration coefficient below (T = 293K) (a) and above (T = 371K) (b) the Curie temperature.

Figure 22b shows the results of measurements of *dc* magnetic field dependences of penetration coefficient r_m at the temperature of T = 371 K, i.e. when the sample is in a paramagnetic state. It is important that this temperature exceeds the Curie by 66 K. A is seen from the data presented in Figure 21, at these temperatures the long-range magnetic order is absent.

Nevertheless, strong variations of penetration coefficient in a magnetic field are experimentally observed. For the frequency of f = 0.05 MHz with increasing field r_m also monotonously increases and reaches the values of 60 % in the field of 10 kOe. In the dependences recorded for higher frequencies a maximum appears, however in strong fields the value of r_m becomes negative. It is necessary to note, that no saturation is observed on the $r_m (H)$ dependences in the range of fields from 6 to 10 kOe.

For the analysis of the experimental dependences obtained we can use the formula (1.11).

As is seen from Figure 22b, for the frequency of f= 0.05 MHz the value of r_m is positive, that is, variations of the penetration coefficient are mostly determined by changes of magnetic permeability. The values of r_m become negative with increasing frequency and mostly determined by magnetoresistance. From the dependence of Figure 22b, corresponding to the frequency of f = 0.05 MHz, we obtain r_m (H=10 kOe) = 0.6. This means, that $\left|\frac{D(H=10\kappa Э)}{D(0)}\right| = 1.6$, whence, taking into account the formula (1.11) and that MR (10 kOe) = - 7.6 %, we obtain that $\mu(0) \approx 1.7$. Since the data presented in Figure 22b correspond to far paramagnetic region, more exactly to the temperature, by 66 K exceeding T_c, the fact of difference of the differential initial magnetic permeability from unity should be treated as the experimental proof of preservation of short-range magnetic order in some regions of the sample while the long-range magnetic order in the whole volume of the sample is absent. Such treatment corresponds to the phase separation model, which was proposed to explain the physical mechanism of the "colossal" magnetoresistance in doped lanthanum manganites [46]. The existence of ferromagnetically ordered clusters in the range of temperatures by 100 K exceeding T_c was experimentally observed in [47].

The temperature dependences of the signal in the receiving coil, measured at different frequencies in the *dc* magnetic field of H = 0,8 kOe are presented in Figure 23. Data were recorded upon heating. The data of Figure 23a correspond to the frequency of f = 0.1 MHz and that of Figure 23b – to the frequency of f = 10 MHz. The field of 0.8 kOe approximately corresponds to the anisotropy field of the sample material. The shapes of the dependences in Figs. 23a and 23b differ substantially: the value of the signal at the frequency of f = 0.1 MHz decreases with increasing temperature while the opposite behavior is observed at the frequency of f = 10 MHz. Such difference can be explained by two prevailing contributions which in different ways depend on frequency: the magnetic permeability for low frequencies (Figure 23a) and magnetoresistance for high frequencies (Figure 23b).

The basic result of this section is in the experimental proof of preservation of a local dynamic magnetic ordering in the $La_{0.60}Er_{0.07}Ba_{0.33}MnO_3$ manganite in the range of temperatures, by 66K exceeding the Curie temperature. The magnetic permeability has the greatest value at frequencies of the order of tens kilohertz. This corresponds to an average lifetime of existence of local dynamic magnetic ordering of about 10-100 μs. According to measurements of static magnetic properties long-range magnetic ordering is absent in this temperature range.

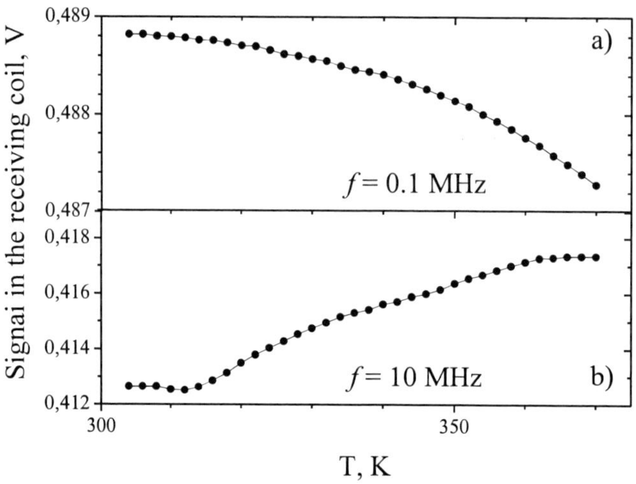

Figure 23. Temperature dependence of the signal in the receiving coil in the *dc* magnetic field of H = 0.8 kOe at the frequencies of f = 0.1 MHz (a) and f = 10 MHz (b).

1.7. Radiofrequency Electromagnetic Properties of Doped La-Sr Manganites. Compositions with Basic Ferro- and Antiferromagnetic States

Though the basic laws of correlation of magnetic and transport properties of doped manganites have been explained long ago within the double exchange model [24], more detailed theoretical and the experimental investigations, carried out in recent years, required introduction of such new concepts, as electronic phase separation, charge ordering, correlated polarons and others. By present, the new concept of the CMR phenomenon has been formulated [18, 46]. It consists in the supposition that the electronic phase separation is the basic physical mechanism of the phenomenon of "colossal" magnetoresistance. In this connection the investigations which allow to obtain experimental arguments in favor of this or other physical mechanism are of special interest. The region of phase transition attracts particular attention since at these temperatures the most strongly pronounced variations of magnetic and transport properties are observed.

Bulk polycrystalline manganites with compositions $La_{1-X}Sr_XMnO_3$ ($0.33 \leq X \leq 0.60$) have interesting physical properties. With increasing concentration of doping cation, smooth change of the basic low-temperature state from metallic ferromagnetic (X = 0.33) through metal antiferromagnetic (X = 0.50 and 0.55) to insulating antiferromagnetic (X = 0.60) occurs, see Figure 24. Since earlier (sections 1.2-1.7) the results of investigations of peculiarities of penetration of a radio-frequency electromagnetic field through doped lanthanum manganites, which have only the basic metallic ferromagnetic low-temperature state, were presented, investigations of doped lanthanum manganites with various types of the basic low-temperature magnetic states was especially interesting.

Bulk polycrystalline samples of the $La_{1-X}Sr_XMnO_3$ (X = 0.33; 0.50; 0.55; 0.60) system were investigated. Measurements were made in the range of temperatures from 300 to 400K. The plate-shaped samples of manganites with the thickness $d = 0.4 \div 0.6$ mm were used in experiments. The values of skin depth δ_l for frequency of f = 1 MHz were calculated assuming magnetic permeability $\mu = 1$. For the sample $La_{0,67}Sr_{0,33}MnO_3$ sample, depending on frequency, the conditions $d<\delta$ or $d<<\delta$, can be realized. For other samples the condition $d <<\delta$ was always fulfilled.

Bulk polycrystalline sintered manganites differ from the single crystals of the same composition by the presence of structural microinhomogeneities: they consist from small crystallites (grains), the size of which is defined by composition and regimes of annealing, and intergrain boundaries. The existence of these microinhomogeneities manifests itself in transport properties: at low temperatures the additional component of magnetoresistance appears [49], which is absent in single crystals. Besides that, the magnetic and transport properties of sintered manganites are influenced by regimes of heat treatment. The question about that, to what extent the physical properties of the investigated polycrystalline sintered samples are close to those of single crystals of the same composition described in [48, 50], was very important.

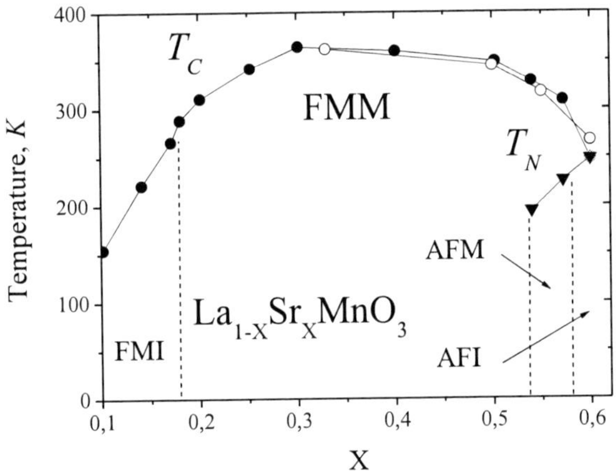

Figure 24. Phase diagram of the $La_{1-X}Sr_XMnO_3$ system. Data for single crystals from [4.23]: ● – Curie temperature T_C, ▼ – Neel temperature T_N. Our data for polycrystalline samples ○ - T_C. FMI – ferromagnetic insulator. FMM – ferromagnetic metal. AFM – antiferromagnetic metal. AFI – antiferromagnetic insulator.

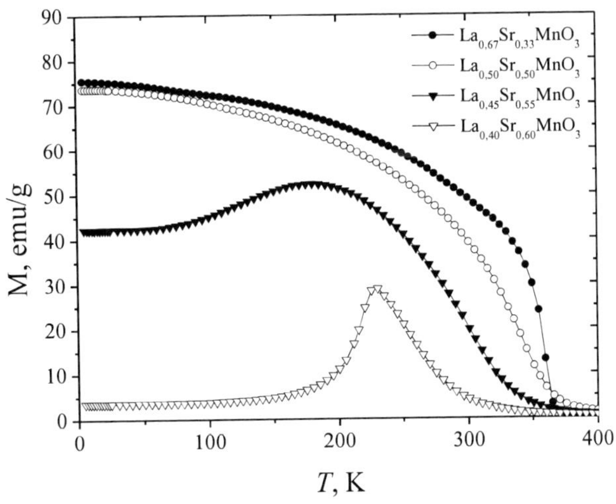

Figure 25. The temperature dependences of *dc* magnetization in the field of 5 kOe.

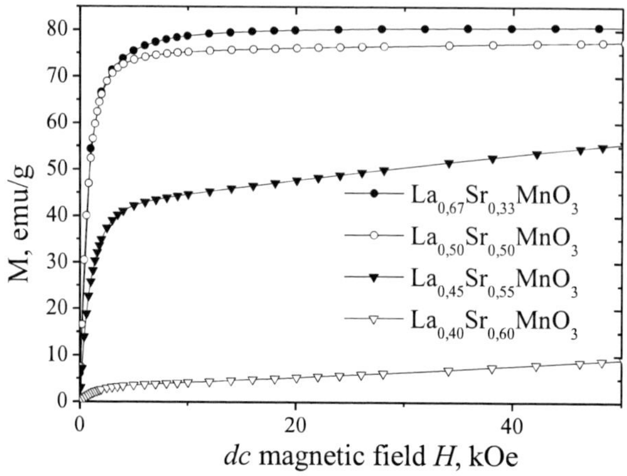

Figure 26. The *dc* magnetic field dependences of magnetization at $T = 5$ K.

For this purpose the temperature dependences of magnetization in the field of 5 kOe (*M (5 kOe)*) (see Figure 25), which can be compared to similar dependences for the single crystals of the identical composition (see Figure 3 in [48]), were measured. As is seen from Figure 25, the shapes of the *M (5 kOe)* temperature dependences are similar for the compositions with $X = 0.33$ and 0.50 and strongly differ for compositions with $X = 0.55$ and 0.60. These differences are caused by the fact that the basic low-temperature state for the first two compositions is ferromagnetic. This is confirmed by the data of Figure 26 where the *dc* magnetic field dependences of magnetization at $T = 5$ K are presented. The value of spontaneous magnetization is close to the corresponding value for single crystals and the susceptibility of paraprocess is small. For the compositions with $X = 0.55$ and 0.60 the basic low-temperature states are, respectively, the metallic and insulating antiferromagnetic. These states manifest themselves by appearance of maxima at $T \sim 190$ and 240 K (see Figure 25) in the *M (5 kOe, T)* temperature dependences sharp decreases of spontaneous magnetization and increase of susceptibility of the paraprocess (see Figure 26). Note that for single crystals in the antiferromagnetic range of temperatures (the data of [48]) the values of *M(5 kOe, T)* is very small. But for our polycrystalline sintered samples in the same temperature range the values of *M(5 kOe, T)* are much higher. Besides that, as it follows from the data presented in Figure 26, for our samples with $X = 0.55$ and 0.60 at $T = 5$ K the finite values of spontaneous magnetisation still exists. At temperatures above T_C the shapes of the *M(5 kOe, T)* dependences were similar both for single crystal and polycrystalline sintered samples. The temperature dependences of resistivity $\rho(T)$ for the investigated compositions are shown in Figure 27a. The shape of the $\rho(T)$ dependence for the composition with $X = 0.33$ practically coincides with similar dependence for the single crystal: in the range of temperatures of 300-420 K the $\partial\rho(T)/\partial T$ derivative is positive. The shapes of the $\rho(T)$ dependences for the compositions with $X \geq 0.50$ are essentially different. First of all, in the investigated interval of temperatures the sign of the $\partial\rho(T)/\partial T$ derivative is negative. It is necessary to note, as for the single crystals of the same composition resistivity slightly varies with temperature (see Figure 3 in [48]), and it is difficult to determine the sign of the derivative of resistivity on temperature upon crossing the magnetic phase transition temperature region. In the case of our polycrystalline sintered samples, as is seen from Figure 27a, the inflection point of the $\rho(T)$ curve can be determined only for the composition with $X = 0.55$. It is located near 340 K.

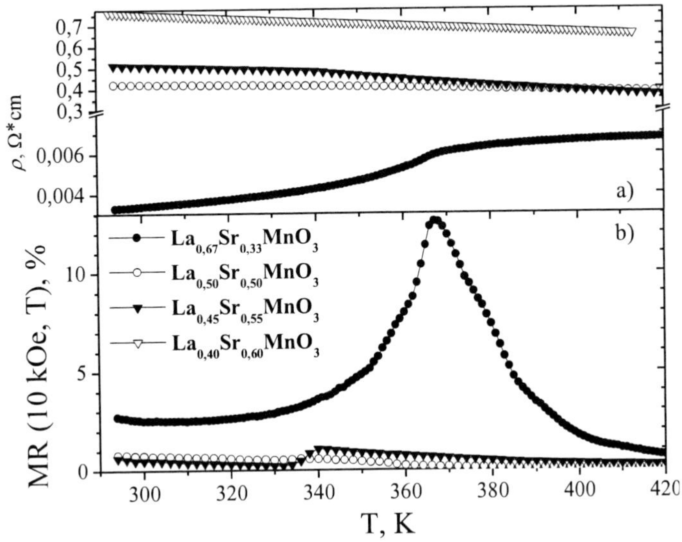

Figure 27. Temperature dependences of resistivity (a) and magnetoresistance in the field of 10 kOe (b).

For the compositions with $X = 0.50$ and 0.60 the $\rho(T)$ dependences have the shapes close to linear. Besides that, the values of resistivity at 300 K (ρ_{300K}) differ substantially: for the composition with $X = 0.33$ the value of $\rho_{300K} = 3{,}3\cdot10^{-3}$ Ω*cm, which is of the same order of magnitude as the corresponding value the single crystal [50], for compositions with $X \geq 0.50$ the values of ρ_{300K} exceed ~ 0,4 Ω*cm which is substantially higher than values reported in [48]. In our opinion this fact points to the essential contribution of intergrain boundaries in the absolute value of $\rho(T)$. Thus influence of intergrain boundaries on magnetoresistance for temperatures near the phase transition temperature is substantially lower. This is illustrated by Figure 27b in which the temperature dependences of magnetoresistance in the field of 10 kOe *MR*(*10 kOe, T*) are presented. The shape of the curves near the Curie temperature strongly depend on composition: if for the composition with $X = 0.33$ a well-defined peak near the Curie temperature can be found in the *MR*(*10 kOe, T*) dependence, for the compositions with $X = 0.50$ and 0.55 the peaks are essentially less pronounced, but for the composition with $X = 0.60$ and the temperatures from 300 to 400 K values of magnetoresistance were found to be lower than the lowest possible measured limit of our experimental setup and consequently are not shown in Figure 27b. This is related to the fact the Curie temperature of this composition is equal to ~267 K and the temperature range of 300 ÷ 400 K is the far paramagnetic region for it. For the compositions with $0.33 < X < 0.55$ the values of magnetoresistance at room temperature in the field of 10 kOe decreased from -2.50 % for $X = 0.33$ to – 0.05 % for $X = 0.55$. For all compositions under consideration the shapes of the *dc* magnetic field dependences *MR*(*H,300K*) were close to linear.

As follows from the analysis of the *dc* magnetic and transport properties given above, contributions from structural inhomogeneities and intergranular boundaries are mostly pronounced in transport properties, namely, resistivity and magnetoresistance at low temperatures. The polycrystalline structure of bulk sintered sampled causes some differences in magnetic properties at low temperatures. Near and above the Curie temperature the static *dc* magnetic properties and magnetoresistance of bulk sintered and single crystalline samples are similar.

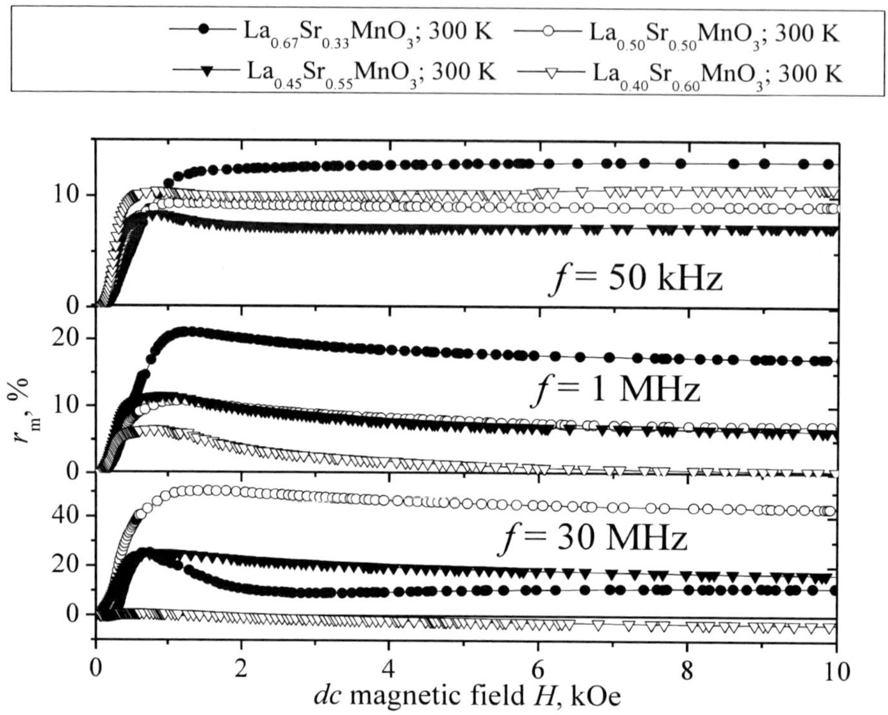

Figure 28. The *dc* magnetic field dependences of penetration coefficient at the frequencies of 50 kHz, 1 MHz and 30 MHz. $T = 300$ K.

The dependence of the module of the penetration coefficient r_m on the strength of external *dc* magnetic field H measured at the frequencies of $f = 50$ kHz, 1 MHz and 30 MHz and the temperature of 300 K is presented in Figure 28. As is seen from Figure 28, a maximum in the region of small fields can be found on the majority of the $r_m(H)$ dependences. Saturation is observed in strong *dc* fields $H \to 10$ kOe in a magnetized condition. For the compositions investigated the temperature of 300 K corresponds to different magnetic states: for the structures with $X \leq 0.55$ this temperature corresponds to a ferromagnetic state, while for the composition with $X = 0.60$ it corresponds to a paramagnetic state. The values of $r_m(H)$ for all compositions and frequencies, except the one corresponding to the composition with $X = 0.60$ and $f = 30$ MHz, are positive.

As it has been shown earlier, see the formula (1.9), the sign r_m it is defined by contributions of relative variations of magnetic permeability and magnetoresistance. Therefore, the positive values of $r_m(H)$, which can be seen in Figure 28, reflects greater contribution from $\frac{\Delta\mu(H)}{\mu}$, in comparison to that from magnetoresistance, since for all compositions, except the one with $X = 0.60$, the temperature of $T = 300$ K corresponds to the ferromagnetic state of the samples. However it is necessary to note, that for the composition with $X = 0.60$, which is in a paramagnetic state at $T = 300$ K, the sign of $r_m(H)$ depends on frequency. As is seen from the data of Figure 28, for the frequencies of $f = 50$ kHz and 1 MHz the values of r_m (H) are positive, while for the frequency of $f = 30$ MHz the values of r_m (H) are negative. Probably, that it is related to the frequency dispersion of magnetic permeability.

Note that for the compositions under investigation the supposition $\left|\frac{\Delta\rho}{\rho}\right| << 1$ is justified, since, as it follows from magnetoresistance measurements, at T = 300 K $|MR(H,\ 300K)|$ < 2.5%, and at T = 400 K $|MR(H,\ 300K)| \leq 1.5\%$ (see Figure 27b). From formula (1.10) it is seen, that under this condition, variations of penetration coefficient between the unsaturated and saturated states of a sample are defined by the initial magnetic permeability and does not depend on direction of the *dc* magnetic field. Therefore, measurements of the $r_m(10\ kOe,\ T)$ dependences for different compositions are of particular interest. The corresponding data are presented for the temperature of T = 300 K in Figure 29. The values of $r_m(10\ kOe,\ T)$ are positive for all compositions with $X \leq 0.55$ for the frequencies up to 20 MHz. For the composition with X = 0.60 they become negative for f > 1 MHz.

With increasing temperature the shape of the $r_m(10\ kOe,\ T)$ dependences is defined by the temperature variations of $\frac{\Delta\rho\ (H)}{\rho}$ and $\frac{\Delta\mu(H)}{\mu}$. Figure 30 shows the $r_m(10\ kOe,\ T)$ dependences recorded for the frequency of f = 50 kHz. For all curves the sign of $r_m(10\ kOe,\ T)$ is positive.

From the presented data it is possible to track variations of $r_m(10\ kOe,\ T)$ upon crossover the temperature of magnetic phase transition: for all compositions except for X = 0.50 the clearly expressed growth of $r_m(10\ kOe,\ T)$ is observed with increasing temperature. For the structure with X = 0.33 with distinctly expressed peak of magnetoresistance curve near the Curie temperature (see Figure 27b) clearly expressed diminution is seen in the $r_m(10\ kOe,\ T)$ dependence in the same temperature range. For the composition with X = 0.55 this diminution is less pronounced, while for the composition with X = 0.60, which is in a paramagnetic state, only growth of $r_m(10\ kOe,\ T)$ is observed. For the two last compositions the values of variations of magnetoresistance in the range of temperature 300 < T < 400 K are substantially smaller, than for the composition with X = 0.33, see Figure 27b. For the composition with X = 0.55 some diminution of the values of $r_m(10\ kOe,\ T)$ upon approaching the Curie temperature is observed, which is probably related to the contribution from the temperature dependence of magnetoresistance. However, above T_C the increase of the $r_m(10\ kOe,\ T)$ values is observed.

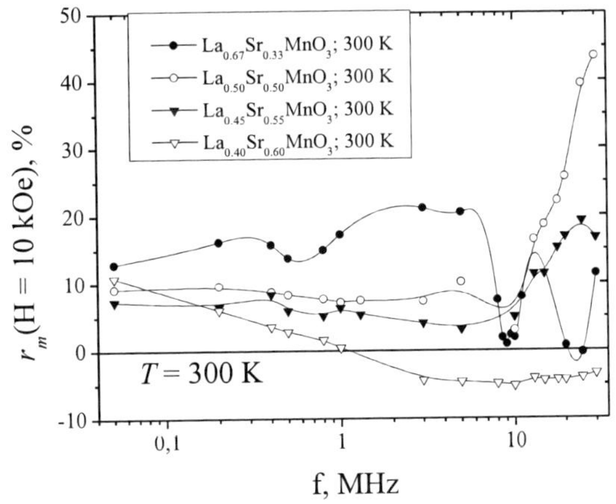

Figure 29. Frequency dependences of the penetration coefficient at T = 300 K in the field of H = 10 kOe.

With increasing temperature to $T = 400$ K all compositions pass to a paramagnetic state. However, as it follows from the data shown in Figure 31, this transition rather weakly influences the shape of the *dc* magnetic field dependences of the module of penetration coefficient at frequencies below 1 MHz: in the corresponding $r_m(H, 400\ K)$ dependences maxima in the weak field still can be found and the curves tend to saturation in the *dc* magnetic fields of an order of 10 kOe. Taking into account the formula (1.9) this can be interpreted as preservation of relatively large value of $\frac{\Delta\mu(H)}{\mu}$. The essential point is that at these frequencies namely the dynamic magnetic permeability is preserved since, as it follows from the data shown in Figure 25, the long-range magnetic order is absent at this temperature. In strong magnetic fields at higher frequencies the contribution from magnetoresistance prevails, and the contribution from variation of magnetic permeability decreases, most likely, because of a frequency dispersion of magnetic permeability. The corresponding frequency dependences of $r_m(10\ kOe,\ T)$ are shown in Figure 32. If for a ferromagnetic state of the compositions investigated the positive sign of $r_m(10\ kOe,\ T)$ was observed for all frequencies (see Figure 29), for a paramagnetic state $r_m\ (10\ kOe,)$ changes sign in the range of frequencies of an order of 1 MHz.

It is interesting to note, that the corresponding dependence for the composition with $X = 0.60$ (see Figure 29) changes sign exactly in the same range of frequencies. From these data and the formula (1.10) it is possible to estimate the value of the initial dynamic magnetic permeability μ (0) at the frequency of 50 kHz for paramagnetic temperature range. These values are equal to 1.084; 1.098; 1.096; and 1.110 for the compositions with $X = 0.33$; 0.50; 0.55; and 0.60, respectively. It is important to note that the values of the initial dynamic magnetic permeability exceeding unity are obtained for all investigated compositions.

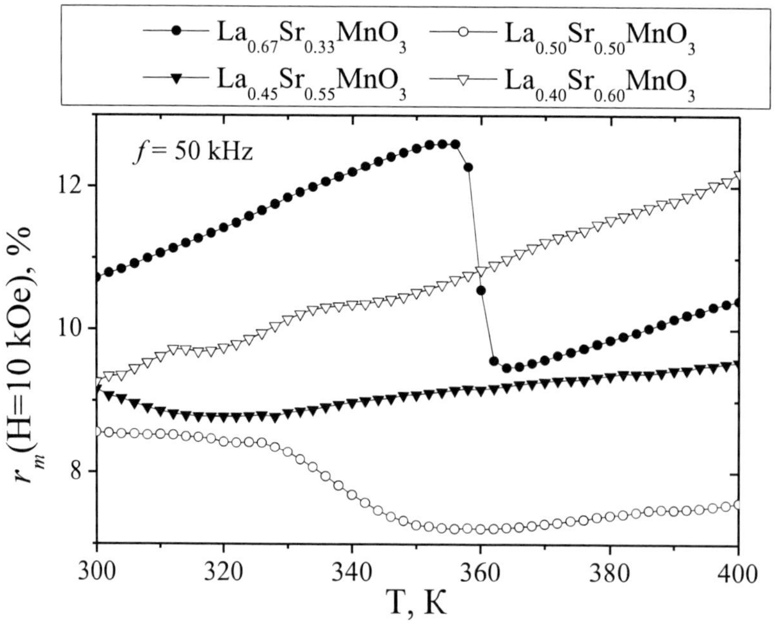

Figure 30. Temperature dependences of the penetration coefficient at $f = 50$ kHz in the field of $H = 10$ kOe.

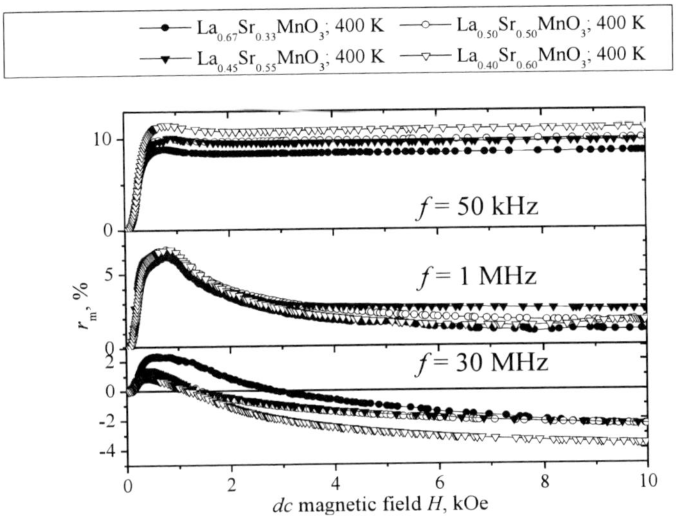

Figure 31. The *dc* magnetic field dependences of the penetration coefficient for the frequencies of 50 kHz, 1 MHz and 30 MHz at $T = 400$ K.

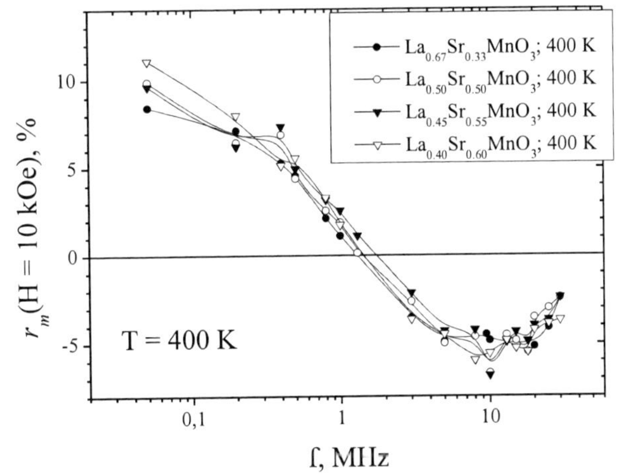

Figure 32. Frequency dependences of the penetration coefficient in the field of $H = 10$ kOe at $T = 400$ K.

In the paramagnetic temperature range just above the Curie temperature unusual magnetic properties were found by several independent methods. In particular, in [51] by the methods of thermal expansion in a magnetic field, measurements of a magnetic susceptibility and small-angle neutronography the local magnetically ordered regions (clusters) were found in doped manganites in the paramagnetic range of temperatures up to $2T_C$. According to this work the size of the clusters was found to be from 0.8 to 2 nm. Similar conclusions have been made as the result of experimental investigations by neutron scattering [52], measurements of heat conductivity [53], oxygen isotope substitution [54], X-ray scattering [55], and EXAFS [56]. As was noted in [57, 58], the unusual magnetic properties are observed in doped manganites in the paramagnetic range of temperatures up to $T/T_C \approx 1.24 \div 1.68$. These features are inherent to the electronic and magnetic subsystems of doped manganites. They are observed in bulk single-crystalline [52], polycrystalline samples [51, 53, 56], and epitaxial thin films [54]. Unlike the low-temperature component of magnetoresistance in weak fields, which is characteristic of bulk polycrystalline sintered samples, these peculiarities of magnetic properties are not related to microstructure of grains or intergrain boundaries.

Theoretically the results obtained can be most consistently considered within the model of electronic phases separation [18, 46]. The basic magnetic state of doped manganites is defined by competition between two kinds of interactions: the ferromagnetic double exchange and the antiferromagnetic superexchange. The ferromagnetically ordered phase has small resistivity while the antiferromagnetically ordered one – large resistivity. The balance between these interactions on a complicated way depends on great number of factors, among which it is necessary to note the structural and magnetic disorder. As it was theoretically shown in [59, 60], within the model, describing the basic properties of manganites, appears the necessity of introduction of additional temperature T^*, exceeding the Curie temperature, below which in a paramagnetic state formation of locally ferromagnetically ordered clusters starts, but a long-range magnetic order is absent in the system. In our opinion, strong variations of the module of penetration coefficient in applied *dc* magnetic field for the frequencies below 1 MHz, and, accordingly, the value of the dynamic magnetic permeability exceeding unity, which are observed in this work for some compositions in the a paramagnetic range reflect the dynamics of variations of orientation of magnetic moments in locally ferromagnetically ordered clusters. From our experimental data the characteristic times, corresponding to average lifetime of these locally ordered clusters, can be estimated about tens and hundreds of microseconds.

1.8. Radiofrequency Electromagnetic Properties of $Pb_{0.4}La_{0.6}MnO_3$ Manganites upon Magnetization with Static Magnetic Field

Potential perspectives of practical utilization of doped manganites with perovskite structure are based on the unique combination of their magnetoresistive and magnetic properties. The possibilities of practical applications of sensors, nonlinear, and other electronic devices with manganites are defined by their properties in a wide range of frequencies. It is possible to control the radio-frequency magnetic parameters using additional *dc* magnetizing field. The parameters of a magnetic material used in an electrotechnical circuit as a transformer core under *dc* magnetizing, have been investigated for a long time. Partial hysteresis loops and magnetization losses at the frequency of 50 Hz for the processes of magnetization of a low-carbon electrotechnical steel by static and high-frequency fields were studied in [61]. The influence of biasing magnetic field on the real part of magnetic permeability and its frequency dependence in soft magnetic nanocrystalline tapes was investigated in [62]. The model for the analysis of work of the transformer under conditions of *dc* biasing with an account for hysteresis and nonlinearity of a magnetization curve was presented and analyzed in [63]. Using this model it is possible to calculate distribution of fields in a core if the results of measurements of static magnetic properties of a core material are known.

Several works [64, 65] were devoted to investigations of the processes of magnetization reversal in soft magnetic polycrystalline materials. Special attention was given to the relation between the reversible and irreversible processes and the formula approximating the *dc* magnetic field dependence of differential magnetic permeability was proposed. The method of penetration of radio-frequency electromagnetic field for investigations of doped manganites was proposed in [31]. In [36] the surface impedance of bulk manganite plate-shaped samples was experimentally investigated for various relations between skin depth and thickness of a sample. Frequency dependences of impedance and magnetoimpedance were investigated for the $La_{0.65}Ba_{0.35}MnO_3$ manganite in [39]. The frequency dependence of magnetoimpedance was found to be nonmonotonic.

Unlike the static *dc* magnetic and transport properties, the physics of the processes of magnetization reversal in manganites under simultaneous action of *dc* and *ac* magnetic fields have been studied insufficiently. The peculiarity of doped manganites, in difference, for example, from metals, consists in the fact that because of their low conductivity the value of skin depth is rather great, and the case of weak skin effect can be realized in a bulk core from manganite even in a wide range of frequencies. It was interesting to investigate the reversible and irreversible processes of magnetic reversal of the $Pb_{0.4}La_{0.6}MnO_3$ manganite in the geometry of a toroidal transformer when the external *dc* magnetic field was applied perpendicular to the radio-frequency *ac* electromagnetic field.

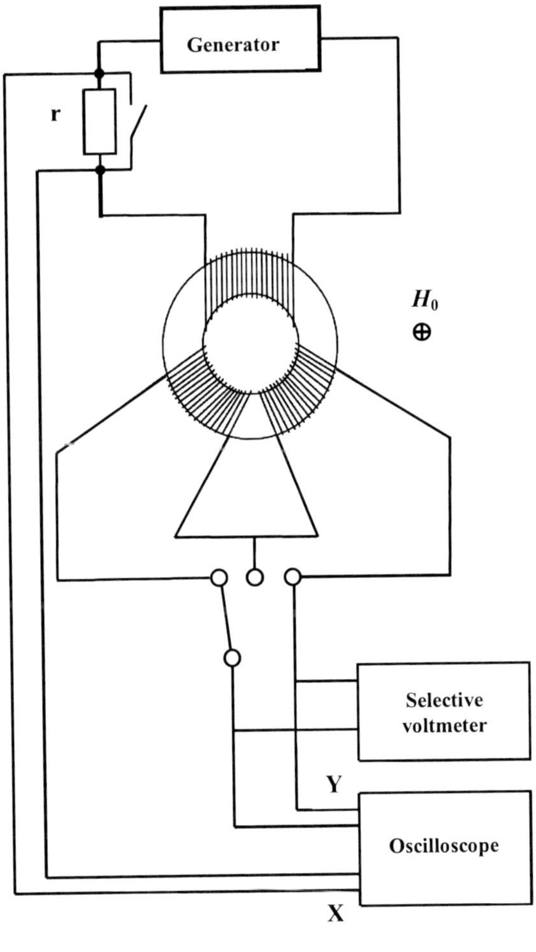

Figure 33. Experimental setup for measurements in the geometry of the toroidal transformer with the core from maganite.

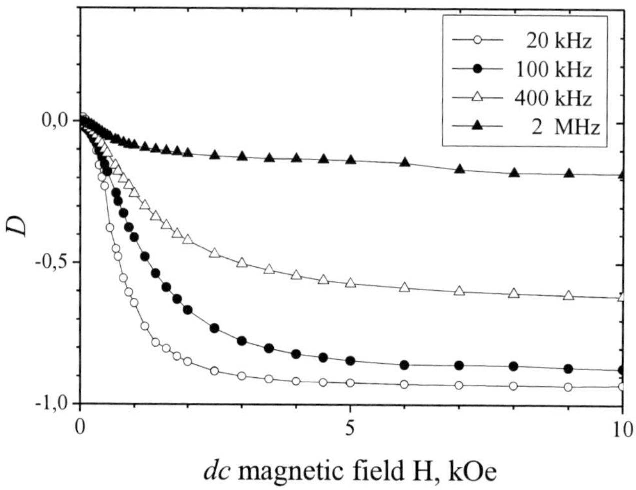

Figure 34. The *dc* magnetic field dependence of the module of transformation coefficient, measured at several frequencies.

The bulk sintered $Pb_{0.4}La_{0.6}MnO_3$ manganite was used as the material of the transformer core. The core had the shape of toroid with the external diameter of 70 mm, internal diameter of 50 mm, and the height of 5 mm. The scheme of measurements is shown in Figure 33. The external *dc* magnetic field was applied perpendicularly to the plane of toroid. Thus, the static *dc* and high-frequency *ac* magnetic fields were perpendicular each other, $H_{\sim}\perp H_0$. The maximum value of the applied *dc* magnetic field was 10 kOe while that of *ac* magnetic field was about 1 Oe. Thus, the condition $|H_{\sim}|<<|H_0|$ was always satisfied. Assuming the relative magnetic permeability μ_r equal to unity, for frequency of $f = 10$ MHz the value of skin depth δ_1 can be estimated to be 0.7 mm. Hence, in the megahertz frequency range the condition of strong skin-effect $d < \delta$ was satisfied both under application of *dc* magnetic field and without it.

The frequency dependence of transformation coefficient of a four-pole network with the core made of manganite is defined by several factors. As shown in section 1.5, in manganites, doped with lead, the frequency dispersion of magnetic permeability is essential at frequencies of a radio range starting from units of megahertz. The imaginary part of magnetic permeability inevitably appears in the presence of frequency dispersion. This is true for our experimental situation when the frequencies from 0.9 to 25 MHz are considered. Further, the primary and secondary windings, with an account for their own inductances and interturn capacitances, form inductively coupled oscillatory contours. This circumstance also must be taken into considering while analyzing the experimental data.

Let us consider first the results obtained in the range of frequencies of 0.02 kHz - 2 MHz. Figure 34 shows the results of measurements of the *dc* magnetic field dependences of the module of transformation coefficient D at several frequencies. It is seen, that variations of D upon application of the *dc* magnetic field are more pronounced at low frequencies. This can be considered as the result of a frequency dispersion of magnetic permeability for the $La_{0.6}Pb_{0.4}MnO_3$ manganite. As was shown in section 1.5, the frequency dispersion becomes essential starting from frequencies of hundreds of kilohertz. The fields of saturation of the radio-frequency dependences in Figure 34 approximately correspond to the saturation field of the manganite composition under investigation.

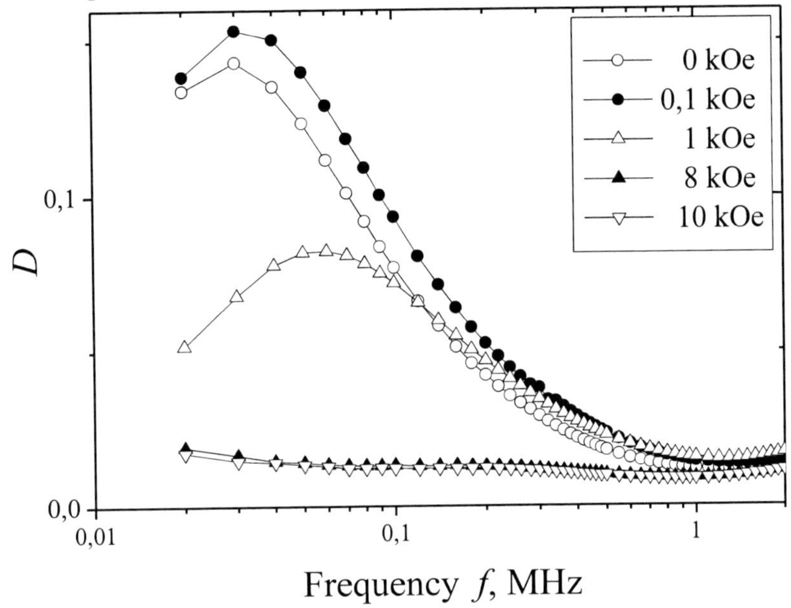

Figure 35. Frequency dependence of the transformation coefficient measured at several values of the *dc* magnetic field.

Frequency dependences of the transformation coefficient, measured in different *dc* magnetic fields, are shown in Figure 35. The general tendency consists in the fact that for the non-magnetized state of the toroid the transformation coefficient starts to decrease abruptly from the frequencies of 100 kHz and above. This confirms the abovementioned supposition about strong frequency dispersion of magnetic permeability in the $La_{0.6}Pb_{0.4}MnO_3$ manganite at the frequencies of hundreds of kilohertz. Magnetization in fields above 1 kOe reduces magnetic permeability which results in increase of the magnetic stray flux and decrease of the transformation coefficient in the whole frequency range under consideration. As is seen from Figure 35, some increase of the transformation coefficient is observed at high frequencies. This is related to realization of resonance conditions in our experimental electrodynamical system.

From a Figure 35 it is seen, that the frequency dependence of the transformation coefficient has a maximum at low frequencies. In Figure 36 the dependence of the maximum value of the transformation coefficient as a function of the *dc* magnetic field is shown. The maximum value of *D* abruptly decreases with increasing *H*. In Figure 37 the dependence of the frequency $f_{1/2}$, at which the value of *D* decreases by twice in comparison with its maximum, is shown as the function of the *dc* magnetic field. It is obvious, that there is no strong *dc* magnetic field dependence of $f_{1/2}$. This means that variations of the transformation coefficient with frequency shown in Figure 35 are really caused by a dispersion of magnetic permeability, and the influence of redistribution of stray fields, with an account of changes of conditions of realization of skin-effect, does not render essential influence.

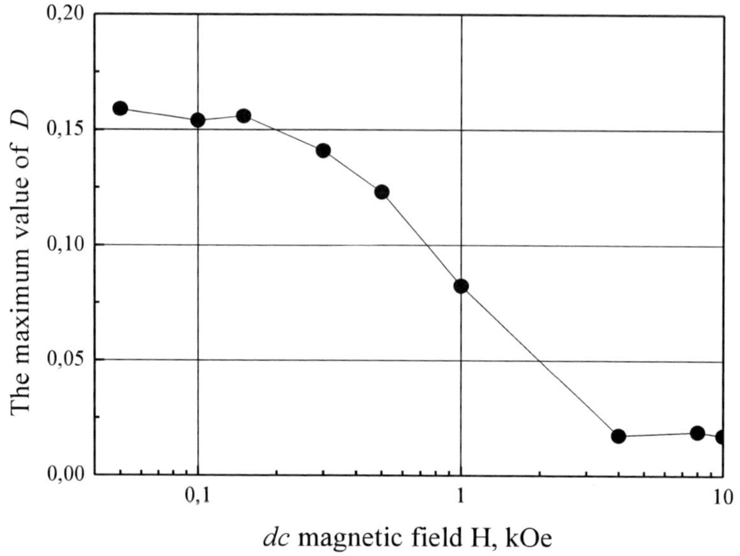

Figure 36. Dependence of the maximum value of the transformation coefficient on intensity of the *dc* magnetic field.

Returning to Figure 35 one can notice, that the dependences of the transformation coefficient obtained for $|H_0| \leq 1$ kOe increase in small frequencies with increasing frequency. This is more pronounced in the range of the lowest frequencies, see Figure 38. This figure shows the frequency dependence of the output voltage of secondary winding in a wide frequency range. Note the logarithmic scales in this figure. Measurements were carried out for $H_0 = 0$ and several values of the exciting voltage. These results show, that linear frequency dependence of the output voltage is observed. Such linear dependence can be expected in the system the magnetic chain of which substantially includes stray fluxes under the condition that in a considered range of frequencies the frequency dispersion of magnetic permeability is not important.

Thus it is shown, that application of the *dc* magnetic field in the range of frequencies 0.02 kHz - 5 MHz results in decrease of the transformation coefficient the four-pole network containing the core made from the $La_{0.6}Pb_{0.4}MnO_3$ manganite. These variations are most essential at frequencies below 100 kHz, and the values of transformation coefficient in a magnetic field, sufficient to saturation, are by 5 - 10 smaller, than without an external magnetic field. Redistribution of magnetic flux over cross-section of the core because of variations of conditions of skin-effect upon frequency variations does not substantially influence the module of the transformation coefficient. Such considerable variations of parameters of the four-pole network containing the magnetic chain from manganite, can be useful for practical applications in magnetoelectronic devices. To achieve the maximum variations of the transformation coefficient it is necessary to choose the manganites with the greatest initial differential permeability.

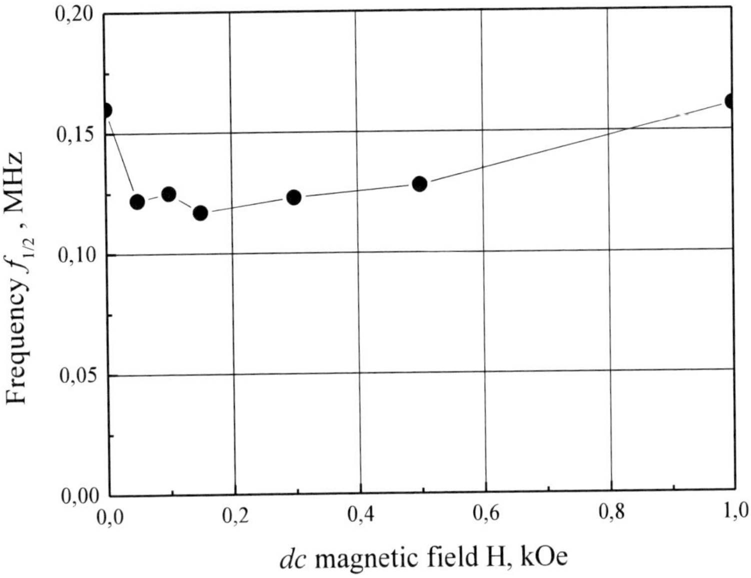

Figure 37. The *dc* magnetic field dependence of the $f_{1/2}$ frequency at which the value of D decreases by twice in comparison to the maximum one.

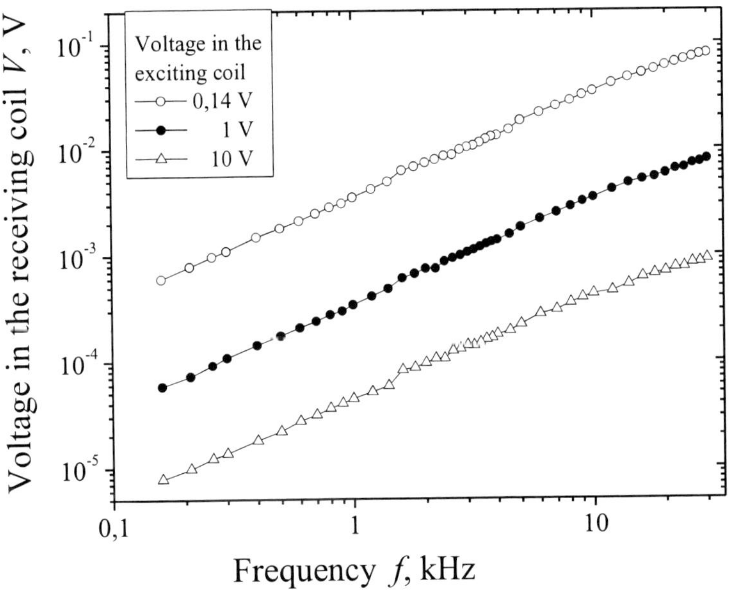

Figure 38. Frequency dependence of the output voltage of secondary winding, measured in a wide interval of variations of frequency at several values of exciting voltage in the primary winding.

Let us consider now the results obtained in the range of frequencies up to 31 MHz. The results of measurements of the *dc* magnetic field dependences of the module of transformation coefficient are shown in Figure 39. Figure Figure 39a, in which the frequency dependence of the transformation coefficient in zero *dc* biasing field is shown, illustrates the presence of dispersion. In the range of frequencies from ~ 0.5 MHz to 2 MHz reduction of the transformation coefficient, which is caused by the frequency dispersion of magnetic permeability of the manganite core, is clearly seen. The resonance of the transformation coefficient starts to appear at frequencies about 10 MHz. The influence of the *dc* biasing magnetic field on the position and amplitude of resonance is illustrates by Figure 39b. With increasing *dc* magnetic field the position of the resonant maximum shifts to higher frequencies, as is seen from Figure 40. The size of the resonant maximum as a whole increases with increasing the *dc* magnetic field, see Figure 41. In the strongest fields of about 10 kOe the resonant line splits and two maxima are seen in the frequency dependence of *D*. For the experimental conditions in which the system shows one resonant maximum, it is possible to introduce the parameter *Q* similar to the quality factor of an oscillatory contour $Q = f_0 / \Delta f_{1/2}$, where f_0 is the resonant frequency, $\Delta f_{1/2}$ is the interval of frequencies at the edges of which the module of the transformation coefficient is equal to the half of its value in the resonance. The value of the quality factor characterizes the losses of the radio-frequency power in a system. The losses contain the contributions from eddy currents and from magnetic reversal in the manganite core. Besides, the value of *Q* is influenced by variations of the fraction of stray fluxes upon *dc* magnetization. In Figure 42, the dependence of quality factor on the *dc* magnetic field is shown.

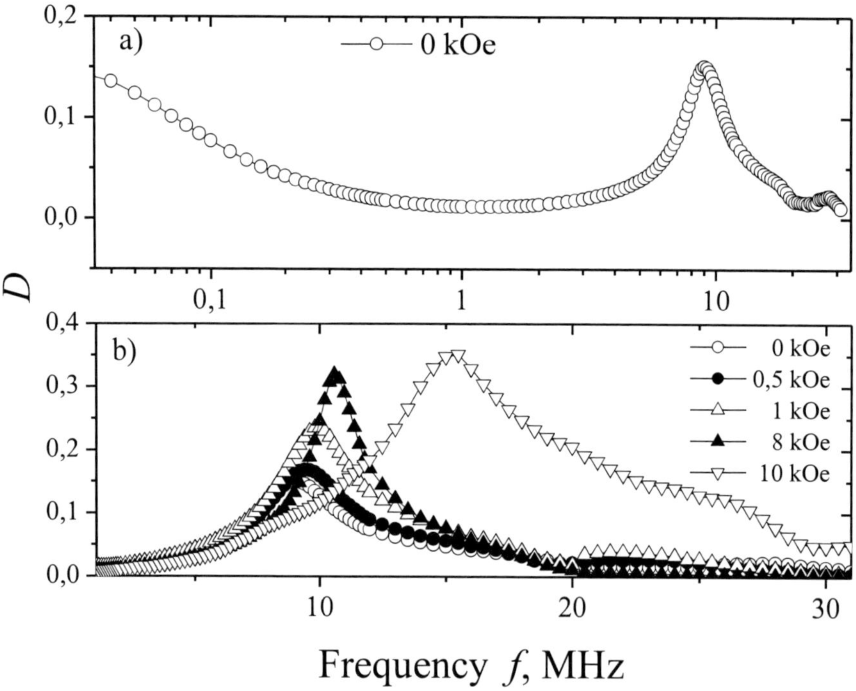

Figure 39. Frequency dependence of the transformation coefficient (a) - the dependence measured at $H = 0$; (b) - the dependences measured at several values of the static *dc* biasing field.

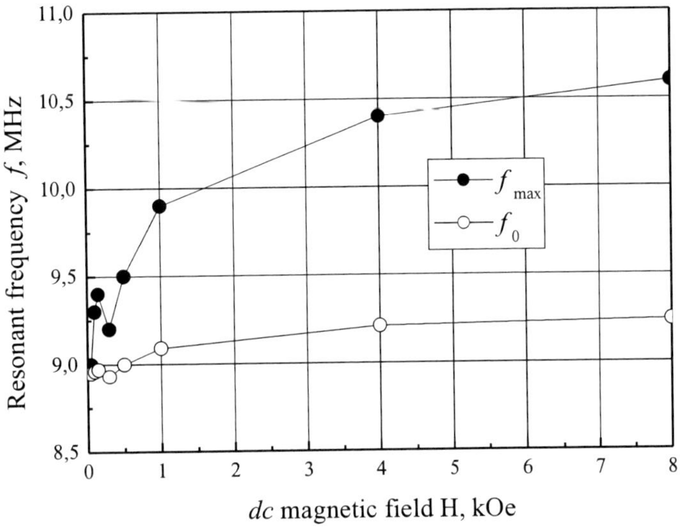

Figure 40. The dependence of the resonant frequency on the *dc* biasing magnetic field.

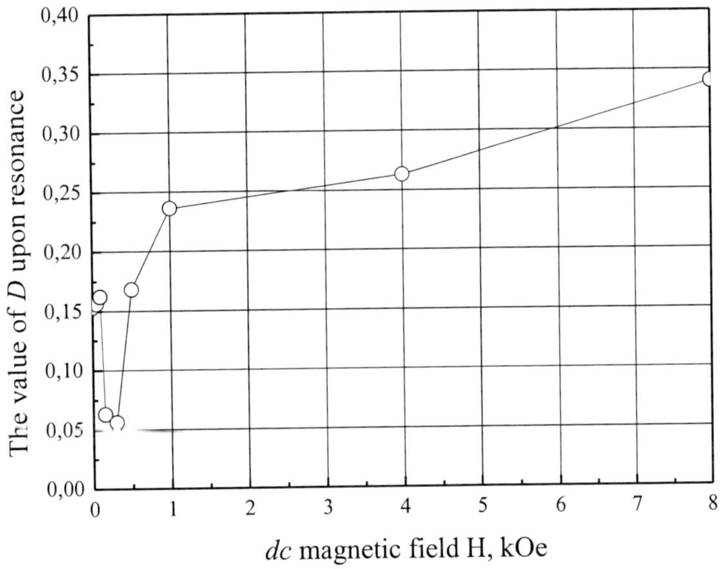

Figure 41. The dependence of the maximum value of transformation coefficient at resonance on the *dc* biasing magnetic field.

The dependence of the transformation coefficient on the dc biasing magnetic field was also investigated. The results of these measurements are presented in Figure 43. In this figure the vertical axis corresponds to the relative variations of the transformation coefficient $r_m = [D(H)-D(0)]/D(0)\cdot 100\%$. As is seen from Figure 43, depending on frequency application of the *dc* magnetic field can either increase, or reduce relative variations of the transformation coefficient. Reduction of *D* occurs at frequencies below 5 MHz and expresses the fact, that with increasing biasing *dc* magnetic field the magnetic permeability decreases and the contribution from stray fluxes increases. The increase of the transformation coefficient in the *dc* magnetic field is of great interest. Such increase occurs at the frequencies where the resonant frequency dependence of the transformation coefficient is realized. As is seen from Figure 43, the increase of D can be rather substantial. For example, at the frequency of $f = 25$ MHz the value of the transformation coefficient as large as 240 % relative to the value corresponding to zero *dc* magnetic field, can be reached. The phase shift ϕ between the signals in the primary and secondary windings was also measured. The results are shown in a Figure 44 as function of the applied *dc* magnetic field. At the frequencies above 5 MHz the *dc* magnetic field dependence of phase shift of becomes much more pronounced. The phase shift can be both positive, and negative. Application of the *dc* magnetic field results in increase of ϕ at some frequencies, while at other frequencies - to its reduction.

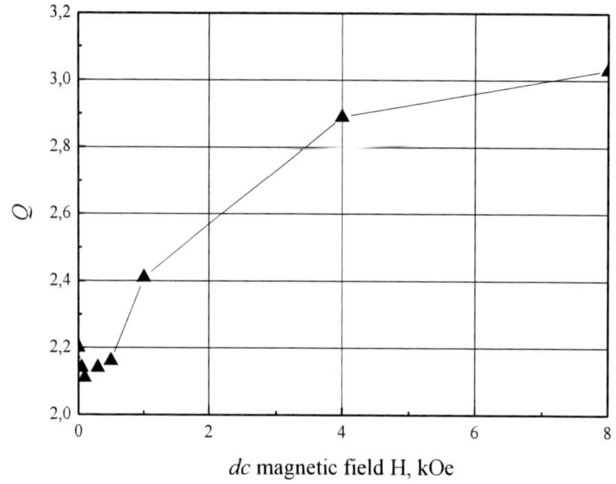

Figure 42. The *dc* magnetic field dependence of the quality factor Q.

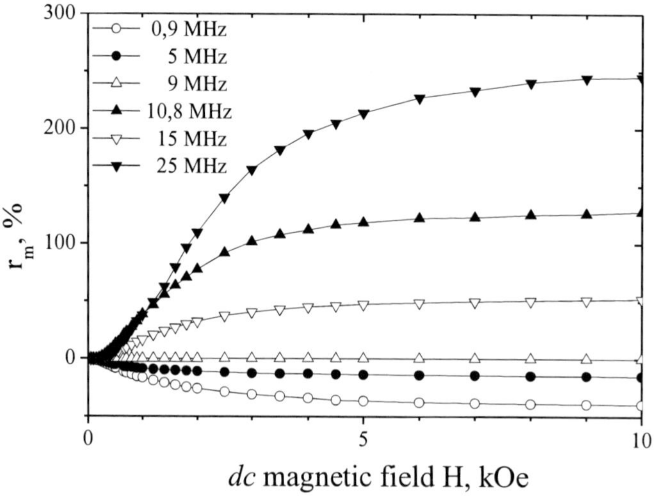

Figure 43. The *dc* magnetic field dependence of the transformation coefficient measured at several frequencies.

As was mentioned above, the reason of the resonant frequency dependence of transformation coefficient of the four-pole network consists in formation of the coupled resonant contours in the chains of the primary and secondary windings.

It is known, that the module of the transformation coefficient of couple contours is expressed by the formula

$$|D| = \frac{k^2Q^2}{\sqrt{\left(1 - a^2 + k^2Q^2\right)^2 + 4a^2}}, \tag{2.1}$$

where $a = 2Q\Delta f / f_0$ is the dimensionless deviation from a resonance, $k = M / \sqrt{L_1 L_2}$ - is the coefficient of coupling of contours, L_1, L_2 are the inductances of primary and secondary windings, respectively, M is the mutual inductance. When $kQ < 1$ the dependence (2.1) represents a resonant curve with one maximum. In a contour with losses the frequency f_{max}, at which the maximum is observed, differs from the resonant frequency f_0 of a contour without losses.

$$f_{\max} = \sqrt{f_0^2 - \frac{1}{2}\left(\frac{f_0}{Q}\right)^2} \tag{2.2}$$

If f_{max} and Q are known from the experiment, using the formula (2.2), it is possible to restore the dependence of the resonant frequency f_0 on the strength of the *dc* biasing magnetic field. It has been made, and the result is shown in Figure 40. The fact that the resonant frequency increases with intensity of the *dc* biasing magnetic field is related to decrease of the L_1 and L_2 inductances because of reduction of the magnetic permeability. It is possible to express the power of losses P_0 through the quality factor Q:

$$P_0 = \frac{2\pi f_0 W_0}{Q}, \tag{2.3}$$

where W_0 is the energy of oscillations in a contour at a resonance. The power P_0 consists from the eddy currents P_e and the power of losses for magnetization reversal P_h. For the unit of volume of ferromagnetic material the power of losses from eddy currents can be expressed as

$$P_e = \frac{1}{2}\sigma k_s^2 d^2 f^2 B_m^2. \tag{2.4}$$

here σ is the conductivity of manganite, k_s is the coefficient of the current shape in a primary winding, B_m is the peak value of the magnetic induction caused by a high-frequency magnetic field. In our experiments the amplitude of the high-frequency magnetic field was kept constant and has the value of H_m = 0.4 Oe. With increasing the strength of the *dc* static magnetic field the amplitude of B_m decreases because of reduction of magnetic permeability. The power of losses for magnetization reversal P_h is connected to the area of hysteresis loop

$P_h = f\oint HdB$ where integration is carried out over a hysteresis loop upon magnetization reversal by a high-frequency magnetic field. Using the harmonious approximation of the process it is possible to write

$$P_h = 2\pi f\mu_0\mu_2 H_m^2, \tag{2.5}$$

where μ_2 is the imaginary part of the complex magnetic permeability $\mu = \mu_1 - i\mu_2$. The increase in quality factor Q with increasing the *dc* biasing magnetic field can be caused both by reduction of the part of magnetic permeability $\mu_1 = B_m/H_m$ and related to it component of power of losses for eddy currents, and also by the imaginary part μ_2 and the power of losses for magnetization reversal.

The phase shift between the voltages in the primary and secondary windings and its dependence on the strength of the *dc* biasing magnetic field is defined by the complex resistance of a magnetic core $Z_m = R_m + i2\pi f L_m$, where

$$R_m = 2\pi f\mu_0\mu_R w^2 S/l,\ L_m = \mu_0\mu_L w^2 S/l \tag{2.6}$$

w is the number of windings of a coil, $S = (D_{max} - D_{min})\ d/2$ is the cross-section area of the core, $l = (D_{max} + D_{min})/2$ is an average length of a magnetic force line. The different magnetic permeabilities $\mu_R = |\mu| + \mu_2$ and $\mu_L = |\mu| - \mu_2$ enter the expressions for the real and imaginary parts of Z_m. They depend in different ways on the strength of the *dc* magnetic field. Thus, the phase shift must change upon magnetization. The difference between the μ_R and μ_L magnetic permeabilities exists only if the imaginary part of the magnetic permeability is present. At frequencies below 1 MHz, where the frequency dispersion of magnetic permeability is not so essential, variations of phase shift with increasing *dc* magnetic field are not expected. The data of Figure 44 confirm this statement.

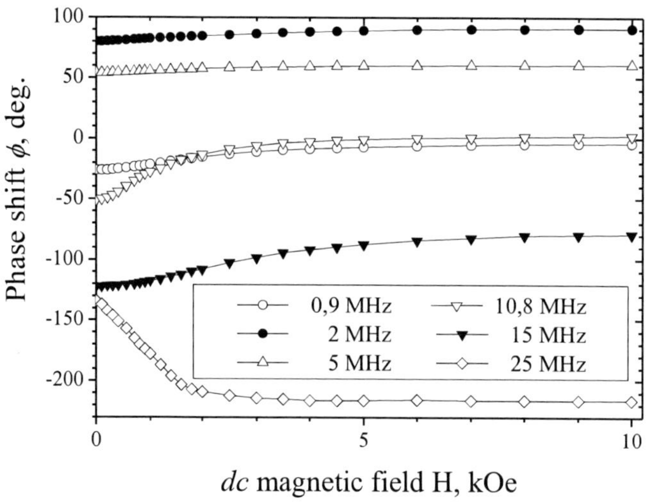

Figure 44. The *dc* magnetic field dependence of the phase shift between the voltages in the primary and secondary windings, measured at different frequencies.

Thus, the frequency dependence of the transformation coefficient of the four-pole network with the circular core made from the $Pb_{0.4}La_{0.6}MnO_3$ manganite has the shape of the curve with resonant maximum at megahertz frequencies. The position of the maximum and its amplitude increase with application of the *dc* biasing magnetic field. The electromagnetic losses in the system, consist of contributions form magnetization reversal and eddy currents, decrease with the *dc* biasing magnetic field. This property distinguishes the core made from manganite from typical cores made from soft magnetic metallic materials with high conductivity.

The shape of the *dc* magnetic field dependence of the transformation coefficient changes with frequency. At low frequencies the values of D decrease with increasing *dc* magnetic field because of reduction of magnetic permeability and increase of stray fluxes. In the range of frequencies of units of megahertz the increase of the transformation coefficient with the *dc* magnetic field is observed. The phase shift between the voltages in the primary and secondary windings of the four-pole network was observed. Its value was found to be dependent on the strength of the *dc* magnetic field. These results are interpreted taking into account the existence the imaginary part of magnetic permeability. Strong dependence of the transformation coefficient on the intensity of the *dc* biasing magnetic field demonstrated the principal possibility of creation of electronic components with the cores made from manganites.

1.9. Complicated Dynamics of Oscillation of Magnetic Flux in Bulk La-Pb Doped Manganite

The dynamics of processes of magnetic reversal in doped lanthanum manganites in crossed radio-frequency and *dc* magnetic fields is very interesting. It was found, that there are conditions when the complicated oscillation variations of magnetization vector are observed. The results reported in this section, were obtained using the same manganite sample which is shown in Figure 33. The core had the shape of toroid to minimize the stray fluxes. The range of frequencies of the *ac* magnetic field from hundreds hertz to tens kilohertz have been chosen in such a way that it was possible to realize the regimes of both weak, and strong skin-effect.

At first we will consider the general problem of magnetization of ferromagnets in two crossed magnetic fields: the *ac* radio-frequency and *dc* static.

Even in the rather simple case when $|H_{\sim}| \ll |H|$ the processes of dynamic magnetization of ferromagnets are studied briefly, not only concerning manganites. The physical essence of the magnetization processes by *dc* and *ac* magnetic fields is described in detail in the monograph [66]. The processes of magnetization by *dc* and *ac* magnetic fields were also considered in [61]. The asymmetrical dynamic hysteresis loops were investigated. The conclusion that even in the limiting case under consideration $|H_{\sim}| \ll |H|$ the value of *ac* magnetization is defined both by H, and the amplitude of the *ac* magnetic field $H_{\sim}$. The frequency dependence of the real part of magnetic permeability of soft magnetic nanocrystalline FeNbBCu material upon *dc* magnetic field biasing was investigated in [62]. The dynamic hysteresis loops of the composite material based on manganese-zinc ferrite were studied in [67] for the frequencies up to 20 kHz. With increasing frequency, the hysteresis loop becomes wider, and its shape approaches to ellipse. In the description of the processes of dynamic magnetization the account of a magnetic hysteresis is the most difficult. The model of an equivalent circuit, which was adapted for the numerical analysis of the processes of magnetization, has been developed in [63]. In this work the scheme of calculation of distribution of magnetization over cross-section of a magnetic core, with an account for hysteresis, have been developed.

The analysis of the equations of movement of magnetization in a film or a particle is given in [68]. It was established, that the phase portrait of a dynamic magnetic system studied in the coordinates, which are the functions of the components of magnetization vector, is rather complicated. It contains the limiting cycles describing periodic movement, saddle point through which a separatrix passes, the points of the "focus" type, to which the helicoids-like phase trajectories of a dissipative system converge. In [68] the method of solving the Landau-Lifshitz equations for magnetic systems under consideration was proposed. However, this consideration was limited only by small values of damping parameter. Besides, in these calculations the magnetic hysteresis was not taken into consideration. This factor can essentially complicate a phase portrait.

The magnetic state of a ferromagnet under simultaneous magnetization by the magnetic fields of different frequencies was considered in [69]. If one of the fields is the *dc* one, while the other is the *ac* one and their vectors are perpendicular each other, the magnetic induction B and the field H at some point of a media are related as $\boldsymbol{B} = f(\boldsymbol{H})$. This relation expresses the magnetization curve, written with the account for the fact that the directions of a magnetic field and induction may not coincide. If the strength of an external *dc* magnetic field is much higher than the strength of the *ac* magnetic field, $|H_{\sim}| << |H|$, then

$$\boldsymbol{B} = f(\boldsymbol{H}_0) + \sum_{i=1}^{3} \boldsymbol{e}_i \sum_{k=1}^{3} \mu_k^i H_{\sim k},$$

where e_i is the unit vector along the i-th axis, $H_{\sim k}$ is the k-th component of the *ac* strength of magnetic field. The values μ_k^i are expressed by the formulas $\mu_k^i = \dfrac{\partial f_i}{\partial H_{\sim k}}$, where f_i is the magnetization curve written down for the i-th components of a magnetic induction. The values of μ_k^i form the tensor of second rank. If without application of the *dc* magnetic field a ferromagnet is isotropic, and the vector of H is directed along the x axis ($i = 1$) the component $\mu_1^1 = \dfrac{\partial f_1}{\partial H_{\sim 1}}$ is the longitudinal reversible permeability, and $\mu_2^2 = \mu_3^3 = \dfrac{\partial f_2}{\partial H_{\sim 2}} = \dfrac{\partial f_3}{\partial H_{\sim 3}}$ is the transversal reversible permeability.

The dynamics of magnetization reversal of the $La_{0.65}Sr_{0.28}MnO_3$ manganite and nickel-zinc ferrite upon simultaneous application of the low-frequency *dc* and the radio-frequency *ac* magnetic fields was considered in [70]. In this work the vectors of all fields were parallel each other. It was revealed, that the component of a signal measured at the frequency of the low-frequency signal, has modulation.

The analysis presented shows, that in the published literature there is no detailed description of the processes of magnetization in *dc* and *ac* magnetic fields, taking into account nonlinearity of the magnetization curve, magnetic hysteresis, and the of influence of skin-effect. The available experimental results show, that realization of dynamic hysteresis loops of the similar to an ellipse, and also asymmetrical loops is possible. Considering the processes of magnetic reversal in the crossed *dc* and *ac* magnetic fields in doped lanthanum manganites with moderate conductivity, it is necessary to consider, in principle, the fact, that the conductivity of these materials can strongly depend on frequency [35].

For the $La_{0.6}Pb_{0.4}MnO_3$ material of the core at a room temperature ($T = 295$ K) and $H = 0$ the value of resistivity is 0.2 Ω*cm. For this value of resistivity, the frequency of $f = 10$ kHz, and assuming the magnetic permeability μ_r equal to unity, one can obtain for skin depth the value of $\delta_1 = 2.3$ mm. This means that at low frequencies of radio range, when $f \ll 10$ kHz, the condition of weak skin-effect is realized in the toroid core and $d < \delta$. At the frequencies in tens and hundreds kilohertz the condition of strong skin-effect $d > \delta$ is realized. The value of the *dc* magnetoresistance, measured in the field of $H = 10$ kOe, is ~4 %. The value of the *dc* coercive force at a room temperature was found to be $H_c = 10$ Oe.

Because of nonlinearity of a magnetization curve the $H(t)$ and $B(t)$ dependences can not be sinusoidal simultaneously, but separately each of these dependences can be sinusoidal. We studied the conditions at which complicated nonsinusoidal oscillations of magnetic flux were observed. One of possible variants of the experiments consists in providing sinusoidal (or close to sinusoidal) variation of the *ac* excitation field. Usually it is reached by inclusion of a ballast resistance in a magnetization chain, as is shown in the scheme of Figure 33.

The primary winding was fed from the high-frequency generator of sinusoidal voltage. At small amplitudes of exciting voltage such connection provides the shape of a current in the primary excitation winding close to the sinusoidal. The maximum strength of a high-frequency magnetic field excited in the core of manganite reached 1 Oe. The voltage from one of the secondary windings was measured by a selective voltmeter, or applied to the *Y* plates of oscilloscope. The voltage from the ballast resistor *r*, which is proportional to the current in a primary winding, and, hence, to the strength if the *ac* excitation magnetic field, was applied to the *X* plates of oscilloscope. The voltage from the secondary winding is proportional to *dB/dt*. Thus, the image at the oscilloscope represented the projection of a phase portrait of oscillations of a magnetic induction to the ($H_{\sim}$, $\dot{B}$). In the other variant of measurements two identical secondary windings were switched subtractively. In this case the first harmonics of the output voltage from these windings were subtracted. Therefore in the output voltage applied to the *Y* plates of oscilloscope, a great fraction of higher harmonics existed as well as of other components of the signal, which originated from the nonlinearity of the dynamic magnetic properties of manganite. In our experiments the condition $|H_{\sim}| \ll |H|$ was satisfied if $|H| \neq 0$, However the module of the strength of *ac* magnetic field $|H_{\sim}|$ can not be consider as small in comparison with the parameters of the hysteresis loop, for example, of the coercive force. Thus at radiofrequencies the magnetic state of the system is defined by both vectors $H_{\sim}$ and H. In Figure 45 the oscillograms of the current I in the magnetization winding, which is proportional to $|H_{\sim}|$, are shown. On each pair of the oscillograms of Figure 45 the bottom signal corresponds to the current. The top signal of the oscillogram is the voltage V from the secondary winding. The voltage V is equal to $-d\Psi/dt$, where $\Psi = w\boldsymbol{B}_{\sim} \cdot \boldsymbol{S}$ is the *ac* part of the flux linkage, S is the vector oriented in the direction, perpendicular to the planes of windings and the size equal to the cross-section area of the toroid core from manganite. The oscillograms were recorded at the generator frequency of $f = 15$ kHz and the excitation voltage of $U = 7.5$ V. Each pair of oscillograms in Figure 45 was measured at certain value of the *dc* magnetic field H_0. In zero *dc* magnetic field the I (t) and V (t) dependences visually were close to sinusoidal. This can be expected since $|H_{\sim}| << |H|$ and when $|H_{\sim}|$ is essentially smaller that the fields, characterizing the limiting hysteresis loop, for example, H_c and H_s, (where H_c is the coercive force and H_s is the saturation field). With increasing the strength of the *dc* magnetic field at first the amplitude of the *ac* current I increases, and the amplitude of the voltage V decreases. The shape of the dependences remains approximately sinusoidal, see the pair of oscillograms recorded at $|H| = 0.83$ kOe. These changes are the obvious consequences of reduction of magnetic permeability with increasing *dc* magnetic field. The detailed analysis the V (t) voltage executed using the selective voltmeter, showed the presence of harmonics in the time dependence. However the value of amplitudes of harmonics is by many times smaller than the amplitude of the first harmonic. At the applied *dc* magnetic field of $|H| = 0.94$ kOe a peculiarity appears in the $V(t)$ dependence, which develops upon further increasing the strength of the *dc* magnetic field. At $|H| = 1.5$ kOe it becomes obvious, that the peculiarity in the voltage oscillogram $V(t)$ corresponds to the process of shock excitation of voltage oscillations. The beginning of the flattened part of the oscillogram corresponds to the moment of shock excitation of oscillations of V (t). Upon further increase of the strength of the *dc* magnetic field the initial amplitude of damping oscillations upon shock excitation also increases. In the fields of ~10 kOe the second region of damping oscillations appears in the V (t) oscillogram. The moment of their occurrence corresponds to the termination moment of the flattened part of the current oscillogram I (t). Attentive investigation of the oscillograms shows, that the abrupt variation of the voltage V occurs at the beginning of oscillations. The frequency of oscillations changes slightly with variation of the excitation. The frequency of oscillations is in the range from 0.2 to 0.5 MHz

The presence of the oscillatory contour formed by the inductance and interturn capacitance of the windings can be considered as the most straightforward explanation of the reason of excitation of oscillations. If this assumption is fair, the frequency of oscillations should be close to the resonant frequency of this contour.

The frequency dependence of voltage V were measured in various *dc* magnetic fields $|H|$ at small amplitude of generator $U = 0.1$ V in order to check this assumption. The results are presented in Figure 35 (the range of frequencies from 0.02 to 2 MHz) and in Figure 39 (the range of frequencies from 2 to 31 MHz). The variations which occurring at small frequencies f<1 MHz, are caused by the frequency dispersion of magnetic permeability. In the process of magnetization of manganite the initial value of D, which corresponding to the lowest frequencies ~ 10 ÷ 100 kHz, decreases. Variations of the resonant nature appear at the frequency dependence in the range of frequencies between 8 - 12 MHz. These frequencies are much higher, than the frequencies at which the shock excitation of oscillations presented in Figure 45 were observed. No abrupt peculiarities can be seen in Figure 35 in the range of frequencies of 0.2 – 0.5 MHz. Figure 34 shows the *dc* magnetic field dependence of the transformation coefficient D measured at several frequencies. With increasing the strength of the *dc* magnetic field value of D decreases. No peculiarities can be found in these dependences. From these experimental data it is seen, that excitation of damping oscillations in Figure 45 is not related to the resonant oscillations of a contour, and is caused by nonlinearity of the process of dynamic magnetization reversal of manganite over the partial dynamic hysteresis loop at a finite amplitude of the excitation *ac* field.

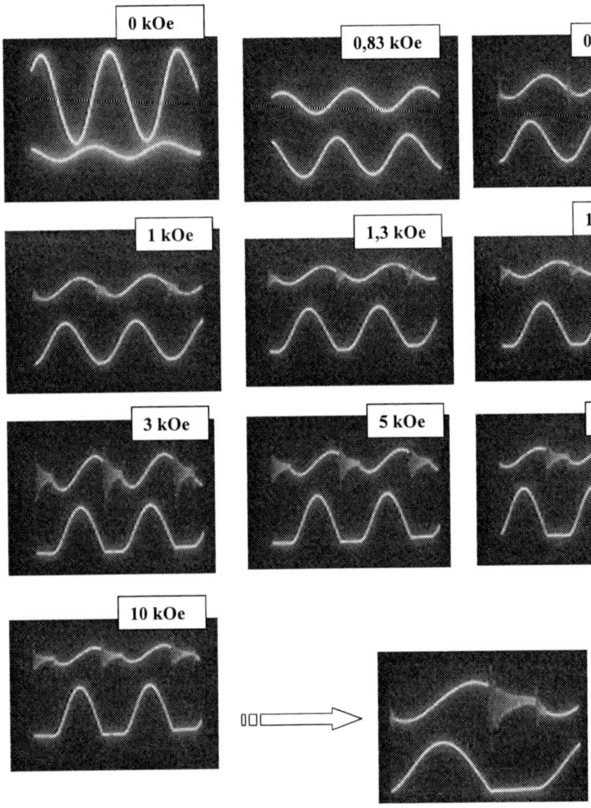

Figure 45. The oscillograms of voltage in the secondary winding (the upper signal) and of the current in the magnetizing winding (the lower signal), obtained at $V = 7.5$ V, $f = 15$ kHz and various values of the *dc* magnetic field.

For explanation of the nature of oscillations let us consider at first the process of magnetic reversal in crossed *ac* and *dc* magnetic fields. The equation of the dynamic system, characterized by magnetic induction B, and the strength of magnetic field H which have variables and constant components, can be written as

$$\frac{d\boldsymbol{B}}{dt} = F(\boldsymbol{B}, \boldsymbol{H}) \cdot \qquad (3.1)$$

The voltage, recorded in the secondary winding, is equal to

$$V = \frac{d\Psi}{dt} = \frac{d(w\boldsymbol{B}_{\sim} \cdot \boldsymbol{S})}{dt} = F'(\Psi, i) \cdot \qquad (3.2)$$

As was mentioned above upon the analysis of oscillograms in Figure 45, at the moment of time, preceding the excitation of shock oscillations, abrupt variation of voltage occurs. Let t_1 denote the moment of the beginning of jump and V_1 the amplitude of the jump. As is seen from Figure 45, in the time interval just after beginning of the jump, the flux linkage varies with time as:

$$\Psi = \int_{t_1}^{t+\Delta t} V(t)\,dt = V_1\Delta t + a_1 e^{-\lambda\Delta t}\sin(\omega\Delta t + \varphi_1), \tag{3.3}$$

where the considered period of time Δt is limited by duration of the flattened part of the current oscillogram I (t). In the formula (3.3) a_1 and φ_1 are the amplitude and a phase of oscillations of magnetic flux, respectively, $\omega = 2\pi f_0$ is the circular frequency of these oscillations, and λ is their damping constant. It is possible to estimate the duration of jump assuming, that the partial dynamic hysteresis loop has the rectangular shape. Then $\frac{d\Psi}{dt} \neq 0$ only at the critical value of the current i_k at which magnetization reversal of the core material occurs, $i_k = H_k \frac{l}{w_1}$. Here H_k is the field at which magnetization reversal of the core material occurs, l is the average length of the core, w_1 is the number of turns in the primary winding. Then $\frac{d\Psi}{dt} = U - i_k r$, from where to estimate the duration of magnetic reversal T_r it is possible to obtain $T_r \approx \frac{\Delta\Psi}{U - i_k R}$, where $\Delta\Psi$ is variation of flux linkage upon magnetization reversal, $\Delta\Psi = 2w_1 B_0 S$, where B_0 – is variation of magnetic induction upon magnetization reversal [66]. For l = 18 cm, S = 0.25 cm^2, w_1 = 100, B_0 = 0.5 kG, i_k = 100 mA, U = 5 V, and r = 50 Ohm we obtain, that $T_r \approx$ 0.2 ms. This value is close to the duration of flattened parts in the current oscillograms in Figure 45 when magnetization reversal of the core material occurs. In [66] the approximating formula relating the voltage in the current winding with the flux was offered assuming the case when the shape of the hysteresis loop can not be considered as the rectangular one

$$V = V_0 + (I - I_k)a - (\Psi - \Psi_0)b, \tag{3.4}$$

where a and b are some material constants. It follows from (3.4), that

$$V = ai - b\int V dt + c,$$

where c is the constant of integration. For a chain containing the magnetic core, application of the equation (3.4) leads to the equation

$$U - ri - L\frac{di}{dt} = V = ai - b\int V dt + c.$$

Substituting the left part of latter equation under the sign of integral and integrating, we obtain

$$U + b\int U dt - c = Ri + L\frac{di}{dt} + \frac{1}{C}\int i dt. \tag{3.5}$$

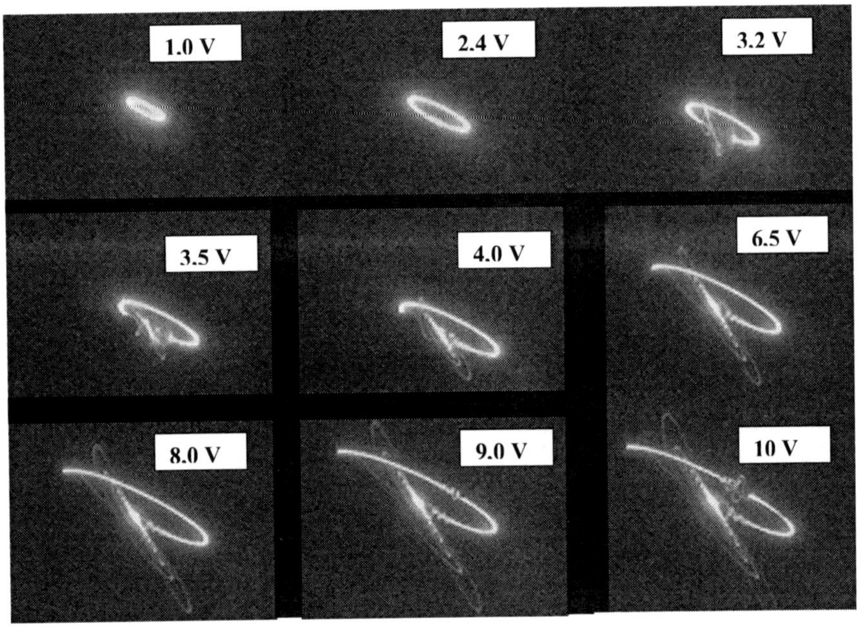

Figure 46. The projection of the phase portrait to the H_m - $\dot{\Phi}$ plane at different values of the excitation voltage U, f= 15 kHz, H_0= 2.5 kOe. For $U \geq 3.2$ V one or two regions of shock excitation of magnetic flux can be seen in the phase portrait.

Here L is the inductance of the coil, $R = r + a + Lb$, $C = 1/rb$. It is important for us, that an account of nonlinearity of the magnetization curve results in the presence of effective capacity C in the electric chain. According to (3.5) the frequency of damping oscillations of the magnetic flux is equal to $f_0 = \frac{1}{2\pi}\frac{1}{\sqrt{LC}}$. An estimation of b under the experimental conditions gives $b \geq U/\Delta\Psi \approx 2 \cdot 10^4$ V/W for U = 10 V. Since r = 50 Ohm, $L \approx 0.5$ μH for the value of the f_0 frequency we obtain an estimate of ~ 0.5 MHz. This value is rather close to the frequencies of oscillations, observed in experiment. These estimations show, that the peculiarities in the oscillograms of a current and voltage are really caused by the process of dynamic magnetic reversal.

It is convenient to study the complicated oscillations of the magnetic flux, considering a phase portrait of oscillations or its projection. The plane with the system of coordinates, with the X axis corresponding to the excitation current in the primary winding, which, in turns, is proportional to the ac component of the magnetic field $H_{\sim}$, and the Y axis corresponding to the voltage in the secondary winding, which is proportional to $dB_{\sim}/dt$. In Figure 46 the series of the experimentally measured projections recorded for the frequency of generator f = 15 kHz in the *dc* magnetic field of 2.5 kOe is shown. These oscillograms differ by the value of the amplitude of generator U. At small amplitudes the projection of the phase portrait of oscillations is close to an ellipse. This corresponds to the oscillations of a magnetic flux close to the sinusoidal. Similar picture of oscillations was observed in many works, including [67, 70]. If the voltage of generator U reaches 3.2 V, distortion appears in the phase portrait which is related to shock excitation of oscillations. At the voltage of U =3.5 V the fluctuations become obviously nonsinusoidal. In the range of voltages U from 4 to 8 V the shock oscillations are close to the damping sinusoidal. Starting from U = 9 V during the time, corresponding to one period of oscillations of the exciting generator, two cases of excitation of shock oscillations occur. Starting from the moment of occurrence, the amplitude of shock oscillations increases with the growth of the voltage amplitude of the excitation generator, *i.e.* with the growth of the strength of the *ac* magnetic field $H_\sim$.

As a whole, it is possible to note, that with increasing the strength of the *ac* $H_\sim$ or *dc* magnetic field H a complicated evolution of phase portrait of oscillations occurs. Two points of bifurcations, that is two points, in which the shock oscillations appear, were fixed. From the point of view of the theory of dynamic systems soft emergence of a cycle occurs in these points. The value of frequency of shock oscillations is far from the resonant frequency of the oscillatory contour formed by inductance and interturn capacitance of the windings. The shape of the dependences in the oscillograms of current in the primary winding and voltage in the secondary one confirms the assumption that shock oscillations are caused by the nonlinear process of dynamic magnetization reversal of the manganite core.

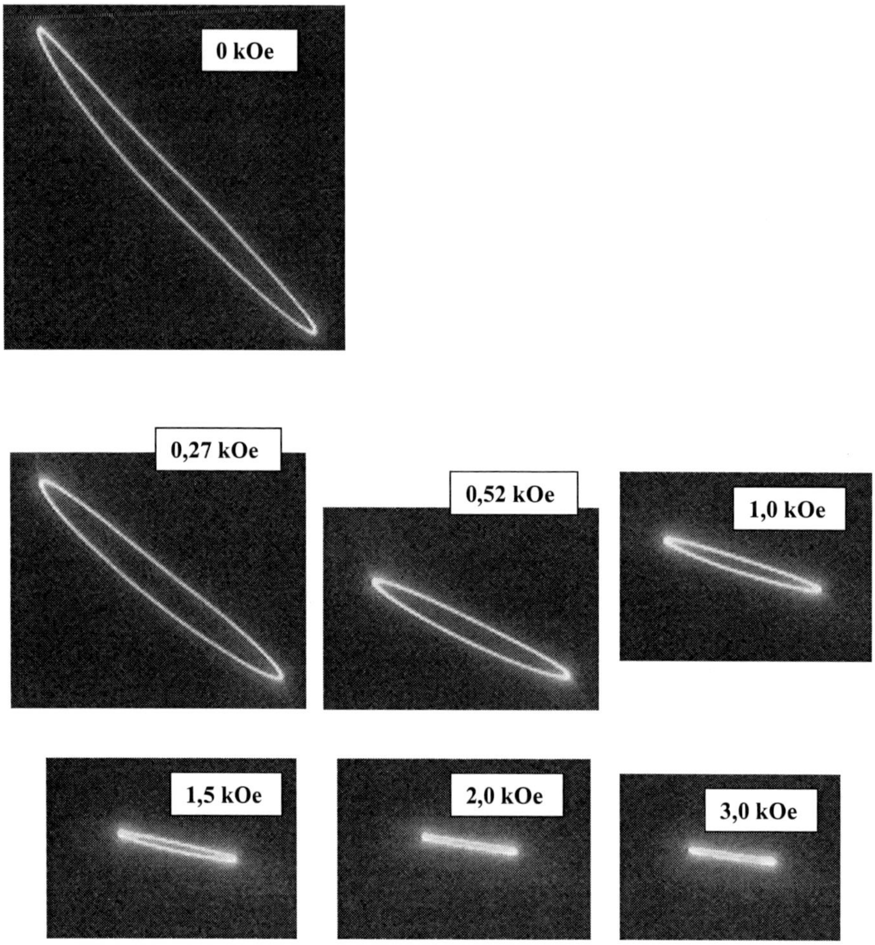

Figure 47. The projection of the phase portrait to the H_m - $\dot{\Phi}$ plane measured at the frequency of $f = 20$ kHz for different values of the *dc* magnetic H_0. The regions of shock excitation of oscillations of magnetic flux are absent.

For understanding of the physical nature of the occurring processes it is important, that this complicated evolution of a phase portrait occurs, under the conditions of our experiments, only if the *ac* magnetic field $H_{\sim}$ has the frequency belonging to the certain interval within 2 and 16 kHz. For an example in Figure 47 the evolution of a phase portrait of oscillations of the output voltage, obtained for the frequency of f = 20 kHz with increasing the strength of the *dc* magnetic field H. In the absence of the *dc* magnetic field or at its small values the projection of the phase portrait is the closed curve. With increasing the strength of the *dc* magnetic field its shape approaches to ellipse. The periodic oscillations exist for such type of phase portrait. The sizes of the axes of the closed curve and the inclination angle relative the X axis decrease with increasing the *dc* magnetic field. Such evolution of oscillations is typical of oscillations of a magnetic flux without bifurcations. It is described by the standard variation of the complex magnetic permeability in a magnetic field and variation of the relation between the real and imaginary parts of magnetic permeability. Such type of evolution is in accordance with the one described in [62, 67, 70]. It is necessary to explain, why in a certain interval of frequencies the other, more complicated type of evolution of the phase portrait of oscillations, is possible.

Let us consider the system of equations describing the distribution of complex amplitudes of the electric $\dot{E}_m$ and magnetic $\dot{H}_m$ fields in a conductive ferromagnetic half-space

$$\frac{d^2\dot{H}_m}{dz^2} = i\omega\sigma\dot{\mu}\dot{H}_m\,, \tag{3.6}$$

$$\frac{d^2\dot{E}_m}{dz^2} = i\omega\sigma\dot{\mu}\dot{E}_m + \frac{1}{\dot{\mu}}\frac{d\dot{\mu}}{dz}\frac{d\dot{E}_m}{dz}\,, \tag{3.7}$$

where σ is the conductivity of a manganite, and $\dot{\mu} = \mu e^{-i\psi}$ is the complex magnetic permeability. We can take advantage of the simplified approach developed in [71]. We can assume, that some distribution of the amplitude of the *ac* magnetic field $H_m = F(z)$, where z is the axis of the coordinate system directed inside the ferromagnetic half-space, is formed in a conductive ferromagnet at given frequency owing to skin-effect. The complex magnetic permeability of a ferromagnet depends on the value of the magnetic field H_m, hence, on the z coordinate: $\dot{\mu} = f(H_m) = f\{F(z)] = \varphi(z)$.

Analysis of the equations (3.6) and (3.7) was carried out in [71]. It does not allow to obtain the detailed characteristics of oscillations with an account for nonlinearity of a magnetization curve and hysteresis. However it gives some concept about the conditions of existence of peculiarities of distribution of a magnetic field in the magnetic conductive core. The existence of such peculiarities serves as the reason for occurrence of complicated oscillatory processes. Therefore we shall use the results of analysis from [71]. While solving the equation (3.6), we can consider, that $\psi \approx \psi'$, where ψ' is the average value of ψ. Certainly, ψ' depends on μ, and, hence, on H_m. The equation (3.6) will become $\frac{d^2\dot{H}_m}{dz^2} = i\omega\sigma\dot{\mu}e^{-i\psi'}\dot{H}_m$. We can assume the dependence of the magnetic permeability on z as

$\dot{\mu} = \frac{\mu_e}{\left(1 - \frac{z}{z_k}\right)^2} e^{-i\psi'}$, where z_k is the z coordinate of the point where the magnetic permeability has a maximum. The allowance for phase ψ' is necessary for inclusion a magnetic hysteresis in consideration. If we denote $k = \left(\frac{\omega\sigma\mu_e}{2}\right)^{1/2}$, then (3.6) will become

$$\left(1 - \frac{z}{z_k}\right)^2 \frac{d^2 \dot{H}_m}{dz^2} = 2ik^2 e^{-i\psi'} \dot{H}_m . \tag{3.8}$$

Solution of the equation (3.8) can be written as

$$\dot{H}_m = \dot{H}_{me}\left(1 - \frac{z}{z_k}\right)^a, \quad a = \frac{2n}{n-1}\left(1 + i\sqrt{\frac{n+1}{2n} - \frac{(n-1)^2}{2n^2} k^2 z_k^2 \sin\psi'}\right) \tag{3.9}$$

In these expressions n is the exponent in the approximation of magnetization curve by parabola $B_m = KH_m^{1/n}$. Solution of the equation (3.9) have various forms in the cases of weak and strong field. In the case of a weak field, when the amplitude of a magnetic field on the surface of a ferromagnet $H_m|_{z=0} < H_{mk}$, where H_{mk} is the field corresponding to the maximum value of the magnetic permeability, the distribution of the *ac* magnetic field inside the core has no peculiarities. In this case $z_k < 0$. In the case of strong field $H_m|_{z=0} > H_{mk}$, inside the sample there exists the point $z_k > 0$ where the *ac* magnetic field has a peculiarities.

It is evident that the distribution of magnetic flux, and, hence, the nature of its oscillations, depends on the relation between skin depth δ and the thickness of a sample d. The peculiarities in distribution of a magnetic flux can occur only if the case of strong skin-effect $d > \delta$ is realized. This means, that there is a lower limit of the frequency interval within which a complicated dynamics of oscillations of magnetic flux can be observed. The upper limit of this interval can be caused by the fact, that on sufficiently high frequency even in a large *dc* magnetic field the skin-effect remains strong, and also because of a frequency dispersion of magnetic permeability. The equation (3.9) is written for the model case when at $H = H_{mk}$ and $z = z_k$ an unlimited increase of magnetic permeability is expected. In the real ferromagnetic material the value of magnetic permeability is finite. With increasing frequency the maximum of magnetic permeability at $H = H_{mk}$ becomes less expressed. Therefore at sufficiently high frequencies the peculiarity of distribution of the magnetic field, which results in complicated dynamics of oscillations of magnetic flux, disappears. All noted characteristic of conditions of existence of complicated dynamics of oscillations of magnetic flux were observed experimentally.

Qualitatively observable peculiarities can be interpreted within the Preisach model for hysteresis [72], which is used for description of the irreversible phenomena in magnetically ordered materials of an arbitrary degree of complexity. Within this model the magnetic state of a material is described assuming decomposition of free energy into an ensemble of elementary bistable fragments, each characterized by two stable states and by two free energy barriers, which inhibit transitions between the states. The height of these barriers is defined by the local fields of anisotropy and ferromagnetic interaction. Utilization of this model for manganites was considered in [73]. Weak variation of the frequency of oscillations f_0 under variation of the excitation conditions within the model [72] can be interpreted as appearance of weak magnetic viscosity upon variation of the states of bistable elements, under *ac* magnetic field excitation.

Thus, while considering the processes of dynamical magnetization reversal of the doped manganite, in the crossed *dc* and *ac* radio-frequency magnetic fields, it was found that the size and dynamics of variations of magnetic flux are defined both by the amplitude and frequency of the *ac* radio-frequency magnetic field, and also by the strength of the *dc* magnetic field. At large amplitudes of the *dc* magnetic field the nonlinear response of the dynamic system, which manifested itself in occurrence of the shock oscillations of magnetic flux, was observed. The moment of occurrence of shock oscillations corresponded to the beginning of the part of oscillogram in which variations of current were minimum. The phenomena observed can be explained by the process of dynamic magnetization reversal of the core on a partial hysteresis loop. The phase portrait of oscillations of the magnetic flux, considered at the H_m - $\dot{\Phi}$ plane, corresponds to an ellipse if the amplitude H_m is small. With increasing H_m, when the shock oscillations of magnetic flux appear, the spiral-shaped trajectories are formed on the phase portrait. Such evolution of the phase portrait corresponds to the soft emergence of a cycle. At large amplitudes of the *ac* radio-frequency and *dc* magnetic fields during one period of the excitation current two points of bifurcations of emergence of a cycle appear.

It is rather important, that the noted elements of complicated dynamics of magnetic flux are realized in the limited interval of frequencies, namely in such interval, within which the skin depth without an external *dc* magnetic field is greater that the thickness of the manganite core, and in stronger enough magnetic field - is less than the thickness of the core. Such excitation conditions in the range of radio frequencies are difficult to realize both in ferrites with high resistivity, and also in metallic ferromagnets with low resistivity. But, as we showed, they can be realized in doped manganites with moderate resistivity.

2. Doped Lanthanum Manganites as Advanced Materials for Sensors of *DC* and *RF* Magnetic Fields

In this section we show, that the static and high-frequency properties of doped lanthanum manganites can be used for development of magnetic sensitive sensor working in a wide range of fields. The peculiarities of the magnetoresistive properties of bulk manganites, doped with various cations were investigated. This permitted to choose the compositions with the best linearity of response to the *dc* magnetic field in the range of temperatures, near the room temperatures. The prototype of the combined (ferrozonde and magnetoresistive) sensor was made and its characteristics were studied. The variant of utilization of doped manganites in magnetic field sensor containing closed magnetic circuit was proposed. The comparative analysis of characteristics of the developed sensors was carried out.

2.1. Utilization of Magnetoresistive Properties of Bulk Polycrystalline Manganites for Detection of Static Magnetic Fields

The problem of registration of the magnetic fields created in magnetizing devices, or the stray fields from defects in already magnetized objects, appears in various tasks of magnetic nondestructive testing. Different principles of recording magnetic fields, and, respectively, several types of sensors, can be used for solution of this problem The magnetoresistive sensors, in which variation of resistivity is proportional to the applied magnetic field, are most frequently used at present for these purposes. The working material of such sensors, as a rule, consist from an alloy or heterostructure of 3*d* metals.

In doped manganites variation of resistivity under the influence of an external *dc* magnetic field (static magnetoresistance *MR*) can be substantial. However the value of the effect depends on many factors. Large negative *MR* is observed in the ferromagnetic range below the Curie temperature T_C. It is linear with the applied magnetic field and is caused by suppression of spin fluctuations. In a paramagnetic range above T_C the values of magnetoresistance are small, it is nonlinear (quadratic) with the applied magnetic field and rapidly decreases with increases temperature. In turn, the values of T_C and *MR* depend on the type of doping cation and doping level *X*. With increasing the concentration of doping cation the value of T_C quickly increases, reaches maximum for $X = 0.3$, and then gradually decreases [1]. From the point of view of practical applications the basic interest represents the *MR* in the range of temperatures close to room temperature. For the prevailing number of cations the values of T_C are at temperatures below room temperature, and, despite high absolute values of *MR*, practical utilization of such compositions seems to be problematic. Only upon doping of $LaMnO_3$ with the Sr^{2+}, Ba^{2+} and Pb^{2+} cations with concentration $X = 0.3$ it is possible to obtain the values of T_C reaching, accordingly, 347 K, 343 K and 347 K.

With an account for possible practical applications it is desirable to have the magnetoresistive material working in a range of temperatures near to the room temperature, with possibly higher values of *MR* in smaller magnetic field and wider range of magnetic fields. The high sensitivity in the region of weak magnetic fields, which is observed in bulk polycrystalline sintered manganites and thin films at low temperatures is caused by tunneling of spin-polarized carriers through intergrain boundaries and drops down with temperature as $1/T$ [74]. In the range of temperatures near to the room temperature, its values are too small for potential practical applications. However the requirements of high sensitivity in small fields and possibility of operation in a wide range of magnetic fields inevitably come into conflict. Theoretically the working range of magnetoresistive sensors, which are widely used now, is limited by the saturation field H_{sat}. In practice, owing to nonlinearity of the magnetic field dependence of *MR*, those working range appears to be essentially smaller, than H_{sat}. The distinctive feature of magnetic dependences magnetoresistance of doped manganitesis is an absence of saturation relatively large values of applied magnetic fields (of the order of 60 kOe and above). According to [75] for $La_{0.90}Pb_{0.1}MnO_3$ bulk polycrystalline manganites at 320 K the value of magnetoresistance reached $\Delta\rho/\rho \sim 85\%$ in the applied field of $H \sim 60$ kOe. These characteristics, if considered with respect to the value of the effect per unity of magnetic field, are worse, than in the known materials used for fabrication of magnetoresistive sensors. For example, at a room temperature in the case of antiferromagnetically coupled Co/Cu metallic superlattices $\Delta\rho/\rho \sim 40\%$ and $H_{sat} \sim 1$ kOe [76]. For the widely used permalloy alloy of the composition $Ni_{0.81}Fe_{0.19}$ the value of $\Delta\rho/\rho$ is about 2.5% and the saturation field is H_{sat}~5 Oe [77]. In comparison with materials based on the 3*d* metals and their alloys, manganites have higher resistivity, the improved linearity of magnetic field dependence of magnetoresistance, higher saturation fields, and, therefore are suitable for measurements of magnetic fields in wider range.

We investigated the magnetoresistive characteristics of bulk polycrystalline sintered $La_{1-X}A_XMnO_3$ ($0.15 < X < 0.55$; A=Sr, Ba, Pb) manganites in the range of temperatures near the room temperature.

The dc magnetic field dependences of magnetoresistance for the $La_{0.60}Sr_{0.40}MnO_3$ composition measured at the temperatures of 5 and 300 K are shown in Figure 48. The distinguishing feature of both curves is the absence of saturation in the field of 50 kOe. In the dependence recorded at 5 K an abrupt increase of the magnitude of magnetoresistance upon application of relatively small (~2 kOe) *dc* magnetic field is clearly seen. With subsequent increase of the applied field the absolute value of magnetoresistance increases almost linearly. The existence of two regions in the *dc* magnetic field dependence of magnetoresistance is caused by two different physical mechanisms. The abrupt variations of MR in weak magnetic field in the case of bulk polycrystalline manganites samples is related to the presence of grains and intergrain boundaries and caused by the processes of tunneling of spin-polarized charge carriers through the boundaries between grains. The contribution of this process essentially depends on temperature and, as it is seen from the dependence, recorded at 300 K, it can be neglected in the range of temperatures near to the room temperature. The magnetoresistance in the range of strong fields is caused by suppression of spin fluctuation by an external magnetic field. This mechanism works in the whole temperature range. In the case of the *MR (H, 300K)* curve, high enough linearity of the magnetic field dependence of magnetoresistance is attractive and it can be of interest for possible practical applications. This property of the *MR (H,T)* dependences, in our opinion, can be considered as the basic advantage doped manganites as materials, the magnetoresistive properties of which can be used for registration of magnetic fields.

In Figure 49 we compare the room-temperature magnetic field dependences of magnetoresistance *MR (H, 300K)* for the Sr, Pb, and Ba manganites with the doping level of $X = 0.33$. The absolute value of magnetoresistance is maximal for the Ba doped manganite. This is explained by the fact that, since the peak of the temperature dependence of magnetoresistance is close to the Curie temperature, and the measurements were executed at the fixed temperature of 300 K, the difference between these temperatures is minimum for the Ba doped manganite for which $T_C = 343$ K. The Curie temperatures for the Pb and Sr doped manganites have higher values, 347 K and 365 K, respectively. In order to characterize quantitatively the degree of nonlinearity of the *dc* magnetic field dependences of magnetoresistance *MR (H, T)*, they were approximated by linear functions ($MR(H, T) = A + B*H$) and the root-mean-square deviation (*SD*) from the linear dependence was calculated. These dependences for different types of doping cation are presented in an insert in Figure 49. The greatest nonlinearity was found for the Ba doped manganites, while the lowest – for the Sr doped manganites. The dependence of nonlinearity on the doping level was also analyzed. These results are presented in Figure 50. Irrespective of the type of doping cation, with increase of its concentration the following general tendencies were observed. Nonlinearity is maximal for small levels of doping, however with increasing X the value of *SD* first passes through a minimum, and then start to increase gradually. The nonlinearity is minimal for the Sr doped compositions, and is maximal for the compositions doped with Pb. Among the factors, which define such shape of the *SD(X)* dependence, one can note variations of the number of free carriers directly defined by the doping level, variation of the Curie temperature (with increasing X from 0.15, T_C quickly grows, reaches a maximum at X=0.33, and further gradually decreases), and also variation of the susceptibility of paraprocess.

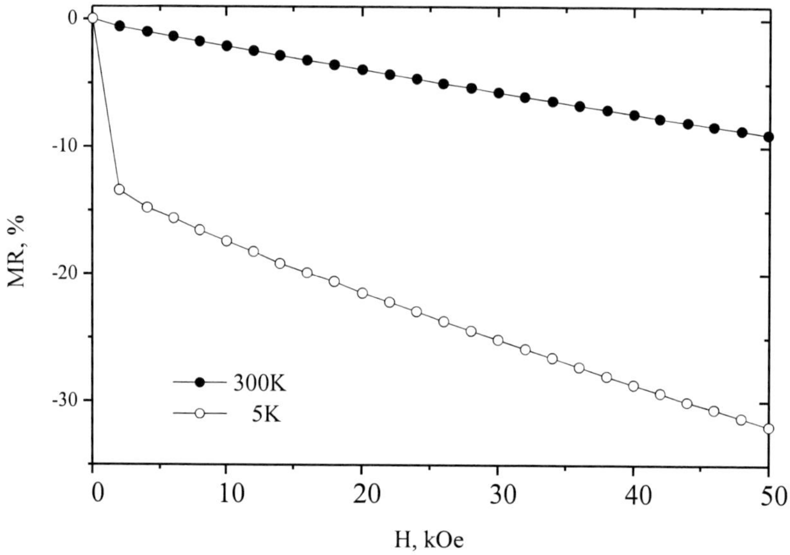

Figure 48. Magnetoresistance of the bulk polycrystalline $La_{0.60}Sr_{0.40}MnO_3$ manganites at 5 and 300 K.

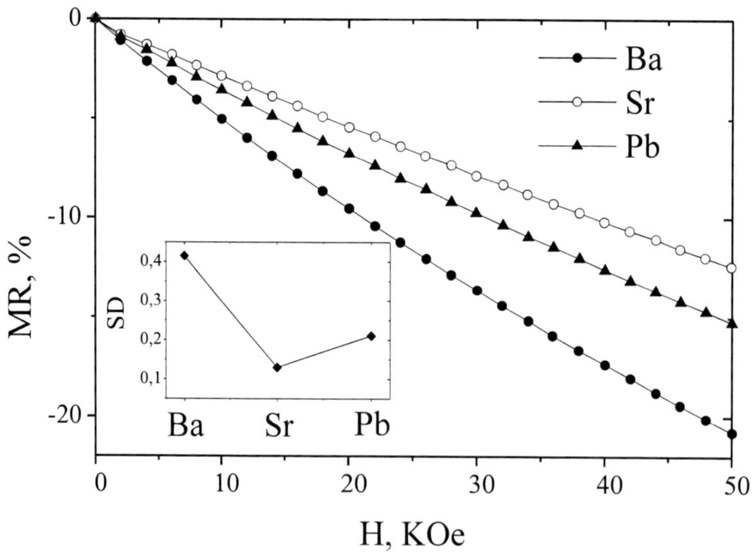

Figure 49. Magnetoresistance of the bulk polycrystalline $La_{0.67}A_{0.33}MnO_3$, (A = Sr, Ba, Pb) manganites at 300 K. Inset: the dependence of nonlinearity *SD* on the type of doping cation.

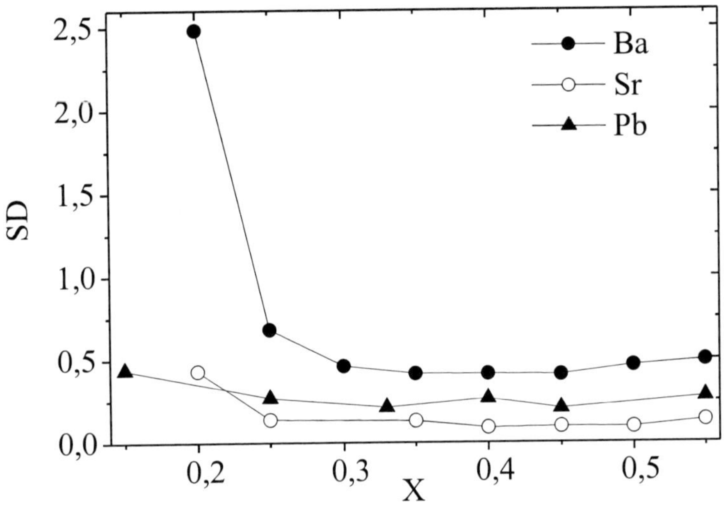

Figure 50. The dependence of nonlinearity of magnetoresistance on the composition of bulk polycrystalline $La_{1-X}A_XMnO_3$ (*A*=Sr, Ba, Pb, and 0.15 <*X* <0.55) manganites at 300 K.

Temperature variations of both nonlinearity and sensitivity for the La-Sr system are shown in Figure 51. It is seen that the $La_{0.60}Sr_{0.40}MnO_3$ composition has the minimum nonlinearity and maximum sensitivity to the *dc* magnetic field. If we take for sensitivity of magnetoresistance for this composition the value of $B \approx 0.2$ %/kOe as an average for the range of temperatures 260-300 K, it is possible to estimate the upper limit of the magnetic field, which can be measured by the sensor made from this material. Estimations give the value of 250-300 kOe. The results obtained can be useful for the choice of a material for manganite magnetoresistive sensor of the *dc* magnetic field with an optimum relation between linearity and sensitivity which can be made, for example, by thin [78] or thick-film [79] technology for applications in devices and systems of the magnetic nondestructive control.

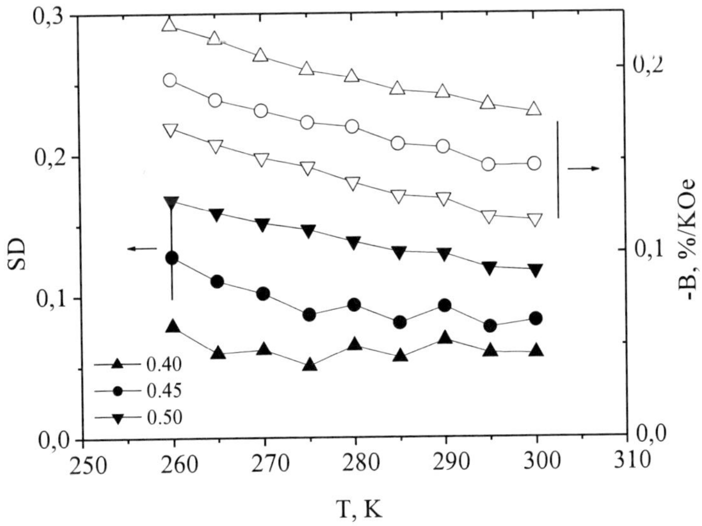

Figure 51. Temperature dependences of nonlinearity *SD* and sensitivity to magnetic field *B* for the $La_{1-X}Sr_XMnO_3$ (*X*=0.40, 0.45, 0.55) manganites.

2.2. Combined Ferrozonde-Type Sensor Based on Bulk Doped Lanthanum Manganites

As it was shown above, using only the magnetoresistive properties of bulk doped manganites, it is possible to obtain the sensors of magnetic fields operating only in the range of strong (above 2 kOe) magnetic fields. However from results of investigations of the radio-frequency magnetic characteristics of manganites, which were considered in section 1, follows, that the reversible differential magnetic permeability of these materials has the large value in the field of weak fields. This circumstance can be used for development of the sensors of magnetic field in configurations of the opened magnetic chain, namely the ferrozonde type of sensor.

The ferrozonde sensor of magnetic field is the type of the sensor which operation is based on nonlinearity of a magnetization curve of the core material. The output signal is measured at the second harmonic of the excitation frequency and is proportional to the external *dc* magnetic field [80].

The bulk polycrystalline manganite of the $La_{0.60}Pb_{0.40}MnO_3$ composition was used for fabrication of the core of ferrozonde sensor. The static magnetoresistive and high-frequency properties of this material were described in detail in section 1. The core had the dimensions 2×2×18 mm. Two identical coils (the exciting and the receiving) containing 285 turns each has been wound on it. The wire with the diameter of 0.08 mm was used.

The scheme of the experimental setup for measurements in the regime of ferrozonde is shown in Figure 52. The excitation coil was connected to the generator of sinusoidal voltage with the frequency of 40 kHz. The signal in the receiving coil was measured at the second harmonic of the excitation frequency by a selective voltmeter after narrow-band filtering. The *dc* magnetic field was applied along the axis of the core. The silver contacts were made on the core for recording the *dc* magnetoresistance by four-probe technique. All measurements were carried out at room temperature.

The *dc* magnetic field dependence of the differential magnetic permeability μ_{dif} is shown in the inset in Figure 2. It has the maximum value of an order of 40 in the field of $H \cong 30$ Oe. As was shown in section 1.6, in doped manganites the value of the differential magnetic permeability differs from unity at temperatures more than by 60 degrees above the Curie temperature. This fact allows to assert, that the proposed variant of the magnetic field sensor with manganite core can operate in such interval of temperatures.

Figure 53 shows the results of measurement in the ferrozonde mode: the dependence of the signal in the receiving coil E_ϕ, measured at the secondary harmonic of the excitation frequency, on the *dc* magnetic field *H*. For the amplitude of the excitation current I_{sup} = 80 мA the dependence $E_\phi(H)$ is almost linear in the range of fields 0÷120 Oe. The stronger magnetic fields can be recorded by the four-probe technique using the magnetoresistive properties of doped manganites and the almost linear dependence of magnetoresistance on the strength of the *dc* applied magnetic field, as was described in detail earlier. In our opinion, the basic advantage of such sensor is the possibility of registration of magnetic fields in very wide range. This can be used for solution of various problems of nondestructive control related to registration of stray magnetic fields from defects in magnetized objects, robotics, sensors of angle of rotation, proximity sensors, and other applications.

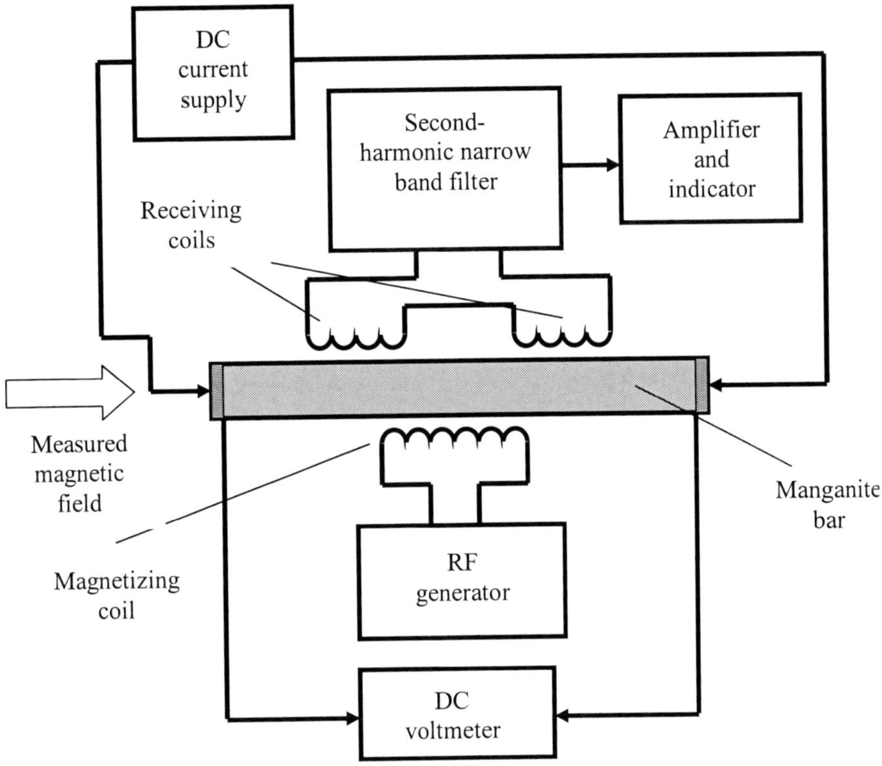

Figure 52. The scheme of experimental setup for measurements in the ferrozonde mode.

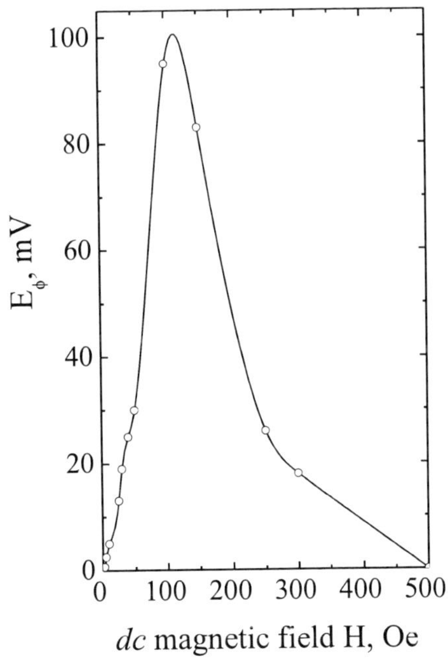

Figure 53. The *dc* magnetic field dependence of the signal in the receiving coil E_{ϕ}, measured at the secondary harmonic of the excitation frequency.

2.3. Sensor of Magnetic Field with Closed Magnetic Circuit made of the La-Pb Manganite

As was shown in section 1, strong variations of the transmission coefficient of radio-frequency electromagnetic waves owing to the *dc* magnetic field dependence of differential magnetic permeability are observed in doped manganites. Earlier in this section the possibility of fabrication of magnetic field sensor using manganite as the material for the opened magnetic circuit was demonstrated. However in this geometry the flux leakage is rather great. Therefore, the possibility of development of a sensor using the closed magnetic circuit made of manganite was of interest.

The transformer with the toroidal core made from the $La_{0.6}Pb_{0.4}MnO_3$ manganite was fabricated. The toroid had the external diameter of D_{max} = 70 mm, the internal diameter of D_{min} = 50 mm, and the thickness of d = 5 mm. Two windings, 100 turns each, have been wound on it. From the *dc* measurements the value of the coercive field for this manganite was found to be H_c = 10 Oe and the value of resistivity – 0.2 Ω*cm. For this value of resistivity the value of skin depth δ_l = 3.6 mm was calculated taking the frequency of f = 1 MHz and assuming the relative magnetic permeability $\mu_{r,}$ equal to unity. From this value of δ_l it follows, that at low frequencies of a radio range the conditions of weak skin-effect ($d < \delta$) were realized in a toroid while at the megahertz frequency range the condition of strong skin-effect ($d > \delta$) were fulfilled.

Radiofrequency measurements were carried out in the frequency range from 20 to 500 kHz. One of the coils was excited by the *ac* signal from generator and the voltage, induced in the secondary (receiving) coil was recorded. The amplitude of the excitation voltage was kept at the level of 0.1 V. The module of transformation coefficient, defined as the ratio of the voltage in the receiving coil to the excitation voltage, was measured. The external biasing *dc* magnetic field was applied perpendicular to the plane of a toroid. The *ac* and *dc* magnetic field were oriented perpendicular to each other, $\boldsymbol{H}_{\sim}\perp\boldsymbol{H}$. The maximum value of the biasing *dc* magnetic field was 10 kOe. Thus, under application of the biasing *dc* magnetic field the condition $|\boldsymbol{H}_{\sim}| \ll |\boldsymbol{H}|$ was satisfied.

The transformation coefficient D of thus obtained four-pole network was measured as the function of the excitation frequency and strength of the biasing *dc* magnetic field.

Figure 54 shows the *dc* magnetic field dependence of the parameter $r_m = [U_{out}(H)-U_{out}(0)]/U_{out}(0)$ 100 %, which characterizes relative variation of the voltage of the secondary coil upon application of magnetic field. With increasing the strength of the *dc* magnetic field the amplitude of the voltage of the secondary coil abruptly decreases. At the frequency of 20 kHz application of the *dc* magnetic field of 2 kOe reduces the amplitude almost by ten times in comparison with amplitude at H=0. The external *dc* magnetic field reduces magnetic permeability, increases the stray fluxes and, as the result, lowers the voltage in the secondary winding. As is seen from Figure 54, with increasing frequency the sensitivity of the device to the *dc* magnetic field decreases. In such scheme of measurement of the *dc* magnetic field the manganites, doped with lead, are efficient up to the excitation frequencies of an order of 500 kHz. The device under considered cannot be considered, certainly, as the variant of sensor ready for practical application. The results of measurements illustrate the basic possibility of utilization of doped manganites in the sensors of the *dc* magnetic field with the configuration of a closed magnetic circuit.

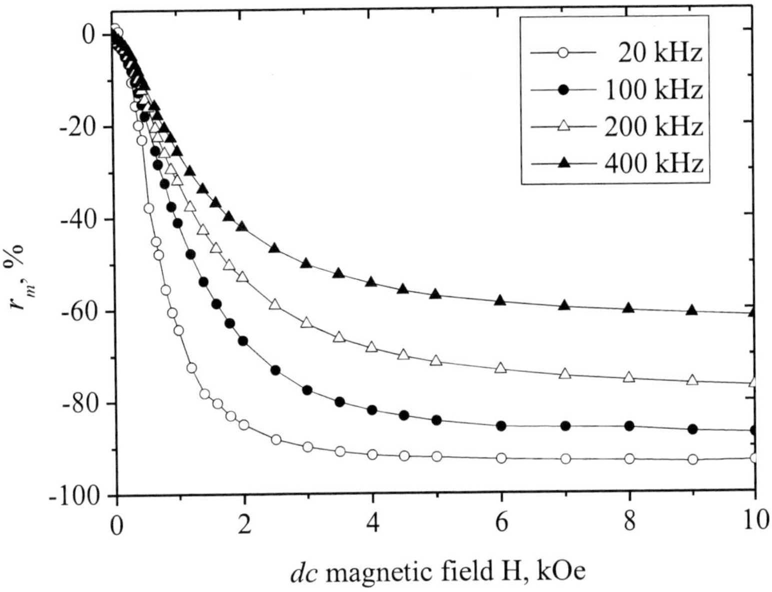

Figure 54. The *dc* magnetic field dependence of the parameter r_m (see definition in the text) for the core from $La_{0.6}Pb_{0.4}MnO_3$ manganite measured at several excitation frequencies.

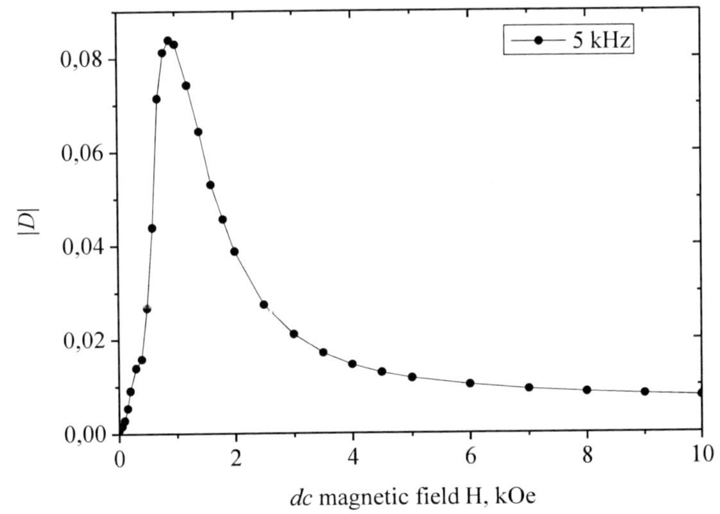

Figure 55. The *dc* magnetic field dependence of the transformation coefficient at the second harmonic $|D_2|$ for the transformer with the core from $La_{0.6}Pb_{0.4}MnO_3$ manganite. The excitation frequency is 5 kHz.

The core made from a ferromagnetic material, magnetized simultaneously by the *dc* and *ac* magnetic fields, can be considered as a nonlinear element. Due to nonlinearity one can expect, the harmonics of the excitation frequency can be detected in the secondary winding of such element. The second harmonic, in principle, can be used for measurement of intensity of the *dc* magnetic field. The dependence of the transformation coefficient at the second harmonic $|D_2|$ on the strength of the *dc* magnetic field is shown in Figure 55. The excitation frequency was 5 kHz. The nonmonotonous dependence in small fields can be eliminated by utilization of the differential scheme of switching of two identical coils. This will permit to ensure linearity of the device up to 1 kOe with high sensitivity and large amplitude of the output signal. The upper limit of working fields almost by 10 times exceeds the corresponding value for the sensors using the configuration of the opened magnetic circuit. However the sensitivity is essentially lower and is about $8 \cdot 10^{-3}$% per Oersted.

CONCLUSION

As a result of experimental investigations of properties of bulk polycrystalline sintered manganites the strong *dc* magnetic field dependence of parameters of penetration of radio-frequency electromagnetic radiation through them was established. It is shown, that the reason of this dependence is the variation of magnetic permeability, but not of magnetoresistance. The frequency dependence of the penetration coefficient at high frequencies can be explained by the frequency dispersion of magnetic permeability. It is experimentally shown, that the physical mechanisms of the magnetoimpedance and penetration effects in doped manganites are different.

It is shown, that for doped manganites at radio frequencies in the range of temperatures up to 60 K above the temperature of magnetic phase transition the dynamic magnetic permeability can essentially exceed unity. This fact is interpreted as the proof of existence of local magnetic dynamical ordering above the temperature Curie for manganites with various basic types of magnetic ordering states. The duration of existence of this local dynamical magnetic ordering is about 10-100 μs. It is established, that at the frequencies of units of megahertz variations of the transmission coefficient in strong *dc* magnetic field can become negative. In this case they are defined, mainly, by the magnetoresistance of a manganite.

In a plate of a conductive ferromagnetic lanthanum manganite in the geometry of the Cotton-Mouton effect the phenomena of the magnetic birefringence and dichroism were observed. They manifested themselves in rotation of the polarization plane and ellipticity of the radiofrequency electromagnetic field. These changes of polarization are caused, basically, by the anisotropy of the dynamic magnetic permeability in an external polarizing *dc* magnetic field.

The frequency dependence of the transformation coefficient of the four-pole network with the core from the La-Pb manganite has the resonant maximum in the range of megahertz frequencies. With application of the *dc* magnetic field its amplitude increases and the maximum frequency shifts to higher frequencies. The electromagnetic losses in the system, consisting from the losses for magnetization reversal and eddy currents, decrease upon application of the *dc* magnetic field. The shape of the *dc* magnetic field dependence of the transformation coefficient changes with variation of frequency. At low frequencies the value of the transformation coefficient decreases with growth of the *dc* magnetic field owing to the decrease of magnetic permeability and increase of stray flux. In the range of frequencies of units of megahertz the transformation coefficient increases with increasing the *dc* magnetic field. The *dc* magnetic field dependent phase shift between the voltages in the primary and secondary windings of the four-pole network was observed.

The processes of dynamical magnetization reversal of manganite, doped by lead, in the crossed *dc* and radio-frequency *ac* magnetic fields were investigated. At large amplitudes of the *dc* magnetic field the nonlinear response of the dynamic system is observed. It consists in appearance of shock oscillations of magnetic flux. The phenomena observed are explained by the process of dynamical magnetization reversal of the core material over a partial hysteresis loop. At great values of amplitude of the *ac* radio-frequency and *dc* magnetic fields during one period of a current two points of bifurcations of emergence of a cycle appear.

The magnetotransport properties of bulk polycrystalline $La_{1-X}A_XMnO_3$ (A = Sr, Ba, Pb, $0.15 < X < 0.55$) manganites were investigated in the range of temperatures 260-300K. The degree of nonlinearity of the *dc* magnetic field dependences of magnetoresistance at room temperature was compared. It is shown, that among the investigated materials in the strong *dc* fields the $La_{0.60}Sr_{0.40}MnO_3$ composition has the maximum sensitivity at the minimum nonlinearity.

The basic possibility of development of the magnetic sensitive element based on combination of two characteristics of doped manganitesis - magnetoresistance and reversible magnetic permeability - is shown. Using this concept the prototype of the sensor of magnetic field was worked out. In the ferrozonde mode of measurements with the opened magnetic circuit the sensitivity of 0.7÷2% per Oersted in the range of fields from zero to 0.1 kOe was obtained. The linearity of characteristic in the magnetoresistance mode is preserved in the range of magnetic fields from 1 to 50 kOe.

The comparative analysis of characteristics of magnetic sensitive elements using the *dc* magnetic field dependence of reversible magnetic permeability of doped manganites in the configurations of open and closed magnetic circuits was performed. Frequency dependences of the output signal are analyzed. It is shown, that advantage of the sensor with closed magnetic circuits is the greater (up to 1 kOe) linearity of the output characteristic relative the *dc* magnetic field, but lower sensitivity, which is about $8 \cdot 10^{-3}$% per Oersted in this range of magnetic fields.

REFERENCES

[1] Jonker, G. H. & Van Santen, J. H. (1950). *Physica*, *16*, 337-349.

[2] Volger, J. (1954). *Physica*, *20*, 49-66.

[3] Jonker, G. H. & van Santen, J. H. (1953). *Physica*, *19*, 120-130.

[4] Jonker, G. H. (1956). *Physica*, *22*, 707-722.

[5] Baibich, M. N., Broto, J. M., Fert, A., Nguen van Dau, F., Petroff, F., Etienne, P., Creuzet, G., Friederichs, A. & Chazelas, J. (1988). *Phys.Rev.B*, *61*, 2472-2475.

[6] Grunberg, P., Schreiber, R., Pang, Y., Brodsky, M. B. & Sowers, H. (1986). *Phys.Rev.Lett.*, *57*, 2442-2445.

[7] Dieny, B., Speriosu, V. S., Parkin, S. S. P., Gurney, B. A., Wilhout, D. R. & Mauri, D. (1991). *Phys.Rev.B*, *43*, 1297-1300.

[8] Monsma, D. J., Lodder, J. C., Popma, T. J. A. & Dieny, B. (1995). *Phys.Rev.Lett.*, *74*, 5260-5263.

[9] Miyazaki, T. & Tezuka, N. (1995). *J.Magn.Magn.Mater.*, *139*, L231-L234.

[10] von Helmolt, R., Wecker, J., Holzapfel, B., Schultz, L. & Samwer, K. (1993). *Phys.Rev.Lett.*, *71*, 2331-2333.

[11] Shapira, Y., Foner, S., Oliveira, N. F. & Reed, T. B. (1974). *Phys.Rev.B.*, *10*, 4765-4780.

[12] Ramirez, A. P. (1997). *J. Phys.:Condens.Matter*, *9*, 8171-8199.

[13] Bratkovsky, A. M. (1997). *Phys. Rev.B*, *56*, 2344-2347.

[14] Soulen, R. J., Osofsky, M. S., Nagorny, B., Ambrose, T., Broussard, P., Cheng, S. F., Byers, J., Tanaka, C. T., Nowack, J., Moodera, J. S., Laprade, G., Bary, A. & Coey, J. M. D. (1999). *J.Appl.Phys.*, *85*, 4589-4591.

[15] Irkhin, V. Yu. & Katsnel'son, M. I. (1994). *Physic.-Uspekhi.*, *37*, 659-676.

[16] Ji, Y., Chien, C. L., Tomioka, Y. & Tokura, Y. (2002). *Phys.Rev.B*, *66*, 012410.

[17] Zutic, I., Fabian, J. & Das Sarma, S. (2004). *Rev.Mod.Phys.*, *76*, 323-410.

[18] Nagaev, E. L. (2002). *Colossal magnetoresistance and phase separation in magnetic semiconductors;* Imperial College Press UK.

[19] Imada, M., Fujimori, A. & Tokura, Y. (1998). *Rev.Mod.Phys.*, *70*, 1039-1263.

[20] Coey, J. M. D., Viret, M. & von Molnar, S. (1999). *Adv.Phys.*, *48*, 167-293.

[21] Izyumov, Yu. A. & Skryabin, Yu. N. (2001). *Phys.-Usp.*, *44*, 109-134.

[22] Kagan, M. Yu. & Kugel', K. I. (2001). *Phys.-Usp.*, *44*, 553-570.

[23] Zener, C. (1951). *Phys. Rev.*, *82*, 403-405.

[24] Mizokawa, T. & Fujimori, A. (1997). *Phys.Rev.B.*, *56*, R493- R496.

[25] Pickett, W. E. & Singh, D. J. (1997). *Phys. Rev.B*, *55*, R8642-R8645 ()

[26] Livesay, E. A., West, R. N., Dugdale, S. B., Santi, G. & Jarlborg, T. (1999). *J.Phys.:Condens.Matter.*, *11*, L279-L285.

[27] Bebenin, N. G. & Ustinov, V. V. (1998). *J.Phys.:Condens.Matter*, *10*, 6301-6309.

[28] Lofland, S. E., Kim, P. H. & Dahiroc, P. (1997). *J. Phys.: Condensed Matter*, *9*, 6697-6709.

[29] Schwartz, A., Scheffer, M. & Anlage, S. M. (2000). *Phys. Rev.B.*, *61*, R870-R873.

[30] Dominguez, M., Bhagat, S. M. & Lofland, S. E. (1995). *Europhys. Lett.*, *32*, 349-353.

[31] Owens, F. J. (1997). *J.Appl.Phys.*, *82*, 3054-3057.

[32] Geinrich, B. & Mescheryakov, V. F. (1996). *JETP Letters.*, *9*, 618-622 (in Russian).

[33] Celinski, Z., Urquhart, K. B. & Heinrich, B. (1997). *J.Magn.Magn.Mater.*, *166*, 6-26.

[34] Semyonov, R. A. (1973). *Technical electrodynamics*; Moscow, "Svyaz'", (in Russian).

[35] Yates, K. A., Cohen, L. F., Watine, C., Tay, T. N., MacManus-Driscoll, J., Freitas, R. S., Ghivelder, L., Haines, E. M. & Gehring, G. A. (2000). *J.Appl.Phys.*, *88*, 4703-4708.

[36] Lofland, S. E., Dominguez, M. & Tyagi, S. D. (1996). *Thin Solid Films*, *288*, 256-261

[37] Solin, N. N., Naumov, S. V. & Camokhvalov, A. A. (2000). *Fizika tverdogo tela (Soviet journal of solid state physics)*, *42*, 899-904 (in Russian).

[38] Pomortsev, R. V. (1996). *The Physics of Metals and Metallography*, *82*, 43-44.

[39] Jifan, Hu., Hongwei, Qin. & Yaming, Zhang. (2000). *Materials Science & Engineering B*, *77*, 280-281.

[40] Atkinson, D. & Squire, P. T. (1998). *J.Appl.Phys.*, *83*, 6569-6571.

[41] Pisarev, R. V. (1974). "Magnetic ordering and optical phenomena in crystals", in the book *Physics of magnetic dielectrics*; Ed. G. A. Smolenskii, Nauka, (in Russian).

[42] Arkadiev, V. K. (1936). *Electromagnetic processes in metals*; State editorial board for scientific and technical literature, (in Russian).

[43] Lofland, S. E., Ray, V., Kim, P. H., Bhagat, S. M., Manheimer, M. A. & Tyagi, S. D. (1997). *Phys.Rev.B*, *55*, 2749-2751.
[44] Schwartz, A., Scheffer, M. & Anlage, S. M. (2000). *Phys.Rev.B*, *61*, R870-R873.
[45] Akther Hossain, A. K. M., Cohen, L. F., Kodenkandeth, T., MacManus-Driscoll, J. & Alford, N. McN. (1999). *J.Magn.Magn.Mater.*, *195*, 31-36.
[46] Dagotto, E. (2002). *Nanoscale phase separation and colossal magnetoresistance;* Springer-Verlag, Berlin.
[47] Manella, N., Rosenhahn, A., Booth, C. H., Marchesini, S., Mun, B. S., Yang, S. H., Ibrahim, K., Tomioka, Y. & Fadley, C. S. (2004). *Phys.Rev.Lett.*, *92*, 166401.
[48] Moritomo, Y., Akimoto T., Nakamura, A., Ohoyama, K. & Ohashi, M. (1998). *Phys.Rev.B*, *58*, 5544-5549.
[49] Coey, J. M. D. (1999). *J. Appl.Phys.*, *85*, 5576-5581.
[50] Urushibara, A., Moritomo, Y., Arima, T., Asamitsu, A., Kido, G. & Tokura, Y. (1995). *Phys.Rev.B*, *51*, 14103-14109.
[51] De Teresa, J. M., Ibarra, M. R., Algarabel, P. A., Ritter, C., Marquina, C., Blasco, J., Garcia, J., del Moral, A. & Arnold, Z. (1997). *Nature*, *386*, 256-259.
[52] Adams, C. P., Lynn, J. W., Mukovskii, Y. M., Arsenov, A. A. & Shulyatev, D. A. (2000). *Phys.Rev.Lett.*, *85*, 3954-3957.
[53] Kim, K. H., Uehara, M. & Cheong, S. W. (2000). *Phys.Rev.B.*, *62*, R11945-R11498.
[54] Zhao, Guo-meng; Wang, Y. S., Kang, D. J., Prellier, W., Rajeswari, M., Keller, H., Venkatesan, T., Chu, C. W. & Greene, R. L. (2000). *Phys.Rev.B.*, *62*, R11949-R11952.
[55] Nelson, C. S., Zimmermann, M. V., Kim, Y. J., Hill, J. P., Gibbs, Doon; Kiryukhin, V., Koo, T. Y., Cheong, S. W., Casa, D., Keimer, B., Tomioka, Y., Tokura ,Y., Gog, T. & Venkataraman, C. T. (2001). *Phys.Rev.B.*, *64*, 174405.
[56] Downward, L., Bridges, F., Bushart, S., Neumeier, J. J., Dilley, N. & Zhou, L. (2005). *Phys.Rev.Lett.*, *95*, 106401.
[57] Salamon, M. B., Lin, P. & Chun, S. H. (2002). *Phys.Rev.Lett.*, *88*, 197203.
[58] Salamon, M. B. & Chun, S. H. (2003). *Phys.Rev.B.*, *68*, 014411.
[59] Burgy, J., Mayr, M., Martin-Mayor, V., Moreo, A. & Dagotto, E. (2001). *Phys.Rev.Lett.*, *87*, 277202.
[60] Burgy, J., Moreo, A. & Dagotto, E. (2004). *Phys.Rev.Lett.*, *92*, 097202.
[61] Enokizono, M. & Takeshima, Y. (2000). *J.Magn.Magn.Mater.*, *215-216*, 704-707.
[62] Xu Feng; Qin, Wen; Peng, Kun; Gao, Wenli; Du, Youwei. (2004). *J.Magn.Magn.Mater.*, *272-276*, 1449-1451
[63] Esposito, N., Musolino, A. & Raugi, M. (1996). *IEEE Trans. on Magnetics*, *32*, 1094-1097.
[64] Bertotti, G., Fiorillo, F. & Pasquale, M. (1991). *J.App.Phys.*, *69*, 5930-5932
[65] Bertotti, G., Fiorillo, F. & Pasquale, M. (1992). *J.Magn.Magn.Mater.*, *112*, 11-13.
[66] Polivanov, K. M. (1957). *Ferromagnetics*; State energetic publishing company, (in Russian).
[67] Slama, J., Vicen, R., Krivosik, P., Gruskova, A. & Dosoudil, R. (1999). *J.Magn.Magn.Mater.*, *196-197*, 359-361
[68] Serpico, C., d'Aquino, M., Bertotti, G. & Mayergoyz, I. D. (2004). *J.Magn.Magn.Mater.*, *272-276*, 734-735
[69] Langvagen, E. N. (1970). *Izvestia Vuzov. Ser. Electromechanics.*, *3*, 250-256 (in Russian).

[70] Tcherkasov, A. N. & Dmitrienko, V. Yu. (2006). *Zurnal tekhnitcheskoi fiziki (Soviet journal of technical physics)*, *76*, 62-67 (in Russian).
[71] Neiman, L. R. (1949). *Skin effect in ferromagnetic objects*; State energetic publishing company, (in Russian).
[72] Song, T., Roshko, R. M. & Dan Dahlberg, E. (2001). *J.Phys.: Condens. Matter.*, *13*, 3443-3460.
[73] M.Roshko, R. & Xi, L. (2003). *J.Appl.Phys.*, *93*, 6653-6655.
[74] Hwang, H., Cheong, S., Ong, N. & Batlogg, B. (1996). *Phys.Rev.Lett.*, *77*, 2041-2044.
[75] Mahendiran, R., Mahesh, R., Raychaudhuri, A. & Rao, C. (1995). *J.Phys.D:Appl.Phys.*, *28*, 1743-1745.
[76] Parkin, S., Li, Z. & Smith, D. (1991). *Appl.Phys.Lett.*, *58*, 2710-2712.
[77] Brug, J., Anthony, T. & Nickel, J. (1996). *Mater.Res.Soc.Bull.*, *21*, 23-27.
[78] Horwitz, J., Chrisley, D., Dorsey, P., Knauss, L., Pond, J., Wilson, M., Osofsky, M., Qadri, S., Caulfield, J. & Auyeung, R. (1997). *Nuclear Instr.Meth.B*, *121*, 371-377.
[79] Balcells, L., Enrich, R., Mora, J., Calleja, A., Fontcuberta, J. & Obrados, X. (1996). *Appl.Phys.Lett.*, *10*, 1486-1488.
[80] Primdahl, F. (1970). *IEEE Trans. Magnetics*, *6*, 376-380.

In: Bulk Materials: Research, Technology and Applications ISBN: 978-1-60692-963-6
Editors: T. Frias, V. Maestas, pp. 75-139

Chapter 2

SINTER-ALLOYING AND PROPERTIES OF MANGANESE STEELS – STATE OF THE ART

Andrej Šalak and Marcela Selecká
Institute of Materials Research of Slovak Academy of Sciences,
Watsonova 47, 040 01 Košice, Slovak Republic.

ABSTRACT

Manganese is the cheapest alloying element with high hardening effect commonly used in wrought steels. Nevertheless, manganese was as an alloying element studied in powder metallurgy since ~1948 mainly under laboratory conditions. The hard reducibility of Mn oxides due to high affinity of manganese for oxygen, especially of MnO, requiring the purity of the sintering atmospheres not attainable in practice, is regarded for a technical problem. In this chapter are given the data (oxygen partial pressure, dew points of the atmosphere) required for the thermodynamic equilibrium conditions for Mn – O system. Presented mechanical and other properties of manganese steels attained prove successful sintering. The used sintering atmospheres for these steels did not fulfill the thermodynamic requirements for the purity. Material and processing factors applied are summarized. Some representative maximal tensile strength values attained are presented. High vapor pressure characterizes manganese, and its values in dependence on temperature are presented demonstrating in a spontaneous sublimation. Effect of manganese vapor on sintering process and alloying of iron powder matrix in the compacts is explained. The reaction of gaseous manganese with the oxygen in the atmosphere is characterized. The result of this is the reduction ("self-cleaning/protection" effect) of manganese for sintering atmosphere to equilibrium conditions for Mn – O system. This process, given by physical property of manganese, is only possible to ensure the effective sintering of manganese containing steels without other special measures.

INTRODUCTION

Sintered steels with high static and dynamic properties for the production of highly stressed precision components are the focus of a permanent basic and applied research.

Copper, nickel and molybdenum are historically widely used alloying elements in the powder metallurgy (PM) production of structural components. The primary reason for use of these elements was their low affinity for oxygen, allowing sintering in atmospheres with less stringent requirements for reducing potential and, to a lesser extent, for metallurgical reasons [1]. Sintering of metal powders as a basic process for the formation of a solid body from dispersed one, requires that the oxide films on the surface always present on all particles are removed, i.e. reduced during preheating, in order to form in the interparticle contacts the sound metallic bonds at sintering.

Copper is one of the most widely used alloying elements because of its strengthening effect (substitution solid solution and precipitation hardening) and for forming usually transient - liquid phase (melting point at 1085°C) to activate sintering. Copper is soluble in γ-iron up to approximately 9%, but only up to 4% in α-iron. Copper causes growth of the dimensions of a component by liquid phase penetration of the pressing contacts ("copper swelling"), depending on green density and copper content.

Nickel is the second mostly used alloying element in powder metallurgy. Nickel belongs to those elements whose behavior during sintering differs from that of copper. Nickel causes shrinkage of the dimensions of the parts and with iron forms at higher temperatures substitution solid solution. Nickel in iron extends the γ-area and by this as alloying element is similar like manganese. The rate of diffusion of iron into nickel is higher than that of nickel into iron. For this reason, the microstructure heterogeneities in sintered mixed nickel alloyed steels cannot be completely removed under acceptable sintering conditions. The concentration of nickel in regions of interparticle bonds is higher compared to the cores. The microstructure inside the grains is ferritic if the nickel addition is too low to stabilize martensite. Nickel is added to the iron powder in the form of the finest powder, and the material should be sintered minimum at 1100°C, better at 1200°C. Nickel and copper belong also to the main alloying elements in diffusion alloyed powders, nickel (2%) also in prealloyed powders.

Molybdenum is a strong carbide forming element with fairly low thermodynamic stability of the oxides. It is therefore the most commonly used carbide-forming alloying element in sintered steels. Molybdenum increases the stability of the steel at the tempering and by this lowers the temper brittleness. Mo is either admixed to iron powder in the form of elemental molybdenum or ferromolybdenum, commonly in amounts of ~0.5% mostly in combination with copper and nickel except carbon, and also in formation of complex alloyed carbides. The Mo prealloyed powders besides the Cr prealloyed powders form a new perspective group of relatively low-alloyed structurally homogeneous powders [2-6].

The use of mentioned elements for alloying of sintered steels by solid state diffusion clearly shows that the formed structures widely differ compared to that formed in alloyed wrought steels. The main combinations in alloying of sintered steels are based totally on other physical-metallurgical characteristics and requirements, therefore are not comparable to those for the structural wrought steels in term of kind and amount of alloying elements. The characteristic base alloying systems in PM steels as listed in Tab.1. As given, some sintered steels prepared by different methods contain up to ~10% of alloying elements (up to quintanary material systems) for the attainment of required high mechanical and toughness properties under common PM routes. The high price of powder nickel with carcinogenic and allergenic properties [8] and some problems in recycling of copper call out the effort to produce less expensive technically more effective sintered products.

Table 1. Base combinations in alloying of sintered steels [5,7]

Steel group	Composition [mass %]
Fe-xMe (mixed)	Fe-(1,2,4,5,8)Cu, Fe-(3,4)Ni, Fe-2Cu-2Ni, Fe-2.5Cu-2Ni, Fe-0.6Mo-0.5Ni-0.25Mn, Fe-0.5Mo, Fe-0.9Mo, Fe-0.5Ni-0.5Mo-0.7Si
Fe-P-xMe (mixed)	Fe-(0.3,0.45,0.60)P, Fe-0.6P-(2,4)Cu, Fe-0.45P-(2,4,5,8)Cu
Distaloy type (diffusion alloyed)	Fe-(1.75,4)Ni-1.5Cu-0.5Mo, Fe-5Ni-2Cu-1Mo, Fe-2Ni-1.5Mo,Fe-2Cu-1.5Mo, Fe-4Ni-2Cu-1.5Mo, Fe-4Ni-0.85Mo-0.7Si, Fe-2Ni-0.8Mo-0.7Si
Prealloyed	Fe-(0.85,1.5,3.5)Mo, Fe-(1,1.5,3)Cr-(0.2,0.5)Mo, Fe-2Ni-0.5Mo-0.2Mn, Fe-(1,3)Cr-0.3Mo-0.3V, Fe-1Cr-0.7Mn-0.2Mo
Hybrid	Fe-0.85Mo-(1,2)Cu, Fe-0.85Mo-(1,2)Ni, Fe-1.5Mo-(1,2)Cu, Fe-1.5Mo-2Ni, Fe-1.5Cr-0.2Mo-2Cu, Fe-2Ni-0.5Mo-0.2Mn-2Cu, Fe-3Cr-0.5Mo-(1-3)Mn, Fe-(1,3)Cr-0.3Mo-0.3V(1-3)Mn, Fe-(0.85,1.5)Mo-(1-3)Mn

The attention was turned therefore permanently to manganese and chromium, as alloying elements in PM, which are base alloying elements in wrought steels, mostly with Mo, in spite of the much higher affinity of Mn and Cr to oxygen. These elements, individually or collectively, increase the hardenability of steels and, at the same time, Cr and Mo are very strong carbide formers. This effort can be called as "nickelless" sintered steels.

Manganese due to its physical-metallurgical properties has a special position in ingot metallurgy (deoxidation and alloying), which can be valid also in PM. At the same time manganese is the cheapest alloying element, which is possible to find as base (up to 0.4% coming from the ore) and alloying element in most structural wrought steel. Manganese exhibits by a high hardening effect for ferrite compared to other alloying elements, e.g. Ni, especially to chromium. On the other side, due to high hardening effect of manganese for ferrite, the Mn prealloyed powders is not possible to press for reaching the satisfactory green strength as demonstrated by the water atomized Fe-(1.4 and 2.9)% Mn annealed powders, which had the microhardness of 175 and 245 HV0.02 [9]. Sintered manganese steels are therefore produced as mixed only. Metallic manganese (electrolytic) is silver-grey, its density is 7.21 g/cm^3 and the melting temperature point 1244°C. Manganese widens the γ - area, and by this decreases the γ - α and α - γ phase transformation temperatures.

Manganese and chromium possess an inherent disadvantage against nickel and copper for use in PM, even against molybdenum, in view of their very stable, i.e. difficult to reduce oxides. The reduction according to thermodynamics requires a sintering atmosphere with an extremely low oxygen potential. It is virtually impossible to achieve in an industrial environment an oxygen partial pressure in sintering atmosphere in balance with Mn and/or Cr, which complies with the limits given by the Mn and Cr lines in the Richardson - Ellingham oxygen potential diagram. Under this view, many research works were performed concerning with sintering Fe-Mn-(C) without some final statement to this problem, which was not up to present time overcome in approach mainly to the application of manganese steels for production of structural components. In spite of the known advantageous economical and technical characteristics of manganese in cast and ingot metallurgy, it did not find the position in alloying in powder metallurgy.

The findings about the sublimation of manganese due to its high vapor pressure (physical property of manganese) explain the thermodynamic conditions for sintering and alloying of Fe-Mn steels in the atmospheres of common purity [10,11]. They explain also the reason for achievement high mechanical properties of sintered Fe-Mn-steels in atmospheres not fulfilling the thermodynamic requirements for the purity.

In spite of this status of the knowledge, the actual research of sintered manganese steels occurs under some physical-metallurgical uncertainties, which affect negatively the use of manganese containing steels in the production of sintered components. This chapter brings in two parts a complex view on the problem of sintered manganese steels. The first part presents a review of main results attained in research of Mn steels since 1948, mostly regarding the thermodynamics of Mn – O system. The second part explains the effect of manganese vapor formed by sublimation of manganese carrier in the compacts due to its high vapor pressure on alloying and sintering of manganese steels. This finally documents that only under the effect of manganese vapor occurs sintering of manganese steels, i.e. independently on the purity of sintering atmosphere applied under laboratory and industrial sintering conditions, because the process is the consequence of the physical property of manganese. Manganese due to low price, high vapor pressure, and high hardening effect, is a "friendly" alloying element for powder metallurgy.

Sintering and Properties of Manganese Steels, Part I

This part of the chapter focuses on the published processing conditions for preparation of sintered manganese steels, which tried to regard mainly the thermodynamic requirements for Mn – O system. As mentioned before a high oxygen affinity of manganese and the difficulty in reducing of its oxides is problem with which even some latest works dealing with the sintered manganese steels are concerned. According to this there is still the risk of manganese carrier oxidation in the compacts during sintering and thus a possible threat to common industrial production of manganese steels. The problem seemed to be the preparation a protective atmosphere having so high a purity that it guarantees an oxygen potential pressure in balance with manganese. Since the first published paper concerning with sintered manganese steel in 1936 [12], the following research was performed under two aspects. The first one is the "pressure" of thermodynamically determined hard reducibility of manganese oxides by searching the methods to overcome this problem. The second one is the sintering without consideration the thermodynamics of Mn-O system, i.e. the sintering of manganese steels under common sintering conditions as for those alloyed with Ni, Cu, P and Mo. The assumed problem of high affinity of manganese for oxygen in powder metallurgy can be explained on the basis of detailed thermodynamics, which can bring a more exact knowledge about this often conflict problem.

Thermodynamics of Mn–O System

In advance is necessary to introduce that the use of equilibrium data to calculate compositional requirements for the protective atmosphere implicitly assumes that the system in question may be brought to an equilibrium or steady state condition. The steady state differs from a true equilibrium here in that only a partial, or local, equilibrium is established at the metal-vapor (atmosphere) interface. Supplying atmosphere at a rate sufficient to overpower any compositional changes that might occur does this by reaction of the atmosphere with metal surfaces. It is readily obvious that this situation is not possible with porous metal materials since the surface extends well into the interior volume (surface of interconnected pores) and continuous rapid replenishment of atmosphere in contact with these inner surfaces is physically impossible under normal furnace conditions.

Additionally, any minor change of local temperatures can affect equilibrium ratios. The situation inside the parts is very different, indeed, in the regions equilibrium can be established only between local carbon content and in our case also local manganese and local oxygen or residual water vapor formed by reaction of hydrogen with iron oxide. The effect of the atmosphere inside the parts can be important at relatively low densities. For this reason, the ideal stages should be considered almost completely valid only for the "skin" - surface of PM parts. What really happens, which reactions proceed in the core, strictly depends on porosity, pore structure, part volume and part size. It is necessary to stress, that the porosity increases in a large amount the total reaction surface of the sintered compact. For instance, at the total porosity of 12% of a sample the surface area of interconnected pores is around 130 times larger as its geometrical surface area [13]. For a better understanding of actual sintering conditions, definitely different from the classical equilibrium concept, it may be necessary to discuss the different possible atmospheres separately for individual material systems, what the researchers tried to do for Fe-Mn system.

Manganese with oxygen forms the compounds Mn_2O_3, MnO_2, MnO_3, Mn_2O_7, Mn_3O_4 and MnO, which is most important manganese oxide also in ingot metallurgy. The majority of them have also different modifications. MnO_2 and Mn_2O_3 easily decompose to Mn_3O_4 and oxygen and by this reason are easily reducible. The oxides Mn_3O_4 and MnO exist at temperature higher than 800°C, and should be reducible only with carbon at presence of the element with high solubility of manganese [14]. These Mn-oxides prove to be of greatest importance for powder metallurgy. Iron in this case can acts as activator at temperatures higher than the ordinary sintering temperature of 1120°C, and therefore the analysis is always concentrated on MnO regarded the most hardly reducible Mn-oxide. The oxidation of metallic manganese in air starts at ~400°C under the formation of an active bright blue-grey Mn-oxide, which at 500°C transforms to a stable in air inactive green in color MnO. At the increased temperature, over 900°C forms also an inactive dark grey form of MnO of unstoichiometric composition [15].

All requirements concerning the reducibility of manganese oxides during the sintering of mixed manganese steel in PM literature therefore relate to MnO only. The basic reaction for manganese and oxygen expresses the equation:

$$2Mn + O_2 = 2MnO.$$

According to this reaction manganese as element of high affinity for oxygen serves in steel making industry as deoxidizing element. The thermodynamic conditions for reduction of MnO oxide by various hydrogen and/or carbon containing atmospheres or by solid carbon are described below. They only particularize the oxygen potentials in the atmosphere under some sintering condition for MnO and partly for FeO. The free enthalpy changes shown in Fig.1 (Richardson-Ellingham diagram) clearly demonstrate why manganese as a PM alloying element did not find up to this time some observable range of application.

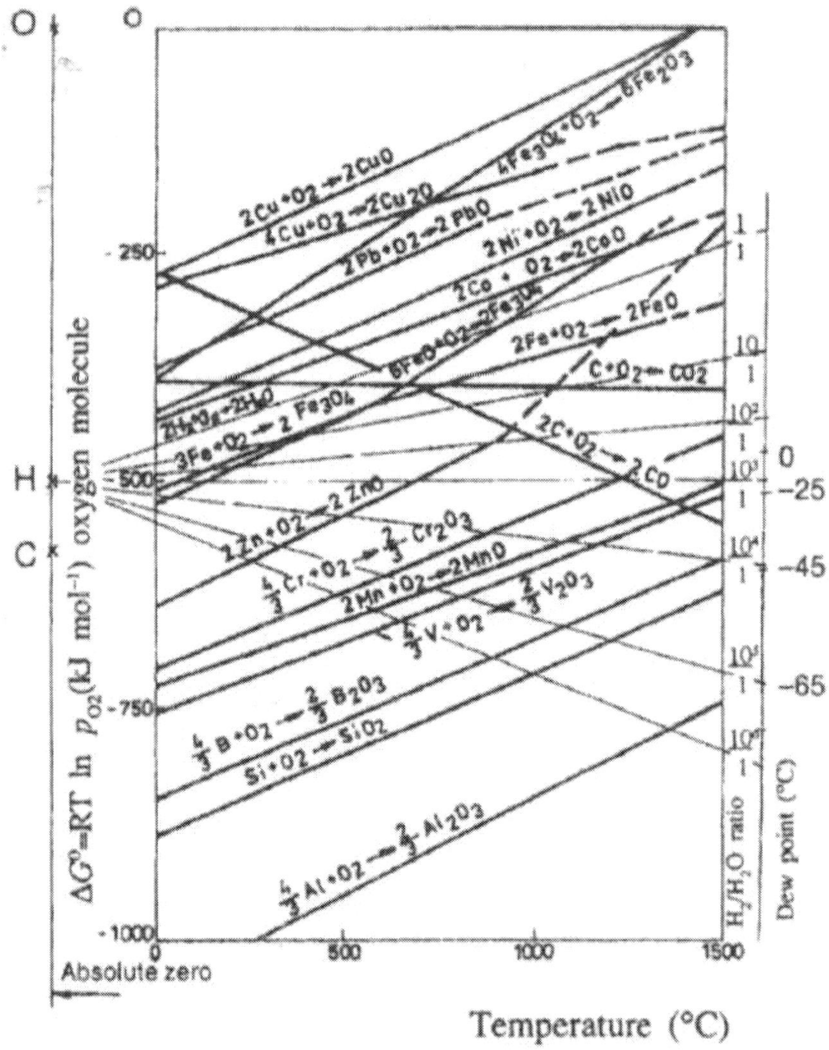

Figure 1. Temperature dependence of enthalpy changes and oxygen potential of metal oxides (Richardson-Ellingham diagram) [4,16].

As can be seen the line for Mn/MnO lies under the lines for other in PM used alloying elements including chromium, which became in the last decade the object of an increased research interest. In Fig.2 are shown the equilibrium dissociation pressures PO_2 for Fe/FeO at 550°C and at 1120°C and 1350°C for manganese oxide Mn/MnO system in H_2/H_2O and CO/CO_2 atmosphere. As can be seen the differences between both reduction agents are for Mn/MnO relatively small but larger in relation to other in PM used alloying elements.

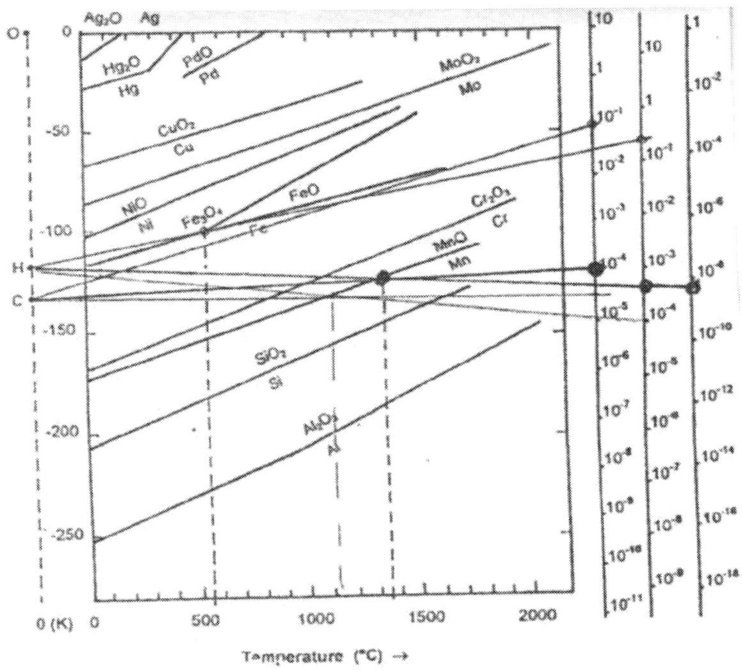

Figure 2. Graphical determination of the equilibrium dissociation pressure PO_2 for Fe/FeO and Mn/MnO systems at various temperatures [4,17].

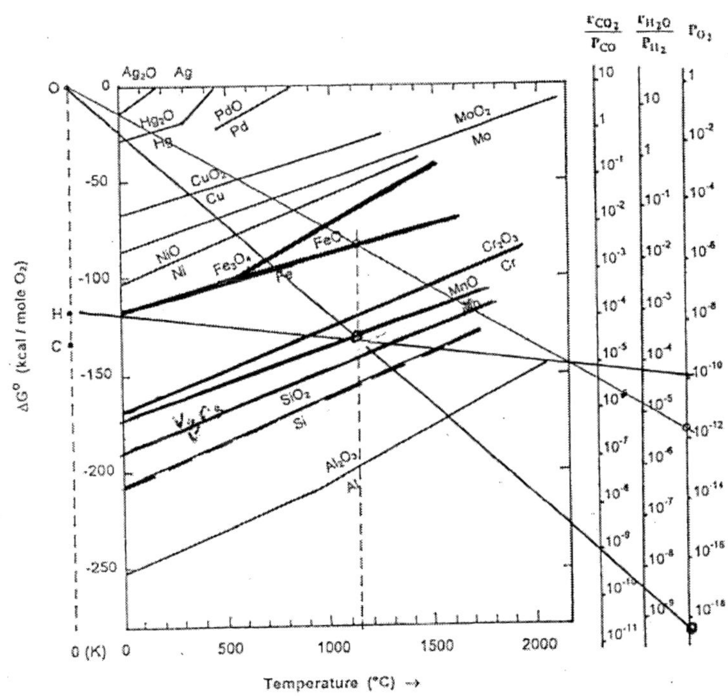

Figure 3. (right) Graphical determination of equilibrium temperatures for Fe and Mn in H_2/H_2O and CO/CO_2 atmosphere [18].

Figure 3 illustrates the equilibrium oxygen dissociation pressure for FeO and MnO at 1120°C, the most common sintering temperature. The difference between these dissociation pressures is larger than by 6 orders of magnitude, i.e. it is very low for MnO. A detailed view on oxygen potential in the sintering atmospheres commonly used in powder metallurgy, i.e. endothermic gas (endogas), dissociated ammonia, hydrogen and on N_2 - based atmospheres at different dew points for Cr and Mn shows Fig.4. It follows from the diagram that H_2/H_2O and temperatures are the critical factors determining whether oxidation or reduction will occur. „None of the analyzed atmospheres according to these calculations can reduce the oxides or avoid further oxidation of Cr and Mn at 1120°C" stated [20]. The equilibrium values of the H_2/H_2O ratio and the dew points in H_2/H_2O atmospheres for selected metal-oxide systems are better shown in Fig.5.

The calculated minimal dew points and oxygen potentials for H_2/H_2O atmospheres in the temperature range of 600 to 1200°C to reduce MnO or to prevent oxidation of manganese are listed in Table 2. It would be very hard up to impossible to fulfill these requirements for the purity of the sintering atmosphere of manganese steel, namely under industrial conditions. The thermodynamic requirements for the purity of a sintering atmosphere should be not constant during the whole sintering temperature range. It is necessary to consider the total preheating period as well under laboratory as mainly under industrial sintering conditions. The preheating period in the sintering process is that in which occurs the dewaxing of the compacts at which the lubricant melts and evaporates followed by burn-off requiring a humid atmosphere yet at the temperature below ~600°C [3]. On the other hand the Mn-containing parts require for elimination the oxidation an atmosphere with extremely low dew point, as for instance of -102°C at 600°C, which is not in practice realizable.

It is necessary also to note that all mentioned reactions for Mn-O-H systems from the thermodynamic point of view are calculated for a unit activity of the system (Richardson – Ellingham diagram). They refer to chemically pure compounds including hydrogen. In the real systems, the chemical activity, e.g. of ferromanganese grades or of some special Mn-containing master alloys, is much lower because they contain different impurities, mainly iron. Accordingly, FeO and vice-versa (unlimited solid solubility) can substitute the places in the lattice of MnO.

Table 2. Equilibrium data of Mn/MnO in O_2 and in gas mixture of H_2/H_2O. D.p. – dew point [4,18]

	Temperature [°C]								
	600	700	800	900	1000	1100	1150	1200	1400
PO_2 [Pa]	10^{-38}	10^{-34}	10^{-30}	10^{-26}	10^{-24}	10^{-22}	10^{-21}	10^{-20}	10^{-16}
H_2/H_2O	$1\cdot10^8$	$1\cdot10^7$	$3\cdot10^6$	$7\cdot10^5$	$2\cdot10^5$	$9\cdot10^4$	$5\cdot10^4$	$4\cdot10^4$	$2\cdot10^3$
D.p. [°C]	-102	-90	-80	-74	-60	-55	-50	-42	-30

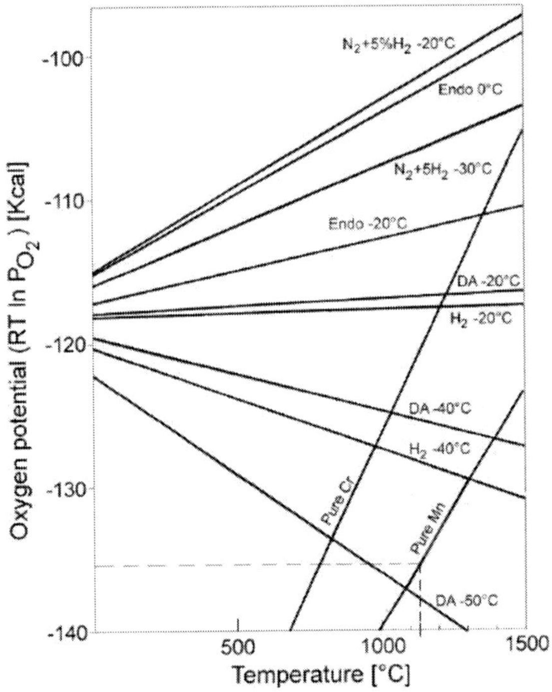

Figure 4. Oxygen potential in endogas, dissociated ammonia, hydrogen and N_2 - based atmospheres at different dew points (indicated temperatures refer to the dew point of the atmosphere) [18-20].

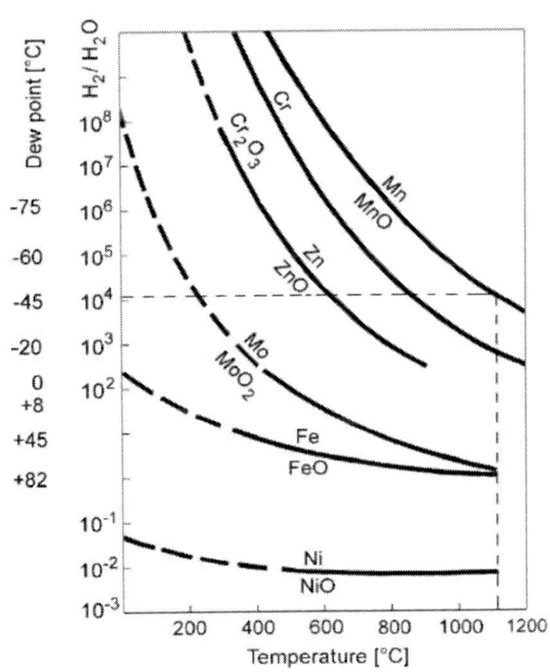

Figure 5. (right) Equilibrium values of H_2/H_2O ratio and the dew points in H_2/H_2O atmospheres for selected metal-oxides systems [4,17].

The phase stability diagrams Mn-O-H and Mn-O-C for 800, 1200 and 1300°C, elaborated on the basis of mentioned thermodynamic analyses for reduction and conversion reactions of manganese oxides with $H_{2(g)}$, $CO_{(g)}$ and $C_{(s)}$ for Mn-O-H and for Mn-O-C system, made possible to deduce comprehensively following results [18]:

- reduction of MnO $\rightarrow$ Mn by neither $H_{2(g)}$, $CO_{(g)}$ nor $C_{(s)}$ occurs below 1280°C
- reduction of MnO is possible only with $C_{(s)}$ at temperature above 1280°C
- conversion of MnO_2 to Mn (MnO_2 is not stable) is possible, the highest step of conversion reaches the reaction with $C_{(s)}$
- reduction both of $Mn_2O_3 \rightarrow Mn$ at >1000°C and of $Mn_3O_4 \rightarrow Mn$ at >1100°C with $C_{(s)}$ can take place
- reduction of $MnO_{(g)}$ in $H_{2(g)}$ and $CO_{(g)}$ can take place.

According to [21] the reduction of MnO to Mn occurs with carbon as graphite (solid) at a temperature over 1200°C evenly only through the Mn-carbide (Mn_5C_2) as follows from the overall reduction reactions for this process:

$$5MnO + 7C = Mn_5C_2 + 5CO$$

$$2MnO + Mn_5C_2 = 7Mn_{(g)} + 2CO.$$

During the sintering of manganese steel, the conditions for the reduction with solid carbon do not form because at the temperature over 900°C, solid carbon (graphite) in sintered steel cannot be present due to its dissolution in austenite.

In term of the mentioned thermodynamic calculations, the presence of MnO in being admixed sintered Fe-Mn compacts is apriori assumed. There are only two possibilities for the presence of some manganese oxides in sintered systems what was in the literature not analyzed. The surface of the starting manganese carrier particles (electrolytic manganese, ferromanganese grades, special Mn-containing master alloys) could be covered with some layer formed at milling in which could be detected some manganese oxide. The MnO, which is green in color and by this easy discernible also by look, is possible to exclude due to its reducibility only at the temperature exceeding the common sintering temperatures as follows from the previous data. Beside it, there are no data in the literature regarding the Mn-oxides on the particles of the manganese carriers. There is only the possibility for oxidation of manganese particles in the compacts during preheating period as well under laboratory as under industrial sintering conditions up to the temperature at which the equilibrium occurs at the dew point of the used atmosphere. For instance, by the use of H/N sintering atmosphere with the dew point -50°C, the equilibrium conditions for Mn-O system corresponds to the temperature of 1150°C. This purity of the atmosphere does not fulfill the requirements for the preheating period over the temperature 400°C, at which manganese starts to oxidize, Table 1.

Properties of Sintered Manganese Steels Attained under Various Processing Conditions

Manganese became the object of systematic interest in PM since 1948 – 1950. Owing to the shortage of nickel in ~1970, there was considerable interest by PM manufacturers in substituting manganese for nickel. An incomplete overview of published works up to 1999 (228) concerning manganese in powder metallurgy [22] confirm a large extent of research activities in this field. A survey of the results by the use of manganese in powder metallurgy with 118 references was also prepared [23].

The following results represent the mechanical properties of sintered manganese steels attained under various material and processing conditions, which make possible to consider in what extent were fulfilled the thermodynamic requirements to avoid the oxidation of manganese or to reduce its oxides at sintering. The presented mechanical and other properties of sintered manganese steels are in following presented only with a short description taken from the corresponding references with the impact on sintering conditions including atmospheres and manganese carriers.

Various Processing Conditions

In the book [24] edited in 1948 are described in that time the processing conditions for the production of sintered alloyed steels in term of physical-metallurgical and technical aspects. Figure 6 shows the tensile strength and elongation of sintered manganese steel in dependence on manganese addition (0-16)% and two carbon levels elaborated on the previous knowledge.

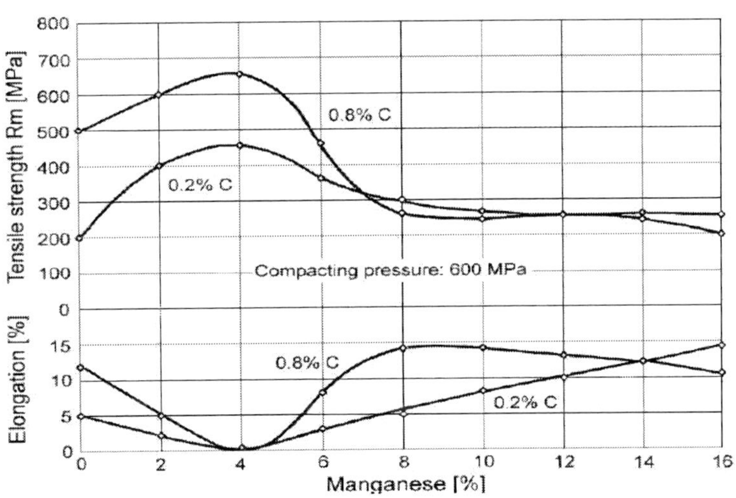

Figure 6. Tensile strength and elongation of sintered steels in dependence on manganese and carbon content. Sintering for 4 h at 1200°C in hydrogen under getter alumina + graphite in closed graphite boxes; Hametag (eddy milled) iron powder (not more available), 600 MPa [24,25].

The sintering for 4 h at 1200°C was performed in hydrogen without an additional cleaning under mentioned getter to minimize the decarburization. Low carbon ferromanganese was used as manganese carrier and natural graphite. The tested range of the compositions includes the manganese contents for structural and partly austenite steels. The maximum tensile strength of 670 MPa attained the Fe-4Mn-0.8C steel.

Authors [26] investigated the effect of sintering temperature and manganese content on tensile strength of carbonless manganese steel (Fe-Mn). Sintering was performed for 1 h in the temperature range of 1000 to 1300°C in hydrogen (purity not declared). Electrolytic manganese <40 μm in size as manganese carrier was used. According with Fig.7, the tensile strength shows a very rapid increase with increasing Mn-content up to 10% with increasing temperature followed by a rapid decline. The Fe-10Mn steel sintered at 1300°C attained the tensile strength of ~440 MPa.

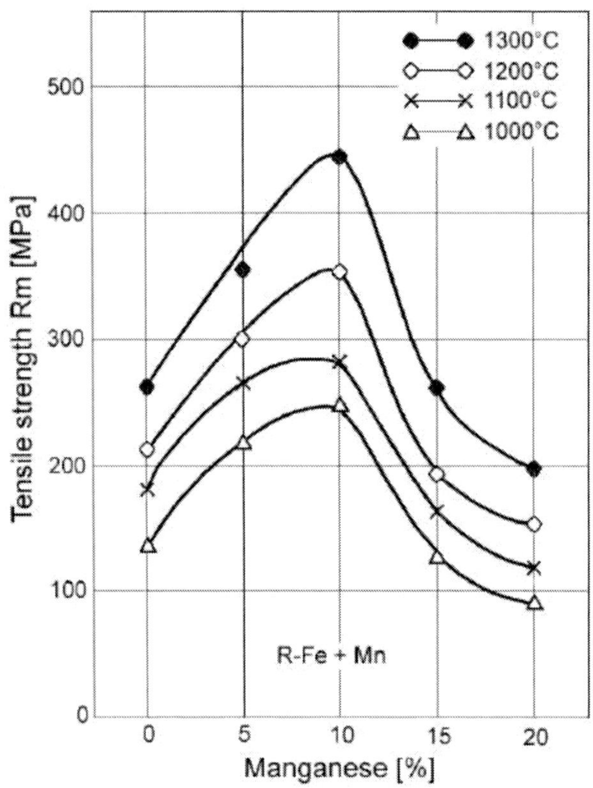

Figure 7. Effect of manganese content (no carbon added) and sintering temperature on tensile strength of sintered manganese steel. Reduced iron powder MH 100.24, 600 MPa [26,27] .

It is possible to note, that the tensile strength was increasing with increasing manganese addition also at 1000°C as the proof of the alloying of the iron matrix by manganese.

Figure 8 shows the tensile strength and hardness of low carbon low-alloyed mixed FeMnCrMoCuC steel sintered at 1120°C after furnace cooling and water and oil quenching in dependence on sintered density. The sintering was performed in cracked ammonia with the dew point -30°C, and manganese was added in form of high carbon ferromanganese (76% Mn, 6% C). This steel with total content of alloying elements of ~3% demonstrated as proper for heat treatment. The water-quenched steel reached the tensile strength in the range of 570 to 690 MPa.

The microstructures of as heat treated steels were bainitic, what demonstrated in impact energy KCV values of 9 to 14 J. The impact energy (unnotched) KC attained the as sintered steel samples in the range of 24 to 61 J.

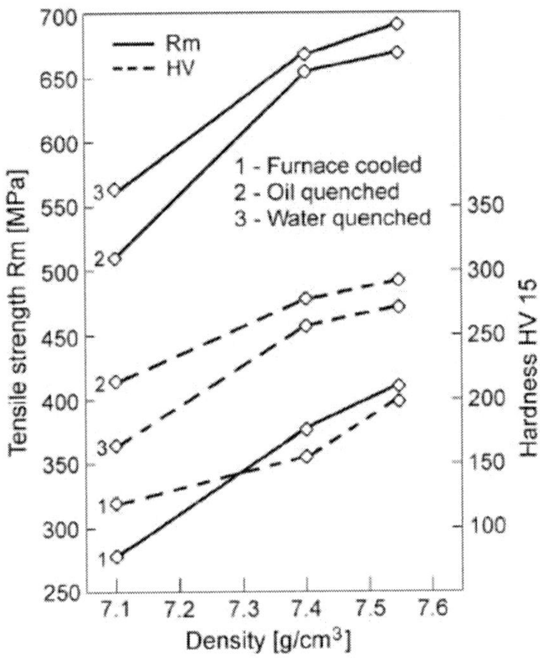

Figure 8. Tensile strength and hardness of as sintered and as heat treated low carbon low alloy Fe-0.1C-1.4Mn-0.5Cr-0.5Mo-0.6Cu steel in dependence on density. Iron powder reduced from mill scale, sintering 2 h at 1120°C, ferrochrome (Fe65Cr), ferromolybdenum (Fe57Mo), electrolytic copper, single, double and triple press-sinter technique. 1 – slow furnace cooled, 2 – oil quenched, 3 – water quenched [28].

In the case shown in Fig.9, the effect of manganese addition in the range of 1 to 8% and manganese carrier grade (electrolytic manganese – EMn (<5 μm) and medium carbon ferromanganese – FeMn (Fe-85Mn-1C, <63 μm) on yield and tensile strength was investigated. Sintering of the samples, based on the previous laboratory sintering, was carried out in industrial walking beam furnace in cracked ammonia under getter alumina + Fe18Al. The maximum tensile strength of 550 – 600 MPa attained the steels with 5-6% Mn. The minimal difference in properties in dependence on manganese carrier was observed. The similar dependence exhibits also the yield strength.

A detailed study performed by [29,30] enlarged the knowledge related to sintered manganese steels shown in Fig.9 also by industrial sintering at the temperatures below 1200°C and above 1280°C (melting temperature of manganese of 1244°C). This temperature was regarded for extreme for the study the sintering process of manganese steels. The mechanical properties of the tested materials points Table 3. The results show that there was no marked effect of varying sintering time and temperature on mechanical properties, except on the elongation at longer sintering time. The sintering of the samples under industrial conditions is necessary to stress. The density values as well as the mechanical properties attained do not enable to assume the sintering at 1280°C under the effect of the liquid phase formed by Mn-carriers.

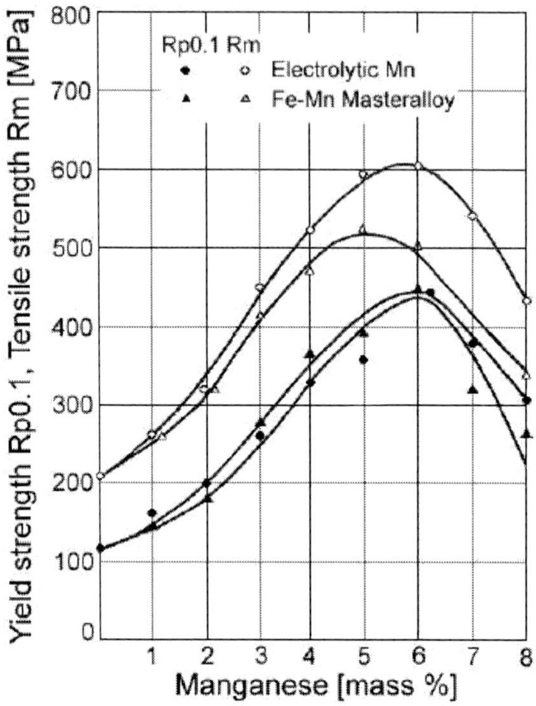

Figure 9. Effect of manganese content on yield and tensile strength for single pressed Fe-Mn steels; water atomized iron powder Atomet 28, 600 MPa, sintering 100 min at 1280°C, walking-beam furnace [23,27,29,30].

Table 3. Effect of sintering time and temperature on density and mechanical properties of single pressed Fe-Mn steels. The preparation of the mixtures as in Fig.9 [29,30]

Steel	Sint. time/ temperature [min/°C]	Sintered density [g/cm^3]	Rp0.1 [MPa]	Rm [MPa]	A_5 [%]	Hardness HRB
Fe-2% Mn	50/1200	7.03	170	300	11.0	41
	100/1200	7.05	220	340	10.0	47
	50/1280	7.06	200	320	10.0	45
	100/1280	7.06	200	330	15.0	44
Fe-4% Mn	50/1200	6.90	340	490	3.5	74
	100/1200	6.91	320	530	4.0	78
	50/1280	6.92	330	530	4.0	76
	100/1280	6.92	330	550	4.0	79
Fe-6% Mn	50/1200	6.79	370	600	3.2	90
	100/1200	6.81	420	650	4.0	86
	50/1280	6.80	450	610	2.5	88
	100/1280	6.87	440	620	4.0	88

The effect of manganese addition, and of two iron powder grades on as sintered, and as heat-treated tensile strength, elongation and hardness of Fe-(2.5-5.5)Mn-C steel values shows Fig.10. Hametag iron powder used is characterized by higher structural activity (eddy milled, fine grained, density of dislocations of 1.3×10^{10} cm^{-2} after annealing at 750°C), and RZ iron powder (air atomized, annealed at~1100°C) is characterized by a coarse grained structure [31-

34]. (Both iron powder grades are not more available). Manganese to the iron powders was added as high carbon ferromanganese (75% Mn, 7.85% C, <45 μm). Sintering was carried out in cracked ammonia with the dew point -30°C. As sintered density values and manganese and carbon contents of the tested steels are listed in Table 4.

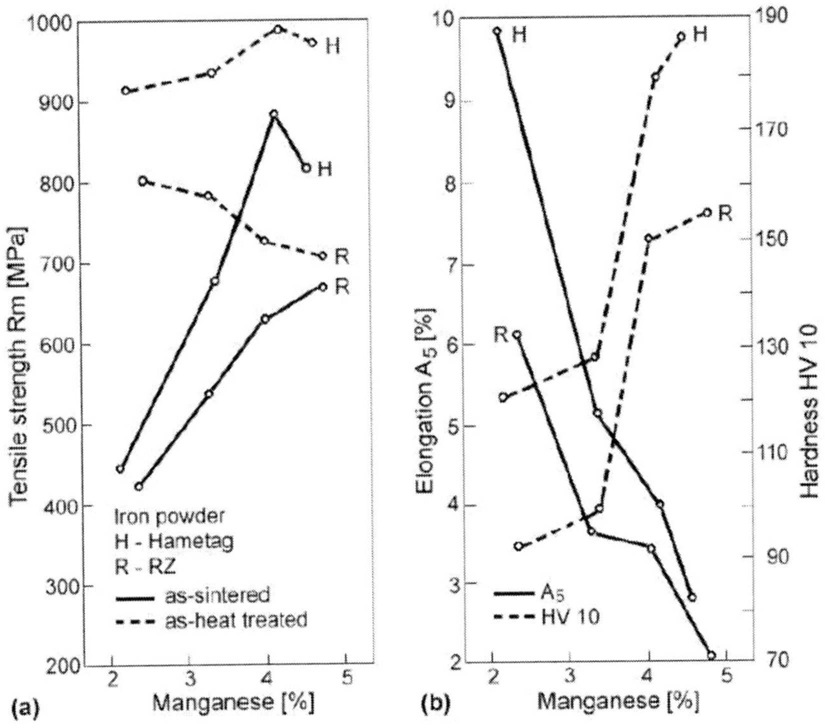

Figure 10. Effect of manganese content and of iron powder grade (H, R) on: a) as sintered and as heat treated tensile strength, b) as sintered elongation and hardness; 600 MPa, 3 h, 1120°C, heat treatment – quenching 850°C, N_2, oil, 200°C, 30 min, air [35-43].

Table 4. As sintered density (ρ) and carbon and manganese content added values (mass %) of the steels as in Fig.10 [36]

Iron powder	Hametag (H)				RZ (R)			
ρ [g/cm^3]	7.26	7.19	7.12	7.08	7.08	7.08	6.95	6.93
Mn added	2.5	3.5	4.5	5.5	2.5	3.5	4.5	5.5
*C added	0.20	0.28	0.36	0.44	0.20	0.28	0.36	0.44
C	0.14	0.15	0.20	0.26	0.11	0.12	0.33	0.35
Mn	2.18	3.39	4.20	4.52	2.43	3.30	4.05	4.79

Remark: *C as combined in the form of FeMnC.

The tensile strength increased almost linearly in both cases with increasing Mn content up to 4 - 4.5%. The highest as sintered tensile strength of 886 MPa attained the Fe-4.2Mn-0.2C steel based on Hametag iron powder. This is the highest tensile strength of single pressed and sintered manganese steel attained under conventional laboratory processing conditions used when compared to other known PM alloyed steels. On the other hand the steel based on RZ iron powder attained the highest tensile strength of 672 MPa. Figure 10b shows the same characteristic effect of iron powder grade and manganese content on elongation and hardness of the tested steels. Both properties show changes (decreasing elongation and increasing hardness) with increasing Mn-C contents. Effect of some physical and technological characteristics of base iron powders coming from the production routes, demonstrated in markedly different mechanical properties of sintered manganese steel.

Heat treatment caused an increase for tensile strength of both steel groups. The steel based on Hametag iron powder attained the tensile strength of 996 MPa and that based on RZ powder of 700 MPa with a different course in relation to the manganese addition.

These Fe-Mn-C materials were investigated also as powder forged and as extruded [44-49] what demonstrated in tensile strength up to 1300 MPa.

Short Time Induction Sintering

Short time induction (high frequency, HF) sintering tests were performed mainly with Fe-4.5Mn-0.36C steel for study the starting states of the formation of the microstructures and corresponding mechanical properties of manganese steels. Induction sintering was performed with individual test samples (ISO 2740). The samples were for comparison sintered also conventionally in the same tube. The mechanical properties of these steels are listed in Table 5. Increase of all properties was observed yet after 1 min HF sintering as the proof of an active alloying by manganese. (Coarse grained microstructure of RZ powder is necessary to regard). Induction sintered samples were tested also as powder forged [51-53].

Table 5. Mechanical properties of induction (high frequency, HF) sintered Fe-4.5Mn-0.36C steel. RZ iron powder, high carbon ferromanganese, 600 MPa, sintering of in a quartz tube in hydrogen, d.p. -30°C, carbon as combined added in form of high carbon ferromanganese, KC – impact energy [50]

Sint. temp. [°C]	Sint. time [min]	C_c [mass %]	UTS [MPa]	Hardness HV 20	KC [J]	*UTS [MPa]
1100	1	0.31	273	160	3.9	-
	**3	0.29	296	159	5.5	246
	10	0.24	315	179	5.7	296
	30	0.19	420	180	5.8	408
1200	1	0.30	238	185	4.4	-
	3	0.27	372	189	5.7	255
	10	0.19	415	177	6.0	432
	30	0.15	427	133	4.8	421
1300	1	0.27	339	137	6.0	-
	3	0.20	377	161	6.8	459
	10	0.15	394	149	8.0	459
	30	0.14	407	152	6.4	460

* conventional sintering **see Fig.11a, C_c – combined.

Characteristic induction sintered microstructures of Fe-4.5Mn-0.36C steels based on RZ (coarse grained) and Hametag (fine grained) iron powder shows Fig.11. The layers alloyed by manganese were formed on the surface of all iron powder particles in the compacts. This demonstrated in strength and hardness values. The difference in alloying of the iron matrix based on both iron powder grades which exhibit markedly different starting structures and substructures demonstrated also in mechanical properties of the samples conventionally sintered shown in Fig.10.

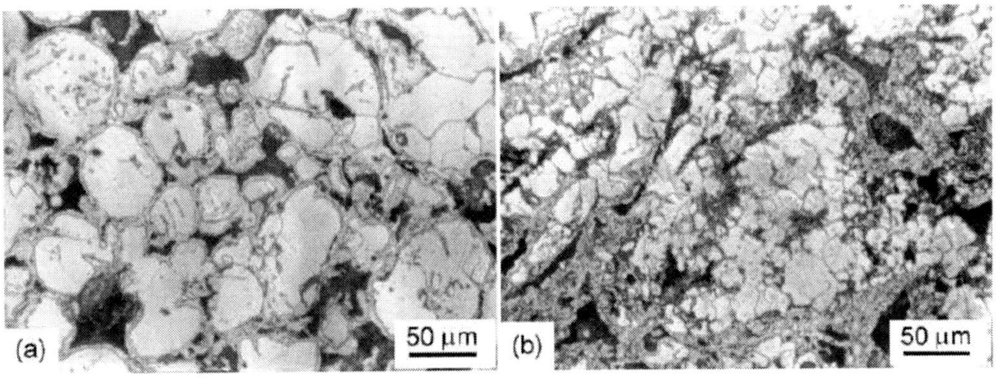

Figure 11. Microstructures of induction (HF) for 3 min at 1100°C sintered Fe-4.5Mn-0.36C steel samples based on: (a) RZ iron powder, (b) Hametag iron powder; optical micrographs. Nital etched.

Effect of Some Sintering Conditions and Base Materials

Effect of many factors affecting the properties of sintered manganese steel studied [54]. Sintering of manganese steels was performed in hydrogen of technical purity under getter alumina + 5% graphite and electrolytic manganese was used as manganese carrier (-10 μm). Effect of sintering time on tensile strength and impact energy of the tested samples is shown in Fig.12. The highest tensile strength of 810 MPa attained the steel Fe-4Mn-0.3C sintered for 180 min. Similarly the highest impact energy and elongation values attained the Fe-2Mn-C steel specimens sintered for 90 min as shown in Fig.13.The decrease of both properties with increasing carbon and decreasing manganese content was observed.

The mechanical properties of the Fe-(1-5)Mn-C steels sintered in hydrogen of technical purity at two temperatures again under getter – alumina + 5% graphite to avoid decarburization gives Table 6. The testing of mechanical properties revealed that their improvement except of elongation by sintering at higher temperature is more pronounced the higher the Mn content. While at 2% Mn the effect of the sintering temperatures is moderate, at 4% a decisive improvement of both and ductility (impact energy) is observed. Relatively high cooling rate of the samples at the evaluation of the properties is necessary to consider.

Investigation carried out with differently compacted Mn-steel compacts showed that in the case of Mn-steels the usual steady increase of the mechanical properties with increasing compacting pressure (some density increase) is found only up to a certain compacting pressure level, at still higher densities as shown in Fig.14.

Effect of manganese content on tensile strength and impact energy of low carbon Fe-(1-5)Mn-xC steels sintered under common laboratory conditions show Figs.15,16. Sintering was carried out in cracked ammonia with the dew point -30°C, and high carbon ferromanganese was applied as manganese carrier without another addition of carbon as graphite. The highest tensile strength of 589 MPa attained the Fe-4Mn-0.32C steel. Impact energy was decreasing from 22.5 J for 1Mn steel up to 5.7 J for 5Mn steel.

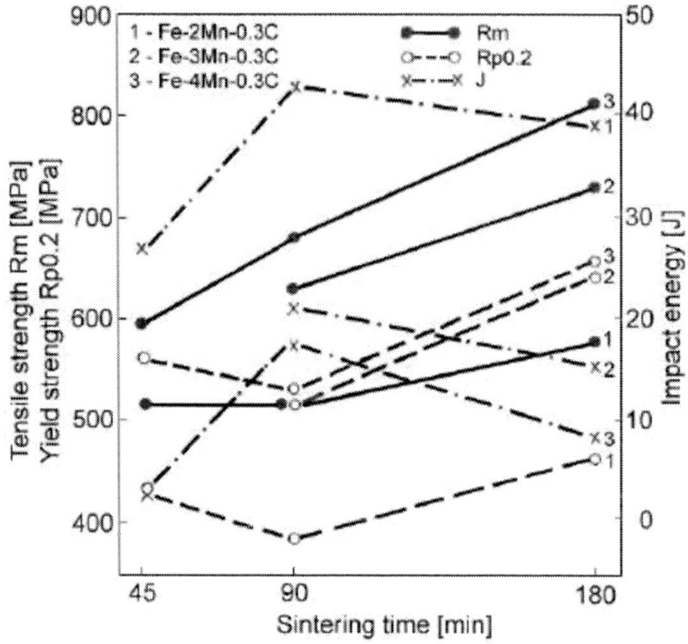

Figure 12. Yield and tensile strength and impact energy of sintered Fe-(2,3,4)Mn-0.3C steels in dependence on sintering time at 1120°C; ASC 100.29 iron powder, 600 MPa [55-57].

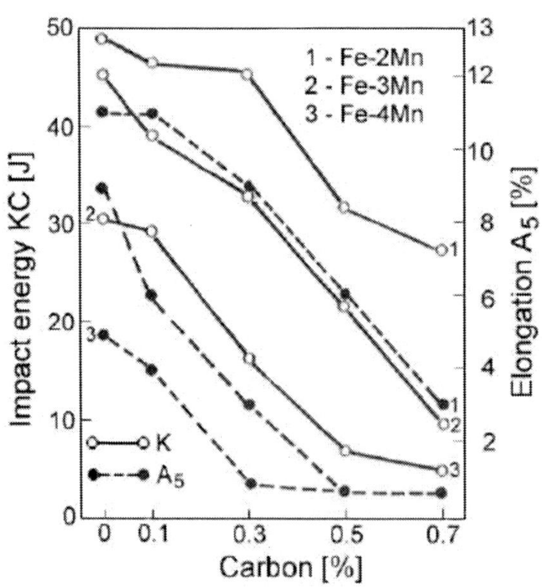

Figure 13. Impact energy and elongation of sintered Fe-(2,3,4)Mn-C steels in dependence on carbon content; sintering for 90 min at 1160°C. Other as in Fig.12 [54-56].

Table 6. Mechanical properties of Fe-(1-5)Mn-0.3C steels. ASC 100.29 iron powder, electrolytic manganese <5 μm, 600 MPa, isothermal sintering for 75 min, cooling rate 50-70°C/min [54-56]

Manganese [mass %]	Sintering temperature [°C]	Sintered density [g/cm³]	Tensile strength [MPa]	Elongation [%]	Impact energy KC [J]	Hardness HV 30
1	1120	7.21	384	11.2	50	93
	1280	7.25	403	12.4	52	103
2	1120	7.11	542	6.2	36	117
	1280	7.16	593	6.1	49	116
4	1120	6.98	594	1.2	8	227
	1280	7.01	966	2.5	23	275
5	1120	6.91	462	0.9	6	258
	1280	6.94	896	1.4	21	269

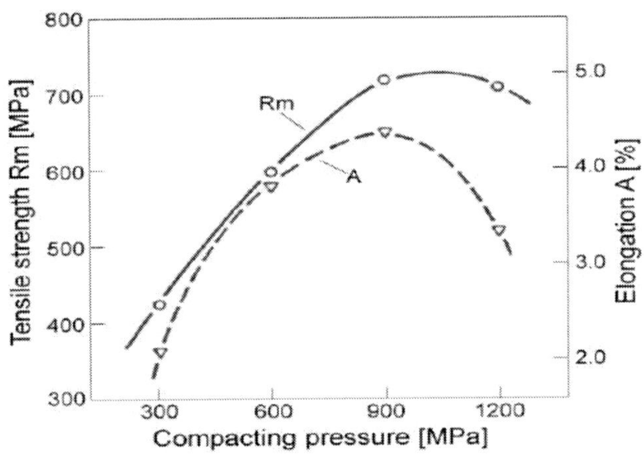

Figure 14. Tensile strength and elongation of Fe-3Mn-0.3C steel in dependence on compacting pressure. Sintering for 75 min at 1160°C, other as in Table 6 [54-56].

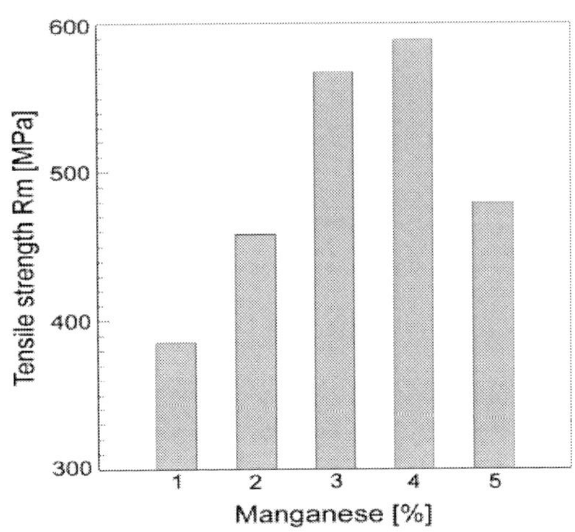

Figure 15. Tensile strength of Fe-(1-5)Mn-(0.08-0.40)C steels. SC 100.26 iron powder, 600 MPa, sintering for 60 min at 1120°C [58]. (UTS of Fe-0.2C steel = ~250 MPa).

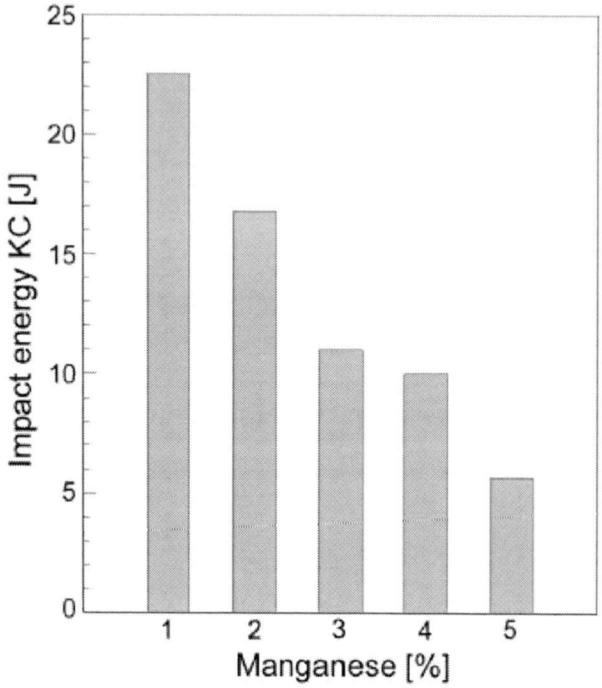

Figure 16. (right) Impact energy of Fe-(1-5)Mn-(0.08-0.40)C steels. Other as in Fig.15.

The sintering of Fe-Mn-C steels with different Mn and C contents in a pusher furnace under conditions commonly used for sintering of the steel components alloyed with Ni, Cu and Mo was performed. No special measures to adapt the sintering conditions for Mn steels according to the thermodynamics were made. High carbon ferromanganese as manganese carrier was used and sintering was carried out in industrial $70N_2/30H_2$ atmosphere with the inlet dew point of -50°C (steels 1-6, 8). The tensile strength values attained are shown in Fig.17. The maximum tensile strength of 589 MPa attained the Fe-4Mn-0.7C steel and the Fe-3Mn-1C steel (555 MPa). Effect of Mo addition demonstrated in increased tensile strength as well after industrial sintering (steel 8), and mainly after sintering at 1200°C (steel 9, 745 MPa) [57,58].

The structural components – rollers used in conveyors (Ø40/16 x 14 mm) industrially compacted and sintered from Distaloy SE (Fe-4Ni-1.75Cu-0.5Mo-0.6C) and Fe-2.5Mn-0.5C powder mixtures were subjected to the reference friction tests. Sintering of the components was carried out for 40 min at 1180°C in $70N_2/30H_2$ atmosphere with the dew point of -50°C. The disc-block method was used for the friction tests. The results are given in Table 7. Markedly lower wear exhibited the rollers prepared of manganese steel under all tested conditions when compared to those prepared of the steel based on Distaloy SE powder.

Regarding a possible effect of cooling rate and of manganese carrier on mechanical properties of Mn steels, [60] investigated the effect of high and medium carbon ferromanganese grades and of two cooling rates, i.e. of 3-4°C/min (furnace, slow) and of ~40°C/min (convective, rapid) on tensile strength of Fe-(3,4)Mn-0.8C steel. The samples were sintered in hydrogen with the dew point better than -40°C. The results show Fig.18. Both ferromanganese grades demonstrated in lower tensile strength of 4Mn steel conventionally (slowly) cooled. The same tensile strength of 460-480 MPa attained both rapidly cooled steels independently on the ferromanganese grade.

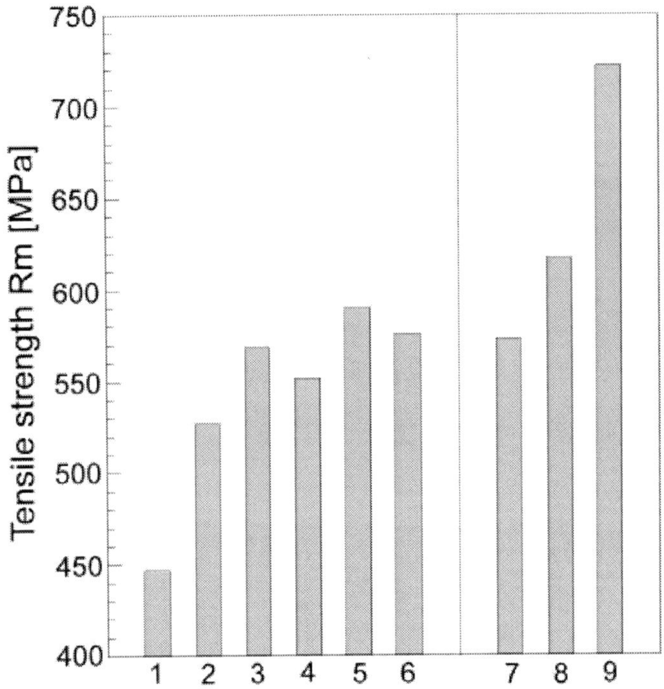

Figure 17. Tensile strength of sintered Fe-xMn-xC steels. SC 100.26 iron powder, 600 MPa, industrial sintering for 40 min at 1180°C. Composition: 1 – 2Mn-0.4 gr., 2 – 2.5Mn-0.7 gr., 3 – 2.5Mn-1.0 gr., 4 – 3Mn-0.7 gr., 5 – 3Mn – 1.0 gr., 6 – 4Mn-0.4 gr., 8 -3Mn-0.2Mo-0.7gr., 7 – 2.5Mn-0.5Mo-0.7 gr., 8 - 3Mn-0.2Mo-0.7 gr., 9 – 3Mn-0.5Mo-0.5 gr., 1 to 6, 8 industrial sintering, 7, 9 – laboratory sintering 1 h, 1200°C, H_2, gr. - graphite [58].

Table 7. Results of dry friction tests of the components – roller made from Fe-2.5Mn-0.5C (Mn) steel (170 HV 10) and Distaloy SE (NiCuMo steel, 206 HV 10). Sliding rate 0.5 m·s^{-1}, sliding distance 1000 m; counterpart: *mild steel, **hardened tool steel [59]

Steel	*Load 140 N			*Load 100 N			**Load 140 N		
	W [mg]	μ	T [°C]	W [mg]	μ	T [°C]	W [mg]	μ	T [°C]
NiCuMo	344	0.40	63	596	0.41	54	2405	0.36	58
Mn	96	0.39	56	104	0.60	64	50	0.19	57

Effect of Additional Elements

Molybdenum

Regarding previous results concerning the effect of Hametag iron powder on mechanical properties of sintered manganese steels, Fig.10, the same iron powder was used for the investigation of the effect of 0.5% Mo addition on sintered properties of Fe-(2.5-5.5)% Mn steels in dependence on compacting pressure. (Effect of Mo addition on tensile strength of the

Mn steels is shown also in Fig.17). Base characteristics of the tested steels compacted at 150 and 600 MPa are given in Table 8. The tensile strength of these materials in dependence on manganese content and compaction pressure shows Fig.19.

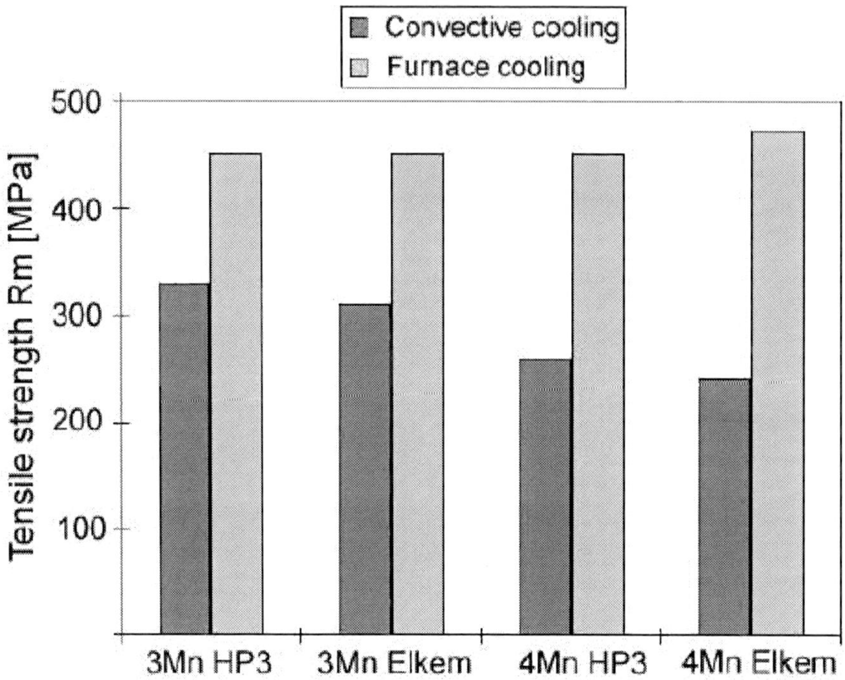

Figure 18. Histogram of tensile strength of Fe-(3,4)Mn-0.8C steels. SC 100.26 iron powder, 600 MPa, sintering for 60 min at 1180°C, HP3 – high carbon ferromanganese (6% C), Elkem – medium carbon ferromanganese (1% C), both <40 μm [60].

Table 8. Manganese and carbon contents (mass %) and microhardness (HV 0.01) values for sintered Fe-Mn-C and Fe-Mn-Mo-C steel samples. Compaction at 150 and 600 MPa, sintering 3 h at 1120°C, c.a., -30°C. Hametag iron powder, high carbon ferromanganese; steel: 1 – 2.5% Mn, 2 – 3.5% Mn, 3 – 4.5% Mn, 4 – 5.5% Mn, 5 – as 1 + 0.5% Mo, 6 - as 2 + 0.5% Mo, 7 – as 3 + 0.5% Mo, 8 – as 4 + 0.5% Mo [46,61]

Compacting pressure	Alloying element	Alloy steel No.							
		1	2	3	4	5	6	7	8
150 [MPa]	Mn	1.92	2.84	3.80	4.60	1.98	2.84	3.85	4.75
	C	0.07	0.08	0.10	0.12	0.07	0.11	0.12	0.14
*600 [MPa]	Mn	2.33	3.00	4.30	4.85	2.25	3.22	4.24	4.93
	C	0.18	0.19	0.23	0.26	0.11	0.11	0.13	0.16
*HV 0.01 (range)		142-167	149-236	148-469	314-789	149-236	175-363	252-394	333-425

Remark: Mn and C added as high carbon ferromanganese, Mo as ferromolybdenum Fe85Mo, *HV 0.01 of samples compacted at 600 Mpa.

The samples compacted at 600 MPa compared to that compacted at 150 MPa had partly a higher as sintered manganese and carbon contents due to lower fraction of open porosity in the material. The ranges of the microhardness values characterize the heterogeneity of the microstructures. The 0.5% Mo addition according to microhardness measurement contributed to the higher homogeneity of the microstructures. This follows when compared the highest microhardness values of the steels No. 3,4 (no Mo) and 7,8 (with Mo) (Table 8). The martensite was found in the steel Fe-5.5Mn-0.44C and contrary in Fe-5.5Mn-0.5Mo-0.44C steel only bainite [46,61].

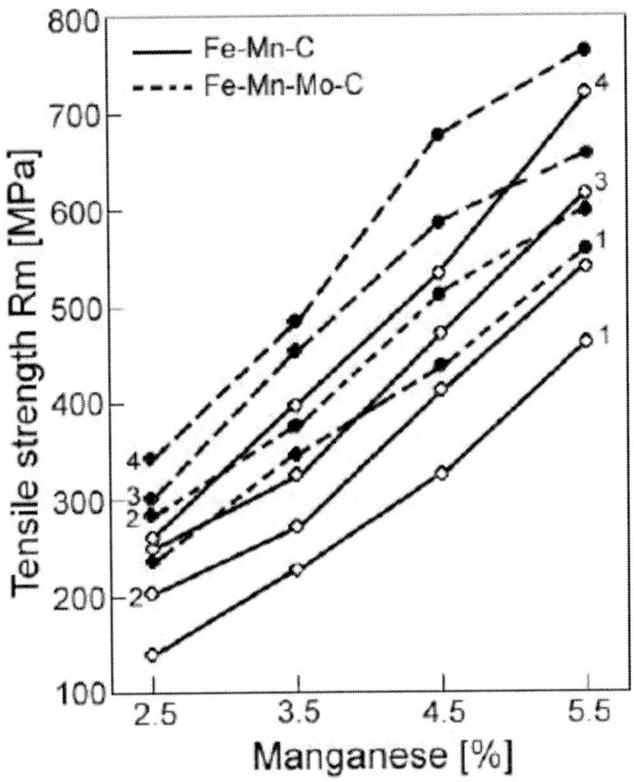

Figure 19. Tensile strength of Fe-xMn-xC and Fe-xMn-0.5Mo-xC steel samples in dependence on manganese content and compacting pressure in the range of 150 to 600 MPa: 1 – 150 MPa, 2 – 200 MPa, 3 – 400 MPa, 4 – 600 MPa; other as in Table 8 [46,61].

The gears for hydrogenerators for operating pressure of 1.5 MPa were prepared from Fe-3Mn-0.5Mo-0.35C steel by industrial compaction at 600 MPa and sintering for 140 min at 1120°C in cracked ammonia with the dew point of -30°C (hardness 178 HV 210) with the aim to replace the case hardened gears prepared from sintered Fe-2Ni-2Cu-0.6C steel or from wrought steel. The as sintered Fe-Mn-Mo-C gears were successfully tested with the varying oil pressure in the range of 3 to 5.0 MPa at 2800 rev./min during 1000 h. The test of FeMnMoC gear was performed also for 5000 h with the pressure of 3 MPa without any change in the tolerances of the dimensions and failure [62,63].

Copper. Effect of copper (melting point 1083°C) addition on properties of Fe-2Mn-(0.6,0.8)C steels lists Table 9 and the mechanical properties of Fe-4.5Mn-0.25C steel with addition of copper sintered at lower temperature as in previous case in cracked ammonia without getter are given in Table 10.

Tin. Tin is an atypical alloying element for alloying of sintered steels. The aim for the use of tin in Fe-Mn steels was to influence the alloying of iron powder matrix by manganese through the formation of a manganese-rich liquid phase and by this to minimize the problem with mentioned high affinity of manganese for oxygen and by this increase the homogeneity of the microstructure. The effect of sintering temperature on tensile strength and elongation of Fe-2Mn-1Sn steel shows Fig.20 and of tin addition on tensile strength and elongation of Fe-(1,4,6)Mn steel shows Fig.21.

Table 9. Mechanical properties of single pressed Fe-Mn-Cu-C steels. Atomet 28 iron powder, 600 MPa, electrolytic manganese, sintering 50 min at 1280°C, cracked ammonia, d.p. -30°C, getter - alumina + Fe8Al [64]

Base alloying [mass %]			Sintered density	Rm	Rp0.1
Mn	Cu	C	[g/cm^3]	[MPa]	[MPa]
2	-	0.6	6.97	540	360
2	-	0.8	6.85	580	400
2	1	0.6	6.97	630	440
2	1	0.8	6.96	660	500
2	2	0.6.	6.94	670	500
2	2	0.8	6.94	710	540

Table 10. Density, tensile strength, elongation and hardness of Fe-4.5Mn-(Cu)-C steels. Hametag iron powder, electrolytic manganese, 600 MPa, 3 h at 1120°C, c.a. -30°C [39,65,66]

Steel	Density [g/cm^3]	Rm [MPa]	A_5 [%]	Hardness HV 10
Fe-4.5Mn-0.25C_c	7.00	745	1.9	186
Fe-4.5Mn-1Cu-0.25C_c	6.94	728	2.3	168

Data given in Table 11 complete the knowledge about the effect of the tin addition on properties of sintered Fe-Mn-(C) steels.

Table 11. Mechanical properties of sintered Fe-Mn-C and Fe-Mn-Sn-C steels. Hametag iron powder, 600 MPa, sintering for 3 h at 1120°C, cracked ammonia, d.p. -30°C [66]

Steel	Density [g/cm^3]	Rm [MPa]	A_5 [%]	Hardness HV 10
Fe-2.5Mn-0.22C	7.10	245	3.2	120
Fe-2.5Mn-1Sn-0.22C	7.10	308	3.9	107
Fe-3.5Mn-0.22C	7.07	418	3.1	199
Fe-3.5Mn-1Sn-0.22C	6.95	399	3.5	172
Fe-4.5Mn-0.22C	7.05	632	1.6	190
Fe-4.5Mn-1Sn-0.22C	6.92	599	1.9	192
Fe-5.5Mn-0.22C	6.97	713	1.3	231
Fe-5.5Mn-1Sn-0.22C	6.90	560	2.0	232

Zinc. The aim for the use the zinc as additional alloying element in sintered manganese steels was the same as in the case of tin, i.e. sintering under the effect of low melting Zn-Cu master alloy (45Zn-35Mn-20Cu, melting temperature of ~850°C, coded as A45Zn). Regarding high vapor pressure of zinc, the investigation of mechanical properties of Fe-2.3Mn-1Cu-1.1Zn-0.35C (6.7% MA45Zn) steel was real only by powder forging at low temperature of 900°C. The samples attained after the quenching and tempering at 500°C the tensile strength of 1050 MPa and hardness of 700 HV 10 [70].

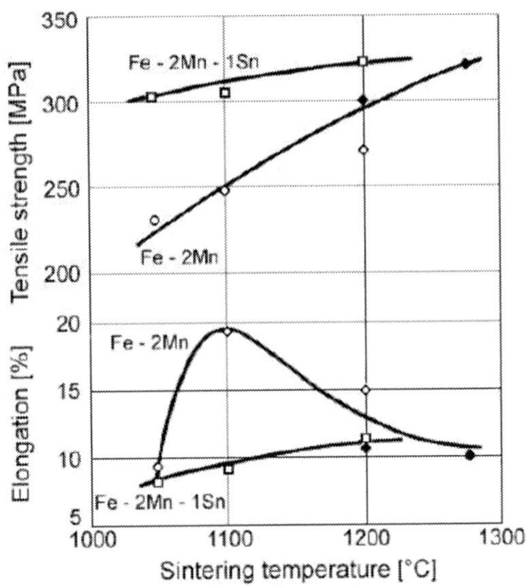

Figure 20. Tensile strength and elongation of Fe-2Mn and Fe-2Mn-1Sn steel in dependence on sintering temperature. SC 100.26 iron powder, EMn, 600 MPa, 1 h at 1280°C, hydrogen, d.p. -70°C [67,68].

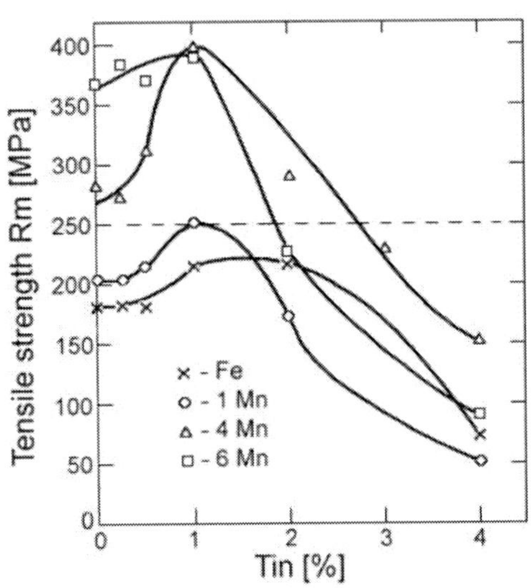

Figure 21. (right) Effect of tin addition on tensile strength and elongation of Fe-(1,4,6)Mn steels. SC 100.26 iron powder, EMn, 600 MPa, 1 h at 1150°C, dry hydrogen [69].

Boron

Addition of boron in form of ferroboron in amount of 0.1, 0.2 and 0.4% on mechanical properties of the Fe-(2,3,4)Mn steel sintered for 60 min at 1150 and 1200°C in hydrogen investigated [71]. Boron addition in amount of 0.1 and 0.2% increased the tensile strength of Fe-2,3)Mn steels to 578 MPa. The highest hardness of 384 HV 10 attained the Fe-4Mn-0.4B steel.

High Alloyed Manganese Steels

The findings in properties of structural sintered low-alloyed manganese steels (Figs. 6 and 7) became an impulse for study the preparation of high-alloyed manganese sintered steels. Microstructure, kinetics of martensite transformations and mechanical properties of powder Fe-(4-40)Mn steels investigated [72-73]. The aim of the works was to study the feasibility to prepare the Mn-steels of those compositions by PM route. Dependence of yield and tensile strength of hot extruded high-alloyed powder Fe-Mn steels on manganese content shows Fig.22. (The nitrogen atomized powders was not possible to cold compact due to high hardness). The high strength properties attained at -196°C and elongation of ~65% for Fe-30Mn steel were regarded for unexpected.

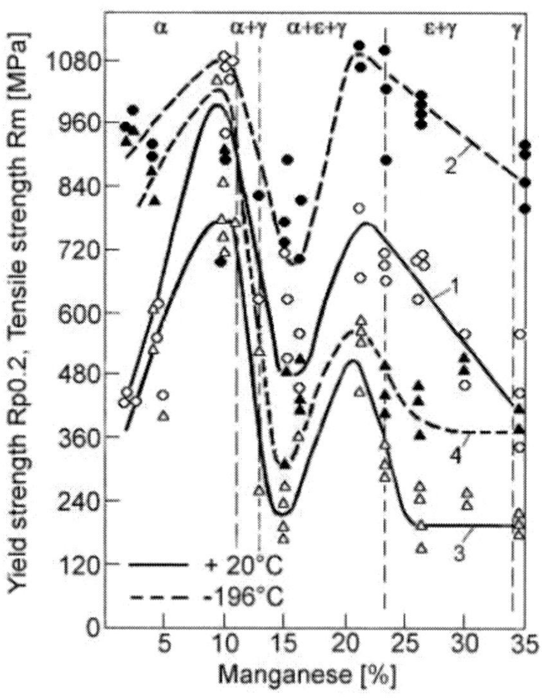

Figure 22. Yield and tensile strength of sinter extruded high-alloyed manganese steels in dependence on manganese content. Nitrogen atomized high Mn alloyed powders, apparent density 4.6-5.6 g/cm^3, direct hot extrusion, water quenching from 1100°C, tempering. Testing of strength properties at +20°C and -196°C, 1,2 – Rm, 3,4 – Rp0.2 [72,73].

[74] studied the sintered austenitic Fe-14Mn-8.5Ni-3Cr and Fe-14Mn-3Cr-2.5Ni-0.2C steels with the aim to increase the stability of the austenite with addition of TiC or Al_2O_3 (dispersion phase) regarding the thermally induced martensite transformation. [55] investigated the properties of sintered Fe-12Mn-1.2C (Hadfield) steel and of Fe-18Mn-1Cr-0.3C steel (ASC 100.29 iron powder, electrolytic manganese, 600 MPa, 30 min at 1280°C, hydrogen + getter). The Hadfield steel reached the tensile strength of 341 MPa, yield strength of 248 MPa, elongation of 11%, hardness of 248 HV 10 and impact energy of 64 J. [75] presented that the sintered Fe-14Mn-1.4C steel with the porosity of 12-13% (sintering for 2 h at 1200°C) exhibited higher abrasive wear resistance compared with conventional high-manganese steel. The tools for mining works were from this steel manufactured. The bushes for locomotives were manufactured by powder forging of high-manganese steels [76].

Liquid Phase Sintering of Fe-Mn Steels

Many low temperature melting master alloys containing manganese were developed and/or determined on the basis of the equilibrium diagrams for an increase the density and the homogeneity of the microstructure of sintered manganese steels by liquid phase sintering. By this was expected also the trouble free sintering of manganese steels due to lower chemical activity of the master alloys. The largest complex of Mn-containing master alloys offers Table 12.

Table 12. Chemical composition of master alloys for alloying of sintered manganese steels in combination with other elements through liquid phase sintering. T_l – liquidus temperature, Ts – solidus temperature [4,77,78]

Chemical composition [mass %]							T_l	T_s
Mn	Ni	Cr	Mo	Fe	Cu	Si	[°C]	[°C]
40	30	15	5	10	-	-	1171	999
44	25	-	11	19	-	-	1204	1166
55	18	3	8	14	-	2	1129	943
56	24	3	6	11	-	-	1157	1032
52	22	8	6	12	-	2.5	1154	1132
47	20	13	6	14	-	2.5	1210	1166
75	25	-	-	-	-	-	1054	982
74	12.5	-	-	-	12.5	2.5	1060	927
36	30	18	6	10	-	2.5	1207	1096
41	25	18	6	10	-	2.5	1216	1077
38	23	18	6	15	-	2.5	1229	1093
64	16	-	10	10	-	-	1232	1068
56	14	-	15	15	-	1	1260	1093
56	14	15	5	10	-	-	1188	1132
56	14	10	10	10	-	-	1227	1102
59	11	15	5	10	-	-	1249	1116
53	17	15	5	10	-	-	1216	1093
56	14	22	8	-	-	-	1335	1049
50	20	15	5	10	-	-	1204	1143
46	24	15	5	10	-	-	1204	1089
72	14	-	-	-	14	2	1043	966

[79] recommended for the preparation of sintered manganese steels to use the 89Mn-11Si (melting temperature of 1000°C), 64.2Mn-35.8Si (melting temperature of 1020°C) and 61Al-39Mn (melting temperature of 880°C) master alloys. The 40Ni-20Mn-18Mo balance Fe extremely fine master alloy enhancing diffusion of alloying elements into iron matrix for sintering at temperatures in excess of 1180°C recommended [80].

The powder low-melting Mn – Ni (40Mn-60Ni) (melting point of 1080°C) master alloy was investigated as manganese carrier for alloying of manganese steels with the aim to minimize the presumed troubles in yet in early research stage [81].

The metallurgical high carbon ferromanganese containing more than 3.8% C melts at the temperature ~1085°C and from this point of view could be regarded also as proper manganese carrier for liquid phase sintering of manganese steel [82]. As mentioned, high carbon ferromanganese was in many cases commonly used as manganese carrier.

The data about the production and use some from mentioned low melting master alloys for alloying of sintered manganese steels are lacking. An exception is Mn-Ni master alloy prepared in powder form by atomization. The attained properties of such sintered Fe-Cu-Mn-Ni steels are listed in Table 13.

Table 13. Properties of sintered Fe-Cu-Mn-Ni steels. RZ iron powder; Mn-Ni low melting mater alloy, 800 MPa, sintering for 1 h at 1120°C, hydrogen, d.p.-32°C [81]

Steel No.	Chemical composition [mass %]			Density [g/cm^3]	Rm [MPa]	Hardness HB
	Cu	Mn	Ni			
1	2	1.6	0.4	7.21	430	145
2	2	1.2	0.8	7.22	420	148
3	2	1.0	1.0	7.39	413	148
4	2	0.6	2.4	7.22	421	150
5	2	1.8	1.8	7.21	399	145
6	2	2.0	2.0	7.17	415	153
7	2	0.8	3.2	7.24	423	148

Alloying of Manganese Steels by Elements of High Oxygen Affinity

The object of the investigation was to improve the mechanical properties of sintered steels by introduction in PM alloying the elements commonly used in ingot metallurgy such manganese, chromium, vanadium, silicon in combination with molybdenum. The high oxygen affinity of Mn, Cr. V and Si presents recently the known problem for their use in PM. The desired alloying element Cr and Mo are known carbide formers, and carbon as was shown in previous results, was studied in many works also in alloying of manganese steels. On the basis of these elements were developed the master alloys for a complex alloying with these elements in the carbide form. Regarding the high affinity of chromium and manganese for oxygen, the master alloys were produced by fusing metallurgical high carbon ferroalloys in vacuum induction furnaces using a susceptor followed by cleaning, jaw crushing, fine grinding (up to 0.15 mm) and attritor milling or milling in a planetary mill under cyclohexane. During fine and ultrafine grinding (average particle size 5 μm) special measures were taken [23,83-86].

The following advantages were attributed to these master alloys containing additional molybdenum: carbon in combined form, at least two of the elements Mn, Cr and/or V, which are stable during the heating up to isothermal sintering, dissociate under sintering conditions, should minimize production costs. (As will be seen later the calculation with thermal stability of manganese during preheating was not correct). Further the amount of added free carbon was minimized and they possess therefore by sufficient oxygen stability. The following master alloys signed as MCM, MVM and MM were developed, and their chemical and phase compositions are presented in Table 14.

Table 14. Chemical and phase compositions of carbide master alloys [23,83,84]

MCM [mass %]					MVM [mass %]					MM [mass %]			
Mn	Cr	Mo	Fe	C	Mn	V	Mo	Fe	C	Mn	Mo	Fe	C
25	23	22	22	7	25.5	23	25.5	20	5	40	20	32	7
80% $(Cr,Mn,Fe,Mo)_7C_3$ 10% $(Fe,Mo,Cr,Mn)_6C$ 5% $(Fe,Mo,Cr,Mn)_3C$ 5% α-(Fe,Mn,Cr,Mo) O - ~0.18%					25% (V,Mo)C 25% $(V,Mo_2)C$ 20-25% α-(Fe,Mn,Mo) 20-25% γ-(Fe,Mn,Mo) MnO (percentage depending on oxygen content), O -~0.2%					Three carbides phases with M_7C_3, little metallic α phase O - ~0.3%			

Mechanical properties of sintered steels alloyed with MCM master alloys shows Fig.23 and those alloyed with MVM master alloy shows Fig.24. The steel with 7% MCM has reached the maximum UTS of 650 MPa and elongation of 2.5%. Similar mechanical properties obtained the steels alloyed with MVM master alloy. By adding free carbon to obtain a total content of 0.6% C, the amount of the master alloy added was reduced. In spite of their optimum tensile strength, the microstructure of the steels containing 4% of MCM master alloy was formed by martensite.

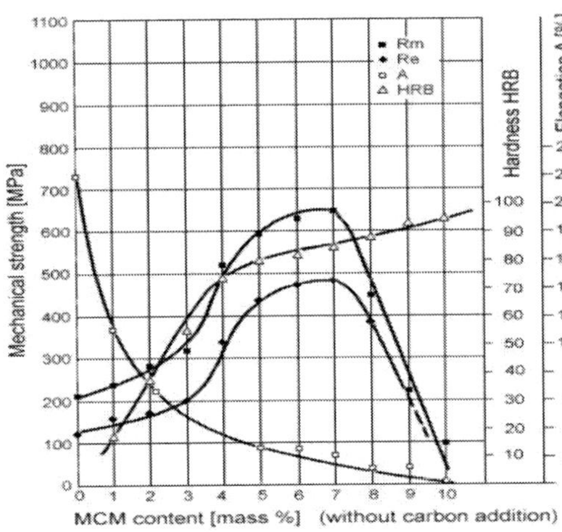

Figure 23. Mechanical properties of sintered steels alloyed with MCM master alloy without carbon addition. Atomet 28 iron powder, 600 MPa, 1h at 1280°C, cracked ammonia, getter – Fe+8% Al, walking beam furnace [83,87,88].

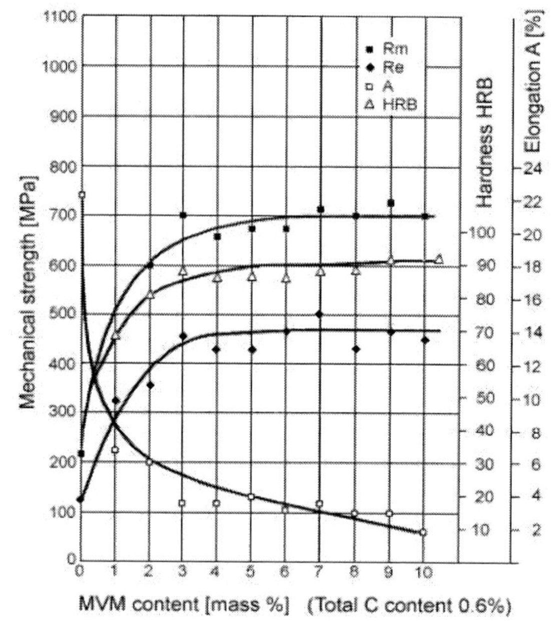

Figure 24. (right) Mechanical properties of sintered steels alloyed with MVM-0.6C (total) master alloy. Other as in Fig.23 [83,87,88] .

The group of special master alloys (MCM, MVM, MM) completes a ternary Fe-Si-Mn master alloy (called SM) developed for the same use consisting of the ternary silicides $(Fe,Mn)_3Si$ and $(Fe,Mn_5)Si)_3$. Figure 25 shows the mechanical properties of Fe-Si-Mn-C steels attained by optimum addition of SM master alloy. Good dimensional accuracy, high strength and wear resistance is expected from these steels. The highest tensile strength of 920 MPa was attained with 0.4% C.

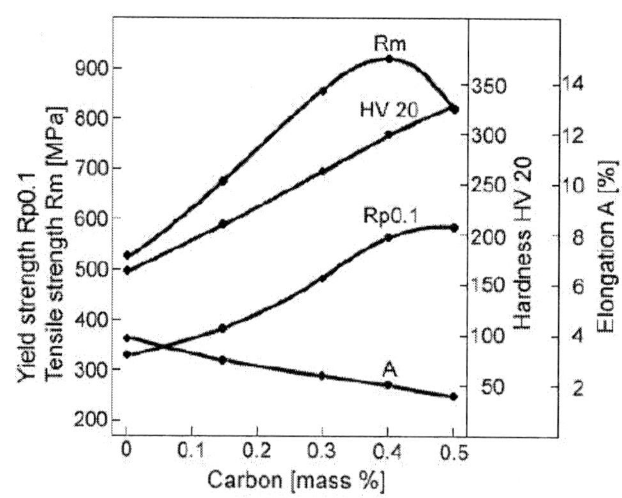

Figure 25. Mechanical properties of Fe-Si-Mn-C sintered steels in dependence on carbon content. Ternary Fe-Si-Mn master alloy, Mn – 3.2%, Si – 1.4%, Atomet 28 iron powder, 600 MPa, 1 h at 1250°C, hydrogen, d.p.-30°C, cooling rate ~ 40°C/min [89-91].

[92,93] investigated the properties of sintered steels alloyed of (1-4)% MCM master alloy with addition of 0.3 and 0.6% phosphorus. According to the results attained, the phosphorus increased the yield and tensile strength. On the other hand [94] investigated the effect of 0.05 and 0.5% P addition on properties of Fe-4.5Mn-0.5C steel. As follows from the data presented in Table 15, phosphorus in both cases caused a significant brittleness of the steels.

Table 15. Mechanical properties of Fe-Mn-C-P steels sintered under various conditions. Atomet 28 iron powder, 600 MPa, cracked ammonia, d.p.-30°C; steel: 1 - Fe-4.5Mn-0.5C-0.05P (red phosphorus), 2 - Fe-4.5Mn-0.5C-0.5P (ferrophosphorus); ρ - density [94]

Steel	ρ [g/cm^3]	Rm [MPa]	HV 10	KC [J]	Sintering
1	6.83	342	152	1.6	3 h, 1120°C
2	6.51	brittle	262	n.d.	3 h, 1120°C
1	6.78	215	191	n.d.	1 h, 1050°C + 2 h, 1120°C
2	6.50	193	235	n.d.	1 h, 1050°C + 2 h, 1120°C

Manganese Carrier - Diffusion Manganese Prealloyed Powders

The preparation of diffusion alloyed manganese powders, which can be called for master alloys, was carried out by following method. The manganese or ferromanganese powder and 1-2% of ammonia chloride (NH_4Cl) were admixed to the iron powder. The mixture was closed in a pot, and heated for 2-4 h at the temperature of 1000-1100°C. (The activation of the process occurred by the addition of NH_42CO_3 in amount of 10%). The iron powder saturated by manganese during this process in form of a cake was ground to powder and used as Mn-carrier for alloying of sintered manganese steels. The content of manganese in the powder is adjustable, usually for 15-30% Mn. The master alloys also for Cr, Ni, Si and Ti and their combinations by this method were prepared [95,96]. Some properties of sintered Mn steels using the iron powders "prealloyed" by the mentioned method as Mn-carrier presents Table 16.

Table 16. Mechanical properties of sintered manganese steels; manganese carrier - diffusion by manganese alloyed powder, sintering for 4 h at 1200°C, hydrogen, d.p. - 30°C, P = 14-16% [95,96]

Mn [mass %]	UTS [MPa]	TRS [MPa]	KC [J]
2	240-260	230-260	2.0-3.0
4	250-270	320-340	2.2-2.4
6	280-300	460-480	1.3-1.8
8	300-310	550-560	0.8-0.9
10	320-330	560-570	1.2-1.4
1-0.40C	284	-	-
1-0.65C	463	756	-
6-1.0C	290	470	-

Sintering in High Purity Hydrogen

Tables 17 and 18 present the mechanical properties of manganese steels sintered with special aim in high purity hydrogen (dew point -60°C). Table 17 presents the properties of manganese steel sintered at three temperatures in high purity hydrogen and cooled at different rates. The effect of cooling rate in the range of 1.4 to 65°C/min on mechanical properties of Fe-(3,4)Mn-0.8C steels sintered in high purity hydrogen at two temperatures investigated [98-100]. Table 18 presents the mechanical properties of the tested steels attained at two cooling rates.

Table 17. Mechanical properties of sintered Fe-3Mn-0.8C steel specimens differently sintered and cooled; NC 100.24 iron powder, 600 MPa, hydrogen, d.p. -60°C, HP – high carbon ferromanganese, Elkem (El) – medium carbon ferromanganese, graphite added to 0.8% C [97]

Steel	Sint. temp. [°C]	Cool. rate [°C/min]	Density [g/cm^3]	UTS [MPa]	HV 10	A_5 [%]
Fe-3Mn-0.8C, HP	1120	40	6.8	516	217	0.2
Fe-3Mn-0.8C, El	1120	40	6.7	559	276	0.3
Fe-3Mn-0.8C, El	1120	10	6.7	480	183	2.0
Fe-3Mn-0.8C, El	1250	10	6.8	590	189	2.5
Fe-3Mn-0.8C, El	1250	40	6.8	614	239	0.5
Fe-3Mn-0.6C, El	1280	7	6.7	360	169	0.5
Fe-3Mn-0.8C, El	1250	40	6.8	614	239	0.5

Table 18. Mechanical properties of Fe-(3,4)Mn-0.8C steels in dependence on low and high cooling rate; NC 100.24 iron powder, density ~6.9 g/cm^3; sintering 60 min, hydrogen, dew point -60°C

Steel	Sint. Temp. [°C]	Cooling rate [°C/min]	YS 0.2 [MPa]	UTS [MPa]	TRS [MPa]	HV 30 core	KC [J]
Fe-3Mn-0.8C (C_c - 0.6%)	1120	4.8	330	478	1270	224	17.2
		55	brittle	535	1000	292	5.2
Fe-3Mn-0.8C (C_c - 0.5%)	1250	4.7	440	595	1330	254	11.5
		57	brittle	491	1090	335	4.9
Fe-4Mn-0.8C (C_c - 0.6%)	1120	4.8	422	669	1021	n.d.	n.d.
		55	brittle	580	1097	n.d.	n.d.
Fe-4Mn-0.8C (C_c - 0.5%)	1250	4.7	434	636	1467	n.d.	n.d.
		57	brittle	428	847	n.d.	n.d.

Summary of Material and Processing Variables Used for Preparation of Sintered Steels Presented in Part I

In following are summarized all in that time known powder materials and processing conditions for the preparation of sintered manganese steels with special respect to sintering atmosphere.

- Base matrix powders: sponge, water and air atomized, eddy milled, carbonyl, prealloyed.
- Manganese carriers: MnO_2, electrolytic manganese, medium and high carbon ferromanganese, low melting manganese containing master alloys, liquid phase forming elements, diffusion on higher content by manganese alloyed iron powders, master carbide powders MCM, MVM and SM.
- Manganese addition range [mass %]: 1 – 6, 1 – 12, 1 – 35.
- Carbon: graphite, carbon ferromanganese grades, master carbide powders (0.2 to 0.9 mass %).
- Additional elements to increase the mechanical properties: molybdenum, copper, boron, phosphorus.
- Compaction: cold single (150 – 900 MPa, mostly 600 MPa), double and triple press – sinter technique, powder forging, hot extrusion.
- Sintering conditions:
  - methods: conventional laboratory, induction (HF), industrial
  - sintering temperature [°C]: 875, 1000, 1100, 1120, 1150, 1180, 1200, 1250, 1280, 1300
  - sintering (hold) time (short and conventional) [min]: 1, 3, 10, 30, 60, 90, 120, 180, 240 (slow furnace cooling rates, fast cooling rates followed by tempering)
  - sintering atmosphere and purity: hydrogen, dew point of -30°C to -70°C; cracked ammonia ~-30°C, in individual cases: getter – alumina + graphite, alumina + Fe8Al, alumina + high carbon ferromanganese
- Heat treatment: through hardening.

The mechanical and other properties of sintered manganese steels, shown in cited and surely many published in uncited references, attained under very various material and sintering conditions, clearly demonstrate a great interest in use of manganese as an alloying element in powder metallurgy. The presented results basically orientated mainly on manganese steel sintering in "cleanest" atmospheres, are in a large extent suppressed by other factors taking part on material characteristics and processing conditions used.

Table 19 presents a survey of some highest tensile strength values, and of related other properties of Fe-Mn-C steels attained under given material and sintering conditions in cited references. It is possible in advance to say that the single pressed and sintered Fe-4.5Mn-0.4C steel attained the highest tensile strength of 886 MPa (No.6).

The presented data prove an effective sintering of steels with manganese content of 1 to 5.5% with carbon content of 0 to 0.9% with the densities around 7.0 g/cm^3 corresponding to compaction at ~600 MPa without a special relation to the processing factors. It is necessary espccially to note that none sintering atmosphere used fulfilled the thermodynamic requirements for the purity for sintering of manganese steels regarding mainly preheating period of the sintering process. All steels, except No.1, were alloyed with the simplest manganese carriers, i.e. with electrolytic manganese and two ferromanganese grades. It is hardly mutually to compare the attained results because of very various materials (iron powder grades, manganese carrier grades and their physical-metallurgical characteristics, carbon content and other), and sintering conditions (temperature, time, atmosphere, dew point, lacking as sintered manganese and carbon contents) including unsystematically determined cooling rates.

Table 19. (a) The highest tensile strength and other properties of as plain Fe-Mn-(C) steels prepared by single press-sinter technique obtained under given experimental conditions presented in mentioned references – (b)

a)

No.	Steel Fe-	Properties					
		Sintered density [g/cm³]	UTS [MPa]	TRS [MPa]	Hardness	A_5 [%]	KC [J]
1	4Mn-0.8C	6.95	670	-	260 HV	-	-
2	5Mn-0C	6.60	270	610	-	-	-
3	5Mn-0C	6.80	360	720	-	-	-
4	6Mn-0C	6.79	600	-	90 HRB	3.2	-
No.	Steel Fe-	Properties					
		Sintered density [g/cm³]	UTS [MPa]	TRS [MPa]	Hardness	A_5 [%]	KC [J]
5	2Mn-0.8C	6.97	585	-	92 HRB	-	-
6	4.5Mn-0.4C	7.12	886	-	179 HV	4.0	21
7	5.5Mn-0.45C	6.93	675	-	155 HV	0.8	17
8	4.5Mn-0.4C	6.62	565	-	157 HV		10
9	3Mn-0.8C	7.10	750	-	215 HV	2.3	-
10	2Mn-0.3C	7.11	495	-	117 HV	6.2	47
11	4Mn-0.3C	7.01	966	-	275 HV	2.5	23
12	3Mn-0.25C	7.00	590	-	190 HV	2.0	-
13	4Mn-0.8C	6.80	480	932	170 HV	-	-
14	3Mn-0.8C	6.66	527	1091	239 HV	0.6	5.6
15	3Mn-0.8C	6.35	445	1016	193 HV	0.3	5.6
16	4Mn-0.3C	7.06	644	1113	162 HV	-	11.6
17	1Mn-0.1C	6.80	335	-	100 HV	-	26
18	3Mn-0.9C	6.94	652	1184	148 HV	-	12
19	3Mn-0.9C	6.92	706	1231	169 HV	-	12
20	2Mn-0.4C	7.29	1570	-	550 HV	2.0	-
21	3Mn-0.8C	6.91	732	1354	-	4.6	-
22	4Mn-0.3C	7.01	966	-	275	2.5	23

b)

No.	Experimental				Ref.
	Sintering [°C/min]	Atmosphere; dew point [°C]	Iron powder grade	Mn - carrier	
1	1200/240	H_2; getter + gr.	Hametag	MnO_2	24,25
2	1100/60	H_2	MH 100.24	EMn	26
3	1300/60	H_2	MH 100.24	EMn	26
4	1200/50	c.a.; getter + gr.	Atomet 28	EMn	84
5	1200/50	c.a.; getter + gr.	Atomet 28	EMn	84
6	1120/180	c.a.; -30	Hametag	FeMnC	40,43
7	1120/180	c.a.; -30	RZ	FeMnC	40,43
8	1300/1, HF	H_2	Hametag	FeMnC	53
9	1150/60	H_2; -40	Carbonnyl	EMn	97
10	1160/90	H_2; getter + gr.	ASC 100.29	EMn	56

b) continued

11	1280/75	H_2; getter	ASC 100.29	EMn	56
12	1180/60	H_2; -30	NC 100.24	FeMn	87
13	1180/60	H_2; <-40	NC 100.24	FeMn	101
14	1200/60	H_2; -40	NC 100.24	FeMn	102
15	1200/60	H_2; -60	NC 100.24	FeMn	102
16	1120/120	c.a.; -33	ASC 100.29	FeMnC	103,104
17	1120/60	c.a.; -33	SC 100.26	FeMnC	104
18	1120/90	c.a.; -33	SC 100.26	FeMnC	105
19	1120/90	c.a.; -33	NC 100.24	EMn	105
20	1250/60	$20H_2/80N_2$	Carbonnyl	FeMn	106
21	1250/60	N_2; -60 (s.h.)	NC 100.24	FeMn	107
22	1280/75	H_2, getter (s.h.)	ASC 100.29	EMn	55-57

Note: Iron powder grade: Hametag - eddy milled; RZ - air atomized. Manganese carrier: FeMnC - high carbon ferromanganese, FeMn - medium carbon ferromanganese, EMn - electrolytic manganese. HF - high frequency sintering; c.a. - cracked ammonia, gr. – graphite, no.21 - MIM heat treated, no. 22 – sinter hardened (s.h.) – cooling rate ~1°C/s.

Alloying and Sintering of Manganese Steels in Term of High Manganese Vapor Pressure, Part II

The experiments with sintering of the Fe-4.5Mn-C steel small cylindrical samples to study in-more detail the microstructure characteristics of this material, carried out by the first author of this work in 1977, started the study the manganese in term of its high vapor pressure not only as alloying element in PM. The samples, casually placed in porcelain instead of commonly used alumina boats, were sintered in a laboratory Mars furnace equipped with quartz tube at the temperature of 1200°C in hydrogen of technical purity. The porcelain boats at this sintering process became black what was not expected compared to the sintering in the alumina (oxal) boats as shown in Fig.26. The being sintered samples were metallic clean. The same became at the treatment of piece-electrolytic manganese, Fig.26c.

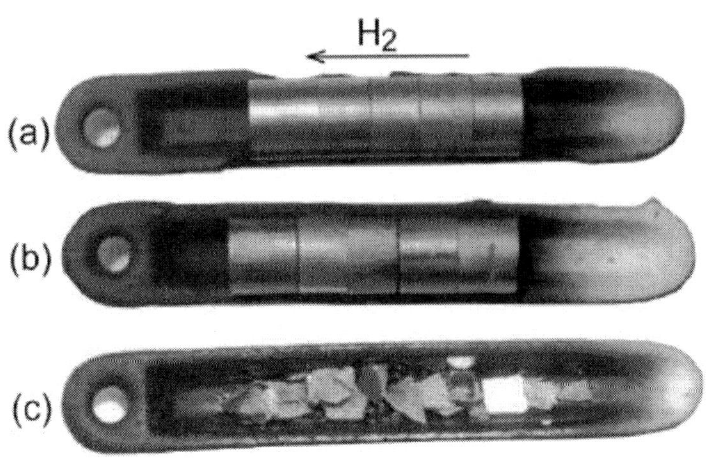

Figure 26. View on porcelain boats with Fe-4.5Mn-0.36C steel samples (Ø10 x 10 mm. Mn as FeMnC, 600 MPa, sintering in at 1200°C for: (a) – 30 min, (b) – 10 min, (c) – piece-electrolytic manganese, 10 min, 1130°C, hydrogen.

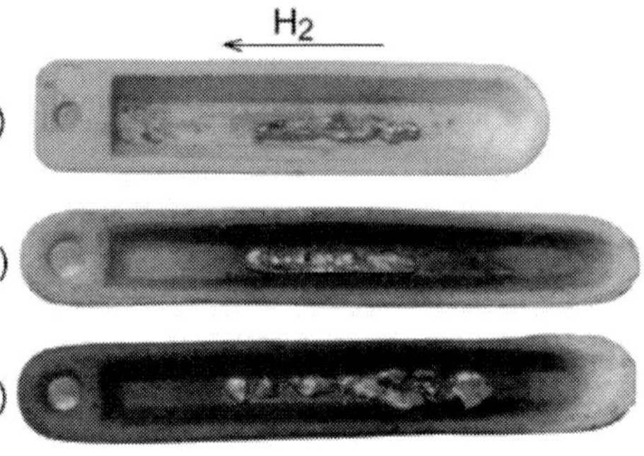

Figure 27. (right) Powder high carbon ferromanganese (FeMnC)(76% Mn, 6.6% C), heat treatment at 1120°C in hydrogen for: (a) –10 min, oxal boat, (b) – 3 min, porcelain boat, (c) – 10 min, porcelain boat.

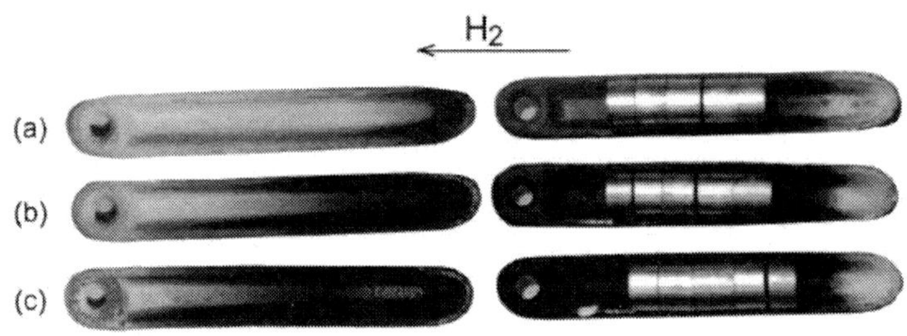

Figure 28. Fe-4.5Mn-0.40C steel compacts. (Ø10 x 10 mm) as in Fig.26, sintered in porcelain boats at 1100°C in hydrogen for: (a) – 10 min, (b) – 30 min, (c) – 60 min, (left) - empty boats in front of active boats.

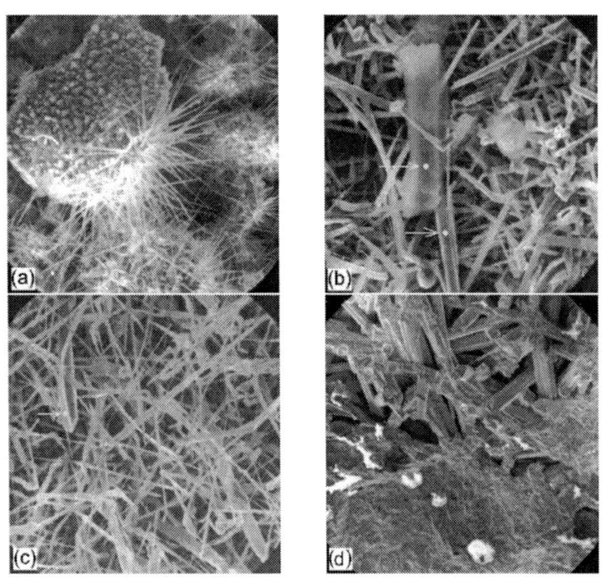

Figure 29. Appearance of: (a) – piece-electrolytic manganese heat treated 15 min at 950°C, (b) – as (a), treatment 60 min at 1000°C (microanalysis of one needle in point, arrow - 98.9% Mn, 0.424% Si, 0.275% Al, balance Cl, K), (c) – powder high carbon ferromanganese, treatment 30 min at 1000°C (microanalysis point, arrow - 95,24% Mn, 0.89% Si, balance Al, K, Cl), (d) – piece-high carbon ferromanganese crushed in production plant, (a),(b),(c) – treatment in hydrogen of technical purity, (d) – air [10,11,108,109]. SEI.

Figure 27 shows the same appearance of an alumina (oxal) boat (a) and of two porcelain boats (b,c) with heat-treated high carbon ferromanganese. The oxal boat did not become black. This proves that the gaseous product – manganese was formed at this treatment which reacted with porcelain only. Besides it, the liquid phase of high carbon ferromanganese occurred in the boats because the applied treatment temperature is higher than that for the high carbon ferromanganese liquid phase formation (1060°C) [11].

A black layer formed, clearly visible on the walls of the boats, is a function of the holding time at the temperature and of their orientation (darker) in relation to the direction of the hydrogen flow. To verify this appearance and to eliminate some accident, further sintering processes in hydrogen were carried out.

Figure 28 shows the blackening of the empty porcelain boats, which were at the sintering placed in the furnace tube in front of each boat with the compacts. All porcelain boats with being sintered compacts became black independently on the sintering time. The further empty boats became also black in dependence on the distance from the sintered samples and sintering time. This proves that the blackening of the porcelain boats was possible in the tested ternary FeMnC systems only by the formation of manganese vapor by sublimation at the temperatures applied. This finding evocated the need of more-detailed study and analysis of this process and raised new questions in sintering of manganese steels.

The appearance of the surfaces of the heat treated piece- and powder electrolytic manganese and ferromanganese and of piece-crushed high carbon ferromanganese shows Fig.29. The manganese vapor formed condensed in needle form. The microanalysis of the needles in Fig.29(b) and (c) confirmed their manganese substance. Figure 29(d) proves the sublimation of manganese and its condensation in needle form also in the caverns of the ferromanganese casting during cooling in air. The shown needles are „pure" metal manganese as confirmed the microanalyses.

Effect of Manganese Vapor Pressure on Sintering

The shown appearances observed in sintering of manganese steel samples in porcelain boats as well as that observed at the heat treatment of piece and powder manganese carriers are the consequence of a physical property of manganese - high vapor pressure when compared to other alloying elements used in PM. Between the alloying elements used commonly in PM steels manganese exhibits by the highest vapor pressure demonstrating at increased temperatures in sublimation, manganese vapor formation. Generally, if a pure solid metal is heated, its atoms sublimate from the free surface. Inasmuch as the sublimation of manganese is the effect of its high vapor pressure, it occurs spontaneously in dependence on temperature and physical-chemical properties of the manganese carrier.

Basic Formulas and Application Conditions for Sintering of Mixed Manganese Steels

To explain the real conditions of manganese sublimation with the impact on sintering process of manganese steels, the vapor pressure, manganese vapor volume and rate and time of sublimation all in dependence on temperature and other factors is necessary to specify. The processes occurring at the sublimation of a solid phase are the same as at the evaporation of a liquid phase, e.g. shown in Fig.27. The difference lies only in the metal surfaces from which the evaporation or sublimation occurs. To calculate manganese vapor pressure was used the relation valid for α and δ manganese including the constants according to [14,16]:

$$\log p = A \cdot T + B \cdot \log T + C \cdot T + D$$

where *A, B, C* and *D* are the constants (-14920, -1.96 , 0, 18.32), *T* is temperature (K), *p* vapor pressure (Pa) for manganese.

The corresponding equation for manganese is:

$$\log p = -14920 \cdot T^{-1} -1.96 \cdot \log T + 18.315 \text{ [Pa]}$$

The vapor volume of manganese generated by sublimation or evaporation is calculated according to the standard equation used for an ideal gas, which with some simplification can be applied even for manganese vapor:

$$p \cdot v = m/M \cdot R \cdot T$$

where *R* is gas constant (9.31441 $J \cdot K^{-1} \cdot mol^{-1}$), *p* is vapor pressure in Pa, *T* is temperature in K, *v* is gas volume in m^3, *m* is mass in g, *M* is molecular mass (54.93 $g \cdot mol^{-1}$ for manganese).

The rate of atom sublimation from a surface unit can be expressed according to equation:

$$\log W = -3.358 + 0.5\log M + \log p - \log T$$

where *W* is the sublimation rate in $g \cdot cm^{-2} \cdot s^{-1}$, *M* the molecular mass in $g \cdot mol^{-1}$, *p* the vapor pressure in Pa, and *T* the metal surface temperature in K.

The sublimation time of certain metal quantities from the surface at a certain temperature can be expressed by the equation:

$$\tau = m/v \cdot F$$

where τ is the sublimation time in s, *m* the metal mass in g, *v* the sublimation rate in $g \cdot cm^{-2} \cdot s^{-1}$, and *F* the free metal surface in cm^2.

Tables 20 and 21 present the data calculated according with the previous equations for manganese vapor pressure, and sublimation rate and time [10,22,108-111]. The sublimation rate and time calculated for one manganese particle cube in size of 15 μm (often used). The calculation refers to the geometrical surface of the particles and to the temperature range of 400 to 1200°C covering the temperatures of ~400°C at which the oxidation of manganese in air starts, the decisive phase - the preheating process in sintering and the sintering temperature range mostly applied also for sintering the manganese steels of 1100-1200°C. It is necessary to expect that at given sublimation rates the sublimation also of larger Mn particles, e.g. of 40-45 μm will finish during heating period in dependence on the heating rate. The longer sublimation time compared to that for pure manganese is possible to expect by using ferromanganese or some special Mn containing master alloys as manganese carriers due to their lower chemical activity (lower Mn content and some impurities). For instance one cube centimeter of a Fe-3Mn sample (0.21 g of Mn) with the density of 7.0 g/cm^3 contains $8.6\cdot10^6$ of particles in size of 15 μm or $4.5\cdot10^5$ of particles 40 μm in size.

It is evident from the given data that the manganese vapor volume, generated by sublimation from Mn particles in some compact, is considerably higher than the pore volume, i.e. the manganese vapor formed in the pores is under pressure.

The influence of the manganese surface area is evident and the fact that by raising the temperature by 100 K the sublimation rate increases approximately by two powers. The manganese sublimation rate from fine particles at high temperatures takes place almost immediately. The evaluation of the properties of sintered manganese steels needs to consider this fact. The calculated data denote that manganese can effectively sublimate even at low temperatures. This proves also that the features shown in Figs.27 and 28 are the effect only of manganese vapor formation at sintering of mixed Fe-Mn steels.

Table 20. Vapor pressure of manganese and of some alloying elements used in powder metallurgy in dependence on temperature

Element	Temperature [°C]				
	900	1000	1100	1200	1300
	Vapor pressure [Pa]				
Mo	$2.03\cdot10^{-17}$	$4.07\cdot10^{-15}$	$3.73\cdot10^{-13}$	$1.85\cdot10^{-11}$	$5.66\cdot10^{-10}$
Si	$2.25\cdot10^{-7}$	$5.40\cdot10^{-6}$	$8.13\cdot10^{-5}$	$8.44\cdot10^{-4}$	$6.49\cdot0^{-3}$
Ni	$3.95\cdot10^{-6}$	$1.17\cdot10^{-5}$	$2.11\cdot10^{-4}$	$2.68\cdot10^{-3}$	$2.25\cdot10^{-2}$
Fe	$2.99\cdot10^{-6}$	$6.47\cdot10^{-5}$	$8.85\cdot10^{-4}$	$8.40\cdot10^{-3}$	$5.93\cdot10^{-2}$
Cr	$1.08\cdot10^{-5}$	$2.36\cdot10^{-4}$	$3.26\cdot10^{-3}$	$3.13\cdot10^{-2}$	$2.24\cdot10^{-1}$
Cu	$4.23\cdot10^{-4}$	$6.10\cdot10^{-3}$	$5.94\cdot10^{-2}$	$4.23\cdot10^{-1}$	$2.33\cdot10^{0}$
Mn	0.38	3.23	19.88	94.69	366.70

Table 21. Manganese sublimation rate and time in dependence on vapor pressure from 1 cubic pure manganese particle 15 μm in size

	Temperature [°C]								
	400	500	600	700	800	900	1000	1100	1200
Vapor pressure [Pa]	$4.0\cdot10^{-10}$	$2.25\cdot10^{-7}$	$2.88\cdot10^{-5}$	$1.3\cdot10^{-3}$	$2.9\cdot10^{-2}$	0.38	3.23	19.88	94.69
Sublimation rate [g/cm^2/s]	$1.0\cdot10^{-14}$	$2.6\cdot10^{-11}$	$3.2\cdot10^{9}$	$1.3\cdot10^{-7}$	$2.9\cdot10^{-6}$	$3.6\cdot10^{-5}$	$2.9\cdot10^{-4}$	$1.3\cdot10^{-3}$	$8.0\cdot10^{-3}$
Time of sublimation [s]	$3.6\cdot10^{10}$	$6.95\cdot10^{7}$	$5.7\cdot10^{3}$	$1.3\cdot10^{2}$	6.1	0.5	0.06	0.01	0.002

Remark: Calculated volume of Mn vapor generated from one particle in size 15 μm is ~$4.3\cdot10^{-5}$ cm^3.

Effect of Manganese Vapor on Alloying of Powder Fe – Mn Compacts During Sintering

The manganese vapor formed from the manganese carrier particles in an iron powder – manganese powder compact during sintering fills the pores and condenses on the surface of iron particles and from there occurs the alloying of their interior by diffusion mechanisms in dependence on temperature. The microstructures of Fe-4Mn-(C) steel samples sintered at

1100 and 1200°C shown in Figs.30-33 demonstrate the uniform alloying by the formation of an alloyed layer on the entire surfaces of individual iron particles in the compacts, and the changes in dependence on time, temperature and iron powder grade (i.e. coarse grained – RZ, fine grained – Hametag). (The boats with the cylindrical samples Ø10 x 10 mm were shifted directly into the isothermal sintering zone of the Mars furnace and after holding time shifted into cooling zone at ambient temperature, i.e. without heating and cooling period. Manganese was added in these cases as high carbon ferromanganese [112,113].

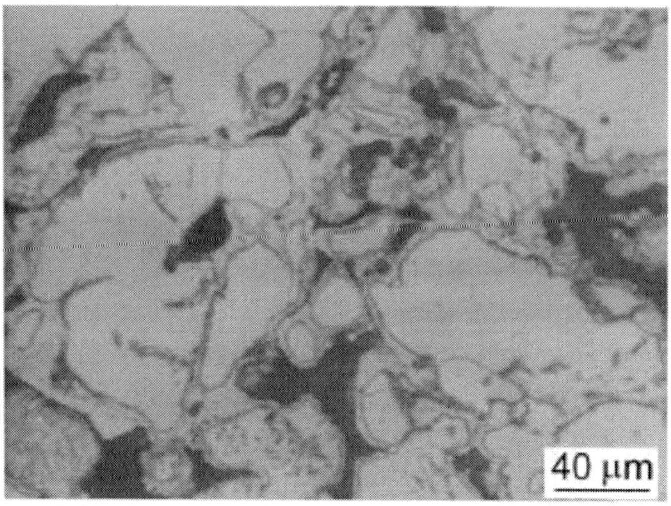

Figure 30. Microstructure of Fe-4Mn-0.32C steel sample. RZ (air atomized) iron powder, 600 MPa, sintering 3 min at 1100°C, hydrogen, d.p. -30°C, optical micrograph. Nital etched.

Figure 31. (right) Microstructure of Fe-4Mn-0.32C steel sample. Sintering for 30 min. Other as in Fig.30.

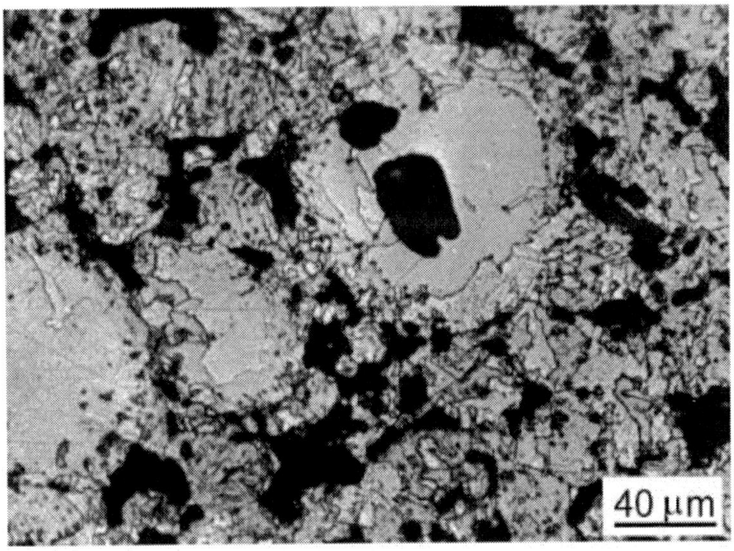

Figure 32. Microstructure of Fe-4Mn-0.32C steel sample. Sintering for 60 min. Other as in Fig.30.

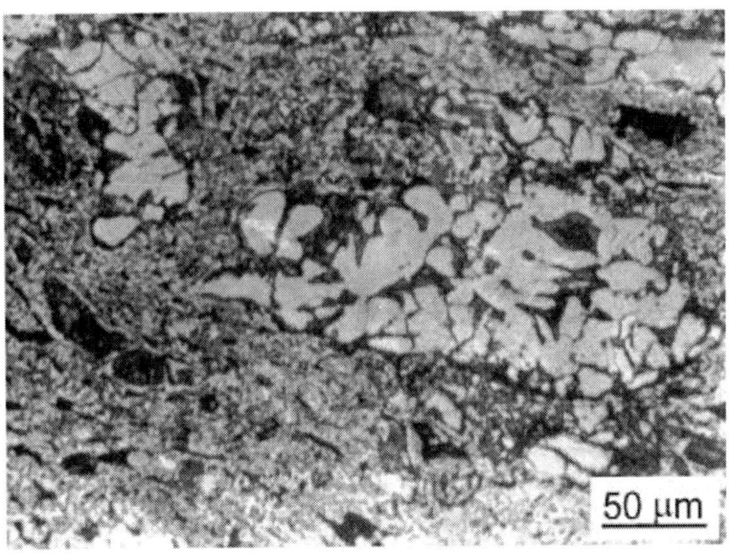

Figure 33. (right) Microstructure of Fe-4Mn-0.32C sample. Hametag (eddy milled) iron powder, 600 MPa, sintering 10 min at 1100°C, hydrogen, optical micrograph. Nital etched.

More uniform alloying shows in a larger extent the microstructure of the steel based on Hametag iron powder (fine grained starting structure and substructure characteristics) compared to RZ iron powder. These differences demonstrated also in mechanical properties, Fig.10. Figure 31 shows the microstructure after sintering for 30 min. The smaller particles are fully alloyed, but the interior of coarser particles remains ferritic (slow volume diffusion). Figure 32 shows the change in the microstructure character of the alloyed layers, they loose the sharpness. This is the effect of a reverse sublimation of manganese from the alloyed structures into surrounding atmosphere due to change in the manganese concentration gradient between the samples (predominantly slow volume diffusion) and atmosphere due to long time at sintering of an individual sample.

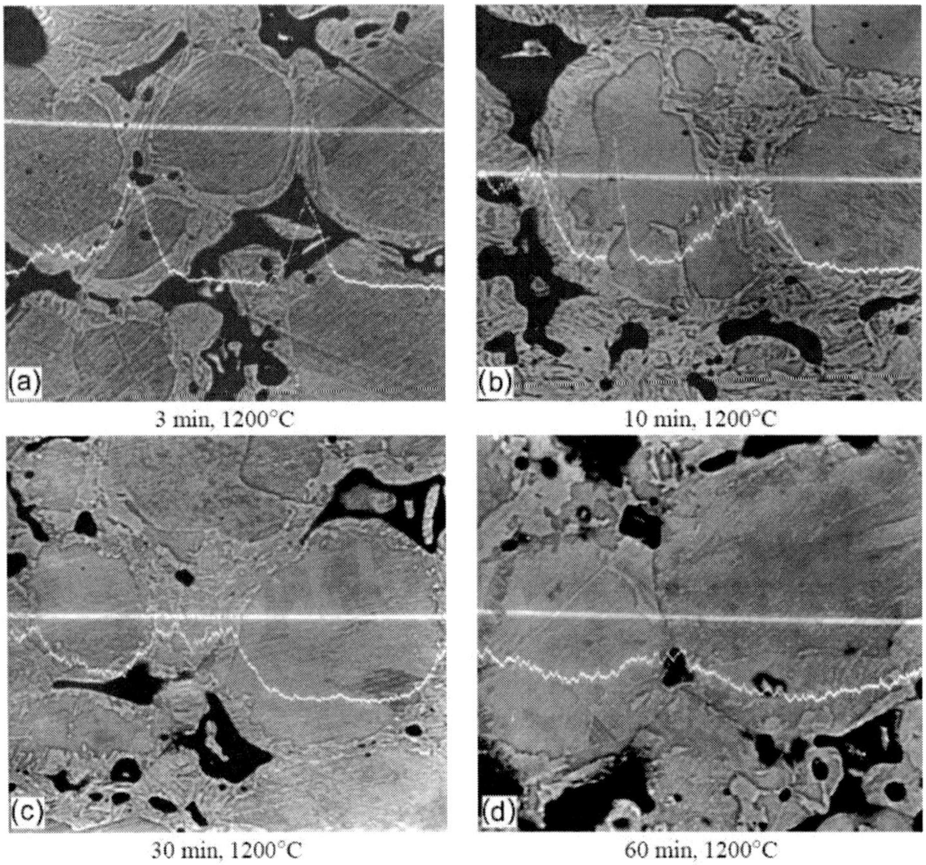

Figure 34. Line distribution of manganese in Fe-4Mn-0.32C steel samples. RZ iron powder, <100 μm, 600 MPa, high carbon ferromanganese, hydrogen. Wave dispersion line analysis, optical micrographs. Nital etched [113].

The changes in line distribution of manganese between the individual starting iron particles in the sintered compacts prove previously shown optical micrographs. The prolonged sintering time at 1200°C demonstrated in more uniform distribution of manganese due to larger alloying of the interiors of the particles and due to mentioned reverse demanganizing, Fig.34(d). (Coarse grained iron powder seemed to be proper for the study the diffusion processes of manganese in iron powder.) The effect of different structural and technological characteristics of iron powder grades demonstrates clearly on the microstructures of manganese steels alloyed through manganese vapor. Figures 35 and 36 show the microstructures of Fe-2Mn-0.3C steel based on water atomized iron powder. Figure 36 shows clearly the alloying of one particle from the surface into the core by volume diffusion and along grain boundaries. Some grains in coarser particles remained ferritic.

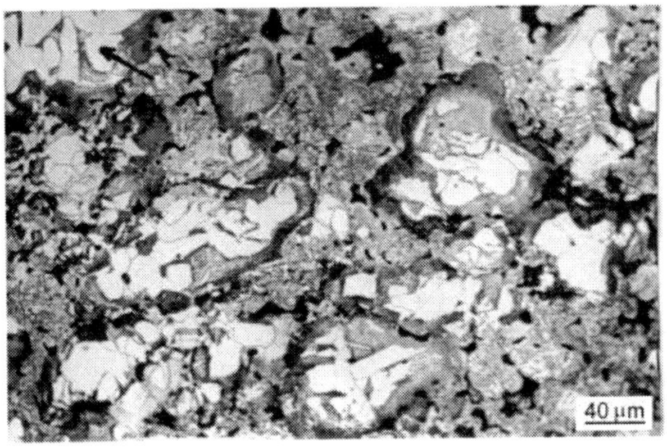

Figure 35. Microstructure of Fe-2Mn-0.3C steel. ASC 100.29 iron powder, 600 MPa, 90 min, 1120°C, cracked ammonia, d.p. -30°C, optical micrograph. Nital etched.

Figure 36. (right) Detail – see arrow in Fig.35 – upper left corner.

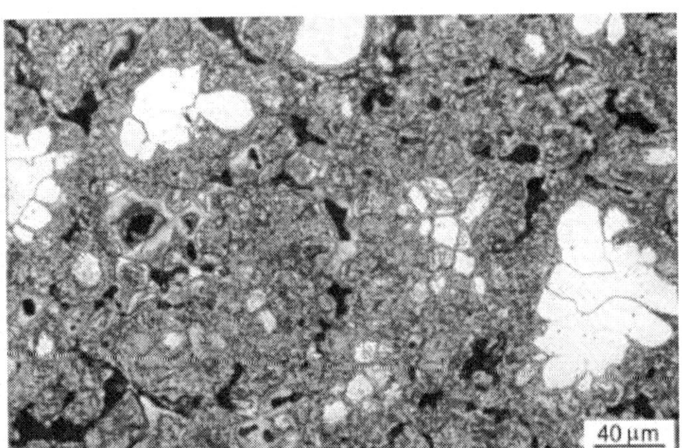

Figure 37. Microstructure of Fe-2Ni-0.5Mo-2Mn-0.3C steel. Prealloyed Astaloy A powder, sintering under getter - alumina + high carbon ferromanganese, 30 min, 1120°C, hydrogen, optical micrograph. Nital etched.

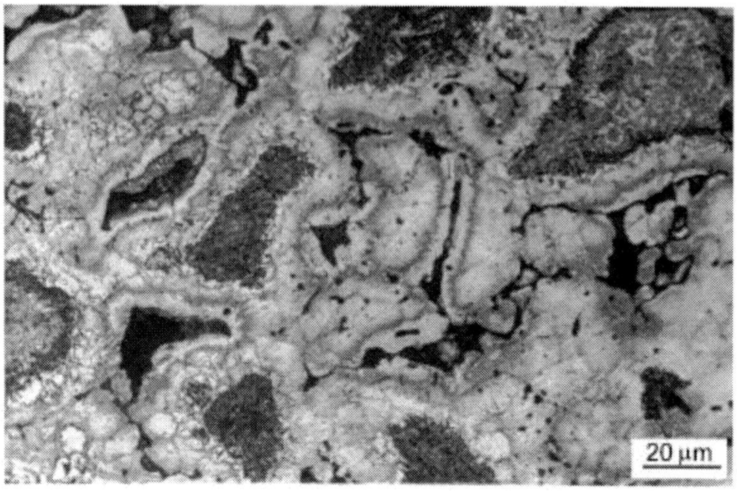

Figure 38. (right) Microstructure of Fe-6Mn-2Ni-1Mo-1.2C steel sintered as in Fig.37. Prealloyed Astaloy A powder, 30 min, 1250°C, hydrogen, 6.0 g/cm^3, optical micrograph. Nital etched [4].

The shown microstructures prove that also the water atomized iron powder particles are structurally not homogeneous. The alloying of them by manganese occurs in dependence on the particles size and structural hetero(homo)geneity. It documents the known heterogeneity of the microstructures of sintered manganese steels based also on sponge iron powder grades. Effect of different base characteristics of iron powder grades clearly are characterized by strength properties shown in Fig. 10. Figure 37 shows the microstructure of Fe-2Mn-0.3C steel based on prealloyed Astaloy A powder sintered under getter – alumina + high carbon ferromanganese. The compact was indirectly through alloyed by manganese vapor coming from the getter. This material exhibited high wear resistance and by this high cutting resistance. Figure 38 shows the microstructure of a Fe-6Mn-2Ni-0.5Mo-1.2C material, based also on prealloyed Astaloy A powder, which was sintered under getter as in previous case. The material was not possible to cut by conventional cutting tools. This manganizing method occurs without halides used in wrought steel manganizing and acts due to pores also inside the compacts. The aim of these experiments was only to prove the alloying of porous material through the gaseous manganese from an outside manganese vapor carrier.

Extreme Sintering Conditions

Regarding the sintering and alloying through the manganese vapor, the sintering at 875°C and during the preheating up to 1120°C only with the effect on mechanical properties of manganese steels seems to be extreme in relation to the conventional sintering conditions especially of the steels alloyed by other elements [114].

The tensile strength of manganese steels, based on different iron powder grades, sintered at 875°C for 80 min shows Fig.39. The sintering in α - region demonstrated in intense formation of interparticle metal bonds in the compact.

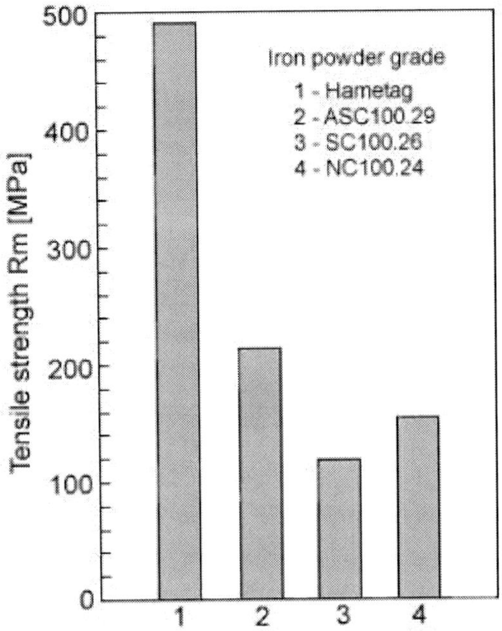

Figure 39. Tensile strength of Fe-4.5Mn-0.4C steels sintered for 80 min at 875°C; c.a., dew point -30°C, heating rate 35°C/min, 600 MPa, high carbon ferromanganese, iron powder grade: 1 - Hametag (eddy-milled), 2 - RZ (air atomized), 3 – ASC100.29 (water atomized), 4 - SC 100.26, 5 - NC 100.24 (both sponge, sintering time: 1,2 for 80 min, 3,4,5 for 30 min [18,115,116].

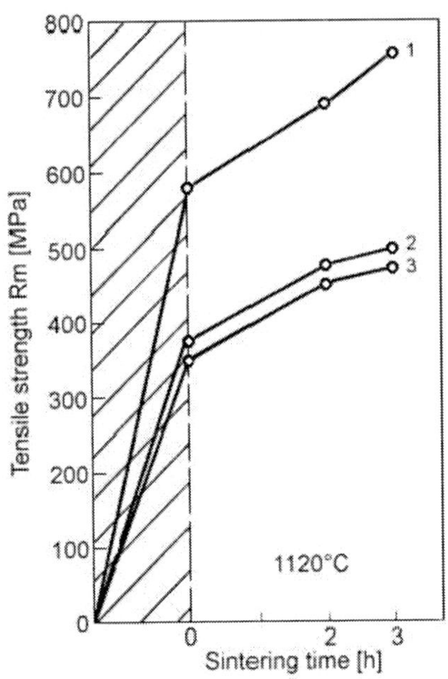

Figure 40. (right) Tensile strength of Fe-4.5Mn-0.4C steel samples in dependence on heating time to sintering temperature of 1120°C; 0 h preheating, 2 and 3 h hold time, heating rate 17°C/min; 1 - Hametag iron powder, 2,3 - RZ iron powder. Other as in Fig.39.

The results prove the manganese sublimation under these sintering conditions and alloying of iron matrices by manganese. The different base structural and technological characteristics of iron powder grades demonstrated on diffusion of manganese from the surface into the cores and also by the formation of the interparticle bonds. The attained tensile strength values document this.

The samples based on Hametag iron powder attained the tensile strength of 496 MPa at the hardness 205 HV 10. The steels based on other powder grades despite of lower tensile strength had the hardness in the range of 153 to 171 HV 10, i.e. all samples were alloyed with manganese without the formation of interparticle necks in the same extent due to their various structural characteristics. The data prove that the oxidation of ferromanganese particles in "insufficiently" clean sintering atmosphere for the tested material system did not occur. The sintering atmosphere should have for these sintering conditions according to the thermodynamics the dew point min. -70°C.

Figure 40 shows the tensile strength of the steel of the same composition as in Fig.39 with special attention to the preheating time to the sintering temperature of 1120°C. The preheating time demonstrated in relatively highest increase in tensile strength caused by manganese compared to that after 2 and 3 h sintering at the preservation of the differences in tensile strength between the materials based on different iron powder grades.

These data show that at sintering of manganese steels during heating period an intense alloying of iron matrix by manganese occurs [40].

The features occurring at sintering of manganese steels, shown in Figs.30 to 40, document their solid phase (iron powder) – gas phase (Mn vapor) sintering and alloying. This character of the alloying is expressed in more homogeneous final microstructures regarding the heterogeneous structures of base iron powders.

Manganese Gas Phase Sintering and Alloying According to Dilatometer Tests

The dilatometer tests enable in powder metallurgy to study the dimensional changes of the compacts caused by various starting materials and alloying formed under selected sintering conditions. The dilatometer tests of manganese steels with the same aims were also carried out. According to Fig.41, the expansion of the length of manganese containing samples clearly starts at low temperature minimum at ~600°C and increases with increasing temperature and manganese content when compared to plain iron curve. It means on the surface of iron particles in the compact formed the layer alloyed with condensed manganese what caused the growth of the dimensions. The larger expansion of the samples over the α - γ phase transformation temperature was caused by increasing solubility of manganese in austenite. These observations prove the previous data shown in Figs.39 and 40 about the alloying of iron matrix by manganese in α - region and during preheating period.

The demanganizing and the activation effect of manganese on sintering caused the shrinkage of the samples length during the isothermal hold when compared to constant dimension of plain iron specimen. The escape of gaseous manganese in some small amount is also a feature caused by high manganese vapor pressure. In the case of the dilatometer samples, the demanganizing confirm the optical micrographs and the microanalysis of Mn.

The range of the demanganizing is in this case a specific feature of small dilatometer samples used characterized by high specific surface area and low density (extreme high reaction surface of open pores) [117,118]. This effect was not expressed on microstructure and properties of the test samples or components when sintered more specimens, parts together in one box as will be explained later.

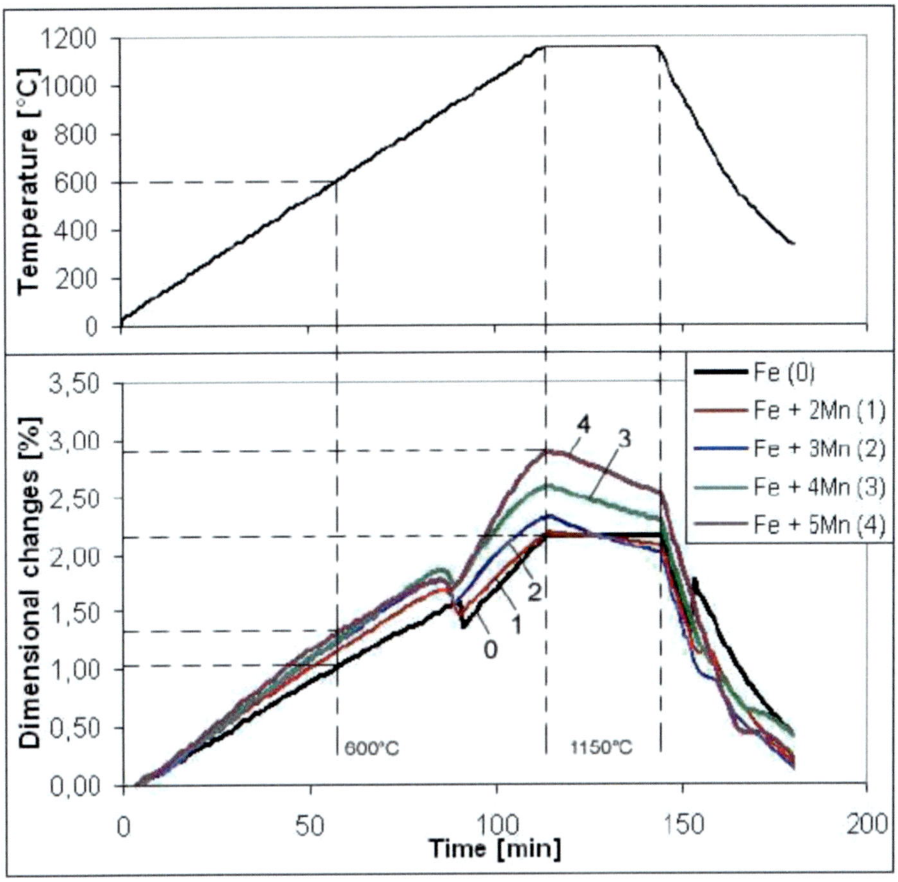

Figure 41. Dilatometer curves and temperature-time profile for Fe-(0,2,3,4)Mn steel samples 4x4x15 mm in size, density 6.6 g/cm^3. Isothermal sintering for 30 min at 1150°C, heating rate 10°C/min, cooling rate 30°C/min, $75H_2/25N_2$, d.p. -30°C, medium carbon ferromanganese [117-120].

The same character exhibit the dilatometer curves for prealloyed Fe-0.85Mo and hybrid Fe-0.85Mo-(0,1,2,3)Mn-(0.5C) steel specimens shown in Fig.42. A more marked expansion of these samples was recorded compared to the samples based on plain iron powder, Fig.41. The prealloyed Fe-0.85Mo powder matrix (alloying and structural heterogeneity) expressed oneself in a more intense through alloying by manganese during heating period due to higher structural activity of prealloyed powder for diffusion. The expansion of dimensions is observed in this case markedly already at ~470°C compared to the plain Fe-Mn samples. Figure 43 shows the effect of carbon addition on the course of the dilatometer curves of plain iron and Fe-3Mn-(0.2,0.5,0.9)C steel specimens. Effect of carbon addition demonstrated markedly in increased swelling of the dimensions over the α - γ phase transformation temperature up to the 1150°C when compared to the carbonless specimens shown in Fig.41 due to the dissolution of carbon in austenite.

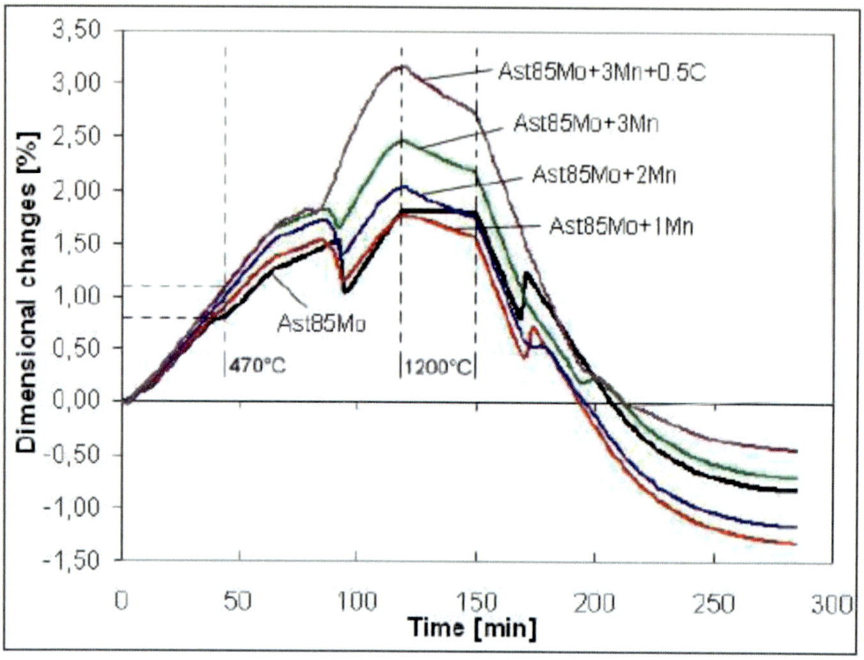

Figure 42. Dilatometer curves of plain Fe-0.85Mo, hybrid Fe-0.85Mo-(0,1,2,3)Mn and of Fe-0.85Mo-3Mn-0.5C steel specimens. Medium carbon ferromanganese, graphite, 30 min at 1200°C, heating rate 10°C/min, cooling rate 15°C/min, $30H_2/70N_2$, d.p.-40°C [119,120].

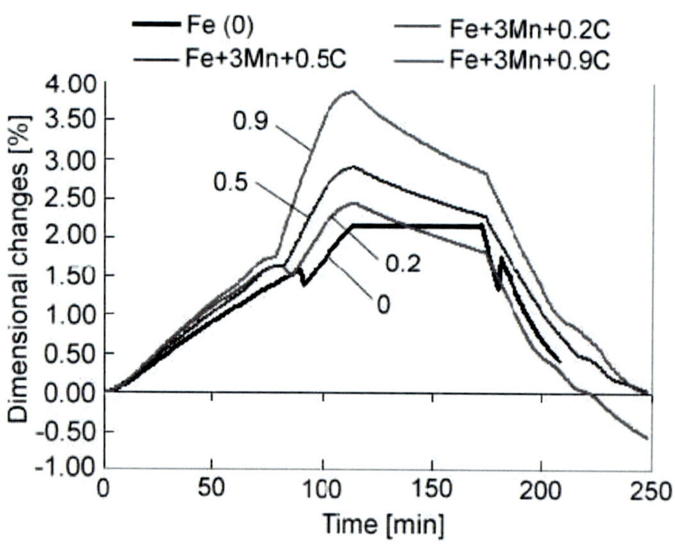

Figure 43. Dilatometer curves of Fe-3Mn-(0.2,0.5,0.9)C steels. C as graphite. Processing as in Fig.41 [117-119].

The final dimensional changes of all manganese steel samples determined by the dilatometer tests do not characterize the effect of base starting composition because of the lower manganese content in final samples due to demanganizing.

Sintering of Manganese Steels under Effect of Manganese Vapor

Figures 26-28 show the blackening of the porcelain boats at sintering of manganese containing iron powder compacts caused by the escape of gaseous manganese from the samples into the flowing atmosphere followed by the reaction of manganese vapor with porcelain. This recorded also the dilatometer curves by shrinkage of the samples length during the hold, Figs.41 to 43.

The manganese vapor escaping from the being sintered samples through the open pores comes into the contact with the flowing sintering atmosphere. Every ordinary atmosphere contains by several orders higher oxygen concentration than the required equilibrium for the system Mn – O for sintering of manganese steels. Now the vaporized manganese atoms, it is necessary in advance again to stress high affinity of manganese for oxygen, react with oxygen molecules being in the atmosphere forming solid MnO. According to the scheme previously illustratively elaborated for evaporation of copper at 1200°C and iron at 1600°C [121,122]. The reaction of gaseous manganese with the humid atmosphere can be explained according to scheme shown in Fig.44.

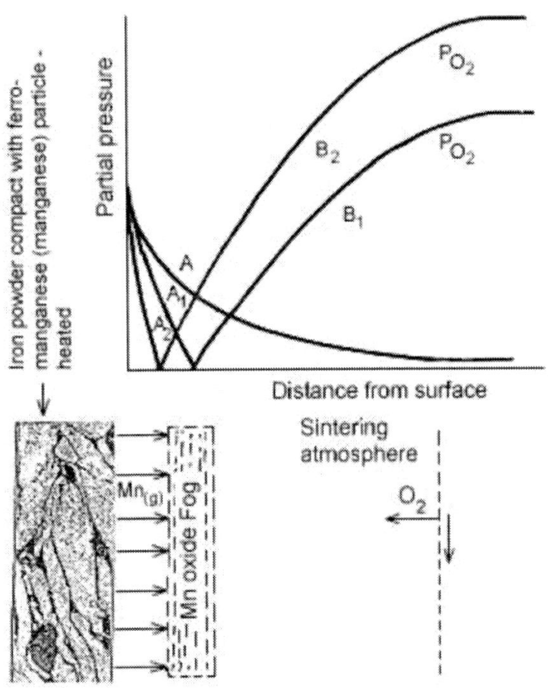

Figure 44. The partial pressures of sublimating manganese in flowing hydrogen under isothermal conditions as a function of the distance from the Fe-Mn specimen surface: A – at the presence of an inactive gas; A1, A2 – at the presence of oxygen in the gas oxidizing the manganese vapor in very fine dispersed oxide MnO; B1, B2 - the corresponding oxygen partial pressure gradients [4,18,112].

Following the fluxes schematically presented beneath the figure, at presence of a neutral gas the manganese gas gradient can follow curve A. However, when a low pressure of oxygen is present, the Mn-vapor oxidizes to solid particles of manganese oxide – involatile MnO, and the manganese vapor and oxygen gradients become steeper as indicated by A1 and B1. If more oxygen is present, the gradients become steeper to A1 and B2. Thus, the reaction zone approaches the compact surface, and free manganese oxide will form on the metal surface. The flowing atmosphere carries the formed volatile MnO away. This reaction starts in the pores in the compact recently with the sublimation and filling the pores with $Mn_{(g)}$ [18,112]. This reaction occurs above the surface of the sintered parts in some restricted space, e.g. in a sintering box.

Now it is evident that at every temperature the equilibrium in the system manganese – oxygen establishes with a reserve of several orders with respect to the sintered manganese containing compacts, and purity of the atmosphere. The result of this analysis is that manganese forms by itself a protection against the oxidation of the manganese particles in the compact during sintering [123,124]. This process, called for "self – cleaning/protection" effect of manganese [112], or reduction effect of gaseous manganese for sintering atmosphere to equilibrium conditions for $Mn_{(g)}$ – O system according to the reaction [18,125,126]:

$$Mn_{(g)} + H_2O = MnO_{(g)} + H_2.$$

The rate of manganese sublimation far exceeds the rate of its oxidation by a hydrogen atmosphere (pure or diluted by a inert gas) with the dew points around -20°C. This finding mitigates very considerably the stringent requirement for the atmospheres with the dew points of -50°C to -70°C, set by purely thermodynamic considerations [127,128]. The high affinity of manganese for oxygen demonstrated by this the mentioned reaction making possible to sinter the manganese steel in the atmospheres of common industrial purity.

The sintering of the Fe-Mn-C parts under industrial conditions in a double steel box under a getter - alumina + graphite to minimize the decarburization documented the formation of MnO as shown in Fig.45. The inner and outer walls of the cover steel box above the getter due to restricted flow of the atmosphere under these conditions covered a condensed green in color film - MnO (pure chemical compound).

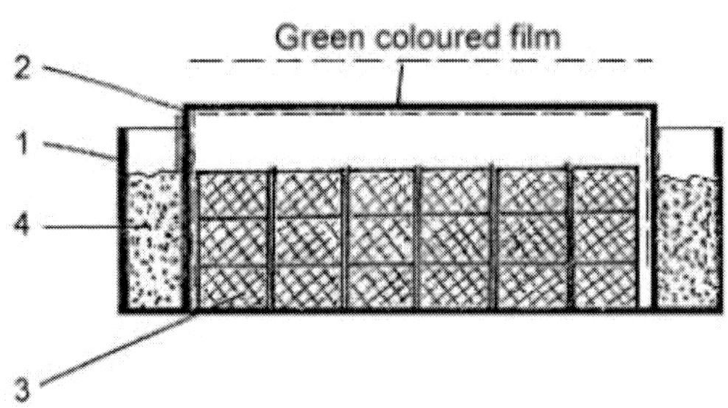

Figure 45. Sintering of Fe-Mn-C steel specimens in a double steel box under getter. Industrial sintering at 1180°C in $70N_2/30H_2$ atmosphere, d.p. -50°C; 1 – base box, 2 – cover box, 3 – parts, 4 – getter - alumina + 5% graphite [18].

Sintering in Continuously Operating Furnaces

The heating up period is a very important period especially for sintering of manganese steels. The start of the formation of the interparticle necks and of diffusion alloying occurs during this period. It is necessary to distinguish between laboratory (stationary) and industrial (continuous) sintering conditions. When sintering in a laboratory sintering furnace, in spite of the mentioned requirement for high purity atmosphere, its flow is during the whole process with constant dew point. The sintering of the parts placed in the boxes in an industrial furnace proceeds with a gas counter flow, the inlet of the atmosphere is at exit of the sintered parts of the furnace and reversely. In term of quality of the sintering atmosphere, the heating up and in a lesser extent the cooling period at sintering in continuously operating furnaces have special function. The first one is the dewaxing (delubing) and binder burn-off function proceeding in the range of 150 to ~600°C requiring the sources providing the oxygen to form carbon monoxide (oxygen, carbon dioxide, water) to form carbon monoxide or carbon dioxide with carbon in the polymer [3]. The sources of oxygen are also the reduction products H_2O and CO_2 coming from the sintered parts. Under these conditions, the manganese (manganese particles) separately or in the compacts without the sublimation must be oxidized. Manganese starts to oxidize at ~400°C, and the alloying of iron through some metal oxide is not possible. In the case of manganese steels, this does not occur only due to manganese sublimation also at low temperatures, Table 19. In the complete cooling zone, this is an inlet of the atmosphere in to the furnace, are situated already alloyed parts, i.e. sintered material with 1-5% Mn, therefore with low chemical activity of the system also in term of possible oxidation, not the manganese carrier particles with high Mn content. The sublimation of manganese regarding low chemical activity of the system occurs also during cooling period. It means the self-protection of manganese against the oxidation occurs also in this sintering period regarding the manganese vapor pressure and chemical activity of the material in relation to the actual temperature.

The as sintered microstructures of Fe-Mn-C and Fe-Mo-Mn-C steels after laboratory and industrial sintering with the margins without some feature of the demanganizing, confirmed also the surface hardness values, shows Fig.46.

Upon box sintering of many compacts, parts, whether in laboratory or industrial conditions, each compact loses a small fraction of manganese in vapor form escaping into the space with the atmosphere surrounding the compacts. This brings a "cloud" of manganese enveloping every compact, i.e. the total batch of the compacts. By this occurs the mutual manganese concentration compensation between the compacts. Admittedly, a small part of manganese is lost by oxidation to MnO, but this loss in relation to the total weight/number of parts sintered in one box, has no adverse effect mainly on the manganese surface content of the specimens/parts as shows Fig.46. This is the consequence of that, as mentioned previously, that the reaction $Mn_{(g)}$ – O occurs in some small distance above the being sintered parts [129,130].

The chemical analyses of manganese and carbon in sintered specimens were carried out with the aim to verify or prove that the demanganizing nor decarburization did not occur as was shown on optical micrographs. The results attained for carbon and manganese contents in hybrid Fe-3Cr-0.5Mo-xMn-0.24C steel samples, sintered under laboratory as well as under industrial conditions stacked in batches of about 100 samples, lists Table 22. Any differences in carbon and manganese contents were determined in relation to the starting ones with the impact on the mechanical properties. The corresponding tensile strength values are shown in following Figs.47,49.

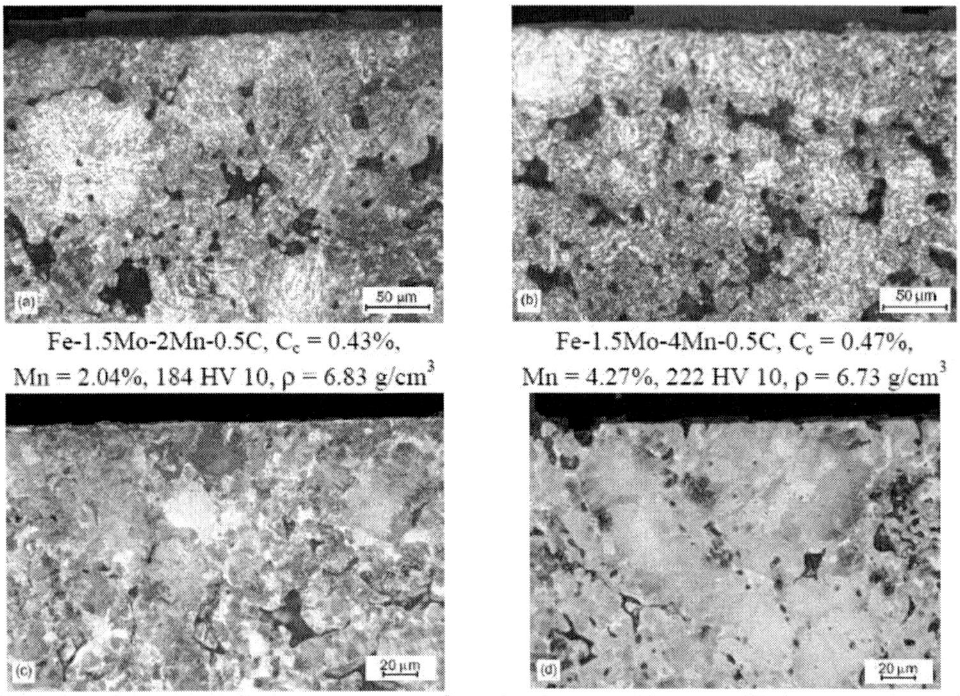

Fe-1.5Mo-2Mn-0.5C, C_c = 0.43%, Mn = 2.04%, 184 HV 10, ρ = 6.83 g/cm^3

Fe-1.5Mo-4Mn-0.5C, C_c = 0.47%, Mn = 4.27%, 222 HV 10, ρ = 6.73 g/cm^3

Fe-2.5Mn-0.8C, 195 HV 10, ρ = 6.88 g/cm^3

Fe-3.0Mn-0.8C, 196 HV 10, ρ = 6.86 g/cm^3

Figure 46. Microstructures of manganese containing steels showing as sintered margins of the specimens without demaganizing. (a), (b) – Fe-xMn-C steels (component – roller mass 125 g), SC 100.26, FeMnC, graphite, 600 MPa, industrial sintering 40 min, 1180°C, $70N_2/30H_2$, d. p. -50°C [129], (c), (d) – hybrid Fe-1.5Mo-xMn-C steels, Astaloy Mo, FeMnC, graphite, 600 MPa, laboratory sintering for 60 min at 1120°C, c.a. -30°C [130], optical micrographs. Nital etched.

Table 22. As sintered carbon and manganese content [mass %] of hybrid Fe-3Cr-0.5Mo-xMn-0.24C steel sintered under laboratory and industrial conditions. Prealloyed Fe-3Cr-0.5Mo powder, EMn - electrolytic manganese, FeMn – medium carbon ferromanganese, FeMnC – high carbon ferromanganese, 600 MPa [131-133]

Mn carrier		EMn			FeMn			FeMnC		
Mn ad. [%]	0	1	2	3	1	2	3	1	2	3
Sintering	Laboratory tube furnace, 30 min, 1250°C, cracked ammonia, d.p. -30°C									
C_c [%]	0.21	0.22	0.23	0.26	0.24	0.22	0.26	0.25	0.26	0.27
Mn [%]	0.10	0.88	2.13	3.04	1.03	2.13	3.30	1.10	2.21	3.11
Sintering	Industrial pusher furnace, 40 min, 1180°C, $70N_2/30H_2$, d.p. -50°C									
C [%]	0.20	0.20	0.23	0.25	0.21	0.23	0.24	0.21	0.24	0.24
Mn [%]	0.10	0.88	2.04	3.09	1.14	2.06	3.11	1.09	2.15	3.05

The standard bars for testing mechanical properties of manganese steels of various manganese and carbon concentrations industrially sintered in a pusher furnace as those shown in Fig.49, obtained the properties on the level corresponding to the compositions and sintering conditions [105]. The structural components "plate" (65 g, thickness 11.5 mm, density ~6.5 g/cm^3),"cube" (90 g, height 24 mm, density 7.0 g/cm^3) and the rollers for conveyors (22, 30 and 40 mm in diameter; G = 14 g for 22 mm, 105 g for 30 mm, 170 g for 40 mm) were prepared of manganese industrially sintered steels. These components tested for bend strength (plate), radial crushing tests (cube)and the rollers for tension and friction characteristics obtained the same or higher values, especially in friction when compared with those, which reached the same components prepared on the basis of Distaloy type powders [134,135].

The as sintered gears for hydrogenerators (outside diameter 38 mm, blank width 28 mm) industrially produced of Fe-3Mn-0.5Mo-0.35C steel were successfully tested with the pressure of 3 MPa for 5000 h [62,63].

Sintering of Hybrid Steels with Addition of Manganese

Sintered hybrid alloyed steels are based on prealloyed powder with admixed some other alloying element. The prealloyed Fe-3Cr-0.5Mo, Fe-1Cr-0.2Mo and Fe-(1,3)Cr-0.3Mo-0.3V powders with addition of manganese seem to be a perspective group of sintered structural alloyed steels for highly stressed parts. Nevertheless, the main alloying elements chromium and manganese are characterized with high affinity for oxygen. This presents perhaps at present time some doubts regarding sintering of these systems.

Figures 47 and 48 show the tensile strength of sintered and sinter hardened hybrid Fe-Cr-Mo-Mn-C steels based on Fe-3Cr-0.5Mo powder (CrM) sintered in cracked ammonia with the dew point of -30°C, all with 0.24% C. As shown, the highest sintered tensile strength of 856 MPa and sinter hardened of 989 MPa with addition of 2% Mn were obtained. By this means an increase in sintered tensile strength by 196 MPa/1% Mn and in sinter hardened by 238 MPa/1% Mn were achieved. The results show that this material system is possible to expose to sinter hardening without addition of some other hardening element.

The effect of individual physical-metallurgical and technical characteristics of the used manganese carriers affected the strength properties of the tested hybrid systems. The optimum properties were obtained added using the medium carbon ferromanganese as manganese carrier.

Figure 49 shows the tensile strength of the same material system as in previous case industrially sintered in a pusher furnace equipped with the delubing zone. In this case the highest tensile strength of 806 MPa was obtained again with 2% Mn. This proves that the shift operation of the box with the samples through the delubing zone of the furnace occurred without the oxidation of manganese particles in the compacts. The presented strength values prove a successful industrial sintering of this hybrid system with two elements of high affinity for oxygen as the consequence of the reduction effect of gaseous manganese for sintering atmosphere.

Figure 50 shows the tensile strength of the steels based on Fe-(1,3)Cr-0.3Mo-0.3V powders (KIP 103V, KIP 30CRV) in dependence on manganese addition and sintering temperature. The highest tensile strength of 929 MPa (~310 MPa/1% Mn) was attained with 1.5% Mn addition [134]. In term of manganese is to seen that the sintering as well as an effective alloying of prealloyed matrix by manganese occurred.

The highest tensile strengths values of both base hybrid Fe-Mn-C systems were obtained with total amount 5.1% and 5.5% of the cheapest alloying elements.

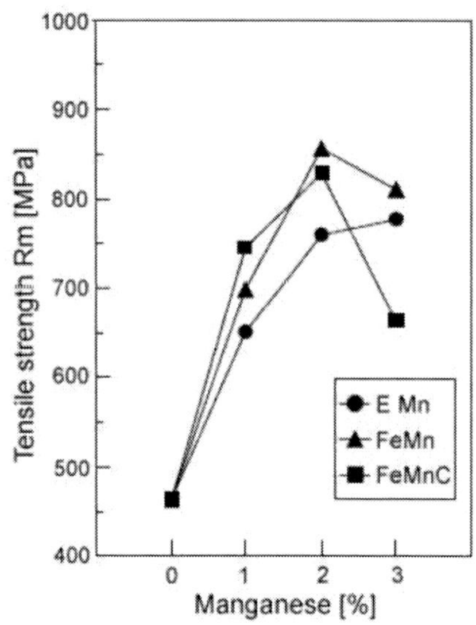

Figure 47. Tensile strength (UTS) of sintered hybrid Fe-3Cr-0.5Mo-xMn-0.24C steel in dependence on manganese addition and manganese carrier grade. 30 min, 1250°C, c. a., d.p. -30°C, cooling rate 0.06°C/s; Mn carrier: EMn, FeMn, FeMnC [131].

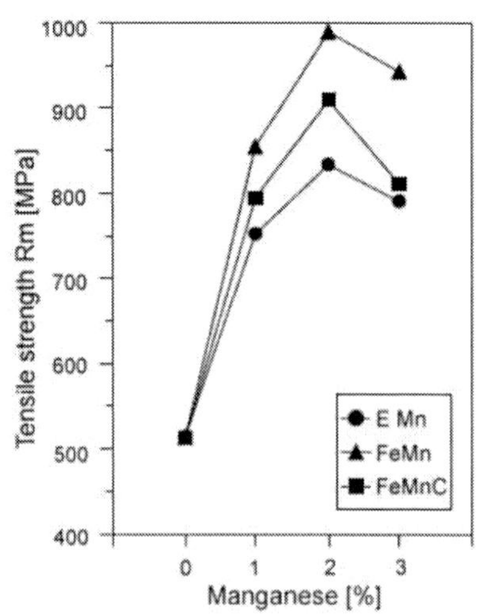

Figure 48. (right) Tensile strength (UTS) of sinter hardened hybrid Fe-3Cr-0.5Mo-xMn-0.24C steel in dependence on manganese addition and on manganese carrier grade. Other as in Fig.47. The cooling rate of ~0.3°C/s in the temperature range of 900°C to 400°C [131].

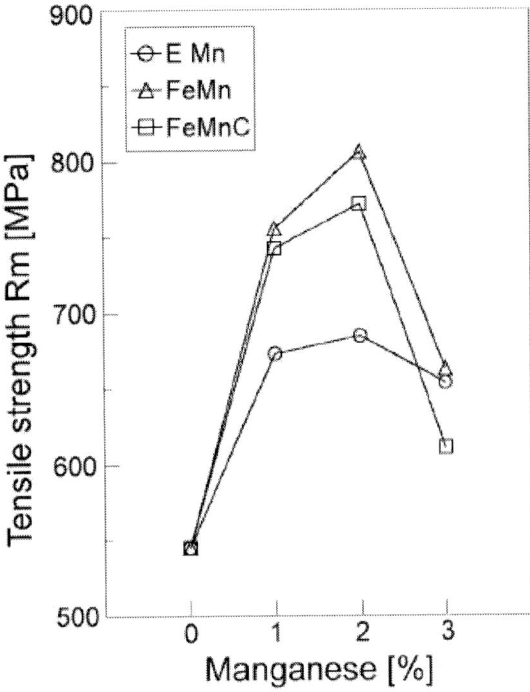

Figure 49. Tensile strength (UTS) of the steels as in Figs.47,48. Industrial sintering 40 min at 1180°C, $70N_2/30H_2$, d.p. -50°C [132,134].

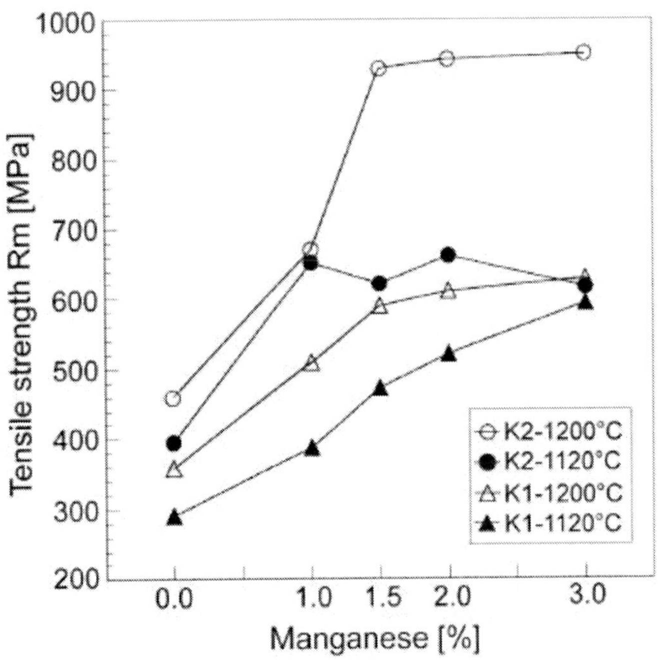

Figure 50. (right) Tensile strength (UTS) of laboratory sintered steels based on Fe-1Cr-0.3Mo-0.3V (Kl,103V) and Fe-3Cr-0.3Mo-0.3V (K2, 30CRV) powders, all with 0.24%C in dependence on manganese addition and sintering temperature of 1120°C and 1200°C, 60 min, c.a., -30°C; Mn as FeMnC [136].

It is necessary to consider in this case possible reaction of Cr_2O_3 ($FeMoCr_2O_3$) with Mn vapor confirmed also by a detailed thermodynamic analysis [125]. A marked effect of higher sintering temperature on tensile strength was observed.

It would be necessary to study in more detail the cause of a stoppage of the strengthening effect of manganese added in amount higher than 1.5% mainly in the material based on 30CRV powder.

Sintering in Semi-Closed Containers

The sintering in double boxes containing getter alumina + graphite is used in laboratory conditions to prevent decarburization of the carbon containing steels. The objective for the sintering of manganese steels in semi-closed containers is according with references:

- the processing to avoid/minimize formation of oxide veins in dry hydrogen-rich atmosphere and sintering in the alpha phase [137]
- the reduction surface oxidation, because the flowing atmosphere can no longer easily interact with the components surfaces at the presence of Mn vapor [125]
- to form the source of Mn vapor in a getter in the semi-closed containers (local dew point), which reacts with the atmosphere and thereby dries it; this enables processing of Mn-alloys at temperatures as low as 1140°C at inlet dew points no better than -30°C [101,102]
- to reduce manganese sublimation and oxidation [138].

The sintering of Mn containing samples in semi-closed steel containers with a labyrinth sea, illustrates Fig.51. To fulfill the above mentioned objective for sintering in semi-closed containers, they did not have any ferromanganese added, but are loaded to nearly full capacity with the specimens (~20), with any remaining space in the container being filled with ferrous ballast. Therefore, the green manganese powder containing compacts are the donors of Mn and C. This arrangement should provide sufficient manganese vapor and free carbon to both improve the local dew point (self-gettering) and minimize the loss of manganese from the compacts [138].

This method uses the previously in detail described findings regarding the manganese sublimation and the reduction of H_2O in the sintering atmosphere by gaseous manganese for improved cleaning the atmosphere, i.e. self-cleaning effect of manganese for sintering atmosphere. Considering the process presented in Fig.44, the reaction of gaseous manganese occurs over the surface of the being sintered samples. No other "cleaner", "drier" atmosphere can come into contact with the components surfaces through the Mn vapor "cloud" (Mn vapor), because the Mn vapor fills the pores and the compacts are enclosed with it.

107] investigated the strength properties of Fe-3Mn-0.8C steel sintered in semi-closed containers with labyrinth seal in hydrogen and nitrogen with the dew point -60°C to produce "Mn rich-microclimate" in dependence on three iron powder grades and two sintering temperatures (1120 and 1250°C). [105] investigated the effect of manganese carrier grade and carbon content on mechanical properties of Fe-3Mn-xC steels in dependence on manganese carrier grade (electrolytic manganese, medium and high carbon ferromanganese). Sintering occurred for 60 min at 1130°C in cracked ammonia with the dew point -23°C in semi-closed containers (getter - alumina + FeMnC + graphite). The effect of various purity of hydrogen (dew points -40, -50 and -60°C) on mechanical properties of Fe-3Mn-0.8C steel sintered in semi-closed containers for 60 min at 1120, 1150, 1180 and 1200°C investigated [102]. Inasmuch as the final mechanical properties of manganese steels sintered in semi-closed containers were affected by more factors, the effect of its alone is hardly to evaluate. The sintering of the same specimens stacked in free containers without iron ballast for comparison is lacking. According to the published data, higher mechanical properties of manganese steels sintered in semi-closed containers compared to that sintered in open boxes were not attained.

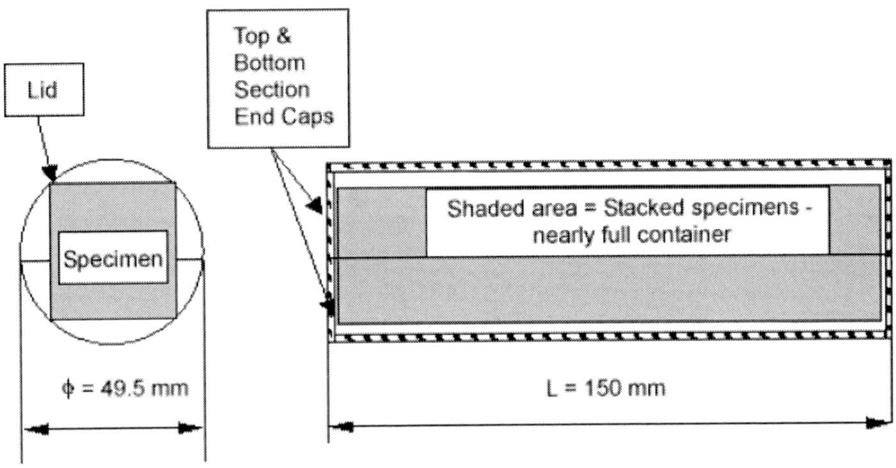

Figure 51. Schematic of sintering in semi-closed container [138].

The industrial sintering of two kinds prototype gears – a) outside diameter 28 mm, blank width 17 mm, b) outside diameter 41.3 mm, blank width 34 mm from Fe-3Mn-(0.6-0.8)C was carried out in semi-closed containers at 1120 - 1150°C in cracked ammonia with the dew point -30°C. The sintered gears were hardened (rapid cooled from 850°C) in a furnace with a protective atmosphere and tempered at 200°C for l h [139,140].

Effect of Manganese on Phase Transformation of Austenite

The evaluation of the properties of sintered manganese – (carbon) steels requires taking into account the special effect of both on phase transformation of austenite known in wrought steels.

The metallographic observations prove the heterogeneous microstructures of sintered manganese steels. Compared to sintered steels alloyed with other elements, in manganese steels martensite is determined on the microstructure also after furnace slow cooling independently on the carbon content. This effect demonstrates in decrease of the strength properties at some manganese and carbon content, Figs.6,7-10,14,15,22,23. This feature is contemporary used in the research for an increase of mechanical properties by faster cooling rates in most cases followed by tempering. The fully martensitic microstructure is not usual.

The cause lies in structural and substructural heterogeneity of the starting iron powder grades according to manufacturing methods. This demonstrates in chemical heterogeneity of course, because individual microregions in the iron powder particles and in the austenite grains, become alloyed by variable concentrations of manganese, i.e. the formation of microregions of various Mn concentrations, Figs.30-33, 35-38. In such case at the phase transformation of austenite in the microstructure can form the different phases corresponding to the manganese concentration in the microregion.

At the analysis of the final microstructure of sintered manganese, steel is necessary to regard that in wrought steels containing 5-6% Mn occurs the diffusionless $\gamma \rightarrow \alpha$ transformation of austenite to martensite also at common slow cooling rates similar to the martensitic transformation in Fe-C steels [141,142]. The final microstructure of such a transformation is an oversaturated α -solid solution of the concentration corresponding to the γ -solution, i.e. manganese martensite without the formation of the tetragonal lattice not requiring the tempering as occurs in Fe-C steels. Similarly, the bainite in wrought Fe-Mn-(C) steels forms in the microregions containing approx. 2% Mn [143].

These factors support the diffusionless transformation of sintered Fe-Mn-(C) in austenite microregions containing 4-6% Mn to martensite and bainite in those containing ~2% Mn. Such transformation of Mn steel austenite corresponds to that occurring at sinter hardening process. The microstructure heterogeneity of sintered Fe-Mn-C steels, characterized by ferrite up to martensite proves the data given in Table 8.

The relatively high Mn concentrations are reaching also the austenite microregions at lower starting manganese addition due to mentioned differences in alloying by diffusion mechanisms and by diffusion induced grain boundary migration (DIGM) [144,145] in relation to the inhomogeneous character of the base powder austenite.

Contemporary is necessary to note that the highest strengthening effect of manganese demonstrates at low content, i.e. one percentage, and at low carbon content. With increasing Mn content its relative strengthening effect is decreasing. Inasmuch as martensite reach the highest hardness with 0.5-0.6% C, higher carbon contents cause the formation of retained austenite in microregions with corresponding Mn and C contents [131].

Manganese besides presented an alloying - sintering effect in manufacturing of manganese steels, in amount ~0.5% improves the machinability of PM steels [146].

Conclusion

Manganese is the cheapest alloying element with high hardenability commonly used in ingot structural and austenite steels. Manganese due to high affinity for oxygen and hard reducibility of its oxides was and in some case is at present time regarded for problematic for use in PM steel production. The main results and knowledge attained in investigation of sintered manganese steels since 1948, are presented in this chapter, and the sintering, alloying and properties are evaluated by two physical-chemical properties of manganese – high affinity for oxygen as well as high vapor pressure.

There were studied the effects of different physical-metallurgical and technological factors on properties of structural and austenite manganese steels with special respect to the purity of sintering atmosphere. These, except of the purity of the atmosphere, are the same as in production of all alloyed sintered steels.

Efficient sintering and alloying of base powders by manganese prove the attained mechanical properties yet after sintering in α - region as well at 1000°C up to 1300°C. The decisive results are those attained by processing of plain Fe-Mn systems unaffected by other factors. Hardening effect of manganese demonstrated in increasing strength properties with increasing manganese content without some interruption in dependence on manganese content and/or sintering conditions.

The final properties of iron-manganese steels attained are affected besides manganese content, compacting pressure and sintering conditions, as base characteristics, also by carbon content and iron powder grade. The highest tensile strength of 886 MPa attained the single pressed-sintered Fe-4.5Mn-0.4C steel based on eddy milled Hametag iron powder.

The comprehensive evaluation of manganese steel sintering processes performed, and of corresponding properties attained in regard to above mentioned assumptions must to regard the thermodynamic requirements for the purity of the sintering atmosphere to achieve equilibrium conditions for Mn – O system. The dew points values for Mn – O system equilibrium in dependence on temperature demonstrate that the most important phase of manganese steel sintering in term of possible oxidation of manganese carrier is the preheating period. Compared to these requirements, it follows from the presented results that none of the used sintering atmosphere fulfilled the mentioned thermodynamic requirements for the purity, nor those with the dew points up to -70°C. In spite of it, all presented results, expressed in mechanical and other properties attained, prove an efficient sintering of manganese steels prepared under very different conditions. It means the sintering and contemporary the alloying of manganese steels occurred almost independently on the atmosphere purity used.

Manganese besides high affinity for oxygen characterizes also high vapor pressure, progressively increasing with increasing temperature which is significantly higher when compared to other alloying elements used in powder metallurgy. The effect of high vapor pressure is manganese vapor formation by sublimation starting already at low temperatures during preheating stage of sintering.

The manganese vapor formed at sintering of manganese steels expresses oneself in the sintering of manganese steels by two effects.

The first one is the condensation of manganese vapor on the surfaces of powder particles in the compacts forming an alloyed surface layer, which forms a source of manganese for alloying of the interior of powder particles in the compact. The alloying of particles cores occurs from the surface layer by diffusion mechanisms according to structural and technical characteristics of iron (prealloyed) powder and temperature. The alloying occurs in solid phase (iron powder) – gas phase (Mn vapor) system.

The second one, as follows from it, is deciding, and only one real sintering process for manganese steels under common laboratory as well as industrial conditions. The manganese vapor formed inside the compacts due to its high vapor pressure (difference in concentration gradient) partly escapes through the interconnected pores into the surrounding humid atmosphere. Here, surrounding the samples, due to high affinity for oxygen, gaseous manganese reacts with the oxygen in the atmosphere according to the equation:

$$Mn_{(g)} + H_2O = MnO_{(g)} + H_2.$$

This is the reaction of gaseous manganese with oxygen in sintering atmosphere forming in gas state the equilibrium conditions for Mn – O system required for alloying and sintering of manganese steel in the atmospheres of common purity (the same also for CO_2). The rate of sublimation of manganese far exceed the rate of its oxidation by a hydrogen atmosphere (pure of diluted by inert gas) with the dew point around -20°C. The use of a protective atmosphere with a low dew point, e.g. ~30°C fulfils the requirements for sintering of manganese steels.

This process expresses oneself in sintering as reduction - "self cleaning/protection" effect of manganese in vapor form against the oxidation with H_2O without the requirements for special measures to prevent the oxidation of manganese carrier. This process does not regard the thermodynamic requirements for the purity of the sintering atmosphere. Self-cleaning effect of manganese is efficient for sintering not only of all mixed Fe-Mn-C steels but also of hybrid steels containing manganese as expressed in properties of sintered and sinter hardened Fe-Cr-Mo-Mn-C and Fe-Mo-Mn-C steels.

The process, called by the authors of this chapter "sinter-alloying" of manganese steels, based on spontaneously occurring sublimation of manganese is only one possible process for explanation of effective sintering of manganese steels since the start of their research interest. This proved the previously presented and other published mechanical and physical properties of sintered manganese containing steels.

Some properties of sintered manganese steels, not fulfilling the expectations according these characteristics, is necessary to look for other causes as, e.g. structural characteristics of base iron or prealloyed powders, manganese carrier kind and particle size distribution, chemical activity of some master alloys (not effective), heating rate, cooling rate, especially carbon content etc. The medium carbon ferromanganese, compared to other previously cited manganese carriers, seems to be from processing and technical point of view the most suitable manganese carrier for the preparation of manganese steels.

Acknowledgments

The work was carried out within the framework of the Scientific Grant Agency of ME SR and SAS (Projects No. 2/6209/26 and 2/0103/09).

References

[1] Bocchini, G. F. Proc. Powder Metall. Summer School, Gothenburg 1999, EPMA, Shrewsbury, UK, 185-204

[2] German, R. M. Powder Metallurgy Science: MPIF, Princeton, NJ, 1994

[3] German, R M. Powder Metallurgy of Iron and Steel: John Wiley & Sons, New York, Inc., 1998

[4] Šalak, A. Ferrous Powder Metallurgy: Cambridge Inter. Sci. Pub., Cambridge, UK, 1995

[5] Höganäs Iron and Steel Powders for Sintered Components: Höganäs AB, Höganäs, SE, 2002

[6] Danninger, H., Pöttschacher, R., Jangg, G., Seyrkammer, J., Šalak, A. Proc. Powder Metall. World Congress, 1994,Paris, EPMA, Shrewsbury, UK, Vol. 2, pp. 879-882

[7] Šalak, A., Danninger, H., Selecká, M. Proc. PM AUTO 05 4th Int. Conf. on Powder Metall. for Automotive Parts, 2005, Isfahan, Mashad Powder Metall. Co., Tehran, IN, , pp.12-24

[8] EU Carcinogenic Directives 90/394/EEC and 91/322/EEC

[9] Sechel, N., Orban, R. L. Proc. PM2001 Eur. Conf. & Exh., 2001, Nice, EPMA, Shrewsbury, UK, Vol. 4, pp. 166-171

[10] Šalak, A. PMI. Vol. 12, 1980, No. 2, pp. 72-75

[11] Šalak, A. Int. J. Powder Metall. Powder Techn. Vol. 16, 1980, No. 4, pp. 369-379

[12] Offermann, E. K. Mitt. Kohl.-Eisenforschung, 1936, No. 1, pp. 55-120

[13] Šalak, A., Rudnayová, E., Selecká, M. Powder Metall. Vol. 42, 1999, No. 1, pp. 45-50

[14] Grethe, K. Mangan - Grundlagen und technische Verwendung: ED. Stahleisen, Düsseldorf, GE, 1972

[15] Tretyakov, Yu. D. Chemistry of Unstoichiometric Oxides: Ed. Moscow University, Moscow, RU, 1974 (in Russian)

[16] Kubaschewski, O., Evans, E, Alcock, C. B. Metallurgical Thermochemistry: Pergamon Press, New York, 1974

[17] Höganäs Handbook for Sintered Components: Höganäs AB, Höganäs, SE, Vol. 2, 1997

[18] Šalak, A., Selecká, M., Bureš, R. Powder Metall. Progr. Vol. 1, 2001, No. 1, pp. 41-58

[19] Rivlin, V. G. Int. Metals Reviews. Vol. 29, 1984, No. 4, pp. 299-303

[20] Tengzelius, J., Grek, S. E., Blände, C. A. Modern Dev. in Powder Metall., MPIF, Princeton, NJ, 1980, Vol. 13, pp. 159-181

[21] Rankyn, W. J., Wynnyckyj, J. R. Metallurgical and Materials Transactions B. Vol. 28B, April, 1997, pp. 307-315

[22] Šalak, A., Selecká, M., Bureš, R. Proc. Höganäs Chair Seminar Sintering Atmospheres for Ferrous Components, TU Vienna, Vienna, A, 1999, pp. 1-21

[23] Upadhayaya, G. S. Manganese in Powder Metallurgy: The Manganese Centre, Paris, FR, 1986

[24] Kieffer, R., Hotop, W. Sintered Iron and Sintered Steel: Ed. Springer Verlag, Wien, A, 1948 (in German)

[25] Benesovsky, F., Kieffer, R. Berg- und Hüttenmän. Monatshefte. Vol. 95, 1950, No. 8, pp. 145-150

[26] Wakabayashi, S., Morioka, Y. Watanabe, T. Report of the Castings Research Laboratory, Waseda University, Waseda, JP, 1965, Vol. 16, pp. 11-16

[27] Romanski, A., Frydrych, H. Metalurgia Proszków Rudy Metale. Vol. 43, 1998, No. 3, pp. 131-139

[28] Šalak, A., Dudrová, E., Miškovič, V., Rudnayová, E. Proc. IV. Int. Conf. Powder Metall. in CSSR, 1974, Štrbské Pleso, SVTS DT, Žilina, SK, Vol. III, pp. 101-122

[29] Zapf, G., Hoffmann, G., Dalal, K. Arch. Eisenhütten. Vol. 18, 1975, No. 4, pp. 287-284

[30] Zapf, G., Hoffmann, G., Dalal, K. Powder Metall. Vol.18, 1975, No. 35, pp. 214-219

[31] Šalak, A., Selecká, M. Pokroky práškové metalurgie VUPM. Šumperk, 1979, No. 1, pp. 15-25

[32] Selecká, M., Chachaľák, M., Šalak, A. Zborník vedeckých prác VŠT Košice. Košice, SK, 1980, No. 2, pp. 195-211

[33] Selecká, M., Chachaľák, M. Kovové materiály (Metallic Mat.). 1980, 1, 126-143

[34] Šalak, A. Practical Metallography. Vol. 17, 1980, No. 8, pp. 390-396
[35] Šalak, A. Practical Metallography. Vol. 17, 1980, No. 6, pp. 273-279
[36] Šalak, A. Proc. Int. Symp. Materials with High Mechanical-physical Properties, 1977, SVTS, Bratislava, SK, Vol. 2, pp.100-105 (in Slovak)
[37] Šalak, A. Proc. 5th Int. Conf. on Powder Metall. in ČSSR, 1978, Gottwaldov (Zlin), SVTS Dom techniky, Žilina, IMR SAS Košice, SK, Vol. I, pp. 425-440
[38] Šalak, A., Leitner, G., Hermel, W. Pokroky práškové metalurgie. VÚPM, Šumperk, 1979, No. 2-3, pp. 41-52
[39] Šalak, A. Modern Develop. Powder Metall., MPIF, Princeton, NJ, 1980, Vol. 13, pp. 183-201
[40] Šalak, A. PMI. Vol. 12, 1980, No. 1, pp. 28-31
[41] Šalak, A. Poroshkovaya metallurghiya. Vol. 215, 1980, No. 11, pp. 100-105
[42] Šalak, A. Proc. UN Seminar on Powder Metall. in Machinery, 1985, Techn. University Minsk, Minsk, BR, R6, pp. 125-132
[43] Šalak, A. Int. J. Powder Metall. Powder Techn. Vol. 18, 1982, No. 1, pp. 11-23
[44] Šalak, A., Dudrová, E. Strojírenská výroba. Vol. 25, 1977, No. 10, pp. 772-776
[45] Šalak, A., Miškovič, V. Proc. VIth Int. Powder Metall. Conf. in GDR, 1977, Tech. University Dresden, Dresden, GE, Vol. 2., pp. 33-1 – 33-15
[46] Šalak, A., Miškovič, V., Maár, K., Špitál, J. Pokroky práškové metalurgie. VUPM, Šumperk, 1979, No. 2-3, pp. 9-15
[47] Šalak, A., Dudrová, E., Maár, K. Zborník vedeckých prác VŠT Košice. Košice, 1978, Vol. 2, pp. 325-340
[48] Šalak, A., Miškovič, V., Maár, K., Špitál, J. Zborník vedekých prác VŠT Košice. Košice, 1979, Vol. 3, pp. 259-269
[49] Šalak, A., Miškovič, V., Maár, K. Zborník vedeckých prác VŠT Košice, Košice. Vol. 2, 1978, pp. 341-354
[50] Šalak, A., Leitner, G., Hermel, W. Pokroky práškové metalurgie. VUPM, Šumperk, 1978, No. 4, pp. 31-44
[51] Šalak, A., Leitner, G., Hermel, W. Pokroky práškové metalurgie. VUPM, Šumperk, 1978, No. 4, pp. 45-54
[52] Šalak, A., Leitner, G., Hermel, W. PMI. Vol. 13, 1981, No. 1, pp. 21-24
[53] Šalak, A.,Leitner, G., Hermel, W. Proc. Vth Int. Conf. Powder Metall., 1979, PAPM, Poznan, PL, Vol.1, pp. 383-409
[54] Fleischer, R. PhD Thesis, TU Vienna, Vienna, A, 1992
[55] Danninger, H., Pöttschacher, R., Šalak, A., Seyrkammer, J. Proc. Int. Conf. TECHNOLOGIA 97, 1997, STU, Bratislava, SK, pp. 16-22
[56] Danninger, H., Pöttschacher, R., Bradac, S., Šalak, A., Seyrkammer, J. Powder Metall. Vol. 48, 2005, No.1, pp. 23-32
[57] Šalak, A. Proc. Int. IIIth Metallography Symp., 1978, High Tatras, TU Košice, Košice, SK, pp. 62-67.
[58] Šalak, A., Selecká, M. Proc. CVD 3rd Int. Powder Metall. Conf. in Turkey, 2002, TTMD, Ankara, TY, pp. 231-246
[59] Šalak, A., Selecká, M., Čajka, A., Franek, Š. Proc. Int. Conf. DF PM 2002, Stará Lesná, IMR SAS, Košice, SK, Vol.1, pp. 285-293
[60] Cias, A., Mitchell, S.C., Wronski, A.S. Proc. 1998 World PM Congress & Exh., Granada, EPMA, Shrewsbury, UK, Vol. 3, pp. 179-183

[61] Šalak, A. Powder Metall. Vol. 24, 1981, No. 2, pp. 64-69
[62] Šalak, A. Strojírenská výroba. 1990, No. 12, pp. 45-50
[63] Šalak, A. Dudrová, E., Miškovič V. Powder Metall. Sci. Techn. Vol. 3, 1992, No. 2, pp. 26-35
[64] Zapf, G., Hoffmann, G., Dalal, K. Powder Metall. Vol. 18, 1975, No. 3, pp. 214-236
[65] Šalak, A. Proc. Vth Int. Conf. Powder Metall., 1979, Katowice, PAPM, PL, Vol. I, pp. 425-431
[66] Šalak, A. Pokroky práškové metalurgie. VUPM, Šumperk, 1983, No. 4, pp. 35-45
[67] Huppmann, W. J., Kaysser, W., Esper, F. Zeitschrift f. Werstofftechnik (J. of Materials Technology), 1976 No. 7, pp. 190-195; 4th Powder Metall. Symp., 1975, Grenoble, EPMA, Shrewsbury, Vol. I, pp. 3-7-1
[68] Chaterjee, S. K., Mackay, C. A. Powder Metall. Vol. 23, 1980, No. 4, pp. 183-188
[69] Anonymous. MPR. 1982, No. 10/November. p. 582
[70] Fischmeister, H., Drar, H. PMI. Vol. 19, 1977, No. 3, pp. 114-118
[71] Selecká, M., Šalak, A., Danninger, H., Parilák, Ľ. Proc. 2000 Powder Metall. World Congress & Exh., 2000, Kyoto, Part 1, JSPM, JN, pp. 16-19
[72] Volynova, T. F., Emelyanova, I. Z., Sidorova, I. B. Poroshkovaya metallurghiya. 1992, No. 5, pp. 29-37
[73] Volynova, T. F., Emelyanova, I. Z. Sidorova, I. B. Poroshkovaya metallurghiya. 1992, No. 6, pp. 1-7
[74] Wilson, F. G., Knott, B .R. The Int. J. Powder Met. Powder Techn. Vol. 14, 1978, No. 2, pp. 121-125
[75] Radomyselskij, I. D., Martjuchin, I.D., Glazkov, V.M., Novomejskiy, Ju. D. Sintered structural materials: Ed. IPM AN USSR, Kiew, UA, 1976, pp. 88-97 (in Russian)
[76] Dorofeyev, Ju. G., Zherditskyi, I. T., Kolesnikov, V. A. Sintered structural materials: Ed. IPM AN UdSSR, Kiew, UA, 1972, pp. 109-116 (in Russian)
[77] Patent USA Nr. 4071354-1978 (Translation of NOYES Data Corp., Pack Ridge, 1980 (in Russian)
[78] Mocarski, S., Hall, D. S., Chernekoff, S. A., Yeager, D. A., McHugh, C. O. Powder Metall. Vol. 39, 1996, No. 2, pp. 130-134
[79] Fedorchenko, I. M., Skorokhod, V. V. Poroskovaya metallurghiya. 1967, No. 10, pp. 29-48
[80] Anonymous. MPR. Vol. 37, 1981, No.4/April, p. 187
[81] Findeisen, G. Problems on Powder Metall.: Ed. SAV, Bratislava, SK, 1964, pp. 189-197 (in Slovak)
[82] Šalak, A. Int. J. Powder Metall. Powder Techn. Vol. 16, 1980, No. 4, pp. 369-379
[83] Zapf, G., Dalal, K. Modern Develop. in Powder Metall., MPIF, Princeton, NJ, 1976, Vol. 10, pp. 129-143
[84] Hoffmann, G., Dalal, K. PMI. Vol. 11, 1979, No. 4, pp. 177-181
[85] Schliepper, G., Thümmler, F. PMI. Vol. 11, 1979, No. 4, pp. 172-176
[86] Banerjee, S., Schliepper, G, Thümmler, F., Zapf, G. Modern Develop. Powder Metall., 1980, Vol. 13, pp. 143-147
[87] Klein, A. N., Oberacker, R., Thümmler, F. PMI. Vol. 17, 1985, No. 1, pp. 13-17
[88] Klein, A.N., Oberacker, R. Thümmler, F. PMI. Vol. 17, 1985, No. 2, pp. 71-75
[89] Retelsdorf, H. J., Fichte, R. M. Proc. 4th Eur. Symp. Powder Metall., 1975, Grenoble, EPMA, Shrewsbury, UK, Vol. 1, pp. 3-5-1/6

[90] Schlieper, G., Thümmler, F. PMI. Vol. 11, 1979, No. 4, pp. 172-176
[91] Klein, A. N., Oberacker, R., Thümmler, F. MPR. Vol. 40, 1984, No. 6, pp. 635-637
[92] Hamiuddin, Md., Upadhyaya, G. S. PMI. Vol. 12, 1980, No. 2, pp. 65-69
[93] Hamiuddin, Md. Int. J. Powder Metall. and Powder Technol., Vol. 20, 1984, No. 4, pp. 343-348
[94] Šalak, A. Proc. Vth Int. Conf. Powder Metall. in Poland, 1979, PAPM, Poznan, PL, Vol. I, pp. 423-431
[95] Radomyselskiy, I. D., Napara-Volgina, S. G., Kazimirenko, S. M., Martjuchin, I.D., Devzhenko, L. D. Powder Metallurgy Structural Materials: Ed. IPM AN UdSSR, Kiew, 1974, pp. 139-146 (in Russian)
[96] Napara-Volgina, S. G. Poroshkovaya metallurghiya. 1984, No. 11, pp. 1-10
[97] Cias, A., Stoytchev, M., Wronski, A. S. Proc. 2001 Int. Conf. Powder Metall. Partic. Mater., New Orleans, MPIF, Princeton, NJ, Vol. 10, pp. 131-135
[98] Sulowski, M., Cias, A. Proc. Euro PM 2003 Eur. Congress & Exh. Powder Metall., Valencia, EPMA, Shrewsbury, UK, Vol. 1, pp. 453-458
[99] Cias, A., Mitchell, S. C. Powder Metall. Progr. Vol. 5, 2005, No. 2, pp. 82-91
[100] Sulowski, M. Powder Metall. Progr. Vol. 7, 2007, No. 2, pp. 80-91
[101] Cias, A., Mitchell, S. C., Watts, A., Wronski, A. S. Powder Metall. Vol. 42, 1999, No. 3, pp. 227-232
[102] Sulowski, M., Cias, A. Inžynieria materialowa. Vol. XIX, 1998, No. 4, pp. 1179-1182
[103] Sinka, V., Šalak, A., Selecká, M., Keresti, R., Wronski, A. S. Proc. Int. Conf. DF PM 99, 1999, Piešťany, IMR SAS, Košice, SK, Vol. 2, pp. 38-41
[104] Šalak, A., Selecká, M., Parilák, L. Naucni Izvestija. Vol. VII, 2000, No. 6, pp. 94-99
[105] Dudrová, E., Kabátová, M., Bidulský, R., Wronski, A. S. Powder Metall. Vol. 47, 2004, No. 2, pp. 181-190
[106] Matsuda, M., Miura, H. P/M Science & Technology Briefs. Vol. 3, 2001, No. 5, pp. 5-8
[107] Cias, A., Wronski, A. S. Powder Metall. Progr. Vol. 7, 2007, No. 3, pp. 139-148
[108] Šalak, A., Ďurišin, J. Pokroky práškové metalurgie. VUPM, Šumperk, 1979, No. 2-3, pp. 29-40
[109] Šalak, A. Hutnické listy. Vol. XXXV, 1980, No. 10, pp. 724-727
[110] Šalak, A. Poroshokovaya metallurghiya. Vol. 214, 1980, No. 10, pp. 103-108
[111] Šalak, A. Hutnické listy. Vol. XXXIX, 1984, No. 2, pp. 124-127
[112] Šalak, A. PMI. Vol. 18, 1986, No. 4, pp. 266-270
[113] Šalak, A. Manganese in Iron Powder Metall. DrSc Thesis, Technical University, Košice, SK, 1995
[114] Šalak, A. PMI. Vol. 16, 1984, No. 6, pp. 260-263
[115] Šalak, A. Pokroky práškové metalurgie. VUPM, Šumperk, 1983, No. 4, pp. 3-14
[116] Šalak, A. Poroshkovaya metallughiya. 1984, No. 8, pp. 58-61
[117] Selecká, M., Šalak, A., Pieczonka, T., Cias, A. Proc. Int. Conf. DF PM 2002, Stará Lesná, IMR SAS, Košice, SK, Vol.1, pp. 93-99
[118] Selecká, M., Šalak, A., Pieczonka, T., Stoytchev, M. Proc. EURO PM2007, Int. Powder Metall.Congr. & Exh., Toulouse, EPMA, Shrewsbury, UK, 2007, Vol. 1, pp. 47-52
[119] Selecká, M., Šalak, A. Powder Metall. Progr. Vol. 8, 2008, No. 1, pp.7-23
[120] Selecká, M., Šalak, A., Pieczonka, T. STEEL GRIPS, Vol. 6, 2008, No. 5, pp. 355-358
[121] Richardson, F.D. Physical Chemistry of Melts in Metallurgy: Vol. 5, Academic Press, London, New York, 1974

[122] Kazenas, E. K., Chizhikov, D. M. Pressure of Composition of Vapor over Oxides of Chemical Elements: Ed. Nauka, Moscow, RF, 1976 (in Russian)
[123] Šalak, A, Ďurišin, J. Method of Removing Oxygen from the gaseous Atmospheres, Patent AO CS 208295/82, SK
[124] Šalak, A, Ďurišin, J. Poroshkovaya metallurghiya. 1985, No. 9, pp. 103-106
[125] Mitchell, S.C., Cias, A. Powder Metall. Progr. Vol. 4, 2004, No. 3, pp. 132-142
[126] Šalak, A. Proc. VIth Int. Conf. Fractography Conf., 1981, Stará Lesná, IMR SAS, Košice, SK, pp. 66-70
[127] Navara E. Proc. VIth Int. Conf. Powder Metall. in Czechoslovakia, 1982, Brno, Dom techniky ČSVTS, Žilina, IMR SAS Košice, SK, Vol. 1, pp. 143-154
[128] Navara, E. Proc. Int. Seminar Sintering 85, Herceg Novi, Plenum Press, New York, 1987, pp. 343-349
[129] Šalak, A., Selecká, M. Proc. Int. Conf. DF PM 2002, Stará Lesná, IMR SAS, Košice, SK, Vol. 1, pp. 284-290
[130] Selecká, M., Keresti, R. Proc. Int. Conf. DF PM 2005, Stará Lesná, IMR SAS, Košice, SK, p. 132-140
[131] Šalak, A., Selecká, M. Powder Metall. Vol. 51, 2008, No. 4, pp. 327-329
[132] Selecká, M., Šalak, A. Proc. PM AUTO 2008 5th Int. Conf Powder Metall. for Automotive Parts, Isfahan, Mashad Powder Metall., Tehran, IN, pp. 1-8
[133] Selecká, M., Šalak, A. RoPM 2009. 4th Int. Conf. of Powder Metall. Book of Abstracts. Craiova. Ed. L. Brandusan. Cluj-Napoca: Risoprint 2009, p.44
[134] Selecká, M., Šalak, A. Powder Metall. Progr. Vol. 8, 2008, No. 2, pp. 121-127
[135] Šalak, A., Selecká, M., Parilák, Ľ., Čajka, A. Proc. PM2004 Powder Metall. World Congr. & Exh., Vienna, EPMA, Shrewsbury, UK, Vol. 3, pp. 237-242
[136] Šalak, A., Selecká, M., Parilák, Ľ. Proc. PM2001 Eur. PM Conf. & Exh., Nice, EPMA, Shrewsbury, UK, Vol. 1, 251-256
[137] Mitchell, S.C., Becker, B.S., Wronski, A.S. Proc. 2000 Powder Metall. World Congress, JSPM, Kyoto, JN, Part 2, pp. 928-926
[138] Mitchell, S.G., Youseffi, M., Abosbaia, A.S., Ernest, J. Powder Metall. Progr. Vol. 8, 2008, No. 2, pp. 91-100
[139] Cias, A., Mitchell, S. C. Powder Metall. Progr. Vol. 5, 2005, No. 2, pp. 82-91
[140] Cias, A. Powder Metall. Progr. Vol. 5, 2005, No. 3, pp. 147-163
[141] Houdremont, E. Handbuch der Sonderstahlkunde: 3rd Ed. Springer Verlag, Berlin, GE, Vol. 1, 1956
[142] Kontorovich, I. E. Heat treatment of steels and cast iron: Book II, Alloyed Steels, Ed. Metallurgizdat, Moscow, RU, 1950 (in Russian)
[143] McPherson, N. A., Baker, T. N. Metal Science. 1976, April, pp. 140-144
[144] Šalak, A. Kovové materialy (Metallic Materials). Vol. 27, 1989, No. 2, pp. 159-170
[145] Šalak, A. Science of Sintering. Vol. 21, 1989, No. 3, pp. 145-154
[146] Šalak, A., Selecká, M., Danninger, H. Machinability of powder metallurgy steels: 2nd ed., Cambridge Inter. Sci. Publ., Cambridge, UK, 2006

In: Bulk Materials: Research, Technology and Applications ISBN: 978-1-60692-963-6
Editors: T. Frias, V. Maestas, pp. 141-194

Chapter 3

SURFACE TREATMENT EFFECTS ON BULK MATERIAL FATIGUE RESISTANCE

Sergio Baragetti and Federico Tordini
Dipartimento di Progettazione e Tecnologie, Università degli Studi di Bergamo, Viale Marconi 5, 24044, Dalmine (BG), Italy.

ABSTRACT

Lightening structural components is becoming more and more important for a lot of applications, particularly in the automotive and aerospace fields. From this point of view, the choice of the most effective surface treatments can play a fundamental role in the enhancement of the fatigue resistance, and hence in the mass reduction, of mechanical components. In this chapter, a review of recent studies on the effects of surface treatments on the fatigue and contact fatigue behaviour of different bulk materials is reported. Fatigue and contact fatigue tests on treated specimens and components made it possible to collect and analyze a large amount of data. Theoretical-numerical models were also developed to verify and justify the experimental results and to predict the fatigue behaviour of surface treated components. Materials ranging from structural steels to light alloys, such as titanium and aluminum ones, are considered. With regard to the surface treatments, particular attention is paid to physically deposited thin hard coatings - i.e. with PVD (Physical Vapor Deposition) and CVD (Chemical Vapor Deposition) techniques. CrN, TiN, WC/C, DLC are some of the coatings which have been studied. Furthermore, the more traditional shot peening treatment is also discussed at the end of the chapter.

Thin hard coatings deposited with PVD and CVD techniques are, for the time being, used over an increasing number of structural applications. They were first introduced in the field of cutting tools and their use has been extended also to the oil, aerospace, transport and biomedical industries. The improvement of the tribological and corrosion-resistant properties of thin hard-coated components has been demonstrated by a number of studies and good results for the fatigue behaviour have been collected too. The surface residual stress field induced by the deposition process is supposed to be one of the main parameters affecting the fatigue behaviour. Both experimental and numerical studies on thin hard-coated components have been carried out. The presence of the cited residual stresses was simulated with finite element models which allowed to foresee the fatigue life of coated components. The features of the fracture surfaces of coated samples, including the possible modifications introduced by the surface treatments, can be highlighted with microscope observations. A collection of scanning electron microscope (SEM) micrographs is shown in the present chapter. The shot peening treatment has been studied with numerical models and by applying the statistical method Design of Experiments (DoE) to optimise the treatment parameters. The proposed procedure could be used to select the best shot peening parameters for each application and bulk material.

INTRODUCTION

The opportunity of lightening mechanical components by broadening the use of light alloys is becoming a crucial task for the next future. In the automotive and aerospace fields, the request for more power with lower CO_2 emission is indeed expected to acquire more and more importance. The fatigue behaviour of light alloys, however, is often worse than that of steels, as well as the wear resistance, the surface hardness and the load bearing capacity. Therefore, suitable surface treatments can be decisive to achieve improved properties, even comparable with the ones of the traditional construction steels. The effects induced on components - made of both steels and light alloys - by treatments like thin hard coating deposition or shot peening have already been investigated, but further research developments are necessary, particularly if the fatigue resistance of the treated components is concerned.

The use of thin hard coatings in a wide range of technological applications is mainly due to the improvement of the tribological and corrosion-resistant properties of the coated components. Coatings deposited with physical vapor deposition (PVD), chemical vapor deposition (CVD) or, in case of decorative applications, electroless nickel plating, are usually not thicker than 10 μm. These surface layers are hence referred to as "thin films". The main benefits of using thin hard coatings versus thick hard coatings and soft coatings are the possibility of achieving high hardness and wear resistance without affecting the dimensional tolerances of the coated components, even in the presence of particular geometries such as sharp edges, the suitability for coating small components and the process capability to coat complex-shaped components. The machining tool and die industry was the first application field for PVD and CVD coatings about thirty years ago, but many other industries have had a huge concern in them from then on – e.g. aerospace, automotive, optical, electronics, chemistry, decorative, biomaterials.

Many research studies on the wear, corrosion [1-6] and mechanical characterization [7-10] of different thin hard coatings are available in the literature. A few studies have also demonstrated that the fatigue resistance of coated components can be enhanced by some coatings [11-18]. In particular, it is well known that fatigue is a phenomenon strictly related to the residual stress values at the component surface. The high surface compressive residual stresses that can be induced by the coating deposition process can prove effective for preventing or retarding fatigue crack initiation, provided that the coating is uncracked and defects-free. On top of evaluating the surface residual stresses, not only induced by the coating deposition but also by the manufacturing process or other surface treatments, the fatigue resistance of coated components also requires the investigation of the microstructure and of the interactions between the residual stresses and the microstructure. Moreover, in case of deposition of a thin hard layer on a mechanical component, the effects induced by the variability of the deposition parameters and the interactions between the base material and the film deposited should be taken into account carefully [19,20]. In the literature, studies dealing with the numerical simulation of the presence of thin hard coatings are available [21-24]. In general, despite the fact that it is not unusual to find studies, both experimental and numerical, on coated components, most of them are focused on the coating characterization and only a few ones on the prediction of the fatigue life with theoretical and/or numerical methods. In fact, further studies are required in the field of crack nucleation and growth simulation for thin hard-coated components in order to develop more and more reliable fatigue life prediction procedures. Suitable models for predicting the damage initiation and propagation in the presence of contact or rolling contact fatigue for coated components are even less frequent in the literature. In [12], Stewart et al. provide a thorough review of studies on the rolling contact fatigue behaviour in presence of thin coatings. With regard to the numerical studies, there are many references on the prediction of the number of load cycles until fatigue damage initiation or of the crack growth rate in mechanical components [25,26] such as meshing gears, but significant efforts have not been made yet to develop numerical or theoretical-numerical models to study the fatigue damage in the presence of thin hard coatings.

The authors have investigated the fatigue behaviour of thin hard-coated components and some of the most important results are reported and discussed in this chapter. Both experimental and numerical studies were carried out on different bulk materials and coatings and theoretical-numerical models enabling a designer to foresee the performance of a coated component under fatigue or contact fatigue conditions were proposed. In [17], a theoretical-numerical procedure for the fatigue life prediction of thin hard-coated components was developed and validated with good results by its application to an electroless nickel-plated component subjected to pulsating fatigue loads. In [14,15], the fatigue behaviour of 2205 duplex stainless steel, H11 tool steel and 6082 aluminum alloy was investigated with four-point bending fatigue tests carried out on both uncoated and CrN (chromium nitride) PVD-coated samples and the experimental results were compared with the numerical ones. Also, the mechanical characterization of the coating was performed with nano-indentation and the residual stress gradient was X-ray measured. With regard to PVD coatings, the effectiveness of thin hard films is related to several microstructural parameters closely connected with the deposition process conditions - i.e. the film texture, the grain morphology and orientation and the residual stress distribution. For the time being, titanium nitride (TiN) is still the most commonly used PVD coating in industrial applications, but also CrN, that shows better anti-oxidative and anticorrosion properties than TiN, has found wide application. Other good properties of CrN coatings are the lower friction coefficient and better toughness than TiN coatings [28]. Furthermore, the possibility of replacing electroplated hard chromium coatings with less environmentally harmful PVD CrN deposits constitutes one more good reason for the recent growing interest in them [29]. In [16], the number of cycles necessary to achieve specified crack depths in coated and uncoated steel and titanium transmission spur gears was numerically evaluated. CrN and TiN coatings applied on, respectively, steel and titanium base materials were considered and the surface micro-hardness and residual stress fields were considered in the models. The possibility of predicting the number of cycles until failure is very important for engine components or structures used in car or motorcycle competition applications or in the aerospace industry and, from this point of view, the proposed procedure could represent a powerful tool for machine designers. In the references [30-33], fatigue tests were performed on both steel and light alloy samples. Different PVD (WC/C and TiN) and CVD (SiO_x and DLC) coatings were deposited on bulk materials whose fatigue behaviour was compared with that without coating. SiO_x was deposited on both steel and aluminium alloys, WC/C and DLC on an aluminum alloy and TiN on a titanium alloy. SiO_x thin coatings, deposited by means of PE-CVD (Plasma Enhanced-CVD) technique in cold plasma, at extremely low pressures and temperatures [30], are suitable for many kinds of substrate and the main applications are those requiring high resistance to scratches and corrosion or a barrier for gas permeation, at both high and low temperatures. PVD WC/C coatings usually show good adhesion and low friction coefficient and can be applied on many materials and employed for structural applications [31]. PA-CVD (Plasma Assisted-CVD) DLC films show extremely low friction coefficient and good wear resistance. For the time being, research studies have been concentrated on the tribological and the mechanical characterization of WC/C and DLC coatings and not on their influence on the fatigue behaviour of aluminum substrates. PVD TiN thin hard-coated Ti-6Al-4V alloy was investigated in [32,33] both numerically and experimentally and the influence of the presence of both the coating and a notch was studied. With regard to the study of the contact fatigue behaviour of coated components, CrN PVD-coated and uncoated case hardened spur gears were investigated in [34] by means of numerical models and experimental tests. A theoretical-numerical procedure able to predict the number of cycles needed to promote fatigue damage initiation at the contact area was developed and the numerical results were compared with the experimental ones.

Shot peening is another surface treatment used to increase the mechanical strength of machine elements and structural components such as shafts, connecting rods or welded joints. In particular, shot peening proves effective in minimising or preventing the material damage from fatigue loading, fretting fatigue, stress corrosion cracking and hydrogen embrittlement. It is well known that such a treatment is performed by impacting a metal surface with hard shots at high speed. This generates surface plastic deformation and, consequently, residual stresses in the surface layers of the impinged material. Such stresses are compressive near the external surface of the target component and tensile in the inner layers. The benefits achieved with shot peening are mainly due to the compressive stresses [35-38]. The study of the relationship between the parameters characterizing the shot peening process and its effects on the treated component – e.g. the residual stresses or the deformation induced – and the development of previsional models to predict the fatigue life of the treated components represent very important issues for the research field. There are few references in which simplifying assumptions, made to achieve reliable solutions, do not enable a realistic simulation of shot peening [39-41]. In the last years, some numerical models able to foresee the residual stress field induced by shot peening were proposed [42-44]. In [44], accurate numerical models considering the collision of one or more spheres against specimens by varying the dimensions of the shot, the impact speed and angle and the target material were developed. Moreover, despite the fact that shot peening is used in a number of applications, empirical considerations or rules are often followed to select the treatment parameters. This can give rise to an improper choice of the levels for the most important factors characterizing the process. Baragetti et al. carried out studies [42,43] to optimise the shot peening parameters by applying an advanced statistical procedure – i.e. Design of Experiments (DoE) [45,46] – to the results obtained with numerical models and by introducing proper non dimensional parameters to achieve the best parameter selection. The use of non dimensional parameters was also made to allow the extension of the results collected to different shot peening applications.

EXPERIMENTAL RESULTS

The authors performed several experimental tests to evaluate the fatigue behaviour of thin hard-coated components and to characterize any difference with respect to the uncoated condition. Results from rotating bending and four-point bending fatigue tests were collected and allowed to evaluate the performance of each coating on the specific substrate which it was deposited on. Suitable micro- and nano-hardness measurements made it possible to characterize both the coatings and the coating-substrate systems after deposition. The residual stress distributions induced by the coating deposition processes were determined with X-ray diffraction technique. Furthermore, scanning electron microscope (SEM) fracture surface observations were carried out and scratch tests were performed to evaluate the coating adhesion. Before reporting the results of the fatigue tests, some details on the characterization of the coatings and the bulk materials investigated are reported.

In [14], PVD-coated 2205 duplex stainless steel specimens were subjected to four-point bending fatigue tests. Finite element (FE) models of the specimens were developed and the experimental results, in terms of crack nucleation and growth from the external surface, were used to validate the effectiveness of such models. A CrN coating was deposited on 2205 duplex stainless steel hot rolled plate (HB = 210) samples using a cathodic arc apparatus. The ultimate tensile strength and the yield strength of the base material were 730 MPa and 515 MPa respectively, the fatigue limit at 2 million load cycles with a probability of rupture equal to 50% was 510 MPa. Each sample was ultrasonically cleaned before coating deposition and the substrate temperature during the treatment was about 573 K. The coating thickness was measured with ball-cratering technique (Calotest) and was verified using SEM. Values in the range 4.5-5.0 μm were detected for this parameter. The nano-hardness values were obtained (ISO 14577) by means of a Vickers indenter with applied loads between 2 and 15 mN so as to minimize the bulk material influence. The coating Young's modulus was between 243 and 247 GPa from nano-indentation measurements and the surface hardness 1500 HV. This value was lower than the published data for PVD CrN coatings [7] and was partially related to the rather high number of droplets (Figure 1(a)) present on the coating surface which acted as discontinuity zones affecting the hardness detection. Microstructural and fractographical analyses were also performed to study the effects of the coating on the fracture mechanism. The coating adhesion was satisfying both during the fatigue tests and for specimens fractured using impact stresses imposed after a short period of immersion in liquid nitrogen (Figure 1(b)).

The residual stress values were determined by means of X-ray stress measurements carried out on a Rigaku diffractometer coupled with a two-dimensional detector using the $\sin^2\psi$ method and a CuKα radiation was employed with a 300 μm point collimator. Compression surface residual stresses ranging from 1.5 to 1.7 GPa were measured. Although residual stresses were not detected below the external surface, possible trends satisfying the force and moment equilibrium equations were assumed – and used in the numerical models too. In fact, regardless of the residual stress profile, the high value of the maximum compressive stresses was found to be the driving parameter influencing the fatigue behavior of coated components (see in the following).

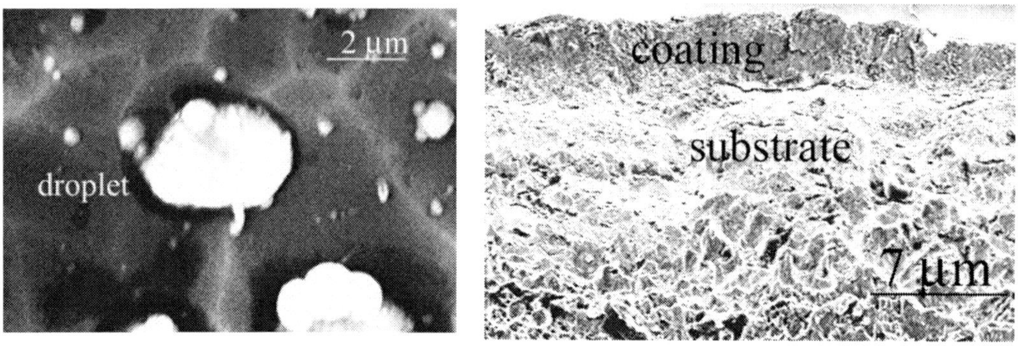

Figure 1 (a) Droplets on the surface of a coated sample and (b) impact stress fracture surface of a coated sample with shot peened substrate. Thcse CrN-coated 2205 steel specimens are the ones studied in [14].

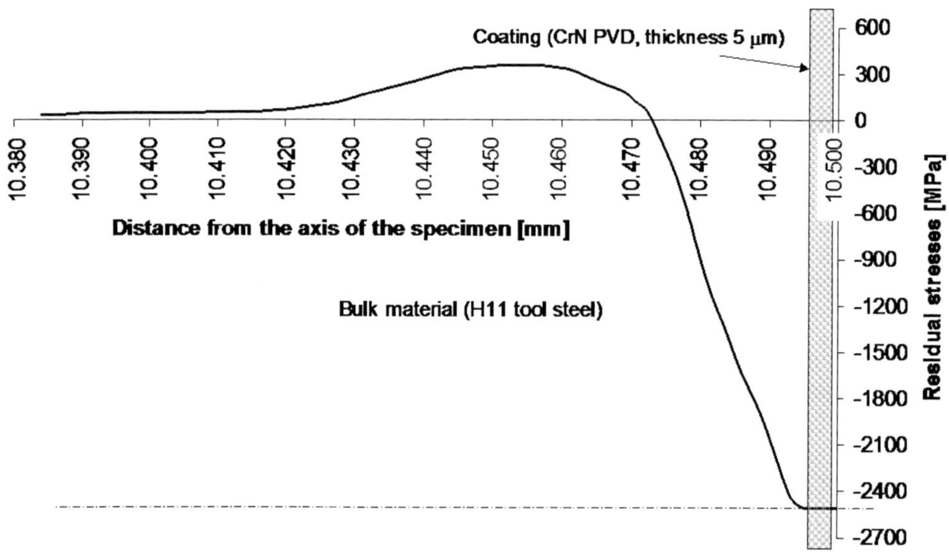

Figure 2. Residual stress trend in a CrN PVD-coated H11 tool steel sample with coating thickness equal to 5 µm [15].

Table 1. Chemical composition (wt.%) of the materials used as substrates in [15]

Material	C	Mn	Si	Cr	Ni	Mo	Mg	Al	Fe
H11	0.43	1.50	0.30	1.99	-	0.21	-	0.031	Bal.
2205	0.019	1.51	0.39	22.45	5.5	3.12	-	-	Bal.
6082	-	0.73	0.99	0.05	-	-	0.78	Bal.	0.46

In the reference [15], the fatigue behaviour of two different steels - i.e. 2205 duplex stainless steel and H11 tool steel - and of the 6082-T6 aluminium alloy was studied. Table 1 reports the chemical compositions of the three materials. Experimental tests on both uncoated and CrN PVD-coated samples were carried out. A cathodic arc apparatus was used for the coating deposition and the standard deposition temperature of 573 K was reduced to 483 K during the deposition on the 6082-T6 alloy. Furthermore, in order to avoid the overheating of the aluminium alloy and the consequent overaging, the film thickness was reduced, and the duration of the deposition process as well, to 2.5 μm – 5 μm-thick coatings were instead deposited on the two steel alloys. The experimental results were interpreted by means of microstructural characterization and residual stress gradient evaluation. Coating micro-hardness measurements gave results practically independent of the substrate material for the steel alloys - 2686 HV for H11 and 2567 HV for 2205, whereas a certain decrease was observed for the coated 6082 samples, i.e. 2238 HV. This variation in the coating hardness was probably due to the load applied with the indenter that, at a penetration depth below the external surface equal to 0.5 μm, resulted to be partially influenced by the hardness of the substrate material [47].

The residual stress distribution from the surface to the in-depth layers of the coated H11 steel was detected with X-ray diffraction measurements. In case of PVD CrN coatings having thicknesses in the order of 5 μm, the film deposition induces high surface residual compressive stresses that are quite constant through the coating thickness and in the range 1.2 to 3 GPa depending on the substrate material [15]. One of the residual stress trends measured by means of X-ray diffraction is depicted in Figure 2. Suitable two-dimensional images were obtained by means of a D/max RAPID Rigaku microdiffractometer in order to measure the texture, the lattice constants and the residual stress values characterizing the coating. With regard to the residual stress analysis, the DRAST method [48] – which allows to measure the residual stresses in the specimen from the evaluation of the displacement of the Debye ring – was used. Details regarding this procedure are also provided in the reference [49]. To evaluate the residual stresses, the following equation for biaxial stress analysis was used [49]:

$$\frac{d_{\phi\psi}-d_0}{d_0}=\frac{1+\upsilon}{E}\sigma_\phi\sin^2\psi-\frac{\upsilon}{E}\left(\sigma_{11}+\sigma_{22}\right) \tag{1}$$

where E is the elastic modulus calculated from nano-indentation measurements and υ the Poisson's ratio of the coating – 303 GPa and 0.2 for CrN respectively [15]. The residual compressive stresses in the coating and in the bulk material surface layers were very high – i.e. more than 2.4 GPa (Figure 2). The micro-hardness was measured to apply literature models able to predict the fatigue crack growth with hardened surfaces [13,50,51]. Figure 3 shows one of the micro-hardness trends measured by means of a nano-indenter equipped with a Vickers indenter in a PVD-coated H11 tool steel sample.

In [30], the authors studied the fatigue behaviour of the quenched and tempered 39NiCrMo3 steel alloy coated with a SiO_x coating deposited by means of PE-CVD technique. The bulk material chemical composition (wt.%) was: 0.395 C, 0.700 Mn, 0.240 Si, 0.018 P, 0.031 S, 0.740 Cr, 0.880 Ni, 0.190 Mo and Fe bal. Standard hourglass-shaped specimens, both coated and uncoated, were addressed to rotating bending tests (R = -1) to observe the changes occurring in the fatigue limit at 10^7 load cycles between the two series. Before the coating deposition process, cleaning operations were performed on the surface of the specimens to remove contaminants like dusts, grease and oxides and therefore to guarantee uniformity and good adhesion of the film. The fracture surfaces of the specimens were observed under a scanning electron microscope to evaluate any differences in the fatigue crack initiation and propagation between the coated and the uncoated samples. Scratch tests carried out on coated plane samples made it possible to evaluate the adhesion of the coating to the bulk material. In [31], similar tests were carried out on the 2011-T6 aluminium alloy both uncoated and coated with WC/C, DLC and SiO_x films deposited with PVD, PA-CVD (Plasma Assisted-CVD) and PE-CVD techniques respectively. The base material had the following chemical composition (wt.%): 5.00-6.00 Cu, 0.20-0.60 Pb, 0.20-0.60 Bi, 0.40 Fe, 0.20 Si, 0.10 Zn, 0.05 Mg, 0.05 Mn and Al bal. The ultimate tensile strength and the yield strength of 39NiCrMo3 were 1050 MPa and 910 MPa respectively, whereas 370 MPa and 272 MPa were found for 2011-T6. The scratch tests on the coated steel alloy (Figure 4), performed with a Rockwell indenter having a radius of 0.2 mm and by applying an incremental load from 0.1 to 10 N and from 0.1 to 20 N, showed some film elasticity. The behaviour of the coating during the tests was quite uniform and its adhesion to the substrate was reasonable – i.e. first and complete delamination loads up to 4.4 and 7.3 N. The results (incremental load from 0 to 30 N) for the aluminium alloy were fairly good only in the presence of the WC/C coating – first and complete delamination at, respectively, 5.6 and 9.6 N. By contrast, the very low hardness of the substrate was detrimental to the DLC film adhesion since the harder and brittle coating was not able to follow the aluminium alloy plastic deformation and soon failed during each test – first and complete delamination at 2.3 and 3.5 N respectively. Harder substrates are supposed to be more effective for this kind of coating [52-54]. The critical load values of the SiO_x film on the aluminum alloy were even lower than DLC – first delamination at 0.9 N and complete at 2.8 N.

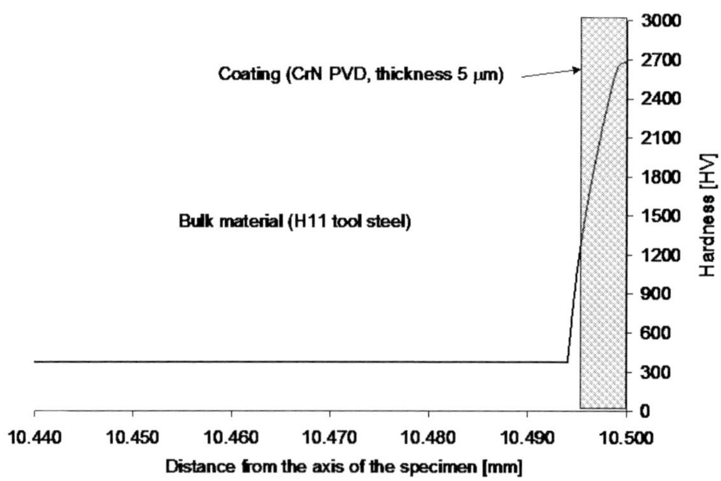

Figure 3. Micro-hardness surface trend in a PVD-coated H11 tool steel sample (coating thickness 5 μm) [15].

As far as the rotating bending tests are concerned, coated and uncoated 39NiCrMo3 [30] was tested at room temperature to calculate the fatigue limit at 10 million load cycles by means of the statistical Stair-case method [55]. The same tests were performed on the 2011-T6 aluminium alloy [31]. The dimensions of the standard hourglass-shaped specimens (ISO 1143) which were used for the rotating bending tests in [30] is shown in Figure 5. The specimens had a wide fillet radius and were accurately polished to achieve smooth surface finishing in the gage length. An average surface roughness below the value of 0.25 μm was achieved to prevent the fatigue behaviour from being influenced by this parameter [56]. The final polishing operations were performed lengthwise not to leave scratches along the specimen circumferential direction.

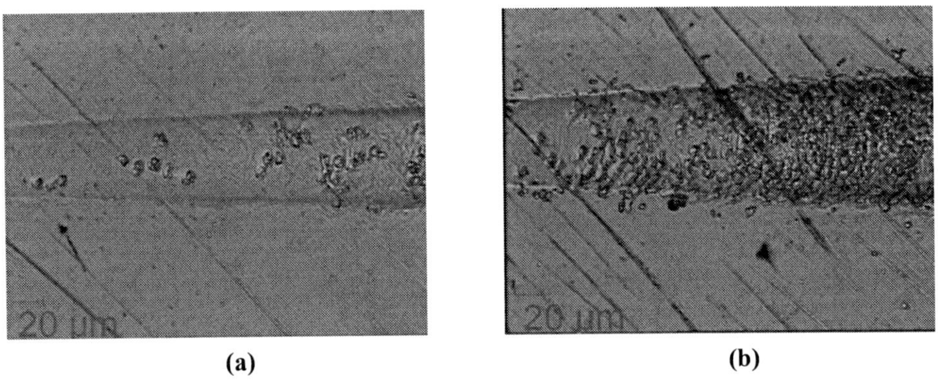

Figure 4. Scratch tests (load 0.1-10 N) on SiO_x-coated 39NiCrMo3: (a) first and (b) complete delamination [30].

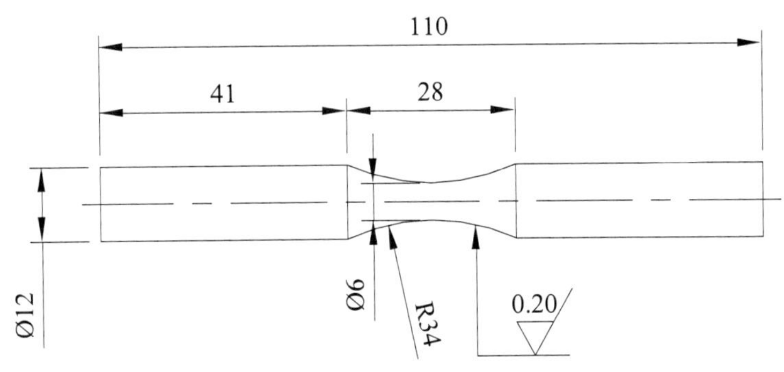

Figure 5 Nominal dimensions of the specimens used in the rotating bending tests in [30].

From the experimental data [30], an average fatigue limit of 538 MPa, with a standard error of 21 MPa, was calculated for the uncoated 39NiCrMo3 specimens, whereas 503 MPa, with a standard error of 23 MPa, was determined for the coated ones. Therefore, the coating was little detrimental to the base material fatigue resistance and the difference between the average values of the fatigue limits was about 10%. Therefore, the fatigue behaviour was not significantly affected by the presence of the SiO_x coating. As a result, for all the applications that require scratch- and corrosion-resistant properties and that must withstand relevant fatigue stresses, the choice of a coating like the one studied in [30] could be effective. The fatigue resistance of the 2011-T6 aluminium alloy base material was slightly increased by the presence of all the three coatings deposited [31]. In particular, the WC/C-coated specimens showed the best behaviour, with an enhancement in the average fatigue limit of about 10% with respect to the uncoated ones. The fatigue limit at 10 million cycles for the uncoated specimens was 154 MPa with a standard error of 6 MPa, 169 MPa and 15 MPa were, respectively, the same parameters calculated in the presence of the WC/C coating, 162 MPa and 6 MPa for the DLC-coated samples and 162 MPa and 7 MPa for the SiO_x-coated ones. The results highlighted that the use of suitably coated light alloys would make it possible to couple the benefits of the wear, friction and corrosion-resistant properties with good fatigue behaviour. The diagram depicted in Figure 6 shows each experimental observation from fatigue tests performed on the 2011-T6 coated and uncoated specimens.

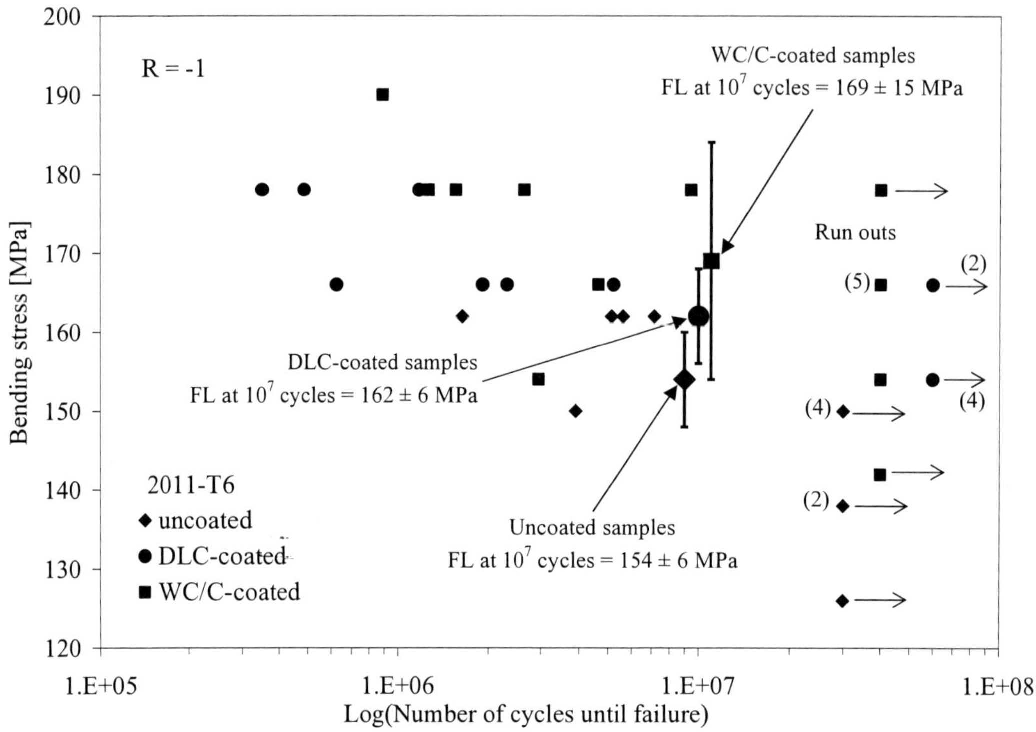

Figure 6. Rotating bending test observations for 2011-T6 coated and uncoated specimens [31].

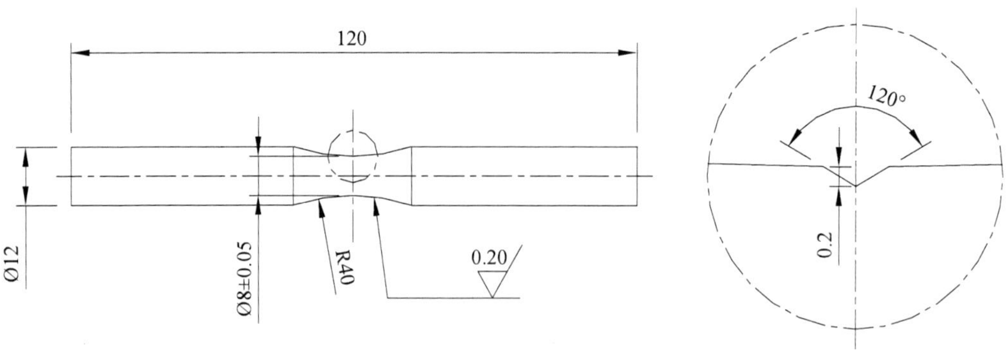

Figure 7. Dimensions of the Ti-6Al-4V specimens tested in [33].

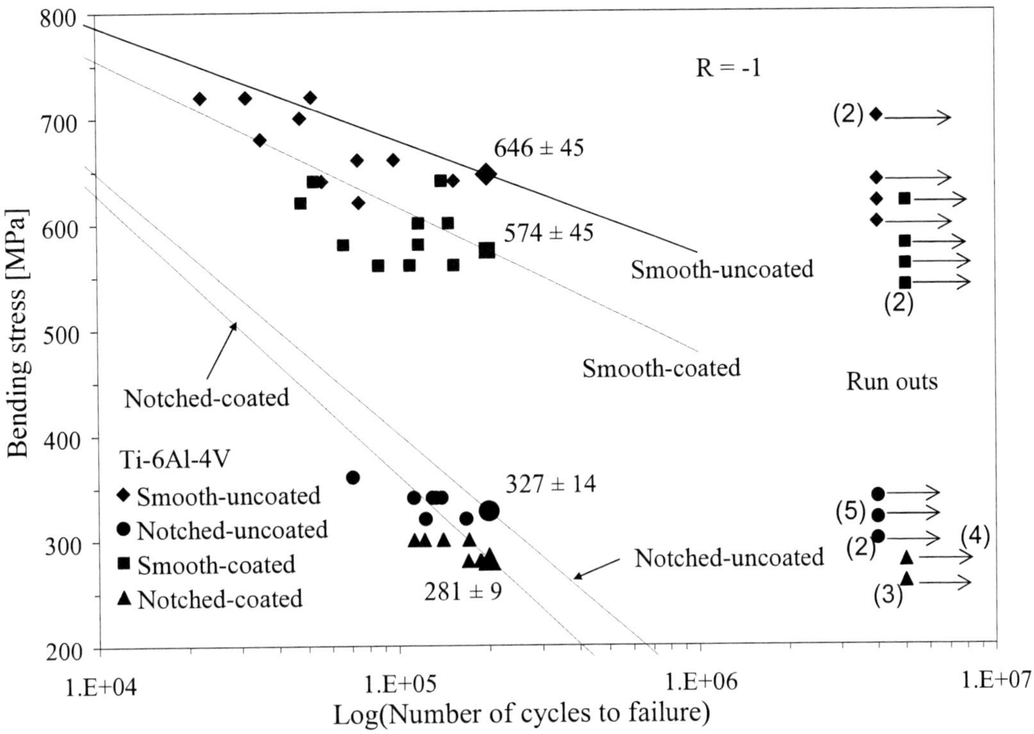

Figure 8. Bending stress vs. Number of cycles to failure diagram for Ti-6Al-4V specimens [33].

Further rotating bending tests were executed by the authors [32,33] on both notched and smooth specimens made of the Ti-6Al-4V titanium alloy. The effect of a PVD TiN coating on the fatigue behaviour of the titanium samples was studied. Also, numerical models were developed in order to calculate the stress distributions below the specimen surface over the minimum cross section and allowed to understand and interpret the experimental results [33]. The substrate ultimate tensile strength and the yield strength were, respectively, 895 and 829 MPa. Figure 7 shows the specimen dimensions and the location of the 120° V-shaped notches – 0.2 mm deep and transversal to the sample axis - which were machined at the minimum cross section in one half of the specimens. A commercial 3.9 μm-thick TiN arc-deposited PVD film was deposited on one half of the notched and of the smooth specimens at a nominal temperature of 420°C. Four series of fifteen specimens each, including smooth-uncoated, smooth-coated, notched-uncoated and notched-coated samples, were addressed to the tests. The bulk material micro-hardness was measured before and after the coating deposition and resulted, respectively, 271±4 $HV_{0.5}$ and 286±4 $HV_{0.5}$ [32]. These results pointed out that the heat treatment due to the deposition process could not have significantly affected the substrate mechanical properties. Rotating bending tests were carried out at a frequency of 50 Hz and at room temperature to calculate the average value of the fatigue limit at $2 \cdot 10^5$ load cycles for each series of specimens. The fracture surfaces were observed by means of an SEM [32] paying particular attention to the crack initiation sites.

The test results highlighted that both the notch and the coating were detrimental to the fatigue behaviour of the bulk material. The average values of the fatigue limit for the smooth-uncoated and the smooth-coated specimens were, respectively 646 and 574 MPa. With regard to the notched specimens, the same parameter was 327 MPa for the uncoated condition and 281 MPa in the presence of the coating. The laboratory observations collected throughout the fatigue tests as well as the fatigue limit calculated for each series of specimens are shown in the diagram of Figure 8.

It is worth underlining that the relatively limited decrease that occurred in the base material fatigue resistance in the presence of the coating was nearly the same between the smooth and the notched material – i.e. less than 12% and 14% for the smooth and for the notched specimens respectively. Therefore, similar behaviour between the smooth-coated and the notched-coated specimens with respect to the uncoated ones was observed through the tests. Previous results [57] confirm this low difference between the fatigue behaviour of uncoated and TiN-coated Ti-6Al-4V. The reduction in the fatigue limit for the notched titanium samples was also quantified with a fatigue stress intensification factor K_f representing the ratio between the average fatigue limits at 200,000 load cycles of the smooth and of the notched specimens. The values calculated for K_f were 1.98 for the uncoated samples and 2.04 for the coated ones. Elastic FE notched specimen models [33] allowed to determine an amplification stress coefficient of 3.70 which, compared to the lower K_f, emphasized that the notch tip plasticization due to the stress redistribution could have been one of the main parameters affecting the fatigue resistance [57,58]. The previously mentioned similarity in the fatigue behaviour between the two types of coated specimens was also pointed out by numerical modeling [33]. SEM analyses (Figure 9) showed that in the coated specimens the fatigue cracks could have initiated slightly below the bulk material external surface and propagated giving rise to a brittle fracture of the coating [32]. The titanium alloy could have developed subsurface fatigue cracks in the maximum tensile stress area resulting from the combination of residual stresses and load application.

As previously said, in [14,15] four-point bending fatigue tests on 2205 stainless steel and H11 tool steel and rotating bending tests on 6082-T6 aluminium alloy were carried out. Both uncoated and CrN PVD-coated samples were addressed to the tests. Suitable prismatic fatigue specimens were produced to test the 2205 duplex stainless steel (110x11x5.5 mm) and the H11 tool steel (110x21x4.4 mm), while standard hourglass-shaped samples were machined for the 6082-T6 aluminium alloy. A sinusoidal pulsating load was used for the four point bending tests and all the tests were carried out at room temperature. Figure 10(a) shows the special device, consisting in a pin connection mounted between the specimen and a universal test machine, assembled to run the four-point bending tests by applying symmetric loads to the specimens (Figure 10(b)). The Stair-case method was applied to the test results to calculate the fatigue limits at $2 \cdot 10^6$ load cycles for the steels and at 10^7 for the aluminium alloy. The average fatigue limit values and the standard errors of the uncoated H11, 2205 and 6082 specimens were, respectively, 730±26, 510±18 and 165±11 MPa. In the presence of the coating the same parameters resulted 845±43, 585±26 and 142±10 MPa. Therefore, the coating was beneficial to the fatigue properties of the two steel substrates – the increase in the fatigue limit was 16% for H11 and 15% for 2205 - and detrimental to the aluminium alloy – with a decrease of 11%. These results were also used to validate an analytical-numerical procedure for the number of cycles to failure prediction (see the next section).

Table 2 summarizes all the fatigue test results reported. It is evident how the choice of the most effective coating for a certain substrate is not trivial and a huge number of further fatigue tests should be performed to assess the best coatings for each base material. Anyway, on top of the tribological and corrosion-resistant properties, the results collected showed that remarkable results in terms of fatigue behaviour improvement can be achieved with coating deposition. It was believed that the compressive residual stress distributions induced in the substrate surface layers by the deposition processes of the most effective coatings – i.e. CrN on steel and WC/C on aluminium – resulted to be particularly effective in increasing the fatigue limits of the coated samples.

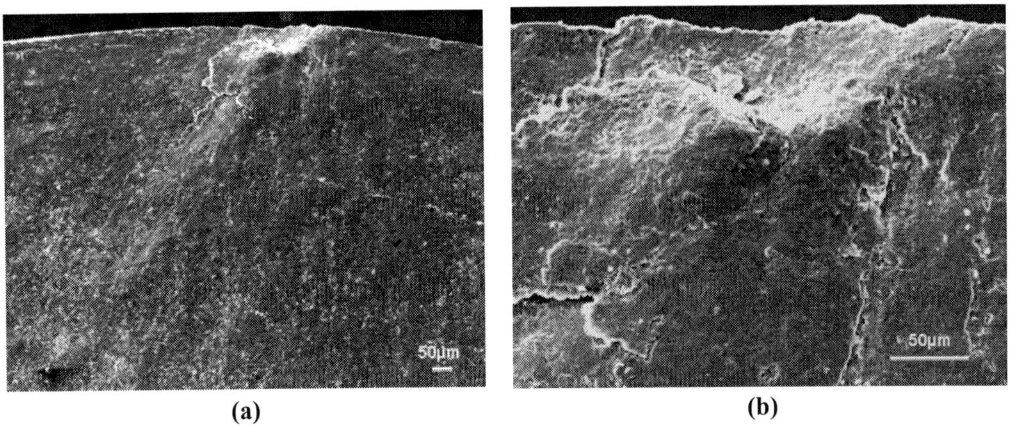

Figure 9. SEM micrographs: (a) fracture surface overview and (b) higher magnification of the crack initiation site of a smooth-coated Ti-6Al-4V sample [32].

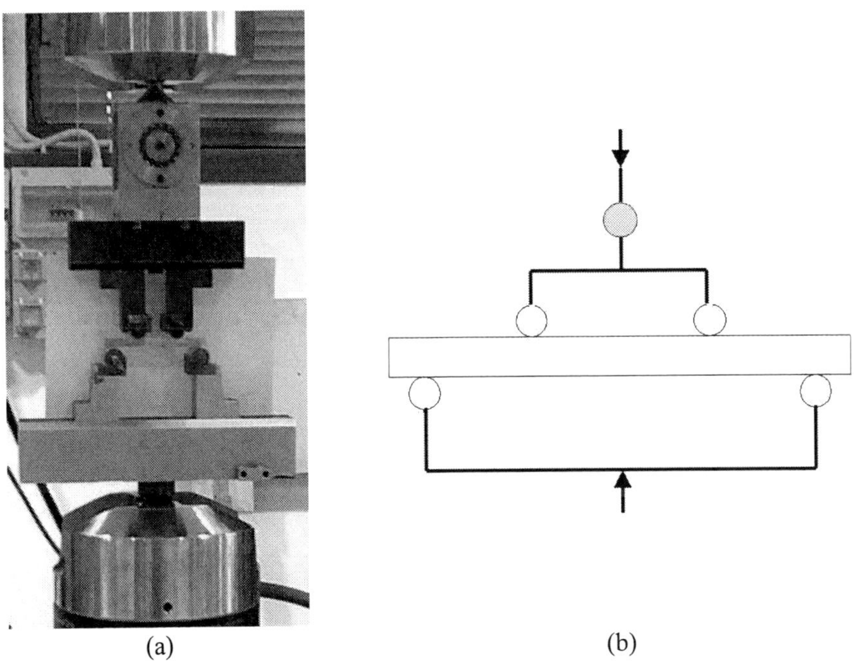

Figure 10. Equipment used for the fatigue tests reported in [14,15].

The possible presence of surface compressive residual stresses was highlighted by SEM observations of the fracture surfaces of WC/C-coated aluminium alloy samples [31]. Figure 11 shows a specimen tested at 190 MPa and failed at 890,178 load cycles. The fracture marking convergence towards the crack initiation site is visible in Figure 11(a), whereas the higher magnification of Figure 11(b) shows that subsurface fatigue crack initiation occurred within a depth range of 10-50 μm below the external surface.

Table 2. Fatigue test results for different coatings and base materials

Base material	Fatigue Limit, MPa			Coating	Δ %	Ref.
	Rotating bending		Four-point bending, $2\cdot10^6$ cycles			
	$2\cdot10^5$ cycles	10^7 cycles				
39NiCrMo3	-	538 ± 21	-	-		[30]
	-	503 ± 23	-	SiO_x	-10%	
2011-T6	-	154 ± 6	-	-		[31]
	-	169 ± 15	-	WC/C	+10%	
	-	162 ± 6	-	DLC	+5%	
	-	162 ± 7	-	SiO_x	+5%	
Ti-6Al-4V	646 ± 45	-	-	-		[32,33]
	574 ± 45	-	-	TiN	-12%	
Ti-6Al-4V (notched)	327 ± 14	-	-	-		
	281 ± 9	-	-	TiN	-14%	
H11	-	-	730 ± 26	-		[14,15]
	-	-	845 ± 43	CrN	+16%	
2205	-	-	510 ± 18	-		[14,15]
	-	-	585 ± 26	CrN	+15%	
6082-T6	-	165 ± 11	-	-		[14,15]
	-	142 ± 10	-	CrN	-14%	

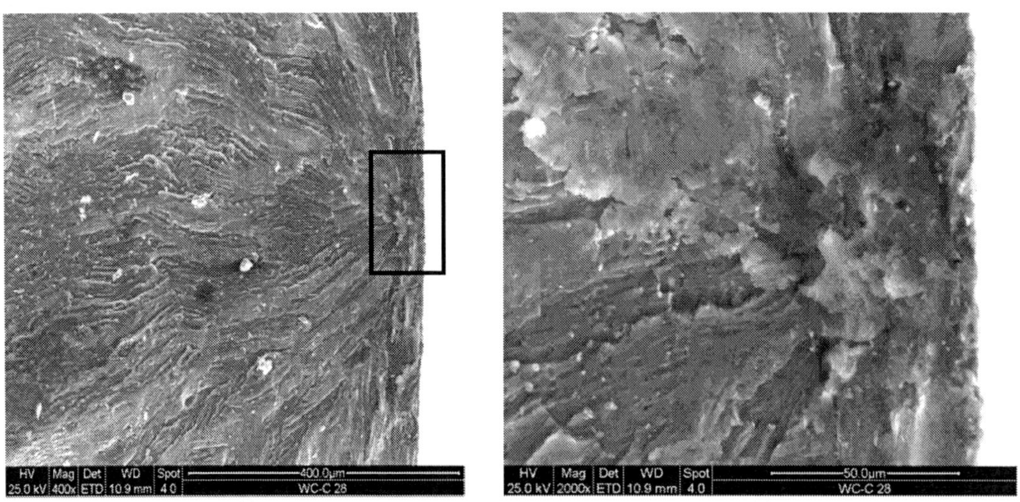

(a) (b)

Figure 11. SEM micrograph of the fracture surface of a WC/C-coated sample tested at 190 MPa and failed at 890178 cycles:: in (a) the crack nucleation site is indicated and in (b) a higher magnification of this area is shown [31]. The crack initiation is located about 10-50 μm below the surface.

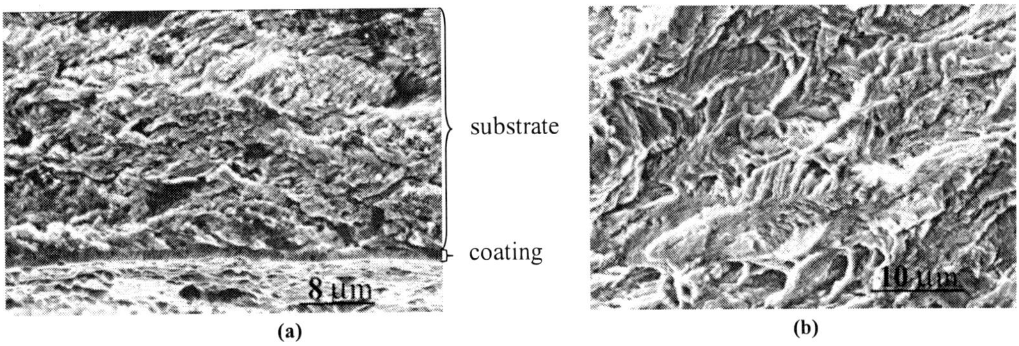

Figure 12. Micrographs of (a) the fracture surface of a CrN-coated 2205 sample and (b) fatigue striations in the substrate [14].

The analysis of the fracture surfaces reported in the reference [14] evidenced smooth fracture without fatigue striations in the coating, which showed compactness and no detachments from the substrate (Figure 12(a)). Figure 12(b) highlights the presence of fatigue striations both in the austenitic and in the ferritic grains of the 2205 duplex stainless steel.

From the test results reported in the reference [15], the CrN-coated steel specimens exhibited a fatigue life increment of 15-16%, whereas, despite the high level of the residual compressive stresses measured in the coating – not lower than 1.2 GPa in modulus, a decrement in the fatigue life was observed for the coated aluminium alloy (Table 2). In fact, it was observed that the nucleation of the fatigue cracks in the H11 steel alloy, which showed quite large non-metallic inclusions, took place only in areas sufficiently away from the interface (Figure 13). In such areas the effect of the compressive residual stresses induced by the coating deposition process was negligible. By contrast, in the coated 2205 duplex stainless steel, being characterized by a higher level of cleanness, the nucleation occurred in the coating. The application of cyclic stresses higher than the yield strength of the substrate was in this case necessary during the fatigue tests. Failure of the coating occurred in the 6082-T6 coated samples: the thin hard film cracked in multiple points due to the impossibility to follow the strains imposed on the overaged aluminium alloy base material. Figure 14 shows a comparison between fracture surfaces of coated 2205 and 6082-T6 samples.

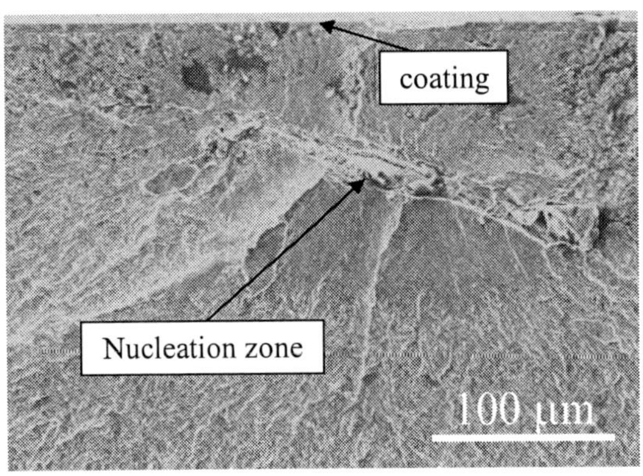

Figure 13. Crack nucleation from a large subsurface non metallic inclusion in an H11 coated sample [15].

Tests on CrN PVD-coated automotive transmission spur gears were carried out [34,60] to study the fatigue and contact fatigue behaviour in the presence of a thin hard coating. Figure 15 shows the device developed to perform these tests. By introducing a cylindrical roller having a diameter of 8 mm between each couple of adjacent teeth, the application of a load perpendicular to the tooth flanks was possible. The device made it possible to keep the gear fixed while the pulsating load (R = 0) was being applied through the roller along the vertical direction. Both CrN PVD-coated and uncoated gears were tested.

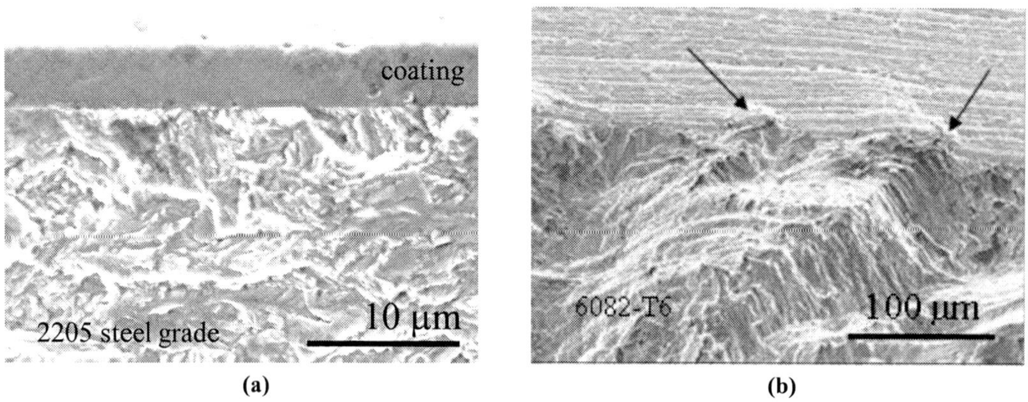

Figure 14. Fracture surfaces of (a) 2205 and (b) 6082-T6 coated samples (the arrows point at fatigue crack nucleation sites) [15].

Figure 15. Device for bending and contact fatigue tests on spur gear teeth [34,60].

The gear bulk material was case hardened 18NiCrMo5 steel alloy and the CrN-PVD arc-deposited coating was about 10 μm-thick. The influence of the coating on the base material fatigue behaviour was studied experimentally and theoretical-numerical models were also developed to predict the number of load cycles necessary for a crack located at the tooth base to propagate from a certain depth to another and for the contact fatigue damage initiation to occur over the contact area. All the tested tooth flanks were observed under optical and scanning electron microscopes. In the reference [34], the spur gears were tested under contact fatigue without lubrication on a computer-controlled universal testing machine at room temperature and at a test frequency of 20 Hz. The tests were also planned to set up the mentioned theoretical-numerical model able to foresee the number of cycles until the first contact fatigue damage on the tooth flank. The bulk material yield strength and ultimate tensile strength were, respectively, 690 MPa and 828 MPa [34]. The case hardening depth was in the interval 0.5-0.6 mm and the surface hardness was 560 HV and 700 HV before and after the coating deposition respectively. The gear module was 2.95 mm, the number of teeth 24, the pitch diameter 70.80 mm and the tooth width about 13.34 mm. The 8-mm cylindrical roller diameter was selected so that the contact with the tooth flanks was very close to the outer point of single teeth pair engagement, which is in the most critical contact loading area during meshing. Figure 16 shows a sketch of the way the load was applied. Table 3 reports the six load levels used in the tests with the resultant normal force on the tooth flanks at each load level and the corresponding maximum contact Hertzian pressure in the case of steel-on-steel contact without friction. The material of the cylindrical roller - i.e. hardened 100Cr6 - had higher strength and hardness than the gear material.

Table 3. Test loads on tooth flanks [34]

Load, kN	F, kN	p_{Hertz}, MPa
0.625	0.56	671
1.25	1.13	949
2.5	2.26	1343
5	4.51	1899
10	9.03	2685
15	13.54	3289

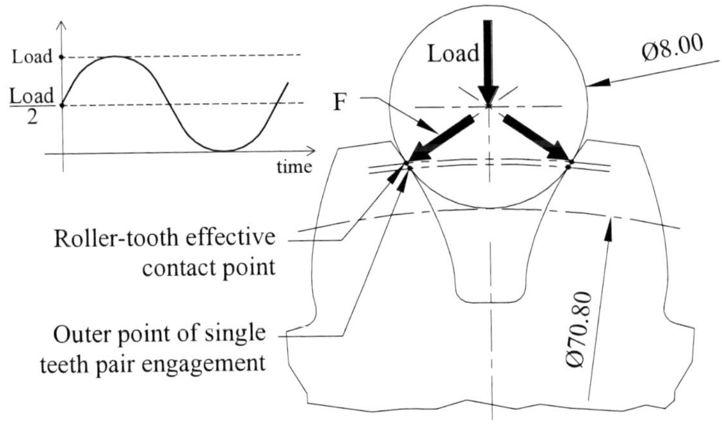

Figure 16. Sketch of the contact load application in the tests reported in [34].

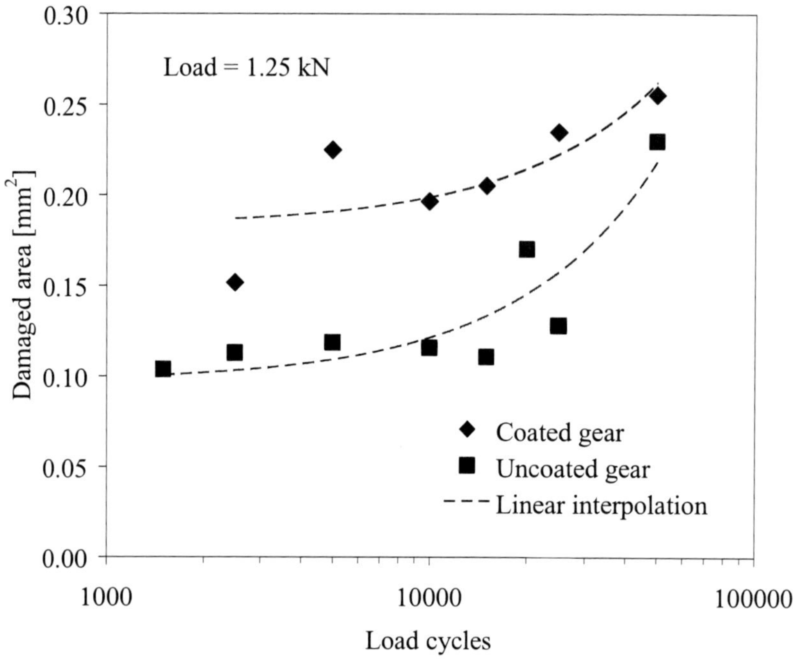

Figure 17. Damaged area vs. Number of load cycles diagram at a constant load of 1.25 kN for the gears tested in [34].

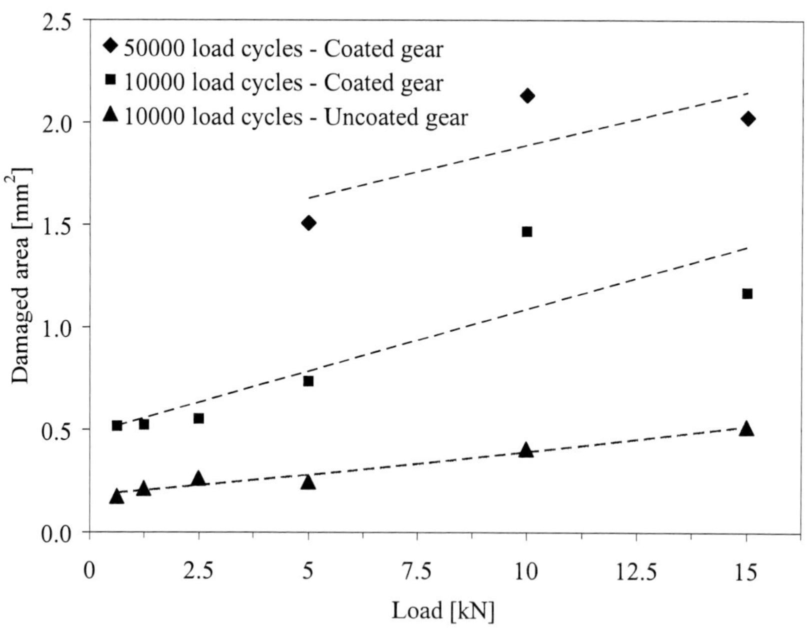

Figure 18. Damaged area vs. Load diagram at constant numbers of load cycles for the gears tested in [34].

Figure 19. (a) Magnification (2x) of a CrN-coated tooth flank tested under contact fatigue (R = 0) at 10 kN and 50000 load cycles and (b) SEM micrograph of an incipient spall formation in a CrN-coated tooth flank tested at 1.25 kN and 50000 load cycles [34].

The damage observed after the tests on the tooth flanks was due to wear, contact fatigue, coating failure and/or delamination. The damaged area on each tooth flank was measured, in terms of mm^2, by image analysis of the optical microscope observations. In Figure 17 the extension of the damaged areas on the tooth flanks vs. the number of load cycles diagram at a test load equal to 1.25 kN for both coated and uncoated gear is depicted, while the damaged areas at constant numbers of load cycles vs. the applied test load diagram is shown in Figure 18. Both the diagrams emphasize that the coated gears had a worse behaviour with respect to the uncoated ones. Furthermore, a certain linearity in the relationship between the damaged areas and both the number of load cycles and the load level was observed. Large coating detachments occurred on the tooth flanks at the highest load levels – namely 10 or 15 kN – and even at low loads and numbers of load cycles the coating showed damage or delamination. This confirmed that lower thicknesses or PVD coatings other than CrN could prove more effective against contact or rolling contact fatigue as reported in [12]. It was also pointed out that the damaged areas in the coated gears were much larger than in the uncoated ones and this could have been due to pure coating failure.

The relationship between the tooth damage and the load level for the uncoated gears (Figure 18) was more uniform than for the coated ones. This suggests that the high residual stresses induced by the deposition process might have been detrimental in the presence of contact fatigue, although it is important to underline that, with regard to the highest loads (namely 10 and 15 kN), the coating could have failed statically (Table 3). The damaged area on a coated tooth flank tested under one of the most severe test conditions – i.e. at 10 kN and 50,000 load cycles – is shown in Figure 19(a). It is evident that the coating delamination occurred with large pieces of coating detached from the bulk material. A spall formation in a coated gear is instead shown in Figure 19(b). The inclination of the spall forming crack suggests that the presence of the CrN coating did not affect the typical damage mechanisms – pitting or spalling – which can take place under rolling contact fatigue in uncoated gears [25,26,61,62]. SEM observations also showed that the initial damage occurred at or close to the external surface for both the uncoated and the coated gears. This could have mainly been due to the high friction from the unlubricated test condition.

(a)

(b)

Figure 20. (a) Appearance of the contact area of an uncoated spur gear and (b) of a CrN-coated one tested under contact fatigue (R = 0) [60].

In the reference [60], the microscope analysis and the liquid penetrant inspection of the tested gears pointed out that fatigue crack propagation under bending loads did not occur at tooth base on both the uncoated and the coated gears. This was probably due to the absence of cracks in the most stressed area of the teeth – i.e. at the tooth base. Figure 20(a) shows an uncoated tooth flank after high load tests, whereas Figure 20(b) shows a damaged surface of a CrN-coated gear. Large coating delamination occurred and, due to a possible lower oxidation, the colour of the coated tooth flanks was darker than that of the uncoated ones.

It is authors' opinion that further studies, both experimental and numerical, on the influence of the coating thickness and elastic modulus values on the rolling contact fatigue could give valuable results for many applications in which gears are used. These studies have already been addressed as part of the authors' future research activities.

NUMERICAL MODELS

The authors et al. have developed numerical and theoretical-numerical models to study the fatigue and the contact fatigue behaviour of both specimens and components under prescribed loading conditions. These models take into account the residual stress distributions induced by the coating deposition process or other surface treatments and allow to predict the number of load cycles necessary for a crack to propagate under fatigue or for the initial damage to occur under contact fatigue.

A theoretical-numerical procedure was proposed in [14-17] with the aim of providing a designer with a powerful tool to predict the fatigue behaviour of thin hard-coated components. By introducing the residual stress field due to the coating deposition process in the finite element models and the hardness values of the surface layers of the coated component in the calculation procedure, a prediction of the number of cycles until failure can accurately be made. The model, once the initial crack depth from the external surface is known, allows to calculate the stress intensity factor, including the effects due to the residual stresses, and to simulate the crack propagation (in a discrete way) – therefore to evaluate the number of cycles necessary to reach specified crack depths until the component failure. The hypothesis of absence of coating delamination – which is quite realistic because of the progress made by the current deposition technique – is assumed to apply the procedure. A commercial code can be used for the calculations and mesh refinement need to be performed at cracked area only. In [17], the suitability of the procedure was confirmed by its application to an electroless nickel-plated (Ni-P) steel component. The model was applied to study the fatigue crack propagation in a load cell, mounted on a double-acting hydraulic actuator exerting pulsating loads up to 100 kN at a maximum test frequency of 1 Hz. An electroless Ni-P coating was deposited on the load cell to increase the substrate corrosion resistance. In order to develop reliable FEM models able to predict the number of load cycles necessary for a certain crack to propagate, suitable methods for the evaluation of the threshold stress intensity factor range and models for describing the crack growth rate above the threshold were applied. Microstructural fracture mechanics (MFM) was used to evaluate the threshold stress intensity factor range in the presence of a crack. El-Haddad et al. [63] proposed the following relationship to calculate the threshold stress intensity factor range:

$$\Delta K_{\mathrm{th}} = \Delta K_{\mathrm{th,Ic}} \sqrt{\frac{a}{a+a_0}} \quad (2)$$

where a is the crack depth, $\Delta K_{\mathrm{th,Ic}}$ the threshold stress intensity factor range for long cracks expressed in MPa√m and a_0 is given by the following equation:

$$a_0 = \left(\frac{\Delta K_{\mathrm{th,Ic}}}{\beta \cdot \Delta\sigma_{\mathrm{w0}}} \right)^2 \frac{1}{\pi} \quad (3)$$

where β is a stress intensity correction factor and $\Delta\sigma_{\mathrm{w0}}$ is the fatigue limit of the material expressed in MPa. The parameter a_0 in equation (3) represents the maximum crack depth for which short cracks are not influenced by microstructural barriers. In the presence of crack depths lower than a_0, microstructural barriers influence the ΔK_{th} value and equation (2) cannot be used. Such a relationship can be used only if a is not higher than $3a_0$-$5a_0$. Another approach was the one proposed by Murakami and Endo [64]. Based on the collection of experimental data for three-dimensional cracks and short cracks, this model allows to calculate the threshold stress intensity factor range as follows:

$$\Delta K_{\text{th}} = 3.3 \cdot 10^{-3}\left(\text{HV} + 120\right)\left(\sqrt{\text{Area}}\right)^{1/3}\left(\frac{1-\text{R}}{2}\right)^{\psi} \tag{4}$$

$$\psi = 0.226 + \text{HV} \cdot 10^{-4} \tag{5}$$

where the parameter √Area (expressed in μm) may be assumed to be equal to the crack depth, HV is the material Vickers hardness and R the ratio between minimum and maximum stress in the fatigue cycle. The model applicability is limited to crack lengths in the range from $3a_0$ to $10a_0$, whereas ΔK_{th} is constant if $a > 10a_0$. The crack growth rate under an MFM regime can then be evaluated by using the following equation [65]:

$$\frac{\text{d}a}{\text{d}N} = A \cdot \Delta\gamma_{\text{p}}^{\alpha}\left(d - a\right) \tag{6}$$

where A and α are material constants, γ_{p} is the plastic component of the shear strain and d is the distance between the microstructural barriers expressed in mm. Paris' [66] or Elber's [67] models can be used to estimate the crack growth rate above the threshold:

$$\frac{\text{d}a}{\text{d}N} = C \cdot \Delta K_{\text{eff}}^{m} \tag{7}$$

where ΔK_{eff} is the effective stress intensity factor range, C and m are material constants. In the present model, different and more recent fatigue crack growth rate equations – derived for spur carburized gears applications [13,50,51] – were used:

$$\frac{\text{d}a}{\text{d}N} = \frac{C}{(1 - \rho^{n})}\left(\Delta K^{n} - \Delta K_{\text{th}}^{n}\right) \quad \text{for} \quad \Delta K_{\text{th}}^{n} \le \Delta K \le K_{C} \tag{8}$$

$$\frac{\text{d}a}{\text{d}N} = \frac{C}{1 - \rho^{n}}\left(\frac{\Delta K^{n} K_{\text{Ic}}}{\Delta K_{\text{Ic}}^{n} - \Delta K^{n}}\right) \quad \text{for} \quad K_{C} < \Delta K < K_{\text{Ic}} \tag{9}$$

The parameters included in equations (8) and (9) are calculated as follows:

$$\rho = \frac{\Delta K_{\text{th}}}{K_{\text{Ic}}}$$

$$K_{C} = \sqrt{\Delta K_{\text{th}} K_{\text{Ic}}}$$

$$A = -\frac{1}{d_2^2}\ln\left(\frac{H_1 - H_3}{H_2 - H_3}\right) \quad \text{for} \quad d^{*} \le d_2$$

$$A = -\frac{1}{(d_c - d_2)^2} \ln\left(\frac{550 - H_3}{H_2 - H_3}\right) \quad \text{for} \quad d^* > d_2$$

$$H = (H_2 - H_3)\exp\left[-A(d^* - d_2)^2\right] + H_3$$

$$\Delta K_{th} = 2.45 + 3.41 \cdot 10^{-3} \cdot H$$

$$K_{Ic} = 141 - 1.64 \cdot 10^{-1} \cdot H$$

$$n = 4.31 - 8.66 \cdot 10^{-3} \cdot H + 1.17 \cdot 10^{-5} \cdot H^2$$

$$\text{Log}(C) = -10.0 + 1.09 \cdot 10^{-2} \cdot H - 1.40 \cdot 10^{-5} \cdot H^2$$

where K_{Ic} is the material fracture toughness expressed in MPa√m, d_c the thickness of the coating in mm, d_2 and d^* the depth, in mm, where the hardness achieves the maximum value and from the external surface respectively, H_1, H_2 and H_3 the surface, the maximum and the core Vickers hardness respectively. Once the hardness trend in the coating and in the bulk material is known, equations (8) and (9) are extended to thin hard-coated components.

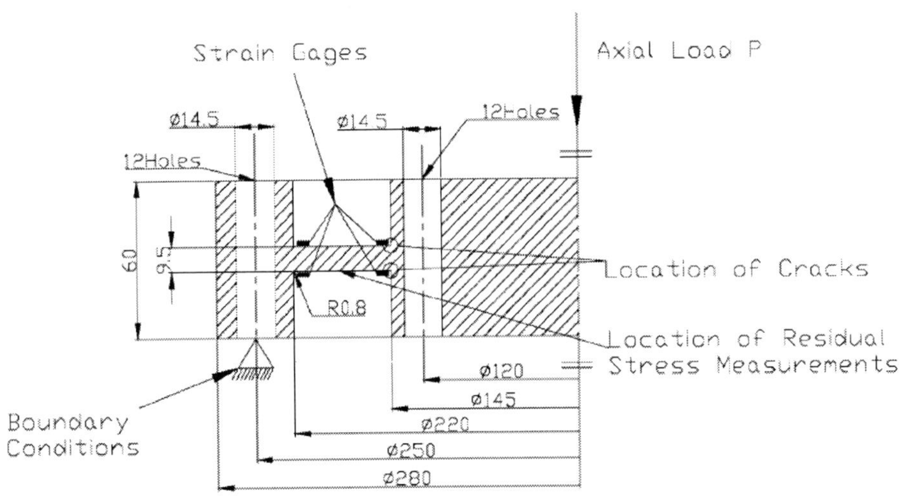

Figure 21. Sketch of the load cell studied in [17].

Figure 21 shows a sketch of the load cell studied in [17], consisting of a circular plate on which the cyclic load was applied in the central part. The outer part is attached to the actuator by means of bolts. Axial and circumferential strain gages allowed to measure the principal strains in the most stressed areas and to check the actuator loads and displacements. The load cell bulk material was forged, quenched and tempered Ni-Cr-Mo steel having the following chemical composition (wt.%): 0.37 C, 0.67 Mn, 0.26 Si, 0.80 Cr, 0.82 Ni, 0.17 Mo, 0.003 S, 0.006 P, Fe bal. The 5 μm-thick coating was deposited by means of an autocatalytic chemical reduction of nickel ions by hypophosphite compounds. The Vickers micro-hardness profile from the coated external surface is depicted in Figure 22.

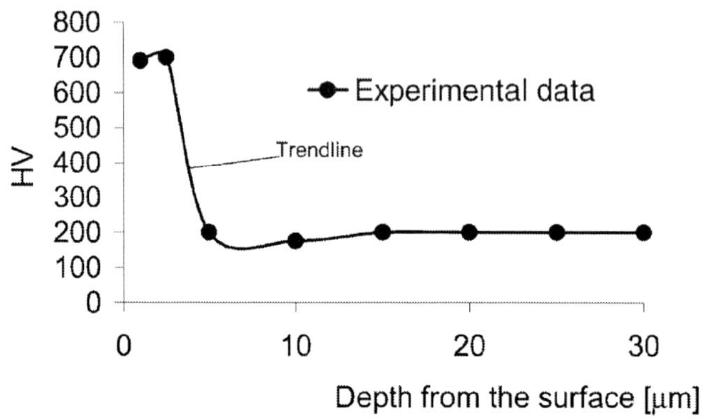

Figure 22. Micro-hardness profile from the surface to the in-depth layers in the load cell coated material tested in [17].

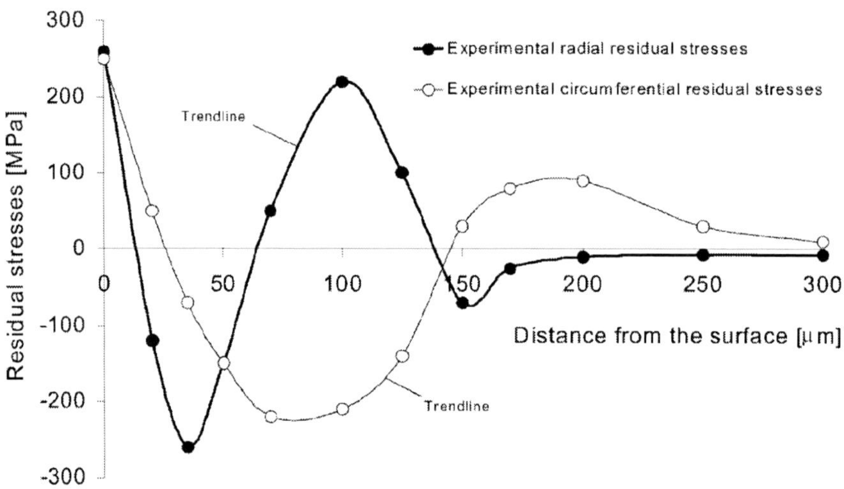

Figure 23. Residual stress trends from the surface to the in-depth layers in the load cell coated material tested in [17].

The circumferential and the radial residual stresses at the external surface of the coating were measured using the X-ray diffraction method, whereas the subsurface distribution was obtained down to 300 μm by using the electropolishing method. The residual stress trends along the circumferential and the radial directions are shown in Figure 23. The surface of the coating was characterized by the presence of tensile stresses higher than 200 MPa for both the two directions. Moreover, it was argued that the whole coating was subjected to residual tensile stresses due to its very low thickness. Figure 23 also shows the steep stress gradients that were measured below the external surface level. The residual stresses became negligible at a depth of more than 200 μm from the external surface. Good adhesion between the coating and the substrate was observed using SEM. High brittleness of the Ni-P layer was pointed out with tensile tests on samples cut from the component far away from the fatigue cracks so as to exclude the presence of strains due to failures occurred during the load cell working. In fact, the deposit was affected by a few cracks even at low applied load levels – i.e. at about 20 MPa. This suggested that the nucleation had started as a consequence of the failure of the brittle coating – reasonably related to the low content of phosphorous in the coating itself. As a matter of fact, a phosphorous content of 11% (wt.) is considered as the threshold above which the residual stresses become compressive [68].

Both axisymmetric and three-dimensional finite element models were then developed to simulate the presence of circumferential and surface defects by modeling semi elliptical cracks. The submodeling technique was also used to simulate, on account of the very low coating thickness, the presence of short cracks in the surface layers of the component material. This technique allowed to refine the mesh by extracting a portion - i.e. a submodel - of the whole (global) model and by imposing on its boundaries the same displacements occurring in the global model. As a result, the accuracy of the calculations was considerably increased. The analyses were processed in the pure elastic field by assuming the limited plasticization at the crack tip to be negligible. Suitable contact elements were used between the crack mating faces to prevent any possible self penetration. The residual stress fields were introduced into the models as initial conditions at the integration points of the external layers of finite elements. A Young's modulus equal to 200,000 MPa was used for the Ni-P coating [17]. The stress intensity factor for each crack modeled was then evaluated by means of the half relative displacements u of the crack faces:

$$K_{\mathrm{I}} = \frac{2E}{1+\nu}\sqrt{\frac{2\pi}{r}}\,\frac{u}{f(\theta)} \tag{10}$$

where:

$$f(\theta) = (2\chi + 1)\sin\left(\frac{\theta}{2}\right) - \sin\left(\frac{3\theta}{2}\right)$$

$$\chi = 3 - 4\nu \quad \text{for plane strain state}$$

$$\chi = \frac{3 - \nu}{1+\nu} \quad \text{for plane stress state}$$

With regard to the load cell axisymmetric model, equation (4) was used to evaluate the ΔK_{th} value and 1.70 MPa$\sqrt{m}$ was for a crack depth equal to the Ni-P layer thickness. To justify the axisymmetry, the presence of the holes in the component (Figure 21) was neglected in this case since they did not have any influence on the stress trends at the notched areas. Bi-quadratic, quadrilateral finite elements were used and mesh refinement was generated at the notched internal area where the modeled fatigue crack was placed. Due to the axisymmetry of the model, this was a 360°-circumfertial crack around the load cell. Such a modeling method for cracks in brittle coatings was supported by the fact that the deposited brittle Ni-P layer was observed to crack after few load cycles at low applied loads around the whole component. A 5 μm-deep crack passing through the thickness of the nickel layer was placed in the model and different analyses were carried out with axial loads ranging from 60 to 100 kN. The relative displacements of the crack flanks – measured on the nodes near the crack tip – and the calculated ΔK values are listed in Table 4 for different applied loads. These results showed that an axisymmetric crack nucleated in the surface nickel coating could propagate for axial loads higher than 72 kN.

Table 4. Numerical crack flank relative displacements (2·*u*) with axisymmetric model and related stress intensity factor ranges (crack depth of 5 μm) in the load cell FEM models discussed in [17]

Load, kN	$2 \cdot u$, mm	ΔK, MPa√m
60	$2.619 \cdot 10^{-5}$	1.23
72	$2.918 \cdot 10^{-5}$	1.55
84	$3.221 \cdot 10^{-5}$	1.80
100	$3.620 \cdot 10^{-5}$	2.03

Table 5. Numerical stress intensity factor ranges with 3D cracks (axial propagation direction, R = 0) in the load cell FEM models discussed in [17]

Crack depth, μm	ΔK, MPa√m			
	60 kN	72 kN	84 kN	100 kN
20	5.30	6.36	7.43	8.84
40	6.29	7.71	9.13	11.02
100	11.10	13.32	15.54	18.49
500	16.90	20.35	23.82	28.42

Table 6. Numerical stress intensity factor ranges with 3D cracks (circumferential propagation direction, R = 0) in the load cell FEM models discussed in [17]

Crack depth, μm	ΔK, MPa√m			
	60 kN	72 kN	84 kN	100 kN
20	4.60	5.52	6.44	7.66
40	4.50	5.40	6.30	7.49
100	10.00	12.00	14.00	16.66
500	9.70	11.66	13.62	16.23

The good bonding between the Ni-P coating and the steel substrate was highlighted with microscope observations. Owing to the brittle failure of the Ni-P deposit, cracks nucleated and propagated in the bulk material in several areas of the notched section of the load cell. It was argued that these defects, probably present at the coating-bulk material interface, propagated because of the steep stress gradient present in the notched area. The simulation of the crack propagation in the base material after the brittle fracture of the coating was then performed by modeling three-dimensional semi elliptical cracks having an aspect ratio equal to 0.5. A suitable portion of the load cell was modeled and symmetrical boundary conditions were imposed on the cutting planes. Cracks having depths equal to 20, 40, 100 and 500 μm were simulated and the calculated threshold stress intensity factor ranges ΔK_{th} – using equation (4) with R equal to 0 – were, respectively, 2.78, 2.89 and 3.20 MPa√m (this last value was the same for the 100 and the 500 μm-deep cracks). The circumferential and axial propagation directions were analyzed and the results, in terms of stress intensity factor range ΔK for different axial loads applied to the load cell, are reported in Tables 5 and 6.

Based on the numerical data, it was pointed out that the stress intensity factor range was nearly the same regardless of the propagation direction – limited differences were calculated for crack depths lower than 100 μm. As a result, the radial and the circumferential propagation patterns were assumed not to produce changes in the crack aspect ratio. It was found that the crack could propagate at each load level, including the minimum axial load to which the load cell was subjected, since the stress intensity factor ranges were always higher than the threshold stress intensity factor ranges. Finally, equations (8) and (9) were used to determine the crack growth rate in the load cell bulk material. These equations require the knowledge of the stress intensity factor range, which was calculated, for each crack depth, by considering in the 3D FEM models both the presence of the residual stresses and the geometrical configuration of the component, and of the Vickers hardness trend from the external surface (Figure 22). Table 6 reports the number of load cycles necessary for the cracks to propagate from a certain depth to another one until the component failure – i.e. until the crack is propagated through the entire load cell thickness along the axial direction. The experimental 2,050,000 cycles until load cell failure with an axial load of 60 kN were in good agreement with the numerically calculated ones at the same load – i.e. 2,150,521 (Table 7). The small difference – less than 5% – between the numerical and the experimental results confirmed the reliability of the proposed procedure. The method could conveniently be extended to other kinds of hard coatings and components, provided that the surface modifications induced by the treatments are known – i.e. the residual stress and hardness distributions – and the coating adhesion is good.

In [14], the model was applied to the four-point bending fatigue tests on CrN-coated prismatic 2205 duplex stainless steel samples previouly discussed. FEM models of the specimens used in the experimental tests were developed and validated. Figure 24 shows a specimen finite element model with the boundary conditions and the applied load. Suitable symmetry boundary conditions were used and half of the sample length was considered along the x direction. Four-node plane stress finite elements were used to build up the mesh and adequate refinement was used at the cracked area. The 5 μm-thick hard CrN coating was simulated at the external surface of the modeled specimen by assigning the coating mechanical properties to the finite elements used to reproduce its presence – Young's modulus equal to 245,000 MPa and Poisson's ratio to 0.23 [28]. Based on the same assumption made for the discussed load cell model about the limited plasticized area at the crack tip, linear elastic behavior was used for both the coating and the base material even in this case. The hypothesis of absence of delamination between the coating and the bulk material, supported by the experimental evidence, was assumed in the models – i.e. same displacements for the nodes lying on the interface line. Self equilibrated residual stresses were applied to the model by means of a suitable temperature gradient imposed from the surface to the in-depth layers. The residual stresses along the x direction from X-ray measurements and the temperature profile used to reproduce these stresses in the models are shown in Figure 25. Contact "gap" elements were used between the nodes lying on the crack flanks in order to prevent any initial overlapping due to the high residual compressive stresses in the specimen surface layers. Figure 26 shows the effect of the presence of the gap elements on the stresses along the x direction.

Table 7. Number of cycles to component failure obtained with the proposed theoretical-numerical model for the load cell studied in [17]

Load, kN	Crack depth intervals				Total number of cycles to failure
	20-40 μm	40-100 μm	100-500 μm	500 μm-failure	
60	881,057	363,416	95,238	810,810	2,150,521
72	105,263	124,378	83,333	398,000	710,974
84	46,981	56,603	49,813	223,325	376,722

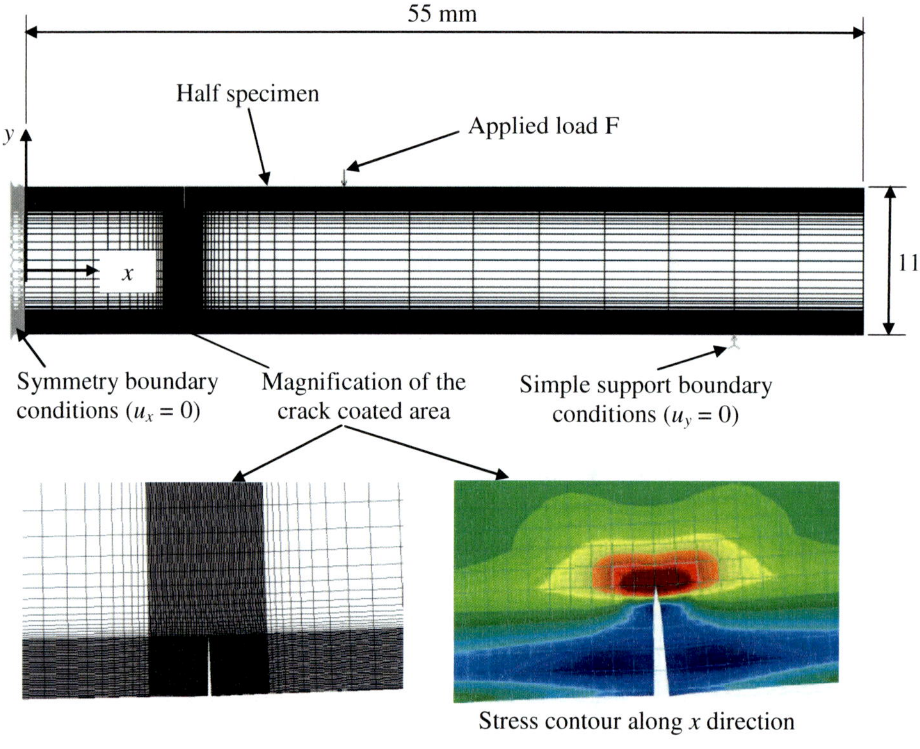

Figure 24. A 2205 duplex stainless steel specimen FEM model [14].

On top of avoiding the crack flank self penetration, the contact elements made it possible to determine the stress intensity factor K_{res} due to the residual stresses before the load application. The effective stress intensity factor range, which takes into account the presence of the residual stresses, was calculated with the following equation:

$$\Delta K_{eff} = K_I - K_{res} \tag{11}$$

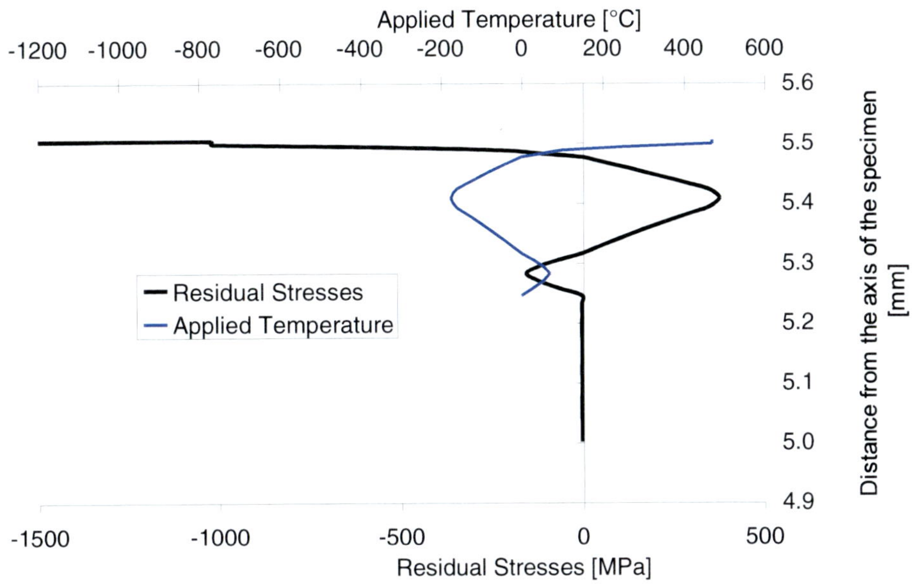

Figure 25. Measured residual stress trend and corresponding temperature profile applied to the 2205 specimen numerical models in [14].

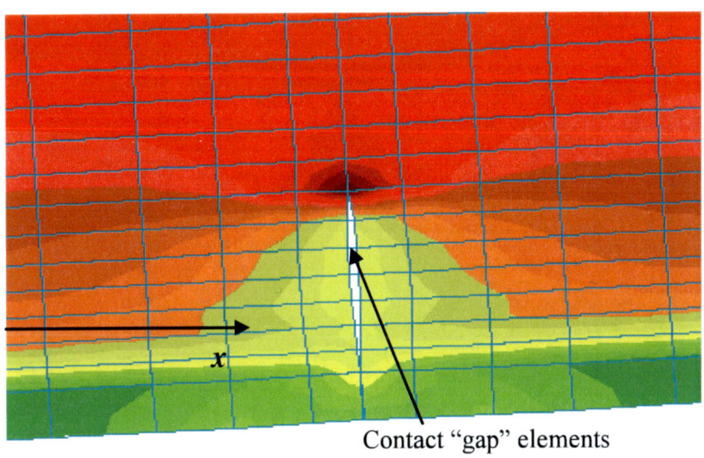

Figure 26. Effect of the gap elements between the crack mating faces on the stresses along the *x* direction in a 2205 specimen numerical model developed in [14].

Furthermore, a reduction of ΔK_{eff} resulting from the pre-load application during the experimental tests was considered by introducing a numerical applied stress intensity factor range ΔK_{app}. Once the stress intensity factor K_I had been numerically calculated for different crack depths, the corresponding stress intensity factor range was determined and compared with the base material threshold stress intensity factor range to verify whether a certain crack could propagate. Several crack depths were considered – i.e. 5, 25, 50, 100, 500 and 1000 μm – and the number of cycles necessary to reach a specified crack depth from another one was calculated by evaluating the threshold stress intensity factor range with the numerical models.

The fatigue crack growth rate from the depth of 1000 μm to the final failure was evaluated by means of the Paris equation [66] using the same material parameters C and n of equation (8).

Table 8 collects, for each crack depth, the half relative displacement value u of the nodes at the crack tip and the corresponding effective stress intensity factor range, the applied stress intensity factor range and the threshold stress intensity factor range at the maximum applied bending stress – i.e. 600 MPa. Table 9 reports the number of cycles necessary for a crack to propagate from a certain depth to another one until failure at the same applied bending stress.

Table 8. Effective stress intensity factor range, applied stress intensity factor range and threshold stress intensity factor range for each crack depth in the CrN-coated 2205 specimen models studied in [14]

Maximum applied bending stress = 600 MPa					
Crack depth, μm	Dist. from crack tip, m	u, m	ΔK_{eff}, MPa $\sqrt{m}$	ΔK_{app}, MPa $\sqrt{m}$	ΔK_{th}, MPa $\sqrt{m}$
5	$5.00\cdot10^{-6}$	0.00	0.00	0.00	-
25	$3.36\cdot10^{-6}$	$8.00\cdot10^{-8}$	3.23	3.05	3.00
50	$4.46\cdot10^{-6}$	$2.60\cdot10^{-7}$	8.99	8.54	3.06
100	$6.81\cdot10^{-6}$	$5.60\cdot10^{-7}$	12.37	11.75	3.10
500	$2.54\cdot10^{-5}$	$2.04\cdot10^{-6}$	29.43	27.96	3.30
1000	$1.28\cdot10^{-4}$	$5.78\cdot10^{-6}$	37.48	35.71	3.30

Table 9. Number of cycles to failure calculated for the coated 2205 specimen models studied in [14]

Maximum applied bending stress = 600 MPa	
Crack depth propagation intervals, μm	
5-25	2,911,346
25-50	8,164
50-100	6,030
100-500	2,812
500-1000	2,436
1000-failure	109
Total number of cycles to failure	2,930,897

The data reported in Table 8 revealed that no fatigue crack growth could take place with a crack depth lower than or equal to 25 μm. The number of cycles until failure calculated with the procedure was 2,930,897, whereas the experimental number of cycles for coated specimens having different surface conditions ranged from 2,086,325 to 4,036,325 [14] (Table 10). The number of cycles – higher than 2,000,000 – determined with the present approach hence confirmed the experimental results. Furthermore, the value of the number of cycles until failure for the rolled specimen, which was free of any surface modifications other than those due to the coating deposition process, was quite close to that obtained numerically. The small difference - less than 14% - between the experimental 3,380,430 load cycles and the numerical 2,930,897 ones confirmed the reliability of the proposed procedure.

Table 10. Experimental fatigue test results for different 2205 bulk material surface conditions [14]

Substrate	Coating	Maximum applied stress, MPa	Number of cycles until failure	
Rolled + polished	No	510	2,000,000	-
Shot peened + polished	Yes	600	4,036,325	Not cracked
Shot peened + polished	Yes	600	2,086,325	Not cracked
Rolled	Yes	600	3,380,430	Cracked
Rolled + polished	Yes	700	177,887	Cracked
Rolled + polished	Yes	650	2,914,391	Not cracked
Rolled + polished	Yes	675	179,841	Cracked
Peening conditions: intensity = 7 Almen, coverage = 100%, glass shots				

With regard to the effect of the CrN coating on the fatigue properties of the bulk material, a good improvement was therefore observed. This increase was justified by considering the high residual compressive stresses – with moduli higher than 1 GPa – induced by the coating deposition process, which were particularly effective against surface cracks having depths lower than 25 μm. Such compressive stresses, even in the presence of surface defects in the coating – e.g. large droplets or microcracks – can prevent the propagation for applied stresses that would cause the rupture of uncoated specimens. As a matter of fact, numerical analyses allowed to estimate that most part of the fatigue life of the coated specimens was spent on increasing the crack depth from 5 μm to 25 μm – namely 2,911,346 out of the total 2,930,897 cycles to failure.

In the reference [15], further models to study the four-point bending tests on CrN-coated H11 tool steel previously discussed were developed. The substrate and the coating mechanical properties were specified in the numerical models by considering, with the same assumption of limited plasticized area at the crack tip of the previous cases, linear elastic behaviour. The temperature gradient method was used again to impose the residual stress distribution (Figure 2) in the numerical models. It is very important to point out that the high residual compressive stress values – i.e. moduli higher than 2.4 GPa – characterizing the coating made the knowledge of the trend of the residual stress field below the external surface not fundamental for the fatigue life prediction. In fact, at different crack depths the effective stress intensity factor range mostly depended on the surface residual stress, as the application of different experimental subsurface residual stress trends did not significantly affect the stress intensity factor range values calculated. Also, on account of the steep residual stress gradient present at the coating-bulk material interface, the presence of irregularities and defects that could generate fatigue crack nucleation and propagation was accurately considered. From this point of view, provided that the crack propagation path is perpendicular to the component surface, the proposed model could be suitable for simulating the presence of subsurface cracks resulting from defects, inclusions or irregularities. The micro-hardness trend measured in the coated H11 tool steel to apply the previsional procedure is shown in Figure 3. On top of the residual stress and micro-hardness data, the surface roughness of the base material should be known in case it exceeds 0.30 μm. In this case, opportune correction coefficients should be assumed to include its effects in the calculations. With regard to the present model, this possibility can be discarded because accurate polishing operations are required before a PVD process to achieve a surface roughness not higher than 0.20 μm to assure good adhesion between the coating and the substrate. The surface roughness of the H11 tool steel samples was lower than 0.15 μm. The fatigue crack propagation from a surface crack having a depth equal to the coating thickness was numerically simulated for the coated samples. Figure 27 shows the FEM models used for stress intensity factor range calculation at different crack depths. Plane stress four-node linear finite elements were used like for the 2205 duplex stainless steel models above discussed.

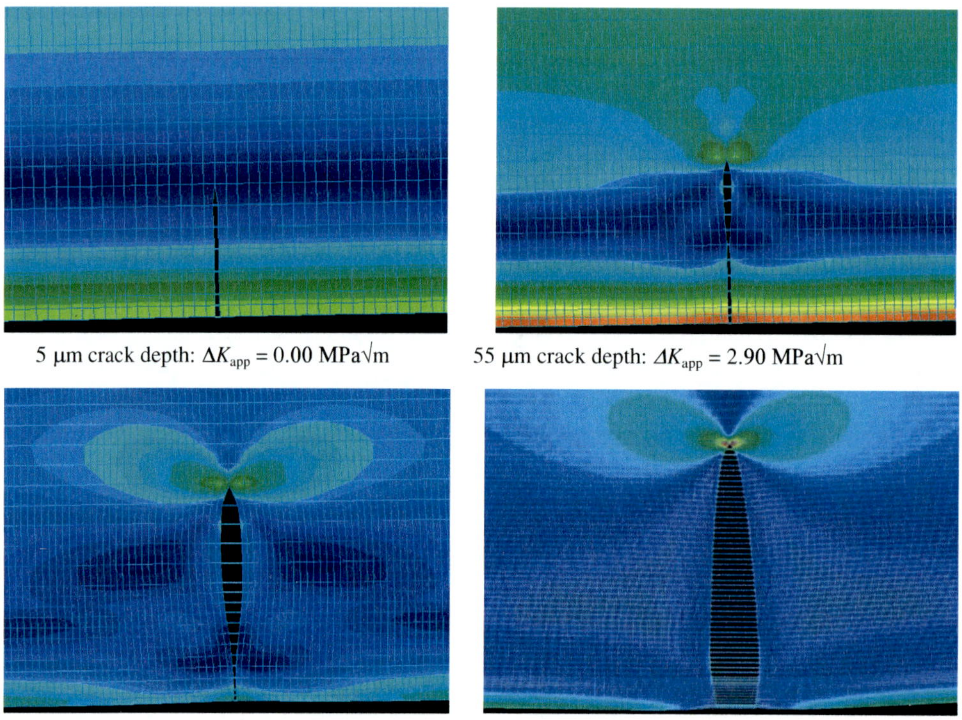

5 μm crack depth: $\Delta K_{app} = 0.00$ MPa√m

55 μm crack depth: $\Delta K_{app} = 2.90$ MPa√m

105 μm crack depth: $\Delta K_{app} = 8.19$ MPa√m

505 μm crack depth: $\Delta K_{app} = 22.40$ MPa√m

Figure 27. Finite element models for ΔK evaluation with different crack depths in CrN-coated H11 tool steel specimens [15].

The applied and the threshold stress intensity factor range values for the coated H11 tool steel calculated with the models with a maximum applied bending stress of 845 MPa – i.e. the fatigue limit from the experimental tests – are reported in Table 11. It is worth underlining that the stress intensity factor range is higher than the threshold stress intensity factor range only for cracks deeper than 55 μm. Table 12 reports the number of cycles until failure. In the presence of, at least, a 55 μm-deep surface defect, the propagation could take place and the predicted number of cycles until failure was equal to 37,670.

Such results were then compared with the ones obtained with the experimental tests carried out on both coated and uncoated H11 tool steel samples. The coated H11 tool steel specimens failed before 2,000,000 load cycles only in the presence of subsurface cracks or inclusions like the one shown in Figure 13. In case of specimens presenting irregularities, surface cracks or subsurface defects at depths lower than 55 μm, the fatigue crack nucleation or propagation for applied bending stresses lower than the fatigue limit should not occur. The numerical results were in good agreement with the experimental ones, as the specimens having initial surface cracks deeper than 55-105 μm either presented a number of cycles until failure higher than 2,000,000 or did not fail at all (stopped tests).

In [16], the procedure was applied to the numerical models of uncoated and PVD-coated transmission spur gears for motorcycle competition applications. Particular attention was paid to titanium as a base material. Accurate numerical models of the spur gears were developed to evaluate the crack propagation in the presence of an initial defect present at the tooth base. The gearbox including the spur gears analyzed – i.e. the pinion and the driven gear constituting the fifth speed – is shown in Figure 28.

Table 11. Applied stress intensity factor range and threshold stress intensity factor range for each crack depth in CrN-coated H11 samples [15]

Maximum applied bending stress = 845 MPa		
Crack depth, μm	ΔK_{app}, MPa $\sqrt{m}$	ΔK_{th}, MPa $\sqrt{m}$
5	0.00	-
30	2.22	4.23
55	2.90	5.18
105	8.19	3.39
505	22.40	3.73
1000	43.80	3.73

Table 12. Number of cycles to failure calculated for CrN-coated H11 samples [15]

Maximum applied bending stress = 845 MPa	
Crack depth propagation intervals, μm	
5-30	No propagation
30-55	No propagation
55-105	31,360
105-505	5,520
505-1000	690
1000-failure	100
Total number of cycles to failure	37,670

Table 13. Main mechanical properties of the gear materials studied in [16]

Material	Condition	Hardness, HV	Density, kg/m^3	Young's mod., MPa	Poisson's ratio	A%	YS, MPa	UTS, MPa	σ_{w0}, MPa
15NiCr13	Carburized	300	7800	206000	0.28	10	883	1128	560
Ti6Al4V	Annealed	350	4500	113000	0.34	14	880	950	510

Figure 28. Gearbox including the gears studied in [16].

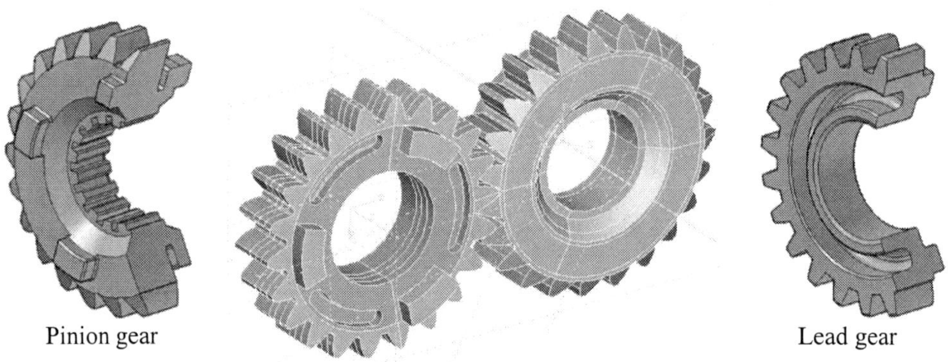

Figure 29. Three-dimensional pinion and driven gear models developed in [16].

Table 14. Hardness values for the coated and the uncoated gears studied in [16]

Material	Condition	Hardness, HV				Depth [mm] of	
		Surface	Maximum	Core	Interface	maximum hardness	interface hardness
15NiCr13	Carburized	720	760	300	550	0.2	0.5
15NiCr13	Coated	2680	2686	300	1652	0.0005	0.005
Ti-6Al-4V	Annealed	350	350	350	350	-	-
Ti-6Al-4V	Coated	2680	2686	350	1652	0.0005	0.005

The modulus of the gears and the tooth thickness were, respectively, 2.5 mm and 13 mm. The number of teeth of the pinion was 23, while the driven gear had 22 teeth. The surface roughness of the tooth flanks was equal to 0.8 μm. The service load applied to the tooth flanks of the meshing gears was 3450 N. The three-dimensional solid models developed for the numerical analyses are shown in Figure 29. The actual material of the gears was the 15NiCr13 steel alloy and a case hardening treatment was carried out producing a penetration depth of 0.7 mm. The alpha-beta Ti-6Al-4V titanium alloy annealed at 750ºC was considered as a very attractive alternative base material for both uncoated and TiN PVD-coated spur gears. The main goal of the research was to compare the fatigue behaviour of steel and titanium spur gears under the same conditions. Therefore, in the calculations the CrN and TiN coatings were assumed to have the same hardness and the same surface residual stress values. As a matter of fact, TiN and CrN coatings can actually show similar hardness and residual stress distributions [69-72]. The previously shown residual stress and micro-hardness trends in the presence of PVD CrN coatings deposited on different steel substrates were assumed. CrN-coated steel alloy and TiN-coated titanium alloy gears were analyzed and the coating thickness was 5 μm in both cases. Table 13 reports the main mechanical properties of the two materials under consideration. The hardness values at the external surface and at the core of the coated and the uncoated bulk materials are listed in Table 14 [16].

The same value of surface residual compressive stress was assumed for both the steel and the titanium alloy spur gears. In fact, the fatigue resistance of PVD-coated components is mainly dependent on the high residual compressive stresses in the coating [14,15], as already said. The assumed residual stress value at the external surface was -2500 MPa in the presence of the coatings [15]. The tensile peak of the self-equilibrated residual stresses applied in the numerical models was placed at a depth of 0.1 mm below the external surface, whereas the inversion point depth of the trend – i.e. where the residual stresses are equal to zero – at 0.04 mm. The elastic modulus and the Poisson's ratio of the CrN and TiN coatings were, respectively, 303 GPa and 0.20 [15] and 600 GPa and 0.25 [16]. Pinion and driven gear numerical models were then developed to simulate the working condition of the meshing gears. A surface crack having a depth of 5 μm – i.e. equal to the thickness of the coating – perpendicular to the external surface profile at the base of the pinion engaging tooth was first modeled in the area where the maximum bending stresses took place. A semi elliptical crack was generated with an aspect ratio of 4:1, which means that the width of the crack was four times its depth. Both the orientation of the plane on which the crack lay and the aspect ratio were supposed not to change during the propagation stage. The numerical models allowed to calculate the number of cycles necessary to reach specified crack depths. A comparison, in terms of fatigue resistance with each different treatment applied, among the carburized steel alloy spur gears, the CrN-coated steel ones and the TiN-coated titanium ones was made. Figures 30(a) and 30(b) show, respectively, the finite element global model and the mesh refinement in the most critical areas – namely at the contact area and at the tooth base. Due to the complex geometry of the spur gears (Figure 29), a hybrid non-structured mesh consisting of linear tetrahedral and hexahedral finite elements was generated. Contact elements were placed between the meshing teeth of the pinion and of the driven gears and between the crack mating faces to prevent them from overlapping before the service load application. A friction coefficient equal to 0.1 was adopted in the numerical models to account for the presence of a suitable lubricant in the real application. The presence of cracks perpendicular to the external surface at the base of the engaged tooth and having depths between 5 and 200 μm was simulated. The submodeling technique was used to refine the mesh at the crack tip until the value of the stress intensity factor had been verified not to significantly change for smaller finite element dimensions. Figures 30(c) and 30(d) show a meshed submodel and a stress contour in the same submodel after load application. The residual stress field induced by the coating deposition process was simulated with the temperature gradient method. The linear elastic material behaviour approximation, from the assumption of a limited plasticized area at the crack tip, allowed to obtain conservative results for the stress intensity factor range, particularly in the case of titanium alloy spur gears. The absence of coating delamination was assumed again.

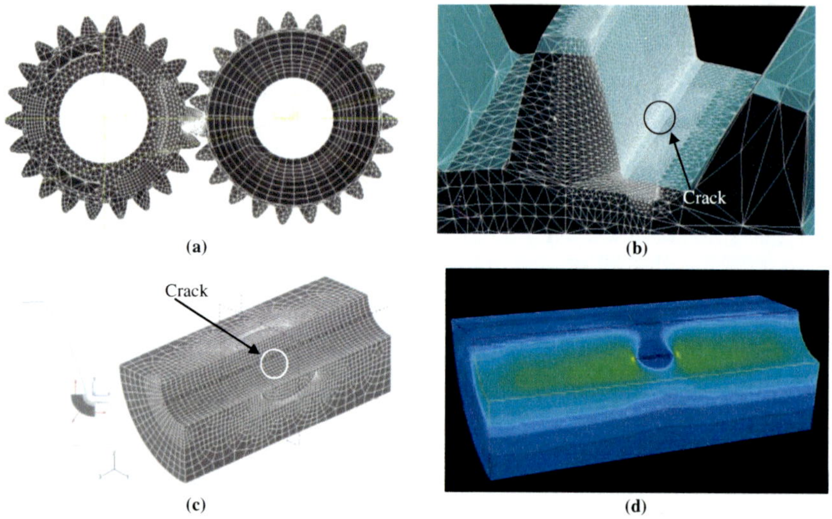

Figure 30. (a) Global model, (b) detail of the refined mesh at the base of a tooth, (c) reference system and mesh of the submodel and (d) stress state around the cracked area in the submodel of the coated pinion and driven gears studied in [16].

Figure 31 depicts, for different crack depths, the stress state in the uncoated non-carburized steel alloy gear model with the corresponding maximum values of the stress intensity factor range at the crack front reported in the caption. Figure 32 shows the stress state in the CrN-coated steel alloy gear in the presence of the deepest cracks studied – i.e. 100 and 200 μm-deep. The high surface residual compressive stresses made it possible to observe crack opening only at depths higher than 100 μm, as far as the coated steel alloy gear is concerned.

Table 15 collects the parameters calculated with the theoretical-numerical procedure to foresee whether the crack propagation occurred at each specified crack depth for the uncoated non-carburized, uncoated carburized and coated steel alloy spur gears. It was pointed out that in the case of the untreated steel gears – i.e. both non-carburized and uncoated – only cracks having a depth lower than 5 μm did not propagate with the service load applied. With regard to the carburized uncoated gears, no crack propagation was possible regardless of the crack depth, while in the coated ones the propagation could take place only with crack depths of at least 100 μm. In this case, due to the high value of the surface residual compressive stress, the crack did not open until a 100-μm defect was present. Furthermore, in the carburized gears the hardened layer was deeper than in the coated ones – namely 0.7 mm against 0.005 mm. Therefore, despite the maximum value of the surface residual compressive stresses in the carburized steel gear was lower than that of the coated steel one, their presence distributed over a deeper layer of material was more effective in preventing the crack propagation.

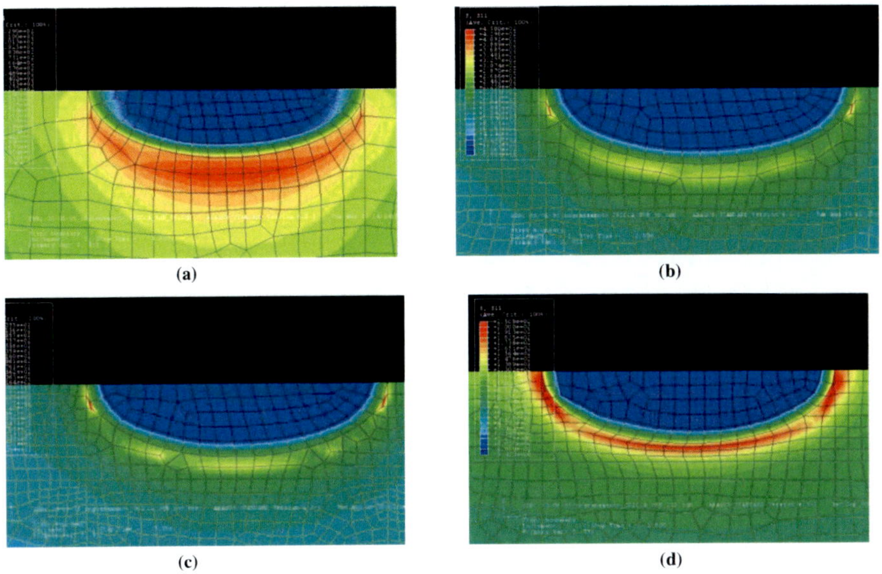

Figure 31. Stress state maps at the crack front in the uncoated non-carburized steel gear model: (a) 5 μm crack depth: $\Delta K_{app} = 1.86$ MPa√m; (b) 50 μm crack depth: $\Delta K_{app} = 5.55$ MPa√m; (c) 100 μm crack depth: $\Delta K_{app} = 6.30$ MPa√m; (d) 200 μm crack depth: $\Delta K_{app} = 5.76$ MPa√m [16].

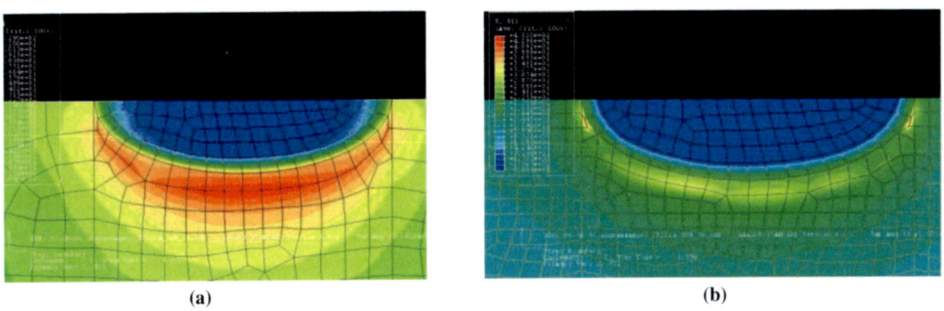

Figure 32. Stress state maps at the crack front in the CrN-coated steel gear model: (a) 100 μm crack depth: $\Delta K_{app} = 5.64$ MPa√m; (b) 200 μm crack depth: $\Delta K_{app} = 9.12$ MPa√m [16].

In Table 16 the same results shown in Table 15 for the steel alloy spur gear are reported for the TiN-coated titanium alloy one as well as the distance r from the crack tip at which the half relative displacement u was assessed. The behaviour was quite similar to that of the coated steel spur gear, since the crack could propagate only with a defect having a depth higher than 100 μm. In the TiN-coated titanium alloy gear the crack did not open until its depth was at least 100 μm. Thus the coatings were beneficial to both the steel and the titanium alloy gears.

Table 15. Crack propagation predictions at specified crack depths for the uncoated non-carburized, the uncoated carburized and the coated steel alloy spur gears. ΔK_{app} and ΔK_{th} are expressed in MPa√m [16]

Crack depth, μm	Non-carburized spur gear			Uncoated carburized spur gear			Coated spur gear		
	ΔK_{app}	ΔK_{th}	Propagation	ΔK_{app}	ΔK_{th}	Propagation	ΔK_{app}	ΔK_{th}	Propagation
5	1.86	2.01	No	1.86	3.87	No	0	7.62	No
25	4.32	3.44	Yes	4.32	6.24	No	0	3.44	No
50	5.55	4.34	Yes	5.55	8.05	No	0	4.34	No
100	6.30	5.46	Yes	6.30	10.55	No	5.64	5.46	Yes
200	5.76	2.99	Yes	5.76	13.73	No	9.12	3.26	Yes

Table 16. Crack propagation predictions at specified crack depths for the TiN-coated titanium spur gear [16]

Crack depth, μm	r, m	u, m	ΔK_{app}, MPa√m	ΔK_{th}, MPa√m	Propagation
5	1.09E-06	0	0	7.62	No
25	2.45E-06	0	0	3.78	No
50	1.07E-06	0	0	4.77	No
100	1.27E-06	2.89E-08	4.00	6.01	No
200	1.76E-05	1.58E-07	5.86	3.00	Yes

Table 17. Number of cycles to reach a 200-μm crack depth for all the spur gear models developed in [16]

Crack depth, μm	Number of cycles			
	Uncoated non-carburized steel	Uncoated carburized steel	Coated steel	Coated titanium
5-25	No propagation	No propagation	No propagation	No propagation
25-50	60,278	No propagation	No propagation	No propagation
50-100	86,254	No propagation	49,976	No propagation
100-200	106,969	No propagation	104,184	432,220
Fatigue life	253,500	-	154,160	432,220

Table 17 collects the number of cycles necessary for the cracks to propagate from a specified depth to another one for all the spur gear models analyzed. These results were obtained by considering the stress intensity factor range variation during the propagation stage. Due to the fact that the continuous variation could not be considered with the discrete propagation model proposed, an average between the values at the initial and at the final crack depths was used in equations (8) and (9) for this parameter. By analyzing the data reported in Table 16 it resulted that for the coated titanium alloy spur gear the number of cycles needed for a crack to grow from 100 to 200 μm was remarkably higher than that for the coated steel alloy spur gear – i.e. 432,220 vs. 104,184 according to the higher value of the applied stress intensity factor range of the coated steel gear with respect to that of the coated titanium gear, say 9.12 vs. 5.86 MPa√m. In fact, by looking at equation (10) it is evident that the stress intensity factor range tightly depends on the elastic modulus of the base material –

206 GPa for the steel alloy and 113 GPa for titanium alloy. Therefore, the higher stress intensity factor range resulted in a higher crack growth rate for the steel gear and so even in a lower number of cycles necessary for the crack propagation from 100 to 200 μm.

The numerical results emphasized that thin hard coatings were beneficial to the fatigue resistance of gears made of both steel and titanium alloys. Therefore, thin hard coatings could represent a good alternative to case hardening, notwithstanding it guarantees a good effect over a greater depth. Figure 33 shows the main results obtained by means of the theoretical-numerical procedure for the coated and the uncoated steel and titanium alloy spur gears. The fatigue resistance of the coated titanium gears was higher than that of the coated steel ones and was comparable to that of the carburized steel gears. The use of titanium instead of steel as a base material for gears would allow a considerable mass reduction, making it possible to have more power available and to reduce the gas emissions.

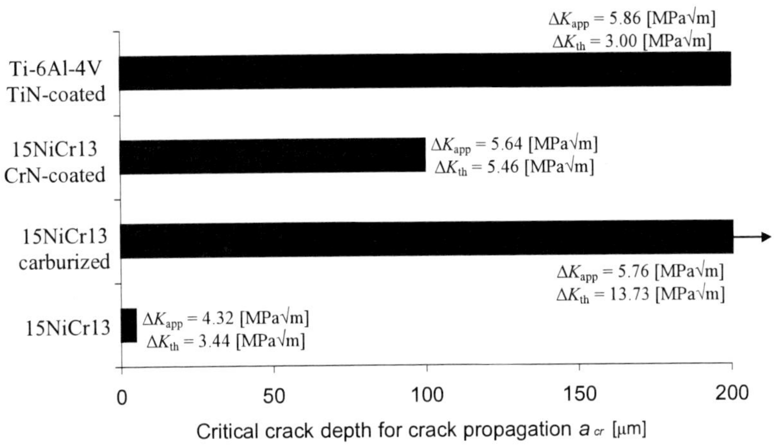

Figure 33. Critical crack depth for the coated and the uncoated steel and titanium spur gears studied in [16].

Models to verify the contact fatigue behaviour of coated gears were also developed by the authors et al. [34] to understand and interpret the experimental results from the contact fatigue tests on uncoated and CrN-coated 18NiCrMo5 spur gears previously reported. The stress and strain distributions on and below the contact surface for both coated and uncoated gears was determined with two-dimensional finite element models. Also, the submodeling technique was used to refine the calculations over the contact area. Quadrilateral four-node bilinear plane strain elements were used to generate the mesh of the bulk material, whereas trilateral three-node linear plane strain elements were employed for the coating (Figure 34). Such models allowed to predict the number of load cycles needed for the fatigue damage initiation to occur by considering a suitable cyclic behaviour for the gear base material [61,73] and by applying the Coffin-Manson law to the numerical results [25]:

$$\frac{\Delta\varepsilon}{2} = \frac{\sigma'_f}{E}\left(2N_f\right)^b + \varepsilon'_f\left(2N_f\right)^c \tag{12}$$

where $\Delta\varepsilon$ is the cyclic deformation amplitude, N_f the number of cycles until the initial damage, σ'_f and ε'_f the fatigue strength coefficient and the fatigue ductility coefficient respectively, and b and c the strength and the fatigue ductility exponents. The following values were used for these parameters for the 18NiCrMo5 base material [61,73]: $\sigma'_f = 1890$ MPa, $\varepsilon'_f = 0.848$, $b = -0.115$ and $c = -0.734$. Both the PVD deposition process [15] and the case hardening treatment [74] were considered by introducing suitable surface residual stress fields into the numerical models. The maximum surface residual compressive stresses were equal to -2400 MPa for the PVD-coated gear models and -150 MPa for the uncoated ones. A friction factor equal to 0.5 was assumed to reproduce the unlubricated test condition. The numerical models were then used to calculate the maximum cyclic deformations and to find their position over the contact area.

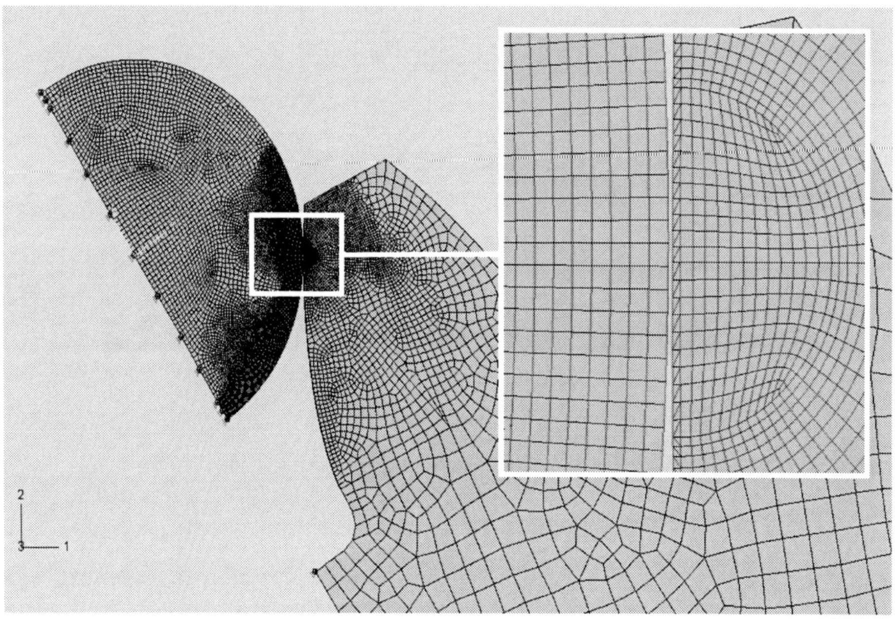

Figure 34. A coated gear numerical model developed in [34].

Figures 35 and 36 show the number of cycles to the initial damage from the external surface to the core of the tooth at each applied test load. The minimum number of cycles was calculated to occur at the bulk material external surface for both the coated and the uncoated gears according to the unlubricated test condition. Also, the number of cycles to the first damage in the uncoated gears was higher than 10,000 only with the load of 0.625 kN, whereas at higher load levels it was very low – less than 300 – for both the coated and the uncoated gears, confirming the experimental evidence from the fatigue tests. Therefore, the theoretical-numerical procedure confirmed that the presence of the 10 μm-thick CrN coating analyzed was unsuitable for this kind of application. The numerical models also showed that the maximum deformation amplitudes promoting the initial damage were at the external surface level for the uncoated gears and at the interface between the coating and the substrate for the coated ones.

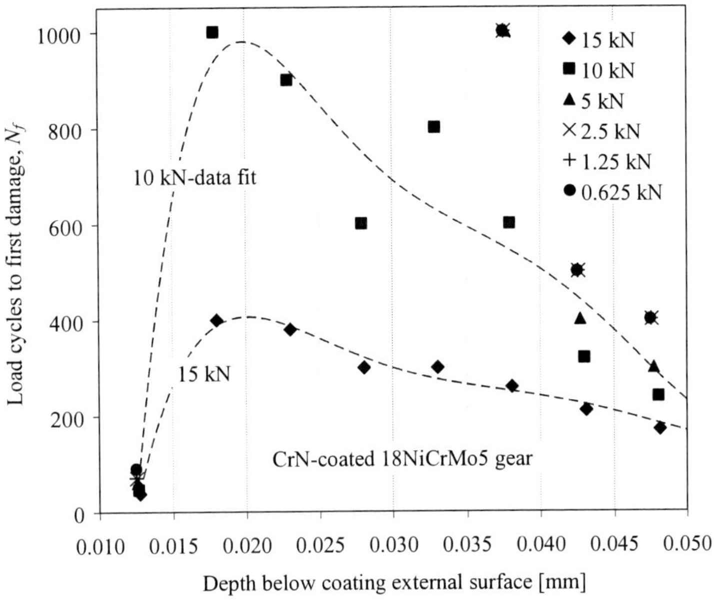

Figure 35. CrN-coated gear models: Number of cycles to first damage vs. Depth below the surface diagram under contact fatigue [34].

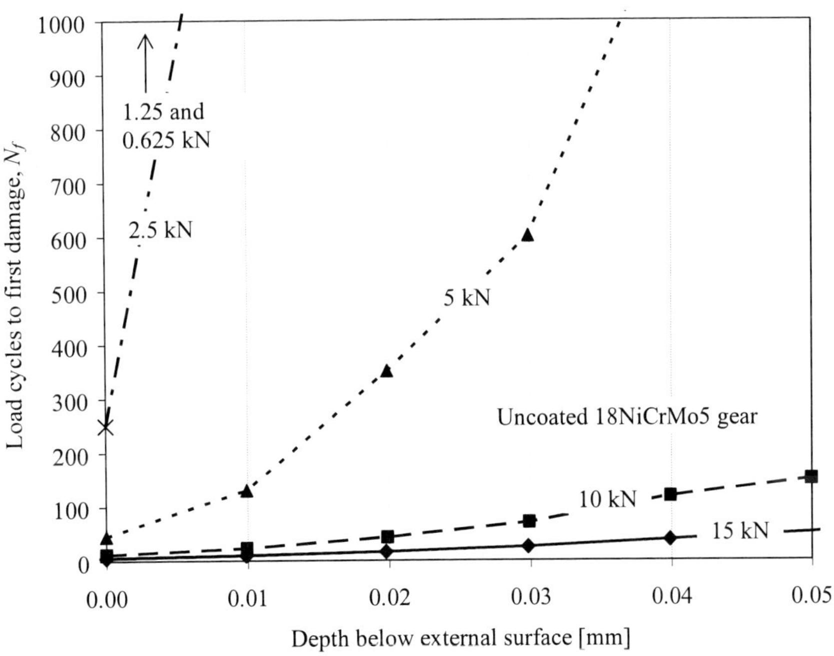

Figure 36. Number of cycles to first damage vs. Depth below the surface diagram under contact loading for the uncoated gear models studied in [34].

With regard to the effect of the shot peening treatment on bulk materials, thorough studies were carried out by Baragetti et al. to improve the results obtainable with the models available in the literature. The most popular of these models had been developed by Al-Obaid

[39-41], who modeled with finite elements a proof plate divided into nine layers – having increasing thickness values from the impact surface to the in-depth layers – impinged with a non-deformable sphere. Each finite element of the top surface was peened in succession with increasing loads and then unloaded. The effects were transmitted to the elements underneath through the connecting nodes. As the surface had been completely peened and the surface elements impinged more than one time, by means of the von Mises' yield criterion the residual stress distribution was calculated. Limits of this model lie in the fact that only the thickness is considered among the geometrical parameters of the target specimen and that neither the material properties nor the process parameters – apart from the shot speed – were specified. In order to develop his model [39], Al-Obaid started from the contact mechanics Hertz's theory, which supposes the contact pressure produced in the specimen surface layers by the contact with a sphere to be hemispherical with the ratio between the average and the maximum pressure values equal to 2:3. The plasticization occurs at a depth equal to 0.47 times the radius a of the mark left by the sphere and with a shear stress equal to 0.468 times the average contact pressure. Equation (13) defines the plasticization depth h_p as a function of the depth $\bar{u}$ of the mark and of the sphere radius R:

$$\frac{h_p}{R} = 2.57 \cdot \left(\frac{\bar{u}}{R} \right)^{\frac{1}{2}} \tag{13}$$

By means of the hypothesis of rigid sphere and elastic-perfectly plastic specimen material behaviour, the depth of the mark can be determined. Furthermore, by supposing the average pressure P generated by the impact of the spherical shot on the specimen to be equal to three times its yield strength and $\bar{u}$ to $a/(2{\cdot}R)$ and integrating the following equilibrium equation:

$$m\frac{dV}{dt} = -\pi \cdot a^2 \cdot P \tag{14}$$

where V and m are the speed and the mass of the sphere respectively, it is possible to express $\bar{u}$ as a function of R, V and the density ρ of the sphere:

$$\frac{\bar{u}}{R} = \left(\frac{2}{3} \right)^{\frac{1}{2}} \left(\frac{\rho \cdot V^2}{P} \right)^{\frac{1}{2}} \tag{15}$$

$$a = \sqrt{2 \cdot \bar{u} \cdot R} = R\sqrt{2} \cdot \left(\frac{2}{3} \right)^{\frac{1}{4}} \left(\frac{\rho \cdot V^2}{P} \right)^{\frac{1}{4}} \tag{16}$$

By substituting equation (13) into equation (15) the following relationship can be written:

$$\frac{h_p}{R} = 2.57 \cdot \left(\frac{2}{3} \right)^{\frac{1}{4}} \left(\frac{\rho \cdot V^2}{P} \right)^{\frac{1}{4}} \tag{17}$$

The residual stresses after the impact are the result of tensile loading and elastic unloading from the interaction between the plastic and the elastic zone produced below the contact area in the target material. Therefore, by means of the theory for beams and thin plates – with the hypothesis that cross sections remain plane in the deformed configuration – the residual stress distribution was described. The residual stress in a thin plate was formulated as follows:

$$\sigma_{res}(u) = \sigma_b + \sigma_a + \sigma_s = \sigma_b + \sigma_a + \frac{E}{1-\nu}\varepsilon(u) \tag{18}$$

where σ_b is the bending stress, σ_a the axial stress and σ_s the source stress. They are necessary for the adoption of the plate theory and are used to balance the bending and the axial stresses arising when the plate tends to recover the undeformed condition. The residual stress distribution was supposed to be sinusoidal with a peak placed at a depth α – expressed as a fraction of h_p – included in the plastic zone:

$$\sigma_{res}(u) = \frac{E \cdot \varepsilon_{\max}}{1-\nu}\left[\frac{12\cdot\lambda}{h\cdot\pi}(1-\alpha)\left(\frac{h}{2}-u\right)C_1 + \frac{2\cdot\lambda}{\pi}(1-\alpha)C_2 - \frac{\varepsilon(u)}{\varepsilon_{\max}}\right] \tag{19}$$

where $\varepsilon_{\max}$ is the maximum deformation, h the thickness of the plate and λ, C_1 and C_2 parameters to determine. Equation (19) allows to evaluate the residual stress profile as a function of the geometry of the plate and of the velocity of the shots. Al-Obaid estimated the three unknown parameters by approximating the deformation trend in the plastic zone with a cosine function:

$$\varepsilon(u) = \varepsilon_{\max}\cdot\cos\pi\left[\frac{u-\alpha\cdot h_p}{2(1-\alpha)h_p}\right] \tag{20}$$

By considering the simple equations for the bending and axial stresses in the pure elastic field in a beam subjected to a bending moment and to an axial load he obtained the following:

$$\sigma_b = \frac{E\cdot\varepsilon_{\max}\cdot 12\cdot(1-\alpha)\cdot\left(\frac{h}{2}-u\right)\cdot\lambda}{h\cdot\pi\cdot(1-\nu)}C_1 \tag{21}$$

$$\sigma_a = \frac{E\cdot\varepsilon_{\max}\cdot 2\cdot(1-\alpha)\cdot\lambda}{\pi\cdot(1-\nu)}C_2 \tag{22}$$

$$\lambda = \frac{h_p}{h} = 2.57\cdot\left(\frac{2}{3}\right)^{\frac{1}{4}}\left(\frac{R}{h}\right)\left(\frac{\rho\cdot V^2}{P}\right)^{\frac{1}{4}} \tag{23}$$

$$C_1 = C_2 - 2\cdot\lambda + \frac{4\cdot\lambda}{\pi}(1-\alpha)\cdot\cos\left[\frac{\alpha\cdot\pi}{2(1-\alpha)}\right] \tag{24}$$

$$C_2 = 1 + \sin\left[\frac{\alpha\cdot\pi}{2(1-\alpha)}\right] \tag{25}$$

The maximum deformation value, considering a curved beam, was calculated as:

$$\varepsilon_{\max} = \frac{2}{3}\cdot\frac{\pi\cdot h\cdot\delta}{\lambda\cdot L^2\cdot(1-\alpha)\cdot C_1} \tag{26}$$

where L is the length of the beam and δ is the height of the arch formed by the deformed beam. By introducing equations from (20) to (26) into equation (19), the trend of the residual stresses can be determined as the sum of the bending, the axial and the source stresses.

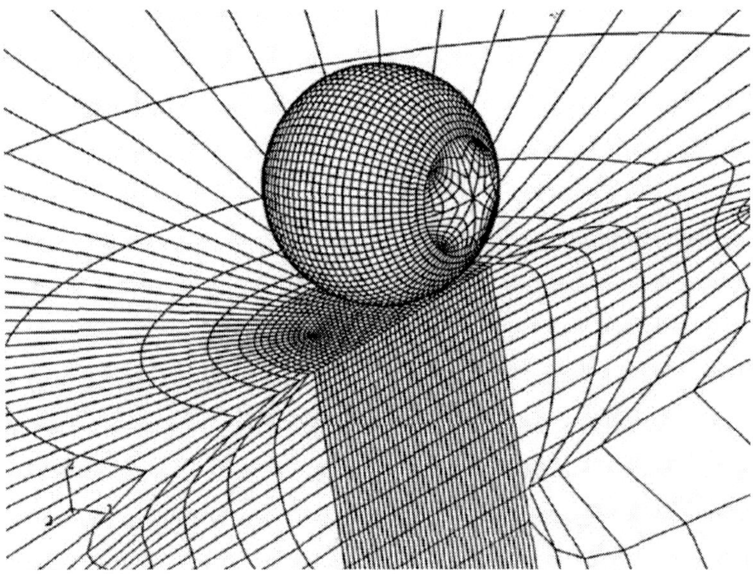

Figure 37. Mesh visualzation in the impact area of a model developed in [44].

In the reference [44], accurate finite element models were developed by Baragetti to evaluate the effects induced by the variation of the process parameters on the residual stress field. Both axisymmetric and three-dimensional numerical models were developed to foresee the surface residual stress distributions with different treatment parameters to point out pros and cons of each one. These studies were carried out to optimise the treatment parameters, since they were commonly chosen with empirical considerations in the practise. Remarkable difficulties were experienced in simulating the non linearities – arising from the contact condition and from the hardening behaviour of the material – of the impact of a rigid sphere on a plastically deformable steel plate. A dynamic explicit procedure [44] was eventually adopted to perform the analyses. The finite element models allowed to investigate the effects of the impact of spheres for different collision angles – namely 0, 30 and 45 degrees with respect to the direction perpendicular to the impinged surface. The sphere was modeled as a rigid body, since further analyses confirmed that the stiffness of the spheres was much higher than that of a treated steel component. The selected shot diameter was equal to 1 mm and the speed ranges were between 80 and 120 m/s. The mechanical properties of the simulated target plate material – i.e. the 35CrMo4 steel alloy – were: ultimate tensile strength and yield strength equal to 764 and 572 MPa respectively, elastic modulus to 206,000 MPa, Poisson's ratio to 0.28 and density to 7800 kg/m^3. A linear hardening stress-strain curve and the von Mises' yielding criterion were assumed.

Figure 37 shows the detail of a mesh of a 3D model in the impact zone. The symmetry of the geometry allowed to model only one half of the whole circular plate and the nodes lying on the symmetry plane – or the symmetry axis for the axisymmetric model – were constrained using suitable boundary conditions. In order to verify the finite element models, impact analyses with both linear elastic and elastic-plastic specimen material behaviours were processed. The results obtained with the linear elastic analyses were then compared with the theoretical ones calculated with the following equations from the extension of the Hertz's contact theory to the impact between solid bodies [75-77]:

$$d_{max} = R^{2/5}\left[\frac{15\left(\dfrac{1-\nu^2}{E}\right)mV^2}{16}\right]^{1/5} \qquad (27)$$

$$F_{max} = \frac{4R^{1/5}\left(\dfrac{15}{16}\pi m V^2\right)^{3/5}}{3\pi\left(\dfrac{1-\nu^2}{\pi E}\right)^{2/5}} \qquad (28)$$

$$P_{max} = \frac{3\cdot F_{max}}{2\pi\cdot d^2_{max}} \qquad (29)$$

where d_{max} is the maximum radius of the contact circle, F_{max} is the maximum force of compression and P_{max} is the maximum contact pressure. A difference less than 1% between the numerical and the theoretical maximum contact pressure values resulted for both the axisymmetric and the three-dimensional models using a shot speed of 100 m/s. This confirmed the reliability of the finite element models developed.

Due to the lack of symmetry in the first results of the elastic-plastic analyses caused by the "hourglass effect" [78] for both the deformed configuration and the residual stress distribution, the hourglass-control parameter was properly adjusted to improve the reliability of the results. The accuracy achieved in terms of stress symmetry was assessed to be acceptable with errors between 5% and 10%. Figure 38 depicts the deformed configuration of the target steel plate with an impact speed equal to 100 m/s obtained by using an hourglass-control parameter equal to three.

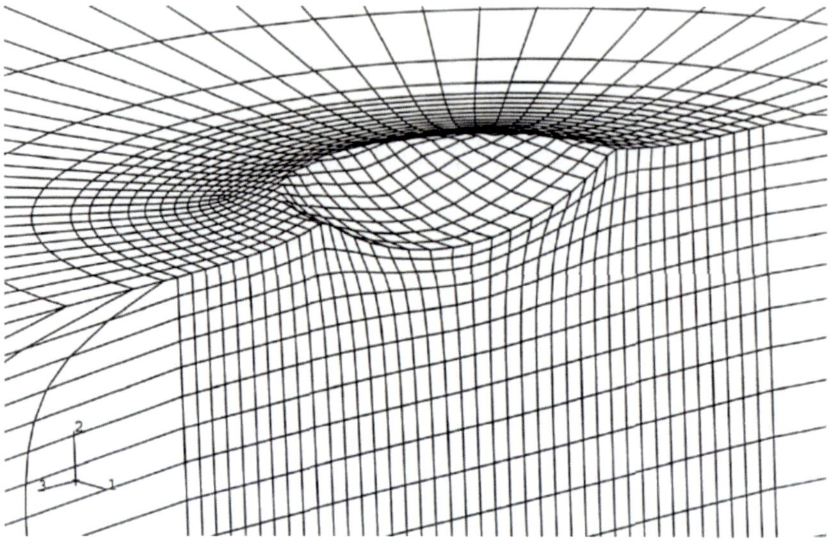

Figure 38. Deformed configuration after the impact obtained with one of the 3D models developed in [44].

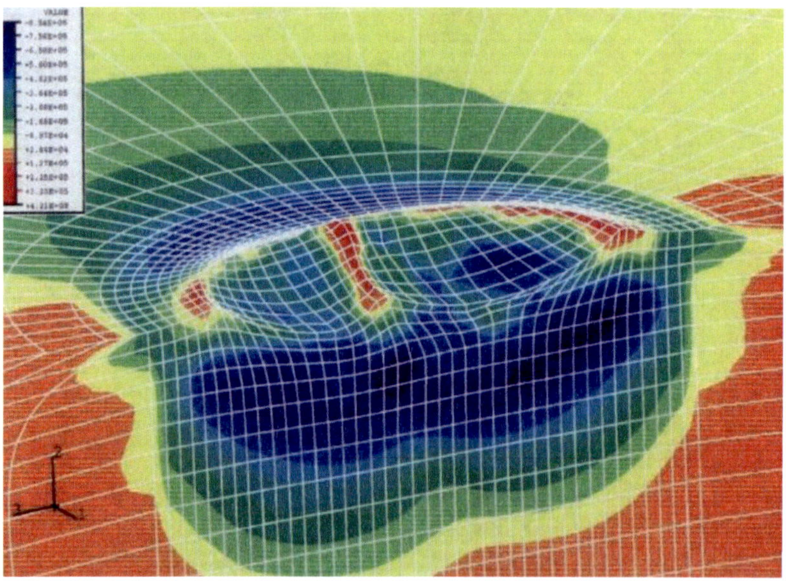

Figure 39. Stress map in the deformed configuration obtained with a model developed for a double impact analysis in [44].

The simulation of multiple impacts was then carried out. Three-dimensional models were only used since an axisymmetric model could not be used with impacts along different axes. A 100% percentage of coverage treatment was considered and each sphere was opportunely positioned at a prescribed distance from the target specimen. After a first impact and the following stabilisation of the residual stresses generated, a second sphere impinged the specimen on the edge of the crater produced by the first impact. Figure 39 shows the stress map in the deformed configuration obtained with a three-dimensional finite element model including two shots having a diameter of 1 mm, an impact speed of 100 m/s and impact axes perpendicular to the specimen surface. The results highlighted that the multiple shot impact, in the considered range of speeds, produced nearly the same effects of the single impact in terms of residual stresses induced. Figure 40 shows the comparison between the residual stress trends after the first and the second impact along the first sphere collision axis. The residual stress trends are very similar and show negligible differences in the maximum values. Only the surface residual stress value decreased after the second impact.

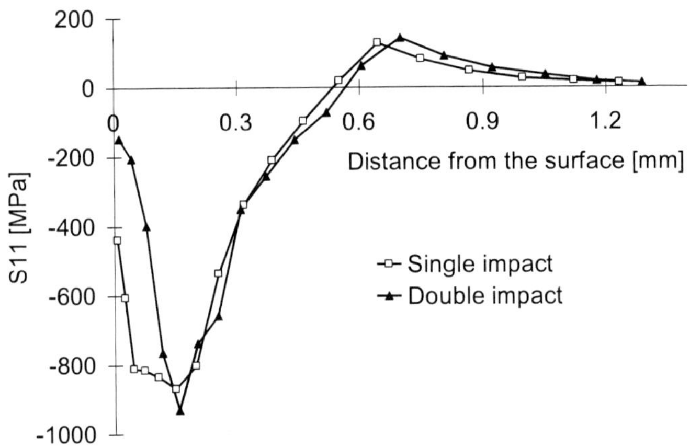

Figure 40. Comparison between the residual stress trends after the impact of the first and of the second sphere along the first sphere collision axis obtained with a model developed in [44].

In the reference [43], Baragetti et al. studied the shot peening process parameters by applying the Design of Experiments (DoE) method [45,46] to the results obtained with numerical models. The procedure allowed to choose the best levels of the treatment parameters in a cheap and fast way. Non dimensional parameters were also defined and used to extend the results obtained for steel targets to allow the choice of the peening parameters for different treatment conditions and several treated materials. DoE made it possible to take into account the effects induced by almost all the factors having an influence on the results of the treatment. Furthermore, due to the negligible effects induced by the variation of the impact angle [42] for the range of speeds of the spheres currently used for steel mechanical components, this parameter was not considered among the influencing factors. Axisymmetric numerical models similar to those previously discussed allowed the collection of the data necessary to apply the procedure. The $L_9(3^4)$ array of Taguchi [45] was used to analyze three non dimensional parameters including the most important factors influencing the shot peening effects on the treated components: the material, the speed and the diameter of the spheres, the material and the dimensions of the target specimen and the number of impacts along the same axis – with a 200% percentage of coverage assumed in case of impact of two spheres along the same collision axis [42]. Such an array enabled the optimisation of four process parameters varying over three different levels each by means of only nine runs, namely nine numerical analyses. In [43], an empty column was used for a fourth non dimensional parameter to verify whether all the non dimensional parameters influencing the results had been considered. Table 17 and 18 list the array parameters and the chosen levels: ρ is the density of the shot material, *k* the number of impacts along the same axis, *S* the ultimate tensile strength of the specimen material, *h* the thickness of the impinged specimen, *Φ* and *V* the sphere diameter and speed respectively. In Table 17(a), the terms P_{ij} are referred to the parameter "i" and the level "j". In Table 17(b), the non dimensional parameter $\rho V^2/S$ represents an important index of the effects induced by the shot peening treatment, particularly if the induced residual stress field and the deformations after the impact are concerned.

Table 17. Non dimensional parameters and selected levels for the Taguchi's $L_9(3^4)$ array used in [43]

(a)				
Non dimensional parameters		Level 1	Level 2	Level 3
P_1	$\rho V^2/S$	0.03824	0.01626	0.02548
P_2	k	1	1	2
P_3	h/Φ	20	2	0.5

(b)			
P_{11}	S = 510 MPa	V = 50 m/s	ρ = 7800 kg/m^3
P_{12}	S = 984 MPa	V = 80 m/s	ρ = 2500 kg/m^3
P_{13}	S = 1300 MPa	V = 120 m/s	ρ = 2300 kg/m^3
P_{21}	k = 1 impact		
P_{22}	k = 1 impact		
P_{23}	k = 2 impacts		
P_{31}	Φ = 0.5 mm	h = 10 mm	
P_{32}	Φ = 1 mm	h = 2 mm	
P_{33}	Φ = 2 mm	h = 1 mm	

Table 18. Taguchi's $L_9(3^4)$ array and parameter values used in [43]

(a)				
Run no.	P_1	P_2	P_3	Empty
1	1	1	1	1
2	1	2	2	2
3	1	3	3	3
4	2	1	2	3
5	2	2	3	1
6	2	3	1	2
7	3	1	3	2
8	3	2	1	3
9	3	3	2	1

(b)									
Run no.	1	2	3	4	5	6	7	8	9
Φ [mm]	0.5	1	2	1	2	0.5	2	0.5	1
S [MPa]	510	510	510	984	984	984	1300	1300	1300
H [mm]	10	2	1	2	1	10	1	10	2
V [m/s]	50	50	50	80	80	80	120	120	120
ρ [kg/m^3]	7800	7800	7800	2500	2500	2500	2300	2300	2300
k	1	1	2	1	1	2	1	1	2

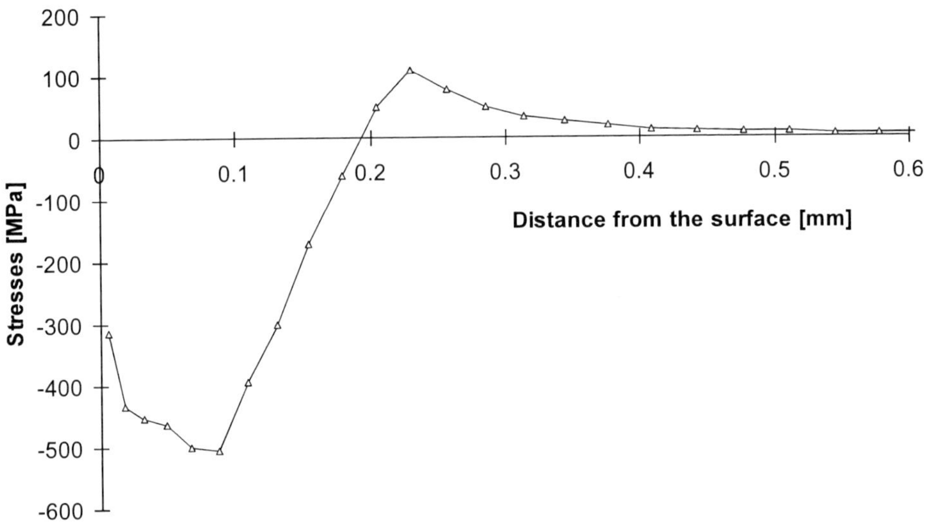

Figure 41. Numerical residual stresses vs. Distance from the surface of the specimen diagram for the Taguchi's $L_9(3^4)$ array first run in [43].

The residual stress field (Figure 41) and the dimension of the mark left by the spheres after the impacts on the specimen were carefully analyzed with the effects. In fact, particular attention was paid to the mark diameter (ϕ_{imp}) and depth (h_{imp}), to the maximum compressive residual stress ($\sigma_{res,max}$) and the depth ($h_{\sigma res,max}$) at which it took place, to the surface residual stress ($\sigma_{res,sup}$), to the compressed area (A-) – i.e. the area of the residual stresses vs. the distance from the surface of the specimen diagram in which the residual stresses are compressive – and to the depth at which the residual stresses turned into tensile stresses (h_{inv}).

The *MSD* parameter [45], which is a measure of the data dispersion, was used to assess the effects according to the two following different modalities: "bigger is better" for ϕ_{imp}, $\sigma_{res,max}$, $\sigma_{res,sup}$, A-, $h_{\sigma res,max}$, h_{inv} and "smaller is better" for h_{imp}. The following equations were used to calculate the *MSD*:

$$MSD = \frac{\frac{1}{Y_1^2} + \frac{1}{Y_2^2} + \ldots\ldots + \frac{1}{Y_n^2}}{n} \qquad \text{(bigger is better)} \qquad (30)$$

$$MSD = \frac{Y_1^2 + Y_2^2 + ... + Y_n^2}{n} \quad \text{(smaller is better)} \tag{31}$$

where Y_i are the values of the effects calculated with the nine runs and n the total number of specimens tested in each run. The knowledge of the *MSD* parameter allowed to determine the "signal to noise ratio" *S/N*, which provides a measure of both the location and the dispersion of the effects, by means of the following equation:

$$\frac{S}{N} = -10 \cdot \mathrm{Log}(MSD) \tag{32}$$

Table 19. Summary of the main results obtained with the analyses performed in [43]

Effect	Most important parameters	Best levels
A-	$\rho V^2/S$	3
	k	3
	h/Φ	3
h_{imp}	$\rho V^2/S$	2
	k	1, 2
	h/Φ	1
$\sigma_{res,max}$	$\rho V^2/S$	3
$h_{\sigma res,max}$	k	3
	h/Φ	3
$\sigma_{res,sup}$	$\rho V^2/S$	3
h_{inv}	k	3
	h/Φ	3
ϕ_{imp}	$\rho V^2/S$	1
	h/Φ	3

A "response table" was then arranged for each effect to analyze the results with suitable data interpretation criteria [45]. The application of the analysis of variance (ANOVA) and the execution of the F-Test [45] allowed the evaluation of the reliability of the results obtained. By analyzing the response table for each effect, the best levels of all the parameters were selected and, furthermore, the analysis of the empty column confirmed that all the influencing parameters had been considered. The F-Test also showed that all the important factors resulted to be significant with a percentage of confidence higher than 75%. Suitable confirmation runs highlighted the reliability of the results and hence the fact that different treatments having the same values of the non dimensional parameters are supposed to produce the same effects on the treated workpieces. In Table 19 the most important parameters and their levels are reported. Therefore, the optimization of each treatment effect was achieved by choosing a combination of the peening parameters, once the best values of the three non dimensional parameters – including the material, the geometry and the characteristics of the treatment and of the component to treat – had been known. Finally, it was remarked to carefully account for the thickness of the impinged target, since the results from the shot peening numerical simulations were obtained with axisymmetric models considering the impact of one or more spheres along the same axis. Such models prove reliable only for thick targets or when the bending stresses induced by the shot peening process are negligible.

CONCLUSION

This chapter reports some of the main results obtained by the authors et al. throughout several years of studies carried out to understand and predict the effects induced by surface treatments on different bulk materials. Among the treatments, thin hard coating deposition with PVD and CVD techniques and shot peening have been particularly investigated. Different bulk materials ranging from traditional construction steels to light alloys for advanced mechanical applications were analyzed. Particular attention was paid to the fatigue behaviour of coated components. It was pointed out that the surface residual stresses induced by the treatments represent one of the most important parameters affecting the enhancement (or the worsening) of the substrate fatigue behaviour. The presence of high surface residual compressive stresses usually proves effective in increasing the fatigue resistance by preventing crack nucleation and propagation. The good results collected for coated light alloys can be really interesting for many applications if the possibilty of lightening mechanical components is addressed.

Based on the experimental evidence from fatigue tests on coated and uncoated specimens, it was pointed out that the choice of the most effective coating for a certain bulk material is not simple and a wider range of substrate-coating combinations should be tested to determine the best solutions. In a few cases the coating was little detrimental to the fatigue resistance of the substrate material. Nevertheless, a limited worsening could even be accepted if one considers the enhancement in the corrosion resistance or in the tribological properties which can be achieved with the presence of a coating. The best results in terms of fatigue resistance were obtained for PVD-deposited CrN on the H11 and 2205 steel alloys and WC/C on the 2011-T6 aluminium alloy. Contact fatigue tests carried out on coated spur gears made of 18NiCrMo5 showed that a 10 μm-thick PVD CrN coating is unsuitable for gear applications. Large coating delamination and detachments were indeed observed during the tests.

A theoretical-numerical procedure was developed to calculate the number of cycles until failure for coated components. By applying this procedure to spur gears used in gearboxes for high speed racing motorcycles it was concluded that PVD thin hard coatings could allow the enhancement of the fatigue resistance of both steel and titanium gears. Furthermore, the coated titanium gears proved better than the coated steel ones and showed performances comparable with those of the carburized steel gears. As a matter of fact, the use of titanium instead of steel as a base material for gears would allow considerable mass reduction. The contact fatigue behaviour of coated gears has instead to be verified with more research activity, as from the results discussed, and paying particular attention to titanium alloy bulk materials. A preliminary theoretical-numerical procedure able to predict the number of cycles necessary for the subsurface contact fatigue damage initiation in gear tooth flanks has been developed by the authors.

Shot peening was studied with numerical models and by applying Design of Experiments to optimise the treatment parameters after their combination to obtain suitable non dimensional factors to use for different treatment conditions and materials. The numerical models were accurately refined to have reliable results – particular attention was paid to the residual stress trends generated by the impacts. The proposed optimisation procedure can represent a powerful tool for the shot peening parameter selection process.

REFERENCES

[1] Merlo, A. M. (2003). *Surf Coat Technol, 174-175*, 21-26.

[2] Vetter, J., Barbezat, G., Crummenauer, J. & Avissar, J. (2005). *Surf Coat Technol, 200*, 1962-1968.

[3] Mendibide, C., Steyer, P., Fontaine, J. & Goudeau, P. (2006). *Surf Coat Technol, 201*, 4119-4124.

[4] Guu, Y. H. & Hocheng, H. (2001). *Mater Sci Eng, A318*, 155-162.

[5] Mendibide, C., Steyer, P., Fontaine, J. & Goudeau, P. (2006). *Surf Coat Technol, 201*, 4119-4124.

[6] PalDey, S. & Deevi, S. C. (2003). *Mater Sci Eng, A342*, 58-79.

[7] Pharr, G. M. & Oliver, W. C. (1992). *MRS Bulletin*, 28-33.

[8] Brotzen, F. R. (1994). *Int Mater Reviews, 39(1)*, 24-45.

[9] Sinnott, S. B., Colton, R. J., White, C. T., Shenderova, O. A., Brenner, D. W. & Harrison, J. A. (1997). *J Vac Sci Technol A, 15(3)*, 936-940.

[10] Shojaei, O. R. & Karimi, A. (1998). *Thin Solid Films, 332*, 202-208.

[11] Kim, K. R., Suh, C. M., Murakami, R. I. & Chung, C. W. (2003). *Surf Coat Technol, 171*, 15-23.

[12] Stewart, S. & Ahmed, R. (2002). *Wear, 235*, 1132-1144.

[13] Inoue, K., Lyu, S., Deng, G. & Kato, M. (1996). *Proc VDI Berichte, 1320*, 357-369.

[14] Baragetti, S., La Vecchia, G. M. & Terranova, A. (2003). *Int J Fat, 25*, 1229-1238.

[15] Baragetti, S., La Vecchia, G. M. & Terranova, A. (2005). *Int J Fat, 27(10-12)*, 1541-1550.

[16] Baragetti, S. (2007). *Int J Fat, 29*, 1893-1903.

[17] Baragetti, S., La Vecchia G. M. & Terranova, A. (2001). *Int J Fat*, 23, S395-S404.

[18] Su, Y. L., Yao, S. H., Wei, C. S., Wu, C. T. & Kao, W. H. (1998). *Materials Letters, 35*, 255-260.

[19] Berrios, J. A. (1998). *Surf Coat Technol, 108-109*, 466-472.

[20] Contreras, G. (1999). *Thin Solid Films, 355-356*, 480-486.

[21] Moinereau, D., Brochard, J., Guichard, D., Bhandari, S., Sherry, A. & France, C. (1996). *ASME-Publications-PVP, 324*, 267-276.

[22] Smith, R. W. (2000). Institute of Physics Publishing, 8(4), 629-648.

[23] Dejun, M., Kewei, X. & Jiawen, H. (1998). *Thin Solid Films, 323*, 183-187.

[24] Souza, R. M., Mustoe, G. G. W. & Moore, J. J. (1999). *Thin Solid Films, 355-356*, 303-310.

[25] Šraml, M. & Flašker, J. (2007). *Int J Adv Manuf Technol*, *31*, 1066-1075.
[26] Glodež, S., Aberšek, B., Flašker, J. & Ren, Z. (2004). *Engng Fract Mech*, *71*, 429-438.
[27] Barata, A., Cunha, L. & Moure, C. (2001). *Thin Solid Films*, *398(399)*, 501-506.
[28] Su, Y. L., Yao, S. H., Wei, C. S., Kao, W. H. & Wu, C. T. (1999). *Mater Scie Technol*, *15*, 73-77.
[29] Cunha, L., Andritschky, M., Pichow, K. & Wang, Z. (1999). *Thin Solid Films*, *355(356)*, 465-471.
[30] Baragetti, S. & Tordini, F. (2007). *Int J Fat*, *29*, 1832-1838.
[31] Baragetti, S., Lusvarghi, L., Bolelli, G. & Tordini, F. (2009). *Surf Coat Technol*, *203*, 3078-3087.
[32] Baragetti, S., Lusvarghi, L., Pighetti Mantini, F. & Tordini, F. (2007). *Key Engineering Materials*, *348-349*, 313-316.
[33] Baragetti, S. & Tordini, F. (2007). *Structural Durability & Health Monitoring*, *3*, 165-176.
[34] Baragetti, S., Terranova, A. & Tordini, F. (2008). *Key Engineering Materials*, *385-387*, 57-60.
[35] Farrahi, G. H., Lebrun, J. L. & Couratin, D. (1995). *Fat Fract Engng Mat Struct*, *18*, 211-220.
[36] Hammond, D. W. & Meguid, S. A. (1990). *Engng Fract Mech*, *37*, 373-387.
[37] Wagner, L., Müller, C. & Gregory, J. K. (1993). *Surf Engng*, 181-186.
[38] Natkaniec-Kokanda, D., Kokanda, S. & Miller, K. J. (1996). *Fat Fract Engng Mat Struct*, *19*, 911-917.
[39] Al-Obaid, Y. F. (1990). *J Appl Mech*, *57*, 307-312.
[40] Al-Obaid, Y. F. (1990). *Comput Struct*, *36*, 681-689.
[41] Al-Obaid, Y. F. (1994). *Comput Struct*, *52(4)*, 693-703.
[42] Baragetti, S. (1997). *Int J Mat Prod Technol*, *12*, 83-109.
[43] Baragetti, S. & Terranova, A. (2000). *Int J Mat Prod Technol*, *15*, 131-141.
[44] Baragetti, S. (2001). *Int J Comput Appl Tech*, *14*, 51-63.
[45] Condra, L. W. (1993). *Reliability Improvement with Design of Experiments*; Marcel Dekker Inc.: New York, NY.
[46] Montgomery, D. C. (1997). *Design and Analysis of Experiments*; 4th ed., John Wiley & Sons: New York, NY.
[47] Ichimura, H. & Ando, I. (2001). *Surf Coat Technol*, *145*, 88-93.
[48] Gelfi, M., Bontempi, E., Roberti, R. & Depero, L. E. (2004). *Acta Mater*, *52*, 583-589.
[49] Gelfi, M., La Vecchia, G. M., Lecis, N. & Troglio, S. (2005). *Surf Coat Technol*, *192*, 263-268.
[50] Inoue, K., Kato, M. & Yamanaka, M. (1989). *Proc Power Transmission Engineering Conf (ASME)*, 663-668.
[51] Kato, M., Deng, G., Inoue, K. & Takatsu, N. (1993). *JSME Int J-Series C*, *36(2)*, 233-240.
[52] Jiang, J., Arnell, R. D. & Tong, J. (1997). *Wear*, *211*, 254-264.
[53] Gadow, R. & Scherer, D. (2002). *Surf Coat Technol*, *151-152*, 471-477.
[54] Nie, X., Wilson, A., Leyland, A. & Matthews, A. (2000). *Surf Coat Technol*, *121*, 506-513.
[55] Dixon, W. J. & Massey, F. J. Jr. (1983). *Introduction to Statistical Analysis*; McGraw-Hill: New York, NY, 386-394.

[56] Juvinall, R. C. & Marshek, K. M. (1991). *Fundamentals of Machine Component Design*; John Wiley & Sons: New York, NY.
[57] Wilson, A. D., Leyland, A. & Matthews, A. (1999). *Surf Coat Technol*, *114*, 70-80.
[58] Haritos, G. K., Nicholas, T. & Lanning, D. B. (1999). *Int J Fat*, *21*, 643-652.
[59] Lanning, D. B., Haritos, G. K. & Nicholas, T. (1999). *Int J Fat*, *21*, S87-S95.
[60] Baragetti, S. (2007). *Proc ISOPE*, 2894-2901.
[61] *ASM Metals Handbook*; 10th ed., ASM International: Materials Park, OH, 1993.
[62] Suresh, S. (1998). *Fatigue of Materials*; 2nd ed., Cambridge University Press: Cambridge, UK.
[63] El-Haddad, M. H., Smith, K. N. & Topper, T. H. (1979). *ASME J Engng Mater Technol*, *101*, 42-46.
[64] Murakami, Y. & Endo, M. (1994). *Int J Fat*, *16*, 519-533.
[65] Miller, K. J. (1982). *Fat Fract Engng Mater Struct*, *5(3)*, 223-232.
[66] Paris, P. C. & Erdogan, F. (1963). *J Basic Engng*, *85*, 528-534.
[67] Elber, W. (1970). *Engng Fract Mech*, *2*, 37-45.
[68] Garcés, Y., Sánchez, H., Berríos, J., Pertuz, A., Chitty, J., Hintermann, H. & Puchi, E. S. (1999). *Thin Solid Films*, 355-356.
[69] Djouadi, M. A., Nouveau, C., Banakh, O., Sanjinés, R., Lévy, F. & Nouet, G. (2002). *Surf Coat Technol*, *151-152*, 510-514.
[70] Ejiri, S., Sasaki, T. & Hirose, Y. (1997). *Thin Solid Films*, *307*, 178-182.
[71] Murotani, T., Hirose, H., Sasaki, T. & Okazaki, K. (2000). *Thin Solid Films*, *377-378*, 617-620.
[72] Suh, C. M., Hwang, B. W. & Murakami, R. I. (2003). *Mater Scie Engng*, *A343*, 1-7.
[73] Higashida, Y., Burk, J. D. & Lawrence, F. V. Jr. (1978). *Weld Res Suppl*, *5083*, 334s-344s.
[74] MackAldener, M. & Olsson, M. (2001). *Int J Fat*, *23*, 329-340.
[75] Johnson, W. (1972). *Impact strength of materials*; Edward Arnold: London, UK.
[76] Timoshenko, S. P. & Goodier, J. N. (1982). *Theory of elasticity*; McGraw-Hill: Singapore.
[77] Johnson, K. L. (1985). *Contact Mechanics*; Cambridge University Press: Cambridge, UK.
[78] Flanagan, D. P. & Belytschko, T. (1981). *Int J Num Meth Engng*, *17*, 679-706.

In: Bulk Materials: Research, Technology and Applications
ISBN: 978-1-60692-963-6
Editors: T. Frias, V. Maestas, pp. 195-244

Chapter 4

DEVELOPMENT OF CONTROLLABLE CHEMICAL SYNTHETIC ROUTES TO ELEMENTAL SELENIUM AND SOME METAL CHALCOGENIDES AS BULK MATERIALS AND THIN FILMS

Biljana Pejova*

Institute of Chemistry, Faculty of Natural Sciences and Mathematics,
Sts. Cyril and Methodius University, POBox 162, 1001 Skopje, Macedonia.

ABSTRACT

The chemical routes to elemental selenium (both the amorphous, *i.e.* red and hexagonal, *i.e.* gray allotropic modifications) and a number of metal chalcogenide materials (such as Bi_2Se_3, CuSe, Cu_3Se_2, NiO, Co_3O_4, Fe_2O_3) which have recently been developed by our group are reviewed. Some of the materials, in particular the elemental Se and Cu_3Se_2, were synthesized for the first time by the chemical deposition method implementing these routes. The developed chemical methods are controllable in the sense that they enable synthesis of the mentioned materials with predefined (*i.e.* designed) properties, as both bulk precipitates and also in the form of thin films on various substrates. The synthesized materials were characterized from structural aspect and also their opto-electrical properties and sensing potentials were investigated. All synthetic approaches were based on controlled homogeneous precipitation reactions in aqueous solutions. Elemental selenium and metal selenides were synthesized using aqueous solution of selenosulfate ions as a precursor of the selenide ionic species, while the synthesis of metal oxides was essentially based on thermally-induced hydrolysis of urea (carbamide) in aqueous solutions. Measurements of the optical and thermal band gap energies of the synthesized Bi_2Se_3 enabled fundamental conclusions concerning the material's microstructure to be derived and also a higher bound to the Bohr's excitonic radius for bulk bismuth(III) selenide to be estimated for the first time in the literature. Optoelectrical investigations of the two copper selenides (CuSe and Cu_3Se_2) indicated the fundamental differences in the bulk band structures of these semiconductors. The synthesized elemental selenium, aside from its substantial engineering importance in

* Corresponding author: E-mail: biljana@iunona.pmf.ukim.edu.mk

xerographic technologies, was shown to exhibit a highly pronounced tendency towards conversion to other metal selenides by very simple chemical treatments. Chemically deposited α-phase of Fe_2O_3, especially when this material is synthesized in thin film form, was shown to exhibit a pronounced sensitivity of its electrical resistivity to moisture. Calibration of the dependence of the α-Fe_2O_3 thin films' resistivity on the moisture content in the atmosphere indicated potential applicability of this material in moisture sensing systems. Important information concerning the reaction pathways to nickel and cobalt hydroxide and oxide phases from the "urea system" has been derived by structural analyses of as-deposited and thermally treated materials ($Ni(OH)_2$, $Co(OH)_2$, NiO and Co_3O_4).

1. INTRODUCTION

In recent years, inorganic semiconductor materials, synthesized either in bulk form or as thin films, have become of considerable interest to various fields of contemporary science and technology. The unique electrical properties of semiconductor materials permit their use in devices (sometimes termed solid-state devices) to perform specific electronic functions [1,2]. The invention of semiconductor devices is to a great extent responsible for extremely rapid growth of semiconducting thin film materials technology. Metal chalcogenide compounds, having a semiconductor nature, are of considerable technical interest in the field of electronics and electro-optical devices. Intensive research has been performed in the past in the course of fabrication and characterization of these compounds in either bulk form or in the form of thin films. Many techniques for fabrication of thin films of these materials have been developed, such as solution growth, electrodeposition, physical vapor deposition, chemical vapor deposition, spray pyrolysis etc. [3]. The solution growth technique (also known as chemical bath deposition, electroless plating or controlled precipitation) is probably the simplest one, in a sense that it does not require sophisticated equipment to actually carry out the deposition process. It is also among the most economical methods for materials synthesis, which is at the same time convenient for both small- and large-area deposition. Any kind of shape or size of the substrates can be used, while the starting chemicals (precursors) are often commonly available and cheap materials.

Following the current trends in microelectronics materials, continuing our work in the field of thin films of semiconducting materials, in recent years we have developed novel solution growth techniques for fabrication of elemental selenium [4], Bi_2Se_3 [5,6], Cu_3Se_2 [7], CuSe [7], iron oxide [8], nickel oxide [9] and cobalt oxide [10] in the form of bulk precipitates and in thin film form. In the present review paper, these synthetic methods developed by our group are reviewed, putting a special emphasis on the chemistry of deposition processes, as well as on the structural, optical and electrical properties of the synthesized materials, from the viewpoint of their potential applicability. In the Introduction section, we will briefly focus on the relevance of the title materials, as well as on previous research which has been devoted to their synthesis and characterization.

Concerning elemental selenium, a number of methods, such as photodeposition, RF sputtering, vacuum evaporation, vacuum deposition by electrochemical decomposition [11-14] have been reported for preparation of this elemental semiconductor in thin film form. However, to the best of our knowledge, prior to our published results [4], the solution growth methodology has not been used for the preparation of thin films of selenium.

Several allotropic forms of elemental selenium are known–monoclinic, hexagonal and amorphous [15,16]. The three red monoclinic phases (α, β and γ) consist of puckered Se_8 rings. Thermodynamically most stable form of the element is the hexagonal *i.e.* the gray, "metallic" selenium, consisting of long helical chains. There are two amorphous forms of selenium-red and black. Both of these forms are characterized with complex and irregular structure. The black selenium consists of large polymeric rings having up to 1000 atoms per ring. The red, amorphous selenium has a deformed chain structure, high resistivity ($\rho = 10^{13}$-10^{16} Ω·cm) [17] and upon heat treatment transforms to stable hexagonal modification. Recently, *cyclo*-Se_6 and *cyclo*-Se_7 red monoclinic forms were synthesized. Concerning the electrical properties of this element, the only modification, which conducts electricity, is gray selenium. In the literature, it is reported that gray selenium is characterized by resistivity $\rho = 10^4$-10^6 Ω·cm [17]. Impurities and defects substantially influence its electrical conductivity. It is known that, bellow its melting point, selenium is a p-type semiconductor.

In recent years, the great interest in various allotropes of selenium and their interconversion is due to its use in electronics and solid-state technology [12,13,15,18]. Selenium exhibits both photovoltaic and photoconductive action. These properties make selenium useful in the production of photocells and exposure meters for photographic use, as well as solar cells. Selenium has been also extensively used in rectifiers, in xerography, in glass industry to decolorize glass and to make rubycolored glasses and enamels. It is also used as a photographic toner and as an additive to stainless steel.

Semiconducting thin film materials, which are based on sulfides, selenides and tellurides, have recently attracted much interest as materials for optoelectronics technology. Among these, bismuth selenide, a member of the family of A^VB^{VI} semiconducting materials, has been extensively studied due to possible application in various fields. This particular material selenide has potential applications as a photographic film in infrared photography [19], in production of optical and photosensitive devices [20] and for modern thermoelectric cooling modules [21].

Thin films of bismuth(III) selenide have been prepared using various techniques such as solution growth [19,22,23], the SILAR method [24], electrodeposition [25] and the molecular jet method [26]. The most exploited method for fabricating bismuth(III) selenide films is the solution growth approach, which is, as discussed before, mainly due to its simplicity and because it is one of the most economical synthetic methods in materials science.

The interest for semiconducting thin films of copper selenides has been motivated by their application in solar cell technology [27-29]. In the literature, a number of methods for preparation of Cu_xSe thin films (where x = 1, 2 or between 1 and 2) have been reported [30-37]. However, thin films of stoichiometrical binary copper selenide Cu_3Se_2 are much less exploited materials. While the revised version of our paper devoted to the developed new method for chemical deposition of Cu_3Se_2 thin films was near completion [7], a paper dealing with chemical bath deposition of Cu_3Se_2 and CuSe films appeared [38]. So, actually, our work was one of the first two parallel researches devoted to Cu_3Se_2 in thin film form.

During the last decade much importance has been attached to the preparation of thin films of various metal oxides as well. This is due to the possibility for application of these materials in the production of technical components for electrical devices, gas sensors, photoelectrochemical solar cells, etc. [39-43].

Gas sensors based on metal oxide semiconducting thin films can be constructed on the basis of conductivity changes upon adsorption and desorption of particular gases on their surfaces [39]. α-Fe_2O_3 thin films have been found to be highly sensitive to some reducing gases, such as methane [39,42] and alcohol [39], without the necessity for application of noble metal catalysts [44]. The gas sensitivity of α-Fe_2O_3 increases upon doping with SO_4^{2-} anions and/or quadrivalent metal cations [45]. In the basic sensing mechanism of α-Fe_2O_3, the grain size has been found to play an important role [46-48].

Thin films of α-Fe_2O_3 are suitable for application in photoelectrochemical solar cells (PEC) [41] as well. Previous investigations have shown that α-Fe_2O_3 is a promising semiconductor anode for the photodecomposition of water in a PEC cell [49], quite stable against photo-corrosion [50,51]. The only problem related to this particular application of this material is the high room temperature resistivity of α-Fe_2O_3 [52], which results in a low PEC efficiency. It has been reported on the basis of the calculated theoretical quantum efficiencies that Cu^{2+}-doped α-Fe_2O_3 could serve even as a better photoanode in PEC cells for water decomposition than the pure α-Fe_2O_3 or Mg^{2+}-doped α-Fe_2O_3 [40].

A number of techniques have been developed for fabrication of iron oxide thin films such as sputtering [53], pulsed laser evaporation [54], chemical vapor deposition [55], spray pyrolysis [56], sol-gel [57], spray-spin-coating [58] etc.

In the context of metal oxide materials, synthesized in bulk or in thin film form, it is particularly important to mention that in recent years, the technology of electrochromic thin films has become more and more attractive for the researchers. The interest for electrochromic materials is based on their potential application to energy efficient "smart windows", glare-free and variable reflectance mirrors, high-contrast non-emissive information displays, switchable displays, devices for thermal control, semiconductor-based sensors etc. [59-72]. Electrochromic devices can change their optical properties under an action of applied voltage pulse [72]. Materials with electrochromic properties are classified in two categories: cathodiccally coloring ones which darken under charge *insertion* and anodiccally coloring ones which darken under charge *extraction* [59]. Among various electrochromic oxides of transition metals, nickel oxide is an interesting material due to its rather high electrochromic efficiency (35-50 cm^{-2} C^{-1}) [73,74], large dynamic range and low material cost [64,75]. The thin nickel oxide films exhibit anodic coloration [59]. It is generally accepted that the mechanism of the electrochromism in the case of nickel oxide material *i.e.* the transition from a colored to a bleached state, is based on the redox process [76,77]:

$$Ni^{3+} + e^{-} = Ni^{2+} \tag{1}$$

The phenomenon of anodic electrochromism of nickel oxide allows potential application of this material as a counter electrode in electrochromic devices where coloring/bleaching changes are increased due to the simultaneous bleaching and coloring of anodic and cathodic films. Other important applications of nickel oxide include preparation of alkaline batteries (as a cathode material), antiferromagnetic layers, p-type transparent conducting films [75,78,79]. Nickel oxide in its lithiated state is also important as a Li^{+} insertion compound for lithium rocking-chair batteries and may have promising electrochromic performance [80,81].

Nickel oxide thin films have been prepared by various methods. Some of the recent works have employed evaporation, electrodeposition, sputtering, oxidation of nickel at high temperatures, solution growth techniques, thermal decomposition, chemical vapor deposition, spray pyrolysis, colloidal deposition, sol-gel route etc. [78,82-100].

The preparative aspects of cobalt oxide thin films have been a subject of investigations by various workers because of their numerous applications in various fields of technology. Thus, they are attractive in application to solar thermal energy collectors as selective absorbing layers [101]. Cobalt oxyhydroxide is electrochromic material, which exhibits anodic coloration [102]. It has been recognized that the electrochromic transition of cobalt oxide is based on the redox process [103]:

$$3CoOOH + e^- + H^+ = Co_3O_4 + 2H_2O \quad (2)$$

where the green product is cobalt(III) oxyhydroxide having a slight stoichiometric excess of oxide [104] and the colored form of the electrochrome is dark brown/black Co_3O_4 (a mixed-valence material that is formally $Co^{II}Co_2^{III}O_4$) [105]. The phenomenon of anodic electrochromism of cobalt oxide allows potential application of this material as a counter electrode in electrochromic devices [102,106-108]. The interest for devices based on electrochromic cobalt oxide-based materials is due to their potential application to energy efficient “smart windows”, glare-free and variable reflectance mirrors, high-contrast non-emissive information displays, switchable displays, devices for thermal control, semiconductor-based sensors [109] etc.

A number of techniques have been developed for fabrication of cobalt oxide thin films such as sputtering [110], chemical vapor deposition [101], spray pyrolysis [111], dip-coating [112], electrolytic method in liquid phase [113-115], electrodeposition methods [107,108], chemical bath deposition [116] etc.

The present paper reviews our previous works in the field of synthesis and characterization of the title materials. It is organized as follows. In Chapter 2, the developed synthetic routes to the mentioned materials are reviewed, with an emphasis on the chemistry of deposition processes. Chapter 3 deals with structural characterization of the synthesized materials, applying the X-ray diffraction methods, while Chapter 4 is devoted to materials' characterization with infrared spectroscopy methods. In Chapter 5, optical properties of the synthesized chalcogenide materials, investigated by the UV-VIS-NIR spectroscopic methods are reviewed and analyzed. Chapter 6, on the other hand, is devoted to the study of electrical properties.

2. Chemical Synthesis of Elemental Selenium and Metal Chalcogenide Materials as Bulk Precipitates and in Thin Film Form

2.1. Pre-Deposition Treatment of the Substrates

The thin films considered in the present review were deposited either on standard microscope glasses or on transparent flexible polyester sheets (Zweck Form, Catalogue No.

3555), which were cut with the dimensions of ca. standard microscope glasses. For some systems, thin films were deposited on both types of substrates. Prior to the deposition process, the used substrates were treated in a diluted solution of tin(II) chloride and than annealed at 200 ^{o}C for about twenty minutes. During this substrate treatment, small and stochastically distributed crystals of tin(II) oxide are created onto the glass surface, which, afterwards, initiate heterogeneous nucleation and improve film adhesion [4-10,117-133]. This procedure is usually called "activation" of the substrates.

2.2. Synthetic Method for Deposition of Elemental Selenium in Thin Film Form or as Bulk Precipitate

The method for selenium thin films solution growth, which was developed by our group [4], is based on hydrolytic decomposition of selenosulfate anions to hydrogenselenides and their successive oxidation to elemental red selenium in acid medium at low temperature. As-deposited material is the amorphous allotropic modification of this elemental semiconductor, which, after post-deposition thermal annealing treatment, transforms into the gray allotrope. The developed method used a solution of sodium selenosulfate as a selenium precursor and citric acid or ascorbic acid as oxidizing agent. Solution of sodium selenosulfate was prepared by dissolving elemental Se in a sodium sulfite solution and heating the obtained mixture for about 1 hour at approximately 90 °C. The excess (unreacted) selenium was afterwards removed by filtration. The same procedure for preparation of selenosulfate species solution was used also for synthesis of metal selenides (Bi_2Se_3, Cu_3Se_2 and CuSe) considered in the present review. It has to be kept in mind, however, that the synthesized solution of sodium selenosulfate is relatively unstable, especially when exposed to light. It therefore had to be prepared prior to thin film deposition process and also kept in a dark container. Selenosulfate anionic species generate H_2Se by spontaneous hydrolysis:

$$SeSO_3^{2-}(aq) + H_2O(l) = H_2Se(aq) + SO_4^{2-}(aq) \qquad (3)$$

H_2Se further dissociates in two steps [15], generating free Se^{2-} ionic species in the solution. Due to the very low value of the second ionization constant of H_2Se [134,135], however, free selenide anionic species are present in the solution in very small concentrations. Despite that, these anions may be relatively easily oxidized to elemental selenium even by the oxygen dissolved in the reaction solution [15,134]:

$$Se^{2-}(aq) + H_2O(l) + 1/2O_2(g) = Se(s) + 2OH^{-}(aq) \qquad (4)$$

Occurrence of this process is prevented in the selenosulfate solution by the sulfite ions, due to the reducing properties of these species, which are enhanced in alkaline medium. As implied by the Nernst equation for the corresponding redox pair, on the other hand, the oxidation properties of the dissolved oxygen become less pronounced in alkaline medium:

$$O_2(g) + 4H^{+}(aq) + 4e = 2H_2O(aq) \quad E^{\circ}_{O_2/H_2O} = +1.23\ \mathrm{V} \qquad (5)$$

$$E_{O_2/H_2O} = E^{\circ}_{O_2/H_2O} + \frac{0.059V}{4}\log a(O_2) - 0.059VpH \quad (6)$$

While in the present case the deposition of elemental selenium films was carried out in acidic medium, as could be concluded from the previous discussion, if one wants to avoid oxidation of selenide ions to Se (e.g. during the synthesis of metal selenides considered further in this review), the pH value should be kept rather high.

The actual deposition solution used for synthesis of elemental selenium in thin film form and as bulk precipitate was prepared by mixing 4 ml 0.1 mol/dm^3 citric (or ascorbic) acid solution and 3 ml selenosulfate solution. Deionized water was further added to the reactor to make a total volume of 100 ml. The film deposition was carried out at temperature of about 0 °C. From one chemical bath, the terminal thickness of the obtained films is about 100 nm. Thicker films could be prepared by re-inserting the initially deposited films into a fresh bath. As implied before, chemical deposition of elemental selenium is based on the process of slow decomposition of selenosulfate in acid solution and subsequent slow oxidation of the produced anions. The process of deposition can be homogeneous (*i.e.* in the bulk of the solution) or heterogeneous (*i.e.* at the substrate surface) [136]. Heterogeneous deposition process leads to the film fabrication. For determination of type of film growth mechanism (*i.e.* crystal growth mechanism) light scattering experiments were carried out on deposition solution. Shining short wavelength laser light (400 nm) through the solution allowed the presence or absence of a colloid in the solution to be seen, by visual observation, due to scattering of the laser beam by colloidal particles. The formation of red selenium can be described by the sequence of the following three processes:

$$SeSO_3^{2-} + H_2O = H_2Se + SO_4^{2-} \quad (7)$$

$$H_2Se + H_2O = H_3O^+ + HSe^- \quad (8)$$

$$C_3H_4(OH)(COOH)_3 + HSe^- + H^+ = C_3H_4(OH)(COOH)_2CHO + Se + H_2O \quad (9)$$

The reduction of carboxylic groups of citric acid may subsequently proceed. Elemental selenium thin films were successfully deposited by the described method on flexible polyester substrates.

According to the light scattering experiments, it was found that the growth of elemental selenium films is based on the cluster mechanism throughout the main part of the deposition. For the most part of the deposition process, the laser beam was significantly scattered by colloidal particles and bulk precipitation was observed.

2.3. Thin Films and Bulk Precipitates of Bismuth Selenide and Two Copper Selenides (CuSe and Cu_3Se_2)

Bismuth(III) selenide films were deposited onto glass and polyester substrates from a solution containing soluble complexes of bismuth ionic species and selenosulfate ions. The proposed route is based on a hydrolysis of selenosulfate to selenide anions in alkaline solution accompanied by parallel slow dissociation of the complexed ions. The following stock solutions were prepared: 0.1 mol dm^{-3} $Bi(NO_3)_3$, 0.1 mol dm^{-3} sodium citrate, 10 % ammonia and sodium selenosulfate. The solution of sodium selenosulfate was prepared as explained in the previous chapter. To prevent the hydrolysis of bismuth(III) ions in aqueous media, $Bi(NO_3)_3 \cdot 5H_2O$ was dissolved in triethanolamine. The final composition of solution for the film growth was 5 cm^3 0.1 mol dm^{-3} non-aqueous solution of $Bi(NO_3)_3$, 20 cm^3 0.1 mol dm^{-3} solution of sodium citrate, 2 cm^3 10 % ammonia, 10 cm^3 solution of selenosulfate and deionized water to a total volume of 50 cm^3. The chemical deposition was carried out at about 60 °C. At the end of the deposition, the substrates were well washed and dried in air. The terminal thickness of the films was *ca.* 200 nm.

Chemical deposition of bismuth(III) selenide involves slow decomposition of selenosulfate in alkaline solution containing soluble complexes of bismuth ionic species, which allows us to control the Bi^{3+} ions concentration.

In alkaline medium, the selenosulfate anion hydrolyzes to hydrogenselenide and selenide ions:

$$SeSO_3^{2-} + OH^- = HSe^- + SO_4^{2-} \quad (10)$$

$$HSe^- + OH^- = H_2O + Se^{2-} \quad (11)$$

As implied in the previous chapter, and discussed in much more details in our previous publications (for a review, see e.g. Refs. [132,133]), the stability of selenide anionic species against their oxidation to elemental selenium is significantly enhanced in alkaline medium. This is due to the enhanced reducing properties of sulfite anions, which are present in excess in the sodium selenosulfate solution, which serve as a stabilizing agent for the selenide-releasing species, and also due to the decreased oxidizing properties of elemental oxygen in alkaline medium. From a thermodynamical viewpoint, these statements could be confirmed by analyses of the Nernst equation written for the corresponding redox pairs. This is shown in more details in our previous review papers [132,133]. In the growth solution, various complexes of bismuth with triethalonammine, ammonia and citrate are present. As a result of the slow dissociation of the soluble complexes, the concentration of Bi^{3+} ions in the solution is kept low.

As noted above, film growth mechanism was determined using coherent light scattering experiments. It is well known that if in the deposition solution bulk precipitation is not observed, then the film growth is based on the ion-by-ion mechanism. On the other hand, if the film growth is based on the cluster mechanism, colloidal species are formed in the deposition solution [136]. Throughout most of the bismuth(III) selenide film growth, bulk precipitation and scattering of the laser beam was observed suggesting the predominance of the cluster mechanism. However, towards the end of the deposition, bulk precipitation and light scattering were reduced. This may be due to a cross over to ion-by-ion growth. It could be concluded that both mechanisms may be involved in film formation towards the end of the deposition process.

According to the ion-by-ion mechanism, Se^{2-} and Bi^{3+} ions react forming bismuth(III) selenide. However, the deposition of Bi_2Se_3 can occur when the ionic product of Bi^{3+} and Se^{2-} exceeds the solubility product of Bi_2Se_3:

$$2Bi^{3+} + 3Se^{2-} = Bi_2Se_3 \qquad (12)$$

According to the cluster mechanism, film growth is based on absorption and coagulation of colloidal particles, which are, most probably, bismuth selenide or bismuth hydroxide species [136,137] in the present case. Selenide anions, generated by a slow hydrolysis of selenosulfate, could replace hydroxyl groups and convert bismuth hydroxide to bismuth selenide. This ability of the selenide anions is due to the lower solubility product of Bi_2Se_3 compared to that of bismuth(III) hydroxide. The grain (*i.e.* crystal) size of bulk precipitates and thin films is dependent upon the relative rates of two processes which occur subsequently in the reactor - nucleation and grain (crystal) growth.

Formation of Cu_3Se_2 and CuSe thin films (and bulk precipitates) was also based on chemical decomposition processes occurring in alkaline media. The proposed methodology used trisodium citrate as a complexing agent for copper ions and aqueous solution of selenosulfate as a selenide anion precursor. For thin films depositions, the following stock aqueous solutions were prepared: 0.1 mol dm^{-3} $CuSO_4$, 0.1 mol dm^{-3} trisodium citrate, 1 mol dm^{-3} Na_2SeSO_3 and 1 mol dm^{-3} NaOH. The solution of sodium selenosulfate was synthesized as explained before. The decomposition of selenosulfate solution and formation of red, colloidal selenium is prevented by excess of sodium sulfite.

For Cu_3Se_2 film growth, the deposition solution was prepared by mixing equal volumes (6 cm^3) of the stock solutions of copper(II) sulfate and sodium citrate and adding 3 cm^3 solution of selenosulfate and deionized water to make a total volume of 100 cm^3. Upon adding the sodium citrate to copper(II) sulfate solution, a clear blue solution is created as a result of formation of complex between Cu^{2+} and Cit^{3-}. pH of the deposition solution was 9.

Previously treated flexible polyester substrates were vertically introduced in the deposition solution. The film deposition was carried out at approximately 60 °C. Upon heating for approximately 15 min, the deposition process starts and the color of the solution turns to dark brown. After the deposition process, the obtained thin films were rinsed with distilled water and dried in air. It takes several hours to deposit Cu_3Se_2 films by the proposed methodology (depending on the required thickness). The terminal thickness of deposited films, from one bath, was about 100 nm.

For copper(II) selenide (CuSe) thin film growth, the deposition bath was composed of 4 cm^3 0.1 mol dm^{-3} solution of copper sulfate, 4 cm^3 0.1 mol dm^{-3} solution of sodium citrate, 3 cm^3 solution of selenosulfate, 1 cm^3 1 mol dm^{-3} solution of sodium hydroxide and distilled water to a total volume of 100 cm^3. pH of the deposition solution was 12.

The treated substrates were vertically introduced in the deposition solutions. The film deposition was carried out at approximately 60°C. After the deposition process, the obtained thin films were rinsed with distilled water and dried in air. The terminal thickness of deposited CuSe, from onc bath, was about 200 nm.

Similarly as in the case of bismuth(III) selenide, chemical deposition of CuSe and Cu_3Se_2 is based on decomposition of selenosulfate in alkaline solution containing a copper(II) salt and a suitable complexing agent which allows us to control the Cu^{2+} ions concentration. Precipitation process can take place either in the bulk of the solution (homogeneous process) or at the substrate surface (heterogeneous process), the second leading to film formation.

In alkaline medium, the selenosulfate anion hydrolyzes to hydrogenselenide and selenide ions according to equations (10) and (11).

According to the ion-by-ion mechanism (*i.e.* growth by successive anion and cation adsorption on the growing crystal), the formation of copper(II) selenide can be written as follows:

$$[Cu(Cit)]^{-} = Cu^{2+} + Cit^{3-} \quad (13)$$

$$Cu^{2+} + Se^{2-} = CuSe \quad (14)$$

Namely, selenide anions are generated by the slow alkaline hydrolysis of selenosulfate, while Cu^{2+} cations are released by dissociation of the complexed ions. It is well-known that, in alkaline medium, copper(2+) ions in presence of organic compounds which contain hydroxyl groups (for example citric acid, tartaric acid etc.) do not precipitate copper(II) hydroxide, but form a blue colored complex. The deposition of CuSe occurs when the concentrations of copper(2+) and selenide(2-) ions are such as to exceed the solubility product of the CuSe.

In the case of Cu_3Se_2 bath deposition, the sulfite anion, having reducing property, can reduce Cu^{2+} into Cu^{+} ions. The deposition of Cu_3Se_2 can occur when the ion concentration product of Cu^{2+}, Cu^{+} and Se^{2-} exceeds the solubility product of the Cu_3Se_2.

$$2Cu^{2+} + SO_3^{2-} + 2OH^{-} = 2Cu^{+} + SO_4^{2-} + H_2O \quad (15)$$

$$Cu^{2+} + 2Cu^{+} + 2Se^{2-} = Cu_3Se_2 \quad (16)$$

According to the cluster mechanism, film growth is based on absorption and coagulation of colloidal particles. The colloidal particles are, most probably, metal chalcogenide or metal hydroxide species [136,137]. Chalcogenide (in our case selenide) anions, generated by a slow hydrolysis of selenosulfate, could replace hydroxyl groups and convert metal hydroxide to metal chalcogenide. This ability of the selenide anions is due to the lower solubility product of both CuSe and Cu_3Se_2 compared to that of copper hydroxide.

Analogously as in the case of Bi_2Se_3, in the case of copper selenides, light scattering experiments have also shown that the cluster mechanism is dominant in the growth processes for both film types. For the largest part of deposition processes, bulk precipitation, accompanied by scattering of the laser beam, is observed. However, at the end of the depositions, bulk precipitation and light scattering reduce. This may be due to the ion-by-ion growth. It can be concluded that both mechanisms may interact leading to the film formation at the end of the deposition process.

2.4. Synthesis of Iron, Nickel and Cobalt Oxides in Thin Film Form and as Bulk Precipitates

The deposition process which was implemented in our studies to synthesize Fe_2O_3 was actually based on a modification of a method previously reported by Chen *et al.* [138]. The film growth was based on thermal treatment of an aqueous solution of iron(II) chloride containing urea. The reaction mixture was set up from 50 cm^3 saturated solution of iron(II) chloride and 50 cm^3 1 mol/dm^3 solution of urea. The pH of final bath was 3. The film deposition was carried out at 90 °C for about 2 hours. Cleaned and activated glass substrates were vertically supported against the walls of a laboratory beaker. As-deposited films were washed in deionized water and annealed at temperature of 350-400 °C for about 24 hours. The post-deposition heat-treated films were uniform, homogeneous and excellently adherent to the substrate surface. The dependence of the film thickness on the deposition time is shown in Figure 1. As can be seen, the deposition is practically completed after about 2 hours. The terminal thickness is about 90 nm. Thicker films can be prepared by re-immersing the initially deposited thin films into a fresh bath.

The overall chemical reaction of the deposition process culminating in the formation of $Fe_2O_3 \cdot H_2O$ may be written as:

$$2Fe^{3+} + NH_2CONH_2 + 5H_2O = Fe_2O_3 \cdot H_2O + 2NH_4^+ + CO_2 + 4H^+ \quad (17)$$

Fe^{3+} cations in the chemical bath are provided by successive dissociation of iron(II) chloride and oxidation of Fe^{2+} ions. A chemical reaction of decomposition of urea is involved in the deposition process. When heated at higher temperatures (90-100 °C), urea decomposes to CO_2 and NH_3 according to the equation:

$$NH_2CONH_2 + H_2O = CO_2 + 2NH_3 \quad (18)$$

The free iron(3+) ions react with the hydroxide ions (released by the dissociation of ammonia) finally forming $Fe_2O_3 \cdot H_2O$ thin films. Upon aging in aqueous medium, $Fe_2O_3 \cdot H_2O$ is converted to γ-FeOOH which spontaneously transforms to thermodynamically more stable α-FeOOH. Upon annealing in air at 300-400°C, the deposited α-FeOOH is converted to α-Fe_2O_3 [139,140]. The described changes upon annealing are evident from the recorded X-ray diffractograms as well as from the FT IR spectra of the as-deposited and the post deposition treated materials, which are discussed in detail further in this review paper.

Since a bulk precipitation is observed during the largest part of the iron oxide film deposition process, accompanied with scattering of the coherent laser light, it seems that the predominant deposition mechanism is the cluster one (not excluding the ion-by-ion one at least at the end of the deposition) [136,141].

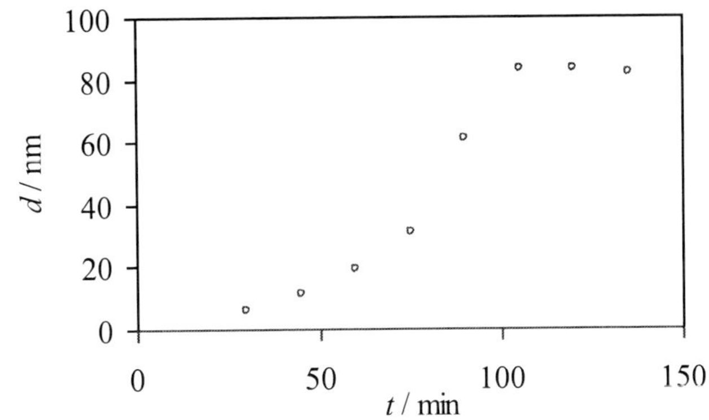

Figure 1. The dependence of the iron(III) oxide film thickness on the deposition time.

The nickel oxide films growth is based on a thermal treatment of an aqueous solution which contains nickel(2+) ions and urea. The deposition bath was prepared by mixing 50 cm^3 1 mol/dm^3 $Ni(NO_3)_2$, 25 cm^3 1 mol/dm^3 urea and adding deionized water up to a total volume of 100 cm^3. The pH of the final bath was 6. Previously prepared substrates were vertically supported against the walls of a 100 cm^3 laboratory beaker. Upon heating at 100°C about one hour, the reaction mixture slowly blurred. Within about two hours it became clear again and a green bulk precipitate of $3Ni(OH)_2 \cdot 2H_2O$ was formed at the bottom of the baker. Both sides of the substrates were covered with the same material as a uniform thin film. As-deposited films were washed in deionized water and annealed at temperature of 350-400 °C for about 48 hours. The annealed films were black/gray, uniform and homogeneous. The dependence of the film thickness on the deposition time is shown in Figure 2.

As can be seen, the deposition is practically completed after about 2 hours. The terminal thickness is about 0.9 μm. Thicker films can be prepared by re-immersing the initially deposited thin films into a fresh bath. Note that the accuracy of the gravimetric method is about 2 %, which may be straightforwardly shown accounting for the relative errors of the measured quantities (the mass and the linear dimensions of the deposited films) [142]. The accuracy of the gravimetric (weighing) method for determination of the film thickness has been the subject of a number of investigations based on comparisons with other, more reliable methods [143]. Such investigations have shown that the agreement between the gravimetric method and the more sophisticated ones is within few percents for films with thickness of about 100 nm, the accuracy being slightly worse for thicker films.

The overall chemical reaction of the deposition process may be expressed with the following equation:

$$3Ni^{2+} + 2NH_2CONH_2 + 10H_2O = 3Ni(OH)_2 \cdot 2H_2O + 4NH_4^{\ +} + 2CO_2 + 2H^+ \quad (19)$$

The Ni^{2+} cations and the hydroxide ions (released by the protolytic reaction of ammonia) react creating finally $3Ni(OH)_2 \cdot 2H_2O$ thin films. Upon annealing in air at 350-400°C, the as-deposited films converted to (hydrated) NiO, accompanied by a color change to black/gray. The described changes upon annealing are established by the recorded X-ray diffractograms as well as from the FT IR spectra of the as-deposited and the post-deposition treated films.

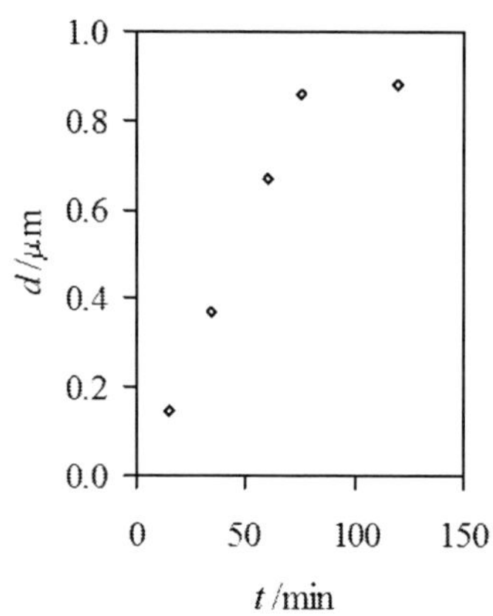

Figure 2. The dependence of the NiO film thickness on the deposition time.

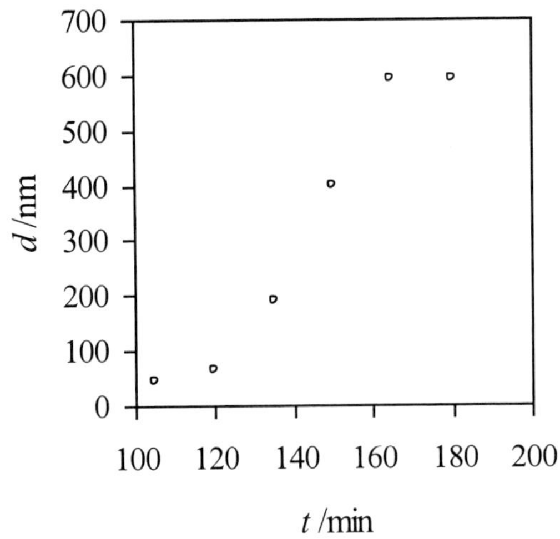

Figure 3. The dependence of the CoO film thickness on the deposition time.

Coherent light scattering experiments indicated that the predominant deposition mechanism is the cluster one (not excluding the ion-by-ion one at least at the end of the deposition) [136,141].

The cobalt oxide films growth is based on a thermal treatment of an aqueous solution, which contains 0.02 mol dm^{-3} cobalt (2+) ions and 0.06 mol dm^{-3} urea. Upon heating at 100°C, the onset of turbidity is achieved after about half an hour. At the bottom of the beaker a pink bulk precipitate of $Co(OH)_2$ was formed and the same material was deposited as a homogeneous thin film on both sides of the substrates. As-deposited films were washed with deionized water and annealed at temperature of 350-400 °C for about 48 hours. The annealed films were dark brown/black, uniform and homogeneous. On Figure 3 the plot of film thickness *vs.* deposition time is presented. As can be seen, the deposition is practically completed after about 3 hours. The terminal thickness is about 0.6 μm. Thicker films can be prepared by re-immersing the initially deposited thin films into a fresh bath.

By analogy with the solution deposition of iron(III) oxide and nickel oxide thin films [8,9], the following reaction is proposed as a basis for the chemistry of solution growth route to cobalt oxide:

$$Co^{2+} + NH_2CONH_2 + 3H_2O = Co(OH)_2 + 2NH_4^+ + CO_2 \qquad (20)$$

The Co^{2+} cations and the hydroxide ions (released by the protolytic reaction of ammonia) react, finally giving insoluble pink $Co(OH)_2$. Upon annealing in air at 350-400°C, the as-deposited films converted to Co_3O_4, which was accompanied by a color change to dark brown/black. It is known that $Co(OH)_2$ decomposes to Co_3O_4 [144] upon annealing in air at 350°C. The decomposition of Co_3O_4 to CoO in air was found to take place above 800°C while this transformation under vacuum occurs at 350°C. The direct decomposition of $Co(OH)_2$ to CoO was possible under vacuum conditions even at 140 °C [145]. The described changes upon annealing are established by the recorded X-ray diffractograms as well as from the FT IR spectra of the as-deposited and the post-deposition treated films.

As in the case of iron and nickel oxides, bulk precipitation throughout the deposition process accompanied with scattering of the coherent laser light, indicated the predominance of the cluster-by-cluster deposition mechanism.

3. Structural Characteristics of the Synthesized Materials

3.1. XRD Studies of Elemental Selenium

Analysis of the XRD patterns of the as-deposited selenium films (and bulk precipitates obtained from the same reactor) clearly indicated that the synthesized material was amorphous. As–deposited material is identified as amorphous, red selenium. The obtained *X*-ray diffractograms of the thermally treated material and the standard are shown on Figure 4 a and b, correspondingly According to the simulated XRD patterns [146], upon annealing at temperature of *ca.* 80°C, as-deposited material is transformed to the most stable modification, hexagonal gray selenium with unit cell parameters a = 4.3662 and c = 4.9536 Å. Annealed material is polycrystalline, it is characterized with random orientation and shows no sings of texturing.

The average crystal size was determined by Scherrer method. The Scherrer formula [147] relates the crystal size to the breadth of peaks in the X-ray diffraction patterns:

$$d = \frac{0.9 \cdot \lambda}{\beta \cdot \cos\theta} \tag{21}$$

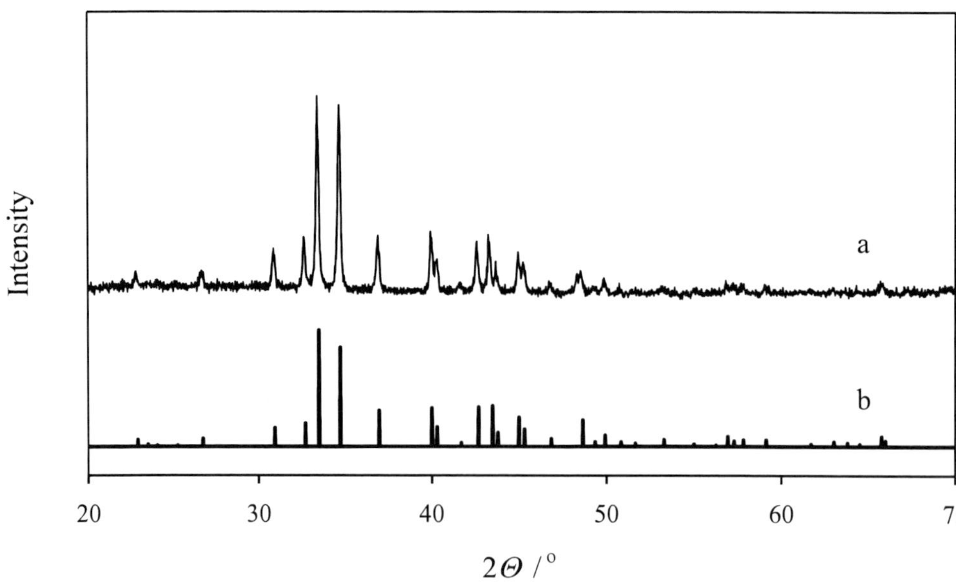

Figure 4. XRD patterns of the synthesized gray hexagonal selenium (a) upon thermal annealing of the as-deposited amorphous red selenium and standard (b).

where d is the average crystal size, θ is the Bragg angle, λ is the X-ray wavelength and β is the full width at half maximum (expressed in radians) of diffraction peaks of the sample. The estimated average crystal size of gray selenium material by this approach is 30 nm.

3.2. XRD Studies of Bismuth(III) Selenide

Figure 5 shows the recorded XRD patterns of the synthesized bismuth(III) selenide films (on glass substrates) and of the corresponding bulk precipitates obtained from the same reactors.

As can be seen, the as-deposited and annealed (at 350 °C) films have a glassy, amorphous structure. A broad 2θ peak, from 20 to 40° is characteristic of the glass substrate. The chemical composition of deposited films was identified by comparing the XRD pattern of the bulk precipitate, as well as that of the film with the standard data [148] and it was concluded that the material is bismuth(III) selenide.

The average grain size of bismuth(III) selenide bulk precipitate was determined using Debye-Sherrer's equation. Calculation based on the peak at $2\theta = 43.693$ ° gave a value of 8 nm.

The degree of purity of the material obtained is relatively high. Thus, the recorded XRD patterns do not show peaks due to the various forms of Bi_2O_3 or bismuth selenium oxide (Bi_2O_5Se) that could be also formed during the deposition process. These impurities, if present at all, appear in concentrations below the detection limit of the XRD technique. The previous statements are further supported by classical chemical analysis of the samples (for details, see the original paper [5]). Infrared spectra of the deposited material also did not show presence of bismuth oxide or mixed selenium oxide phases.

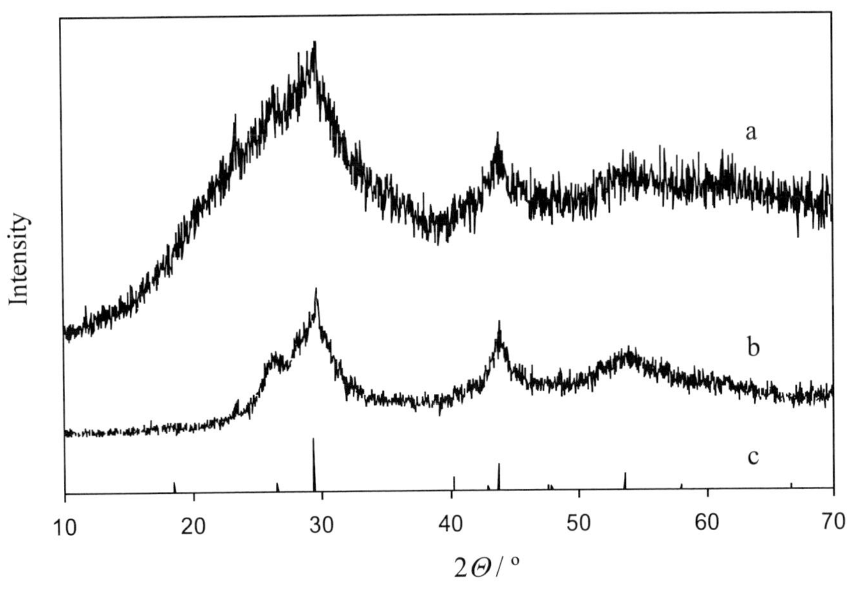

Figure 5. XRD patterns of as-deposited thin film of bismuth(III) selenide (a), bulk precipitate (b) and the standard (c).

3.3. XRD Studies of Cu_3Se_2

The recorded XRD patterns of as-deposited Cu_3Se_2 as bulk precipitate and in thin film form (deposited on polyester substrate) are presented in Figures 6 and 7. All peaks of the diffractograms were found to be characteristic of polycrystalline Cu_3Se_2, modification known as umangite, with a tetragonal structure and crystal parameters a = 6.402 Å and c = 4.276 Å [149]. According to the X-ray study, the obtained films consist of pure Cu_3Se_2 (i.e. the impurities, if present, are below the detection limit of the X-ray diffractometer). They are characterized with random orientation and show no sing of texturing.

The recorded XRD spectrum of annealed material (at *ca.* 120 °C in air) did not show significant improvement in crystallinity upon annealing. The average crystal size, estimated using the Debye-Scherrer relationship, is 42 nm.

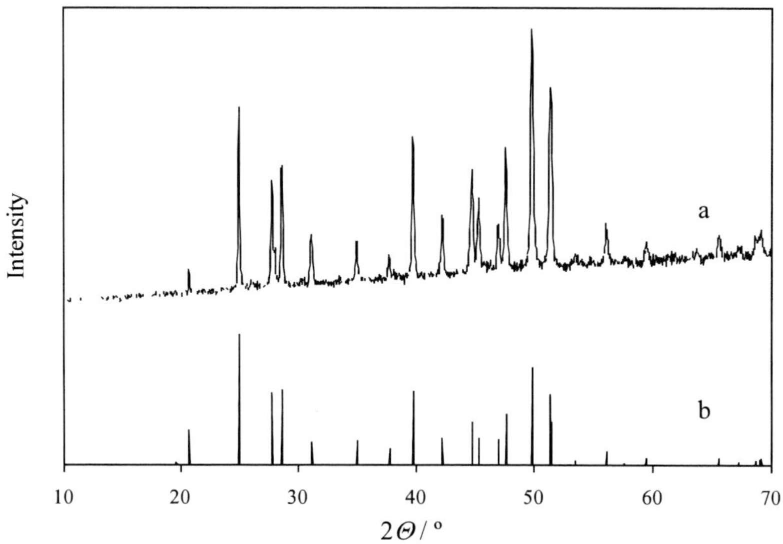

Figure 6. XRD patterns of as-deposited Cu_3Se_2 bulk precipitate (a) and the standard (b).

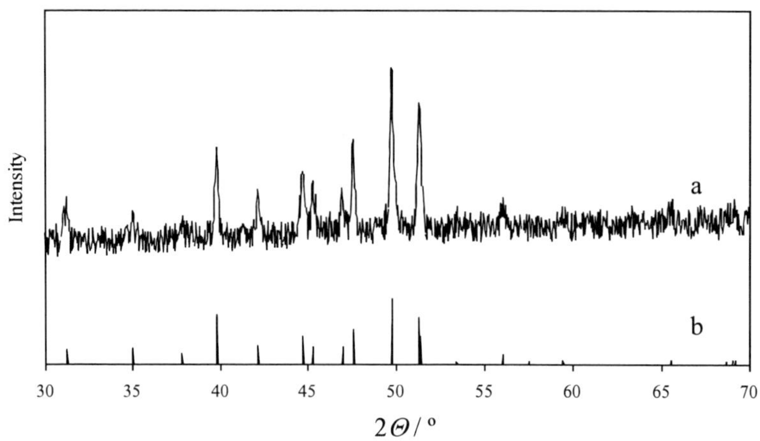

Figure 7. XRD patterns of as-deposited Cu_3Se_2 thin film on polyester substrate (a) and the standard (b).

3.4. XRD Studies of CuSe

The recorded XRD patterns of as-deposited copper(II) selenide bulk precipitates and thin films and are presented in Figures 8 and 9. As it can be seen, the comparison with the simulated XRD pattern [150] confirms that the obtained material is non-textured pure copper(II) selenide, modification known as klockmannite with hexagonal structure and crystal parameters a = 3.939 and c = 17.255 Å. Using the previously mentioned Debye-Scherrer relation, the average crystal size was found to be about 42 nm. In this case also annealing process practically does not improve the film crystallinity.

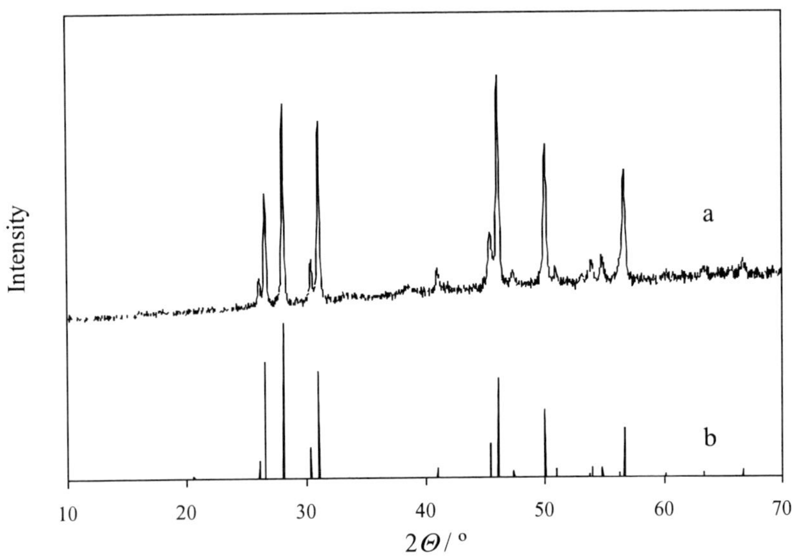

Figure 8. XRD patterns of as-deposited CuSe bulk precipitate (a) and the standard (b).

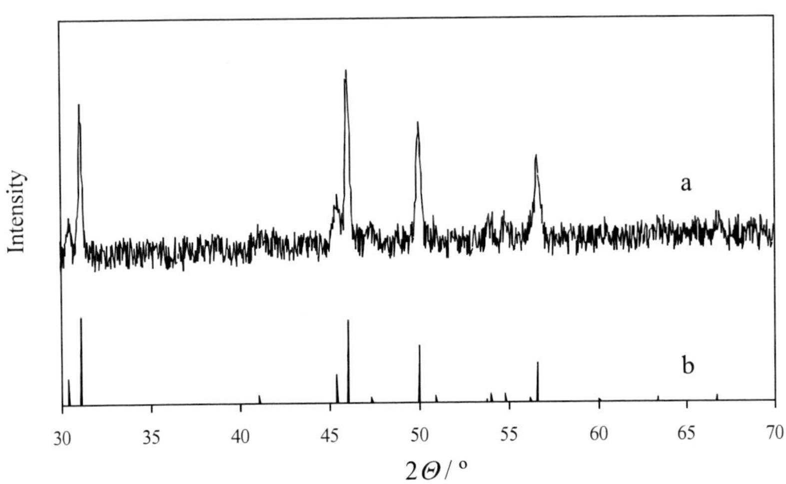

Figure 9. XRD patterns of as-deposited CuSe thin film on polyester substrate (a) and the standard (b).

3.5. XRD Studies of Iron Oxide

X-ray diffractograms for the as-deposited and annealed iron oxide films synthesized by the method described in Chapter 2.4 are presented in Figures 10 a and b correspondingly. As can be seen, the as-deposited (α-FeOOH) films are amorphous in nature, while the annealed ones (α-Fe_2O_3) are crystalline. The comparison of the observed diffraction peaks for the post-deposition heat-treated films with the standards (JCPDS power diffraction data set #201096 from the 84-0306 database) confirmed that the post-deposition treated material is α–Fe_2O_3. The Miller indices in the diffractograms were taken from the previously mentioned database. As can be seen from the X-ray patterns, the obtained α-Fe_2O_3 thin films are non-textured, contrary to those obtained by the previously proposed method [138]. Using the Debye-Sherrer expression [147], on the basis of the full width at half maximum intensity of the XRD peaks, we have calculated an average crystal size of 22 nm for the annealed films.

3.6. XRD Studies of Nickel Oxide

The results from the X-ray investigations of the material synthesized by the route described in Chapter 2.4 are presented in Figures 11 and 12. As can be seen from the X-ray difractograms of the as-deposited and the heat-treated material, the annealing process influences significantly the film structure. Also, an increase of the degree of crystallinity of the deposited material upon annealing is indicated. The observed diffraction peaks for the as-deposited and post-deposition heat-treated films are in a very good agreement with the standard values [151]. It is thus confirmed that the annealed material is $NiO{\cdot}xH_2O$, while the unannealed one is $3Ni(OH)_2{\cdot}2H_2O$. In the bulk, nickel oxide has a cubic (NaCl-type) structure with a lattice parameter of 0.4195 nm [152]. Using the Debye-Sherrer expression [147], on the basis of the full width at half maximum intensity of the XRD peaks, we have calculated an average crystal size of 13 nm for the annealed films. This value is accurate within 4 - 5 %, as may be shown accounting for the uncertainties in the full widths at half maximum intensity.

It is important to note at this point that the Debye-Scherrer equation is itself an approximation, since the particles within a given sample are of course not of a uniform size, but rather some type of probability distribution is in question. Therefore, one can speak only of an *average crystal size* within the sample [153,154].

3.7. XRD Studies of Cobalt Oxide

XRD patterns of obtained cobalt oxide materials (as-deposited and annealed) are presented in Figure 13 a and b, correspondingly. The comparison of the recorded pattern of annealed material (in air at 350-400°C) with the simulated XRD pattern (Figure 13 c) [151] confirmed that the annealed material is polycrystalline Co_3O_4.

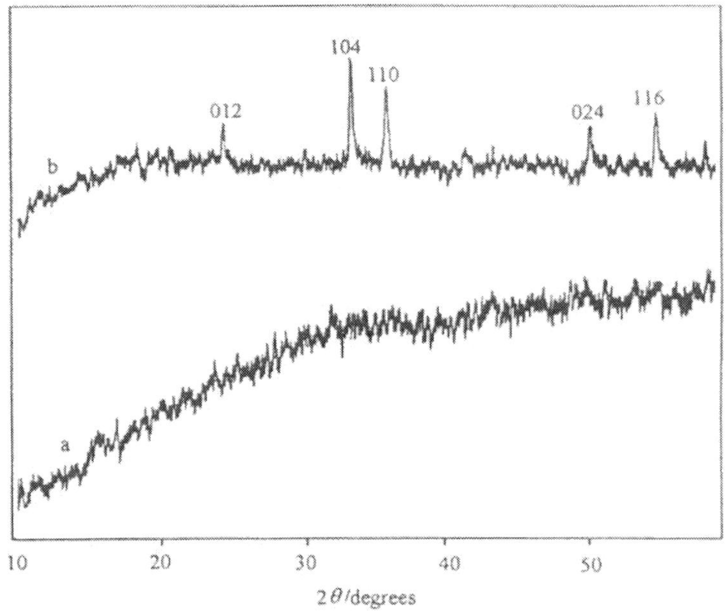

Figure 10. XRD patterns of as-deposited (a) and annealed iron oxide film (b).

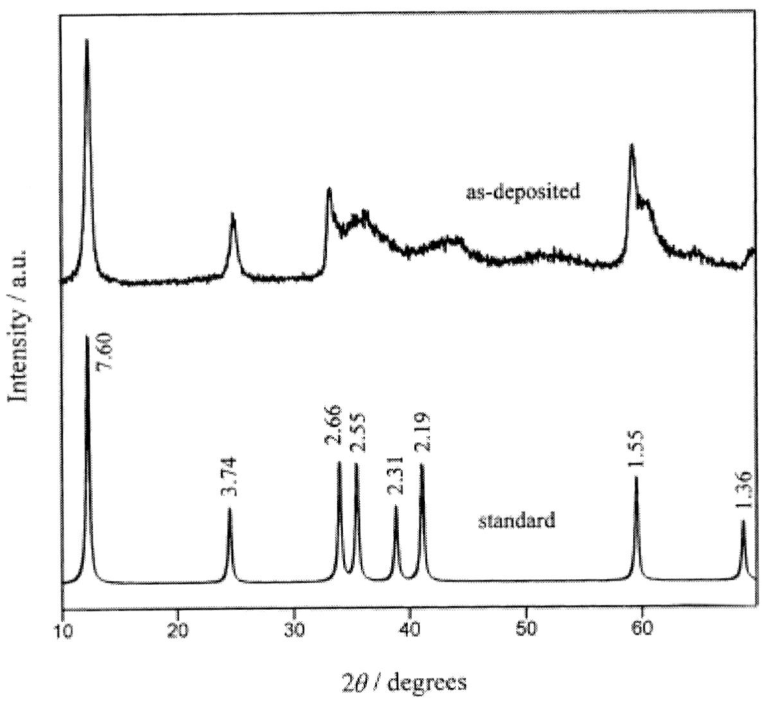

Figure 11. X-ray diffractograms of the as-deposited material ($3Ni(OH)_2 \cdot 2H_2O$) together with the standard (the peaks denoted by an asterisk correspond to the carrier - Al).

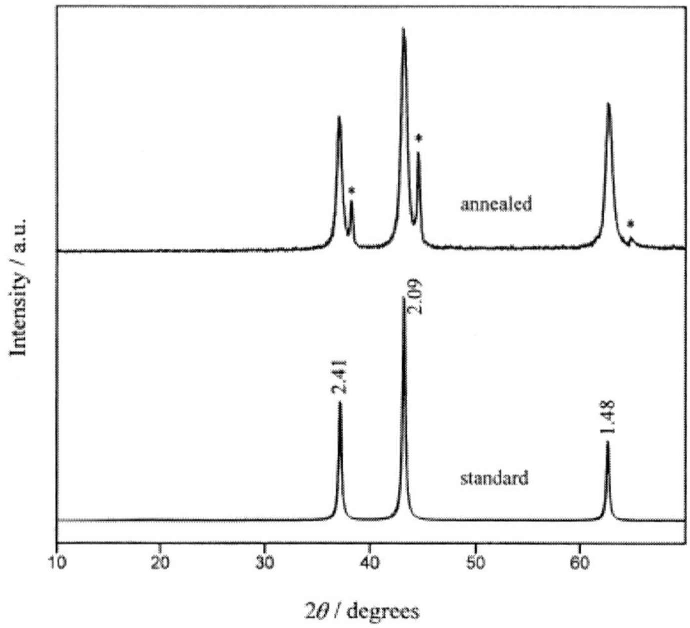

Figure 12. X-ray diffractograms of the post-deposition treated material at 350°C ($NiO \cdot xH_2O$) together with the standard (the peaks denoted by an asterisk correspond to the carrier - Al).

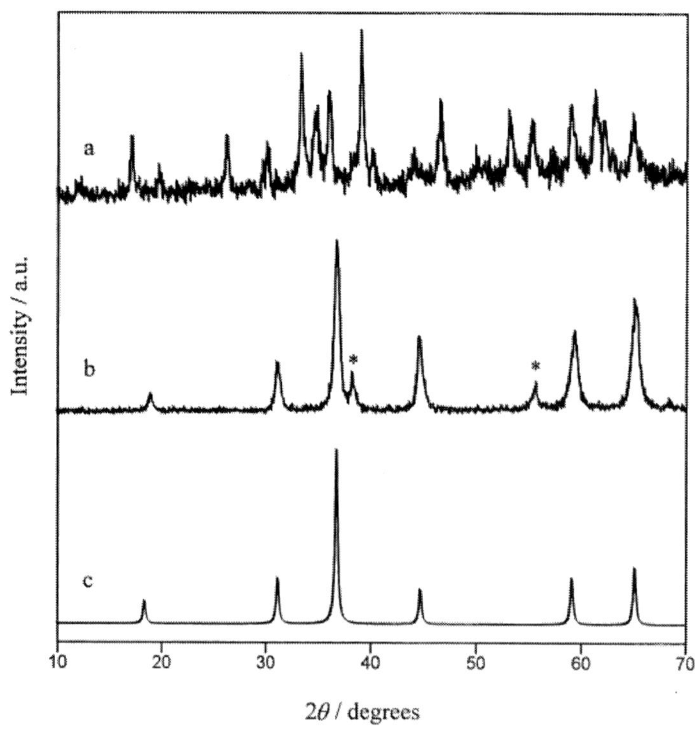

Figure 13. X-ray diffractograms of the as-deposited material (a) post-deposition treated one at 350°C (b) and simulated Co_3O_4 XRD pattern (c).

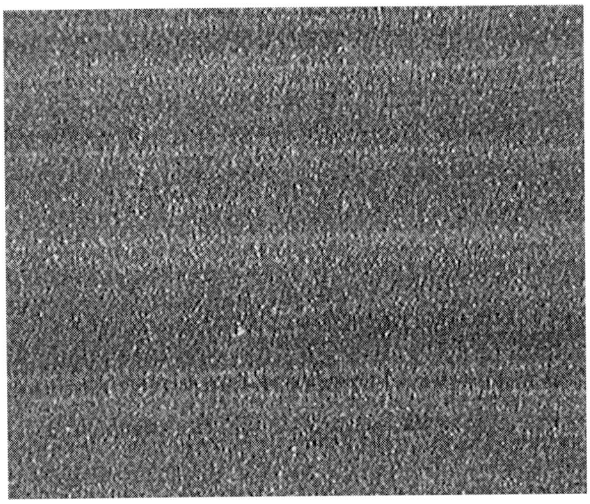

Figure 14. The dark-field microphotograph of a Co_3O_4 thin film.

The as-deposited material was found to be a mixture of different phases of $Co(OH)_2$ [151]. The annealed films are characterized by average crystal size of about 12 nm. This parameter was calculated by the Debye-Sherrer expression [147].

Figure 14 shows the dark – field microphotograph of an annealed thin film. As can be seen from the microphotograph, the obtained films are characterized by uniform and homogeneous microstructure.

4. IR Studies

In order to support the conclusions concerning structural properties of as-deposited and annealed materials, as well as to gain further conclusions on the deposition mechanism and materials structure, Fourier transform infrared (FT IR) studies of the as-deposited and thermally treated metal oxide materials were carried out. FT IR spectra (by the method of KBr pellets) were recorded on a Perkin-Elmer system 2000 FT IR interferometer.

4.1. IR Study of the Synthesized Iron Oxide Phases

Fourier transform IR spectra were recorded for the initially obtained iron oxide bath precipitate and for the post-deposition treated one. They are shown in Figure 15 a and b, correspondingly. The comparison of the recorded spectra with the standards [155-157] confirmed that the as-deposited material is FeOOH and the post-deposition treated one is Fe_2O_3. The band assignments for Fe_2O_3 according to Ref. [155] are given in Table 1. Perhaps the most drastic change in the spectra upon annealing is the disappearance of the IR bands corresponding to O-H stretching and bending modes (appearing at about 3450 and 1600 cm^{-1} respectively in the IR spectrum of FeOOH).

Table 1. Assignments of IR bands for the Fe_2O_3 sample according to Ref. [155]

Wavenumber/cm^{-1}	Assignment
570	ν_1
476	ν_3
444	ν_2

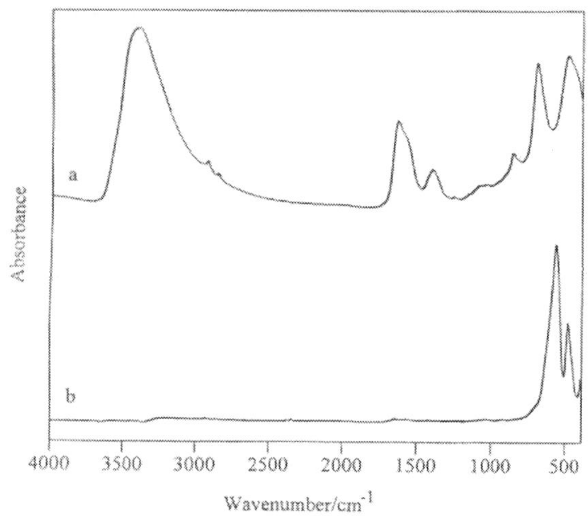

Figure 15. FTIR spectra of the bath precipitate (FeOOH) (a) and post-deposition treated iron oxide material (Fe_2O_3) (b).

4.2. IR Study of the Synthesized Nickel Oxide Phases

Fourier transform IR spectra were recorded for the initially obtained bath precipitate (and film) synthesized from the nickel/urea system, as well as for the post-deposition treated materials at 350-400 °C. They are shown in Figure 16. The comparison of the recorded spectra with the previously published data [76] confirmed that the as-deposited material is $3Ni(OH)_2 \cdot 2H_2O$, while the post-deposition treated one is hydrated NiO. The band appearing at ~ 3648 cm^{-1} in the FT IR spectrum of the as-deposited films reveals the presence of the weakly hydrogen bonded hydroxyde ions (the ν(OH) mode). As can be seen, this band disappears upon thermal treatment. However, the lower-frequency broad band, with centro-frequency of ~ 3440 cm^{-1} (corresponding to the hydrogen bonded ν(HO-H···O) mode) retains even upon annealing (shifting to 3470 cm^{-1}). It is interesting to note that the FT IR study confirmed the presence of the intercalated -NCO species in the as-deposited films. Namely, the very strong band appearing at ~ 2200 cm^{-1} (due to the ν_a(NCO) mode), together with those appearing at 1385 cm^{-1} (due to the ν_s(NCO) mode) and at 640 cm^{-1} (the δ(NCO) mode), all of which disappear upon thermal treatment of the films, prove the presence of these species [158]. Most probably, the isocyanate anions generated in the course of the hydrolysis of urea partially fill the Ni^{2+} coordination sphere, leading to growth of very homogeneous films (preventing the formation of larger agglomerates). Upon annealing the intercalated species are obviously eliminated from the films, which is reflected in the appearance of the

FT IR spectra. It is worth noting that the role of intercalated anions within the nickel oxide films may be very important for their electrochromic properties. However, there is only a limited number of detailed studies devoted to this subject [76]. The appearance of a strong band due to the ν(Ni-O) mode at ~ 430 cm^{-1} in the FT IR spectra of the annealed films, is a clear evidence for the presence of the crystalline NiO. The absence of the hydroxide group stretching mode at ~ 3640 cm^{-1} in the FT IR spectra of the annealed films eliminates the possibility of simultaneous presence of the phase with lower degree of crystallinity.

The summarized band assignments for the as-deposited and annealed films are presented in Table 2.

Table 2. The assignments of the most important bands in the FT IR spectra of $3Ni(OH)_2 \cdot 2H_2O$ and $NiO \cdot xH_2O$ thin films

$3Ni(OH)_2 \cdot 2H_2O$		$NiO \cdot xH_2O$	
Wavenumber/cm^{-1}	Assignment	Wavenumber/cm^{-1}	Assignment
3648	ν(OH)		
3440	ν(HO-H···O)	3470	ν(HO-H···O)
2200	ν_a(-NCO)		
1637	δ(HOH)	1633	δ(HOH)
1385	ν_s(-NCO)		
640	δ(-NCO)		
		430	ν(Ni-O)

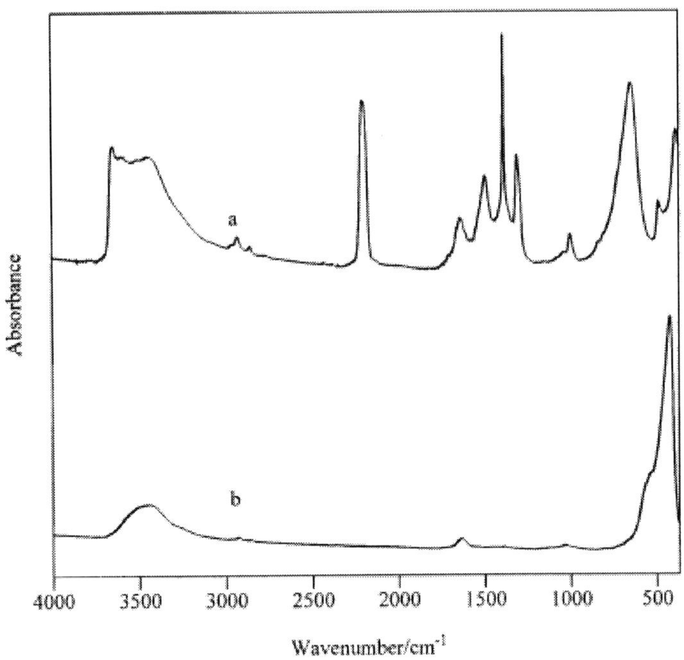

Figure 16. FT-IR spectra of the bath precipitate ($3Ni(OH)_2 \cdot 2H_2O$) (a) and post-deposition treated material (hydrated NiO) (b).

4.3. IR Study of the Synthesized Cobalt Oxide Phases

Fourier transform IR spectra were also recorded for the initially obtained bath precipitate and films from the cobalt/urea system, as well as for the post-deposition treated materials at 350-400 °C (Figure 17). The appearance of the IR spectra of both the as-deposited and post-deposition treated materials fully supports the previously outlined discussion. However, some important aspects regarding the deposition process and the obtained materials are enlightened more deeply. The analysis of the recorded spectra further confirmed that the as-deposited material contains $Co(OH)_2 \cdot xH_2O$ (a mixture of different phases), while the post-deposition treated one is hydrated Co_3O_4. Namely, the band appearing at ~ 3630 cm^{-1} in the FT IR spectrum of the as-deposited films (due to the ν(OH) mode) reveals the presence of the weakly hydrogen bonded hydroxyde ions. This band disappears upon thermal treatment, as can be seen from the spectra. However, the lower-frequency broad band corresponding to the hydrogen bonded ν(HO-H···O) mode (with centro-frequency of ~ 3490 cm^{-1}) retains even upon annealing (shifted to 3440 cm^{-1}). It is interesting to note that, analogously as in the case of nickel oxide, the FT IR study in the present case also indicated the presence of the intercalated -NCO species in the as-deposited films. Namely, the very strong band appearing at ~ 2220 cm^{-1} (due to the ν_a(NCO) mode), together with those appearing at 1385 cm^{-1} (due to the ν_s(NCO) mode) and at 630 cm^{-1} (the δ(NCO) mode), all of which disappear upon thermal treatment of the films, prove the presence of these species [158]. It could be reasonably postulated that the isocyanate anions, generated in the course of the hydrolysis of urea, partially fill the Co^{2+} coordination sphere, leading to growth of very homogeneous films (preventing the formation of larger agglomerates). Upon annealing the intercalated species are obviously eliminated from the films, which is reflected in the appearance of the FT IR spectra. It is worth noting at this point that, similarly as in the case of nickel oxide, the role of intercalated anions within the cobalt oxide films could be very important for their electrochromic properties. However, as mentioned also in the context of NiO films in the previous chapter, there is only a limited number of detailed studies devoted to this subject in general [76]. The appearance of two strong bands (due to the ν(Co-O) modes) at ~ 670 cm^{-1} and ~ 580 cm^{-1} in the FT IR spectra of the annealed films, is a clear evidence for the presence of the crystalline Co_3O_4. The absence of the weakly hydrogen bonded hydroxide group stretching mode at ~ 3630 cm^{-1} in the FT IR spectra of the annealed films eliminates the possibility of simultaneous presence of phases with lower degree of crystallinity. The summarized band assignments for the most important bands in the FT IR spectra of the as-deposited and annealed films are presented in Table 3. As noted before, the previously outlined discussion regarding the FT IR study of the cobalt oxide films is quite similar to that in the case of nickel oxide films. A similar chemical behavior of these species with respect to the chemical bath deposition of the corresponding oxide phases is, however, expected.

5. Optical Properties in the UV-VIS-NIR Spectral Range

The optical (electronic) spectra of chemically deposited thin films of elemental selenium and metal chalcogenides were recorded on a *Cary 50* spectrophotometer in the ultraviolet-visible (UV-VIS) and near-infrared (NIR) spectral range. Also, the attenuated total reflection

(ATR) spectra in the NIR spectral range were recorded on a Perkin Elmer System 2000 FT-IR interferometer. The spectra processing was performed using the MS EXCEL software package [159].

Table 3. The assignments of the most important bands in the FT IR spectra of $Co(OH)_2{\cdot}xH_2O$ and $Co_3O_4{\cdot}xH_2O$ thin films

$Co(OH)_2{\cdot}xH_2O$		$Co_3O_4{\cdot}xH_2O$	
Wavenumber/cm^{-1}	Assignment	Wavenumber/cm^{-1}	Assignment
3630	ν(OH)		
3490	ν(HO-H···O)	3440	ν(HO-H···O)
2220	ν_a(-NCO)		
1637	δ(HOH)	1630	δ(HOH)
1385	ν_s(-NCO)		
630	δ(-NCO)		
		670	ν(Co-O)
		580	ν(Co-O)

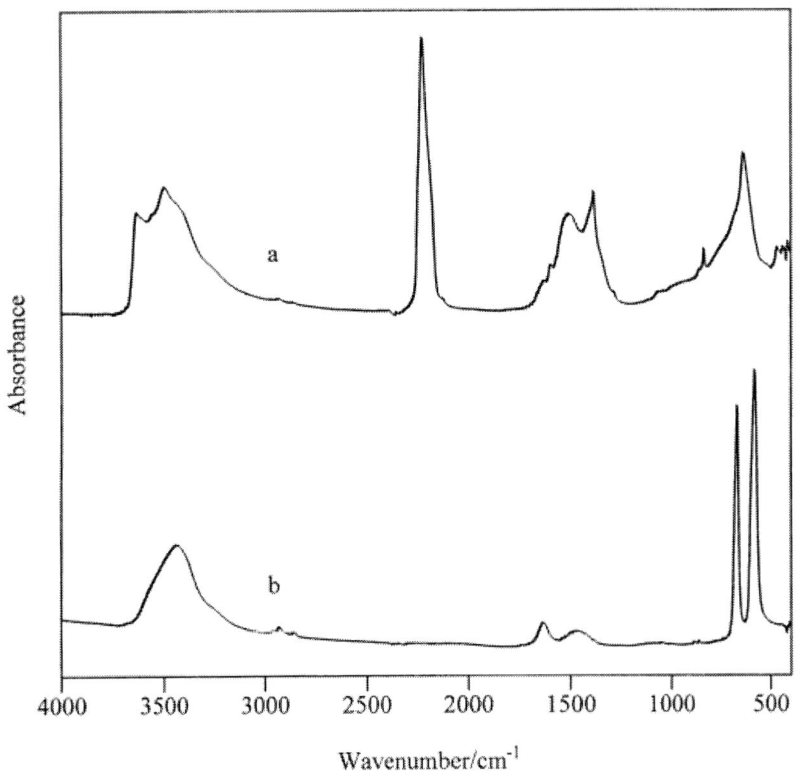

Figure 17. FT-IR spectra of the bath precipitate (a mixture of different phases of $Co(OH)_2{\cdot}xH_2O$) (a) and post-deposition treated material (Co_3O_4) (b).

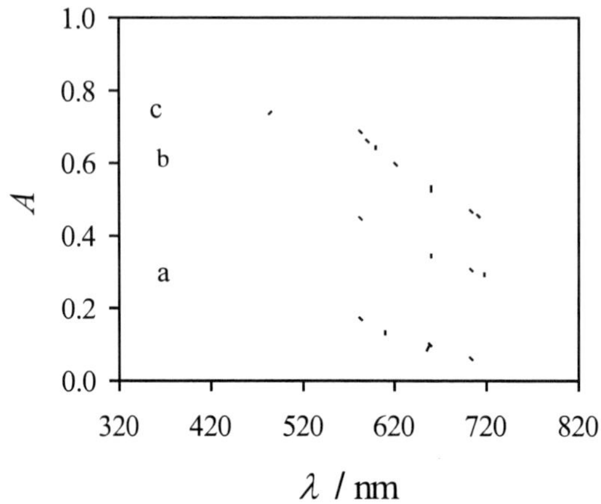

Figure 18. Optical absorption spectra of amorphous selenium thin films with different thickness (23 nm (a), 66 nm (b) and 89 nm (c)).

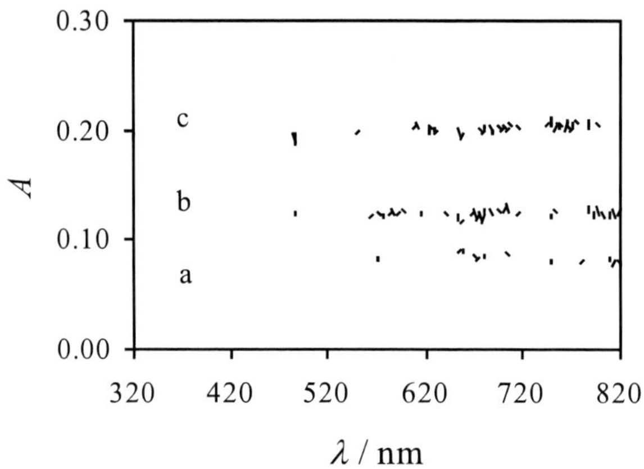

Figure 19. Optical absorption spectra of "metallic" selenium thin films with different thickness (20 nm (a), 58 nm (b) and 79 nm (c)).

5.1. Optical Properties of Elemental Selenium Deposited in Thin Film Form

The study of optical properties of elemental selenium deposited in thin film form showed that gray selenium thin films are characterized with high transparency in the spectral region from 320 to 820 nm. Optical absorption spectra of amorphous and hexagonal selenium thin films (in the previously mentioned spectral range) with different thickness are presented on Figure 18 and Figure 19 correspondingly. The amorphous ultra-thin films (<~50 nm) are highly transparent ($T > 80\%$), with an absorption onset evident at about 600 nm.

The optical band gap energy (E_g), a parameter which is of considerable interest for semiconductor materials, was calculated on the basis of recorded optical spectra using the constructed semiconductor absorption function of the form [160,161]:

$$(\alpha h\nu)^n = C\cdot(h\nu - E_g) \qquad (22)$$

where C is constant, hν is the incident photon energy, while n depends on the nature of the interband transition. For direct interband transitions n = 2 or 2/3, while for indirect ones n = 1/2 or 1/3, depending on whether they are allowed or forbidden in dipole approximation, respectively. As explained in more details in our previous review papers devoted to this subject [132,133], the outlined analysis involving semiconductor absorption functions of the form (22) is based on the expressions arising from Fermi's golden rule for interband electronic transitions, within the framework of parabolic approximation for the dispersion relation. The best fit of $(\alpha h\nu)^n$ vs. the photon energy, in the case of both amorphous and hexagonal selenium thin films, was obtained for n = 2. The extrapolation of the linear part gave band gap energy of 2.0 eV for the amorphous, red selenium thin films (Figure 20), while for gray ones it decreases to 1.8 eV (Figure 21). The optical band gap energy of 2.0 eV for amorphous selenium is very close to the value of 2.1 eV, which has been reported in the literature for vacuum deposited thin films of amorphous selenium [2]. The calculated value of optical band gap of metallic selenium thin films is in agreement with the previously reported one of 1.8 eV for bulk hexagonal selenium [2,15].

5.2. Optical Properties of Bismuth and Copper Selenides

5.2.1. Optical properties of Bi_2Se_3 thin films

The optical band gap energy of bismuth(III) selenide in thin-film form was obtained on the basis of the recorded absorption spectra in the near-infrared spectral region using the method of attenuated total reflection. The absorption spectra were mathematically processed according to the expressions arising from Fermi's golden rule for electronic transitions from the valence to conduction band within the framework of parabolic approximation for dispersion relation.

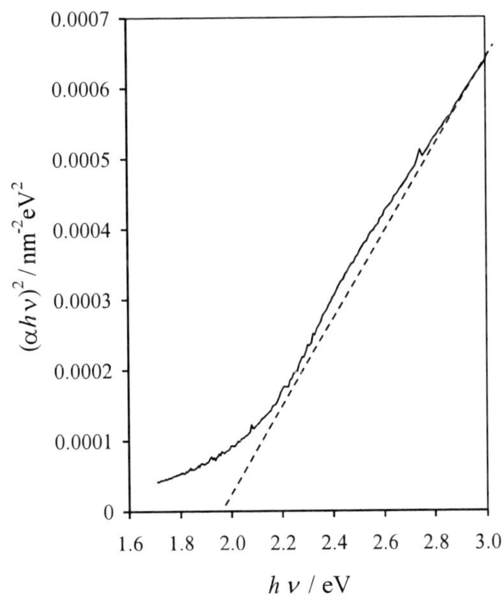

Figure 20. Dependence of $(\alpha h\nu)^2$ on $h\nu$ for amorphous, red selenium thin film.

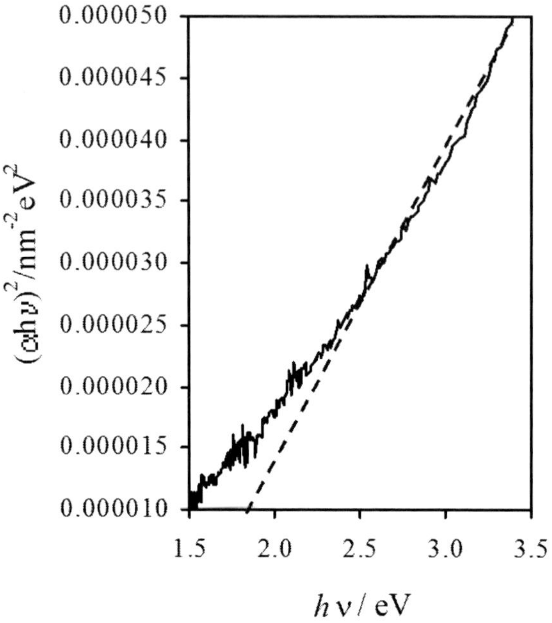

Figure 21. Dependence of $(\alpha h\nu)^2$ on $h\nu$ for gray selenium thin film.

The linear dependence of $(\alpha h\nu)^n$ versus $h\nu$ was obtained for $n = 2$ and this is in correlation with the fact that bismuth(III) selenide is a direct type of semiconductor, *i.e.* the fundamental electronic transitions are direct (allowed in dipole approximation) and do not involve phonons. With linear correlation analysis of the experimental data from the linear trend of the $(\alpha h\nu)^2$ versus $h\nu$ dependence, the value of calculated band gap energy is 0.35 eV at room temperature. The dependence of $(\alpha h\nu)^2$ on $h\nu$ for bismuth(III) selenide thin film is shown in Figure 22. The deviation from the linear dependence of $(\alpha h\nu)^2$ on $h\nu$ is a result of deviations from the parabolic approximation for the dispersion relation. The calculated optical band gap energy of chemically deposited bismuth(III) selenide in thin-film form is in agreement with the literature data for bulk Bi_2Se_3, which is characterized with band gap energy of 0.35 eV [162]. The experimentally obtained band gap energy of Bi_2Se_3 thin films in this study is also in agreement with the results which arise from the theoretically obtained band structure of this system [163]. According to this model, the predicted band gap energy value is 0.24 eV. On the basis of these results, as well as on the basis of resistance–temperature measurements presented in the next chapter, the following conclusions regarding the microstructure of Bi_2Se_3 thin films could be derived. As the optical (and as presented in the following chapter also the thermal) band gap energy is very close to the literature data for bulk Bi_2Se_3, the obtained thin films do not manifest quantum size effects. However, since the estimated average crystal diameter is only 8 nm [5], the films are nanocrystalline. The absence of size quantization effects may be attributed to the very small Bohr's exciton radius for bulk Bi_2Se_3. As mentioned before, to the best of our knowledge, no literature data exist for this value. Thus, on the basis of our results we can conclude that Bohr's exciton radius for bulk Bi_2Se_3 should be smaller than 4 nm.

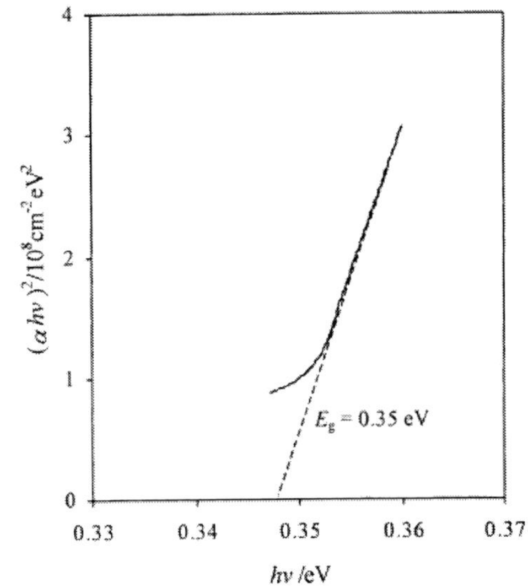

Figure 22. The $(\alpha hv)^2$ vs. hv dependence of a Bi_2Se_3 thin film, together with the extrapolated straight line to $\alpha hv = 0$.

5.2.2. Optical properties of Cu_3Se_2 thin films

For optical characterization, UV-VIS-NIR spectra of as-deposited and annealed Cu_3Se_2 thin films were recorded using polyester substrate as a reference. The absorption spectra of as-deposited thin films of Cu_3Se_2 with different thicknesses (22 nm, 65 nm and 94 nm) are presented in Figure 23. It is seen that the as-deposited thin films are highly transparent in the VIS-NIR spectral region. At about 550 nm an absorption onset is evident. On Figure 24, optical transmittance spectra of both unannealed and annealed Cu_3Se_2 films with equal thicknesses are shown.

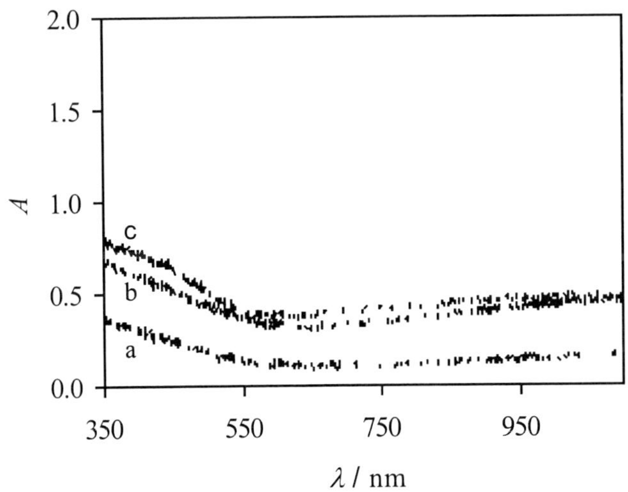

Figure 23. Optical absorption spectra of as-deposited Cu_3Se_2 thin films with thickness of 22 (a), 65 (b) and 94 nm (c).

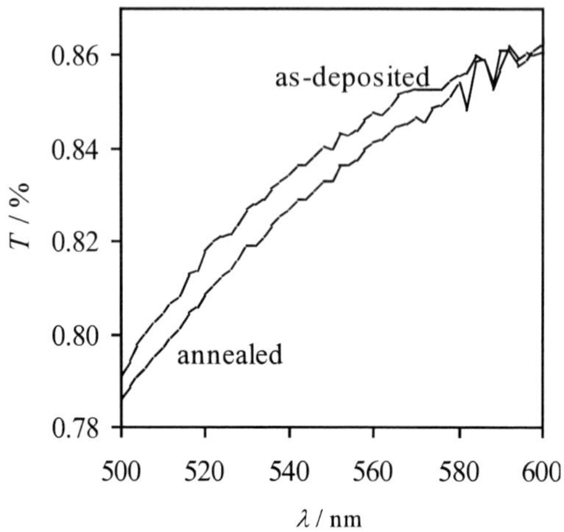

Figure 24. Optical transmittance spectra of both Cu_3Se_2 unannealed (as-deposited) and annealed films, showing a slight red shift of the absorption onset upon annealing.

A slight red shift of the absorption onset is observable upon annealing, indicating a rather small increase in the average crystal size with heat-treatment. The red shift observed in absorption onset upon annealing could be due to increase in grain size [141], and in principle it would be expected to reflect in the decrease of band gap energy as well. However, in the present case the effect is very small and no measurable changes in the last quantity have been observed. Also, no measurable changes in the full widths at half-maximum intensity of XRD peaks for the as-deposited and annealed films could be detected.

The optical band gap (E_g) for as-deposited and annealed films was calculated on the basis of the optical absorption spectra, constructing and analyzing the semiconductor absorption functions of the form (22) within the framework of parabolic approximation for the dispersion relation. The best fit of the experimental curve to a band gap semiconductor absorption function of both as-deposited and annealed films was obtained for $n = 2$. On Figure 25, the dependence of $(\alpha h\nu)^2$ *versus* $h\nu$ is presented for as-deposited thin film of Cu_3Se_2.

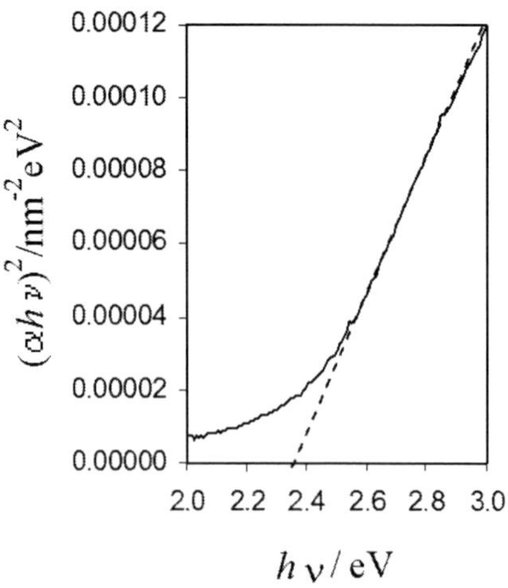

Figure 25. The dependence of $(\alpha h\nu)^2$ on $h\nu$ for as-deposited Cu_3Se_2 film.

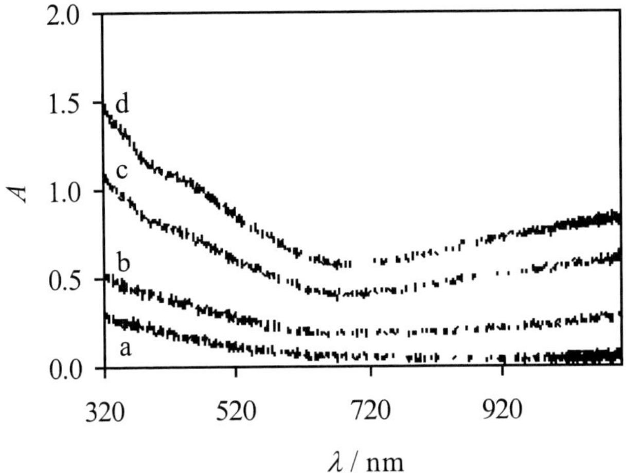

Figure 26. Optical absorption spectra of as-deposited CuSe thin films with thickness 45 (a), 70 (b), 118 (c) and 168 nm (d).

The calculated value of direct energy band gap is 2.37 eV for both as-deposited and annealed Cu_3Se_2 thin films. This value is somewhat lower than the recently published one [38], which may be due to size quantization effect as well as to differences due to various deposition mechanisms and conditions involved (it has been well recognized that the value of the band gap energy is affected by the deposition technique and deposition conditions).

5.2.3. Optical properties of CuSe thin films

The recorded optical spectra for as-deposited thin films of copper(II) selenide with different thicknesses (45 nm, 70 nm, 118 nm and 168 nm) in the UV-VIS-NIR spectral region are presented in Figure 26.

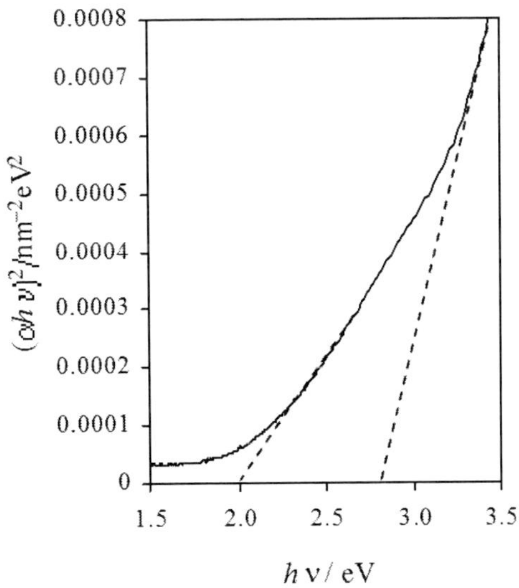

Figure 27. The dependence of $(\alpha h\nu)^2$ on $h\nu$ for as-deposited CuSe film.

The thin films are transparent in the spectral region from approximately 650 to 1100 nm. Upon annealing, the shift of absorption onset is not observed, indicating no changes in the average crystal size and in the band gap energy of the annealed thin films. Using the recorded optical spectra and the relation (22), optical band gap energies of both unannealed and annealed CuSe thin films were calculated. Copper(II) selenide thin films, as-deposited and annealed, are direct semiconductors *i.e.* the best fit of the experimental curve to a band gap semiconductor absorption function was obtained for $n = 2$. The extrapolation of the linear parts of the curves gave two values for band gap energies of 2.0 eV and 2.8 eV corresponding to two direct band-to-band transitions for both unannealed and annealed films. The first value of optical band gap agrees with the previously reported one for copper(II) selenide thin films by A. Mondal and P. Pramanik [30]. On Figure 27 the dependence of $(\alpha h\nu)^2$ *vs.* $h\nu$ for as-deposited copper(II) selenide thin films is presented.

5.3. Iron, Cobalt and Nickel Oxides

5.3.1. Optical properties of iron oxide thin films

The optical absorption spectra in the range from 190 to 820 nm for as-deposited and annealed Fe_2O_3 films (recorded against the glass substrate reference) with thickness of 150 nm are presented on Figure 28. The shift of the optical absorption onset towards the longer wavelength for the annealed film indicates a decrease in the optical band gap upon heat treatment.

For as-deposited films, the best fit of $(\alpha h\nu)^n$ *vs.* $h\nu$ was obtained for $n = 2$. The extrapolation of the linear part gave a band gap of 3.2 eV (the plot of $(\alpha h\nu)^2$ *vs.* $h\nu$ for the unannealed films is presented on Figure 29). In the case of post-deposition treated films, the best fit of $(\alpha h\nu)^n$ *vs.* $h\nu$ was obtained for $n = 1/2$. Thus, the thin films of α-Fe_2O_3 are characterized with an indirect band gap, which agrees with previously reported results [40,41]. The intercept on the $h\nu$ axis is at 2.0 eV as can be seen on Figure 30. The calculated optical band gap for annealed films agrees well with the previously reported values of *ca.* 1.9 eV for sputtered Fe_2O_3 thin films and 1.8 eV for α-Fe_2O_3 ones obtained by chemical vapor deposition [164]. It has been well documented that the band gap energy of particular material depends on the method of fabrication, one of the possible reasons for this being the size quantization effects [117,118,141,164].

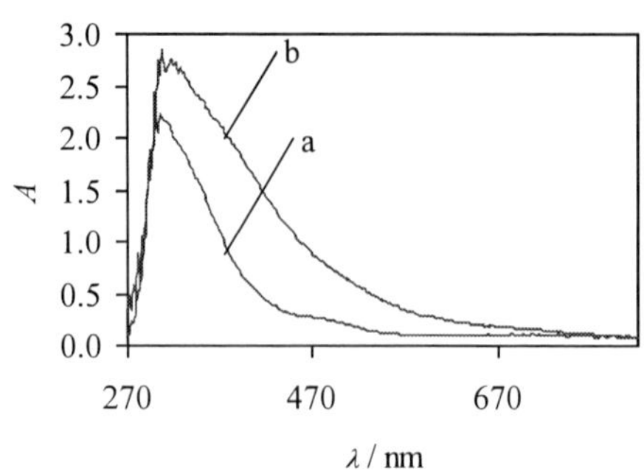

Figure 28. Optical absorption spectra for as-deposited iron oxide film (a) and post-deposition treated one (b) with thickness of 150 nm.

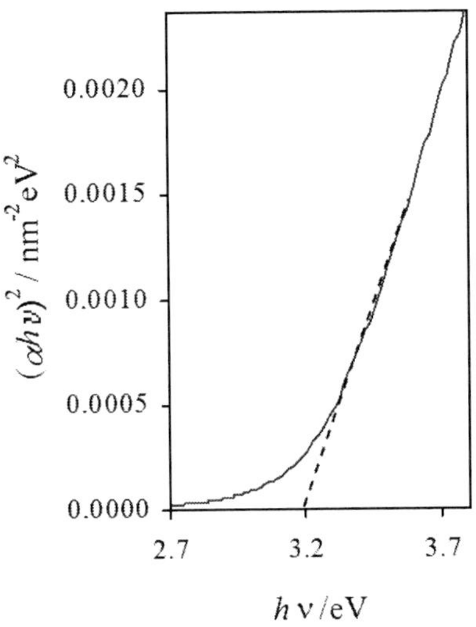

Figure 29. The plot of $(\alpha h\nu)^2$ *vs*. the photon energy for as-deposited iron oxide film.

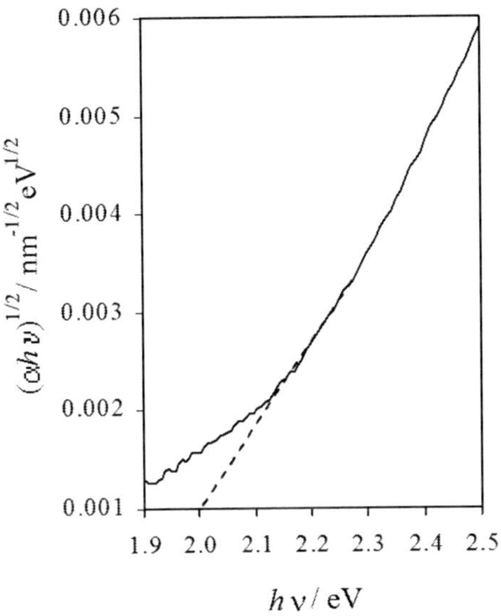

Figure 30. The plot of $(\alpha h\nu)^{1/2}$ *vs*. the photon energy for post-deposition treated iron oxide film.

5.3.2. Optical properties of nickel oxide thin films

The optical absorption spectra recorded in the range from 190 to 820 nm for as-deposited and annealed nickel oxide films (with equal thicknesses, recorded against the glass substrate reference) are presented on Figure 31. As can be seen, the NiO thin films are characterized by a sharp peak at 3.84 eV, while the as-deposited ones have an absorption peak in this energy region as well, but it is much less intensive. The increase of the absorption coefficient upon annealing may be attributed to the structural variations. On Figure 32 the absorption spectra of post-deposition treated films with different thicknesses are presented.

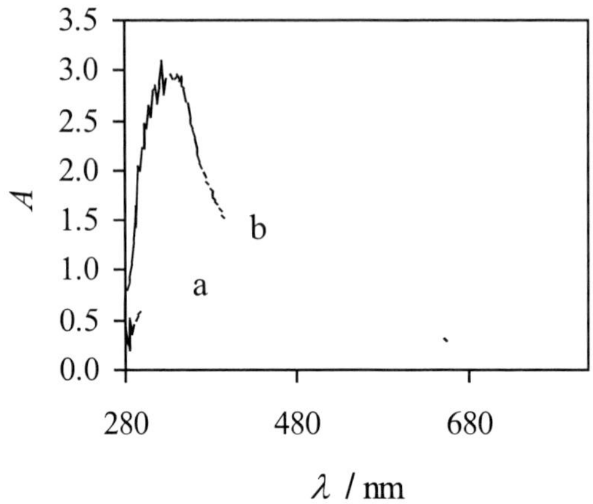

Figure 31. Optical absorption spectra for as-deposited nickel oxide film (a) and post-deposition treated one (b).

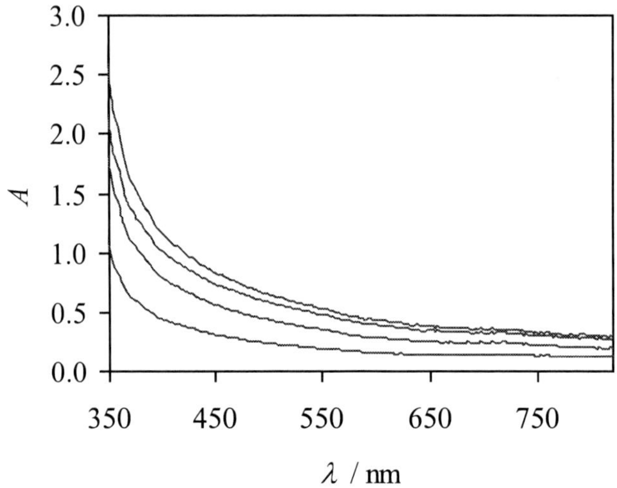

Figure 32. Optical absorption spectra for annealed nickel oxide films, with thicknesses of (from the uppermost curve downwards) 769, 648, 450 and 310 nm.

The optical band gap energy value (E_g) for annealed films, calculated on the basis of the previously outlined analysis of the optical absorption spectra, was found to be 3.60 eV. In the presently studied case of the best fit of $(\alpha h\nu)^n$ *vs.* $h\nu$ dependence was obtained for $n = 2$ indicating a direct band structure of the material in question (Figure 33). The calculated band gap value agrees well with the previously reported ones of *ca.* 3.25 eV [78] and from 3.6 to 4.0 eV [165,166]. Somewhat higher values for the band-gap energy in the case of our films (in comparison with the value of 3.25 eV) may be in principle attributed to size quantization effects in the *ca.* 13 nm nanocrystals [136,164]. However, since the present value is significantly lower than the bulk value (4.0 eV [78]), one has to attribute such findings to other reasons (e.g. existence of impurity levels introduced during the deposition). Thus, the overall lowering of the band gap energy in comparison to the bulk value may be rationalized in terms of a compromise between the quantum size effects and impurity level introducing.

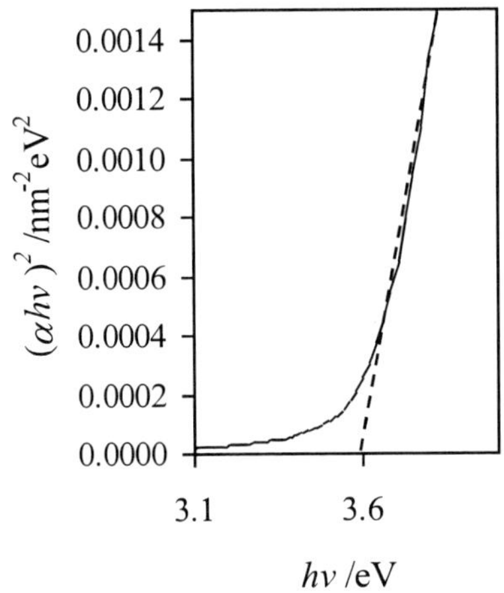

Figure 33. The plot of $(\alpha h\nu)^2$ vs. the photon energy for post-deposition treated nickel oxide film at 350°C.

5.3.3. Optical properties of cobalt oxide thin films

The optical properties of both the as-deposited and the annealed cobalt oxide films were studied in the spectral range from 300 to 820 nm. On Figure 34 a and b, the optical absorption spectra for as-deposited and annealed films (with equal thickness, recorded against the glass substrate reference) are shown. It is clear that upon annealing process, the absorption of the films decreases. The comparison of the absorption spectra of annealed films with different thickness is given on Figure 35.

On Figure 36 the dependence of $(\alpha h\nu)^2$ *vs.* incident photon energy $h\nu$ for annealed film is presented. The best fit of $(\alpha h\nu)^n$ *vs.* $h\nu$ was obtained for $n = 2$. This exponent ($n = 2$), indicates that the obtained Co_3O_4 thin films are semiconductors with dipole allowed direct interband electronic transitions. The calculated value for optical E_g of the obtained Co_3O_4 thin films is 2.2 eV. This somewhat higher value compared to the previously reported one (1.65 eV [116]) may be attributed to the very small crystal size in our case, *i.e.* size quantization effects [141].

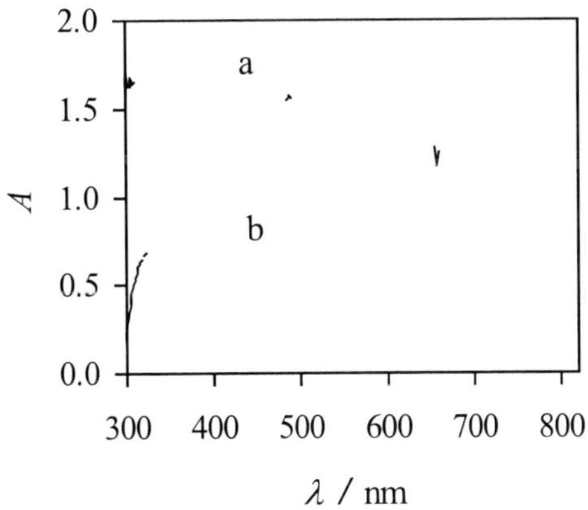

Figure 34. Optical absorption spectra for as-deposited cobalt oxide film (a) and post-deposition treated one (b).

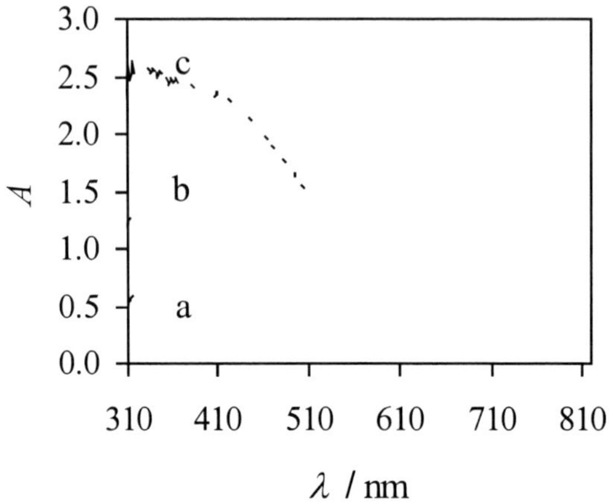

Figure 35. Optical absorption spectra for annealed cobalt oxide films, with thickness of 117 (a), 168 (b) and 194 nm (c).

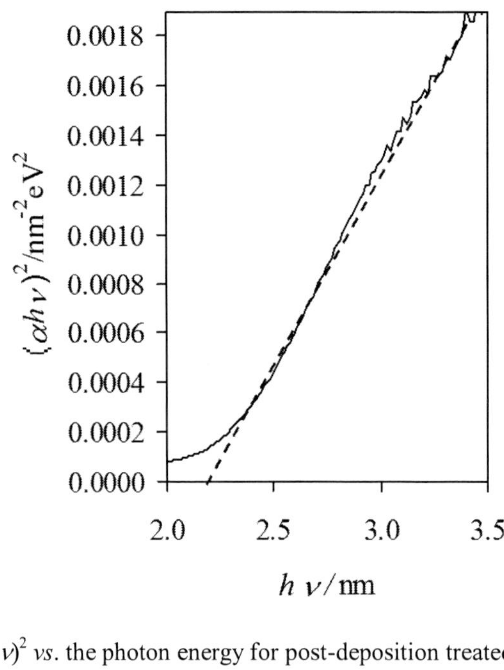

Figure 36. The plot of $(\alpha h\nu)^2$ *vs*. the photon energy for post-deposition treated cobalt oxide film at 350°C.

6. Electrical Properties of the Synthesized Chalcogenide Materials in Thin Film Form

6.1. Electrical Resistivity of the Synthesized Thin Films

Electrical resistivity of the synthesized materials in thin film form was measured by the two-point probe, four-point probe (van der Pauw method) and also by the constant field method [167-169]. During these measurements, injection of charge carriers was allowed through suitable metal contacts. The non-rectifying character of the corresponding metal-semiconductor contacts was proved by measuring their current-voltage characteristics. In general, we have demonstrated that the electrical resistance of the studied thin film materials depends on a number of factors. Among these, the most significant are the experimental conditions during thin film deposition processes, the post-deposition treatment (e.g. annealing) and the average crystal size. The predominant charge carriers' types in the case of the studied materials were determined by Seebeck effect studies. Experimentally, *the sign* of the thermoelectric force of metal chalcogenide-metal (or elemental selenium-metal) thermocouple was measured employing the hot point probe method.

The obtained thin films of red, amorphous selenium are characterized with high room temperature electrical resistivity, which is in agreement with the previously mentioned literature data. To reduce their resistivity and improve their electrical characteristics, as-deposited selenium thin films were doped by some metal ions. Namely, it is well known that the electrical properties of semiconductor materials are extremely sensitive to the presence of even minute concentration of impurities. When the electrical characteristics are dictated by impurity atoms, the semiconductor is said to be extrinsic. Our preliminary qualitative investigations have shown that the electrical resistivity of these films is highly sensitive to impurities. The detailed investigations of the effects of doping on the electrical and optical properties of selenium thin films will be published elsewhere.

The electrical measurements on Bi_2Se_3 films show that its room temperature electrical resistivity (measured in the dark) is about 4 kΩ/cm^2. According to the hot-probe measurements, in the fabricated thin films of Bi_2Se_3 majority charge carriers are electrons, which is in agreement with the previously published data [23].

Unannealed thin films of Cu_3Se_2 are characterized with dark (room temperature) electrical resistivity of the order of few kΩs/cm^2. The effect of annealing process on dark room temperature electrical resistivity of deposited films is practically negligible. At room temperature, the as deposited thin films of Cu_3Se_2 are characterized with *n*-type conduction. The study of the nature of the dependence of dark electrical resistivity *versus* temperature shows that as-deposited thin films of Cu_3Se_2 are semiconducting materials of metallic type.

As-deposited thin films of copper(II) selenide are characterized with room temperature dark resistivity in the range from 100 to 300 Ω/cm^2, depending of the film's thickness. The room temperature dark resistivity of annealed copper(II) selenide thin films is practically equal to that for the corresponding as-deposited films. The obtained films are *n*-type semiconductors.

At room temperature, Fe_2O_3 thin films exhibit a rather high resistivity, as it has already been reported [52,170]. However, our preliminary qualitative investigations have shown that the room temperature resistivity of these films is highly sensitive to moisture, indicating their potential applicability as moisture sensors. Namely, we have found that in atmosphere with relative humidity higher than 50 %, the room temperature resistivity of the films drops to about 200 kΩ/cm^2. The dependence of room temperature (dark) resistivity of Fe_2O_3 thin films on the atmospheric moisture content for relative humidities higher than 50 % is presented in Table 4 and Figure 37. (The constant relative humidities were achieved using appropriate sulfuric acid aqueous solutions [171].) Our further efforts will be directed towards the development of Fe_2O_3-based moisture sensing systems.

Table 4. The dependence of room temperature (dark) resistivity of Fe_2O_3 thin films on atmospheric moisture content for relative humidity higher than 50%

relative humidity / %	R / (kΩ/cm^2)
58	200
70	120
80	80
94 100	70
100	50

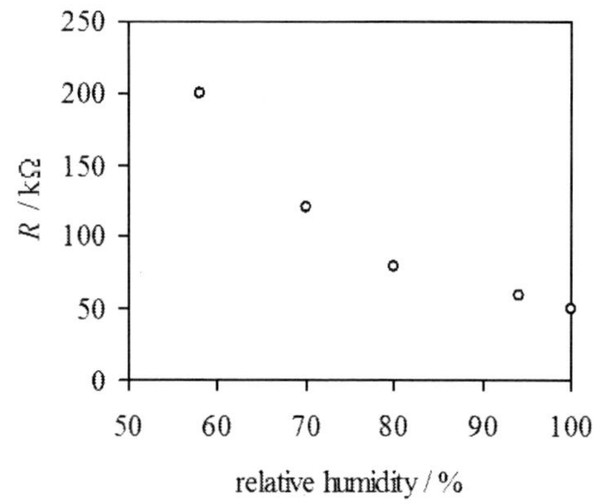

Figure 37. The dependence of room temperature (dark) resistivity of Fe_2O_3 thin films on the relative humidity.

The measurements of electrical resistivity of nickel oxide films showed that as-deposited thin films are dielectric, while post-deposition treated ones are characterized by resistivity of several MΩs/cm^2 at room temperature, as it has already been reported [78,165]. The resistivity of NiO thin films can be lowered by increase of Ni^{3+} ions concentration as result of doping with monovalent atoms or by appearance of nickel vacancies and/or interstitial oxygen in the NiO crystallites [172].

The room temperature dark resistivity of annealed cobalt oxide films deposited by the presented method is of the order of 10^7 Ω/cm^2. The resistivity of the solution grown films may be lowered by adding dopant impurities to the reaction mixture while depositing the films.

6.2. Temperature Dependence of the Dark Electrical Resistivity of the Films (R = f(T) function). Determination of Thermal Band Gap Energies and Impurity Levels Ionization Energies

To measure the temperature dependence of thin films' dark electrical resistivity, essentially the same experimental setup (presented in Figure 38), which has been discussed in details in our previous papers, was used [117-133]. It consists of a special furnace, which has been initially designed for Franck-Hertz-type experiments (FH), a thermocouple with voltmeter (TC), variable transformer (Tr), a d.c. source and a standard resistor (R_s). As explained in more details later in this review, as well as in our previous publications in the field, in order to be able to calculate the thermal band gap energy of the synthesized materials, or at least to obtain data for as many as possible impurity levels on the basis of experimental $R = f(T)$ data, it is often necessary to measure the $R(T)$ dependence in a range of relatively high temperatures. Such high temperatures, on the other hand, could induce changes in the chemical composition of the films. To avoid such changes in the investigated films in the course of these measurements, they were carried out in inert (argon) atmosphere at $P = 80$ kPa. The furnace was heated slowly, by a variable transformer (Tr), monitoring film's temperature by precise measurements, directly on its surface, using a Fe-Constantane thermocouple with a previously determined temperature dependence of thermoelectric force.

The essential circuit on which all calculations were performed on the basis of experimental data is denoted by a dashed circle in Figure 38. It could be easily shown, by mathematical analysis of this circuit, that the following relation holds between the most important parameters obtained by the experimental measurements:

$$R_f = \frac{V_b R_s}{V_s} - R_s \tag{23}$$

where V_b is the test voltage applied in the circuit containing the film, V_s is the voltage drop at the ends of the standard resistor, R_s is the resistance of the standard resistor, while R_f is the resistance of the investigated thin film. In a special, but highly important circumstance, *i.e.* when $R_s << R_f$, it follows that:

$$R_f = \frac{V_b R_s}{V_s} \tag{24}$$

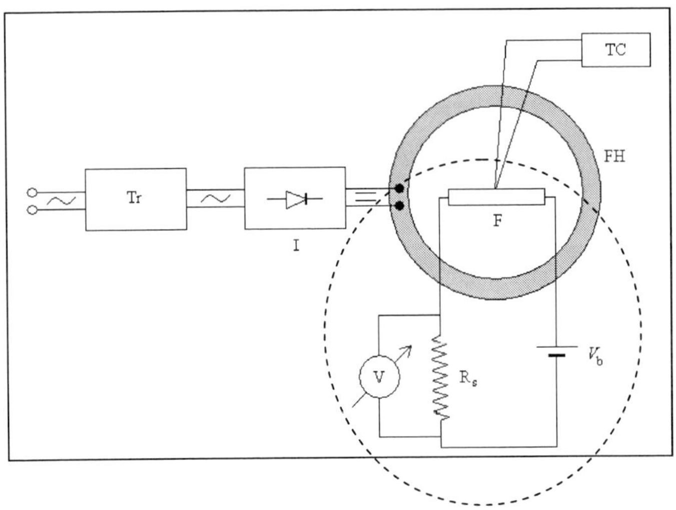

Figure 38. A schematic presentation of the experimental setup used for measurements of the temperature dependencies of thin films' electrical resistivity.

During all measurements of temperature dependence of electrical resisitivity, the condition $R_s \ll R_f$ was provided by controlling the R_s value.

To check for the reproducibility of the $R = f(T)$ measurements in all of the studied cases, the studied samples were successively subjected to multiple heating-cooling cycles. Comparison of the data obtained from various cycles invariably showed that the measurements were highly reproducible, especially with respect to the calculated thermal band gaps (E_g) and impurity level ionization energy values (ΔE_i).

6.3. Temperature Dependence of Resistivity in Bismuth(III) Selenide Thin Films

As was previously published [5], the dark electrical resistance of the thin films of bismuth(III) selenide is of the order of several kΩ/□ (kiloohms per square, *i.e.* the value referring to a thin-film sample with square shape). The measurement of the temperature dependence of the dark electrical resistance of these films was carried out on a sample which is characterized with dark electrical resistance of 13 k Ω/□ at room temperature. The applied voltage and the resistance of standard resistor were 1 V and 10 kΩ, respectively. The experimentally obtained temperature dependence of the dark electrical resistance of Bi_2Se_3 thin film is shown in Figure 39. The dark electrical resistance decreases exponentially with increasing temperature according to the theoretical prediction.

Namely, the cumulative electrical conduction (σ) of a semiconducting substance is determined by the following equation:

$$\sigma(T) = \sigma_0 T^{3/2} \exp\left(-\frac{E_g}{2kT}\right) + \sum_{i=1}^{n} \sigma_{0,i} \exp\left(-\frac{\Delta E_i}{kT}\right) \qquad (25)$$

where ΔE_i is the ionization energy of the i-th impurity level, k the Boltzmann's constant, while σ_0 and σ_0' are constants. As could be inferred from application of the band theory of solid state materials to the electrical conduction problem, the two terms in Eq. (25) correspond to the intrinsic and extrinsic conduction mechanism, respectively. Thermal band gap energy and ionization energy of the impurity levels of bismuth(III) selenide thin films were calculated from the ln(R) versus $1/T$ dependence (Figure 40). As can be seen from this figure two linear trends are observed.

The linear trend which corresponds to the lower temperature region (from room temperature to about 166 °C) is due to the mechanism of extrinsic conduction. The decrease of the dark electrical resistance in this temperature region is due to ionization of the impurity levels, i.e. to the generation of free electrons as a result of the transition from impurity level to conduction band. On the basis of this trend, the ionization energy of the impurity level was calculated. Using a linear correlation analysis, the experimental data from the lower temperature region were fitted with the following equation (Figure 41):

$$\ln R = \frac{\Delta E}{k} \frac{1}{T} + \ln R_o \qquad (26)$$

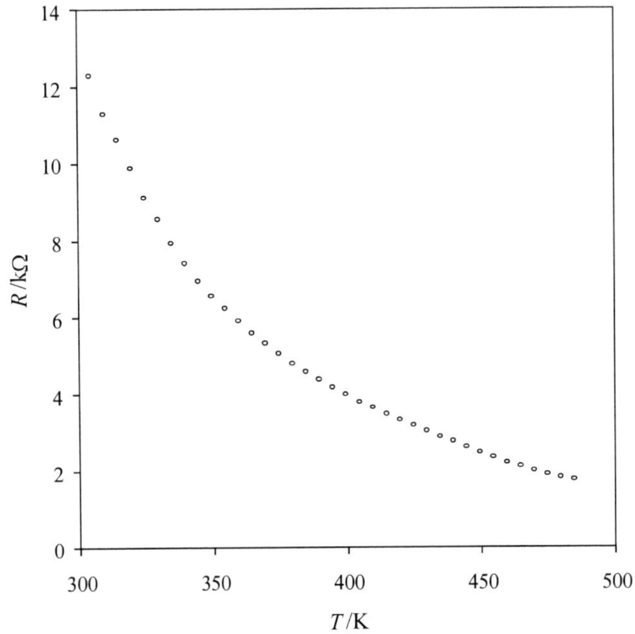

Figure 39. Temperature dependence of the dark electrical resistance of Bi_2Se_3 thin films.

where R is the dark electrical resistance of investigated semiconductor and R_0 is a constant. The calculated ionization energy of the impurity level is 0.13 eV at 0 K. Having in mind that the obtained bismuth(III) selenide thin films are characterized by n-type conduction [5], the presented impurities behave as electron donors and the corresponding energy level is situated near to the bottom of conduction band.

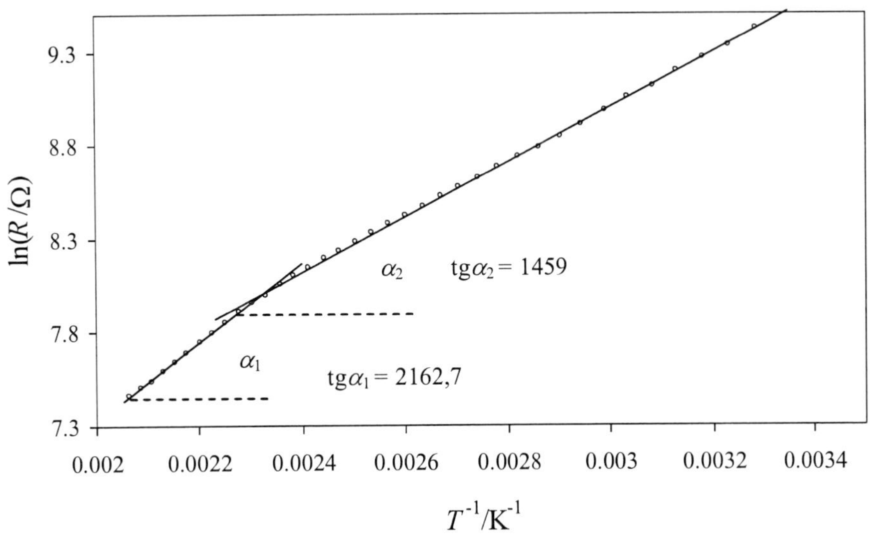

Figure 40. Dependence of ln R vs. $1/T$ of Bi_2Se_3 thin films in the whole temperature range.

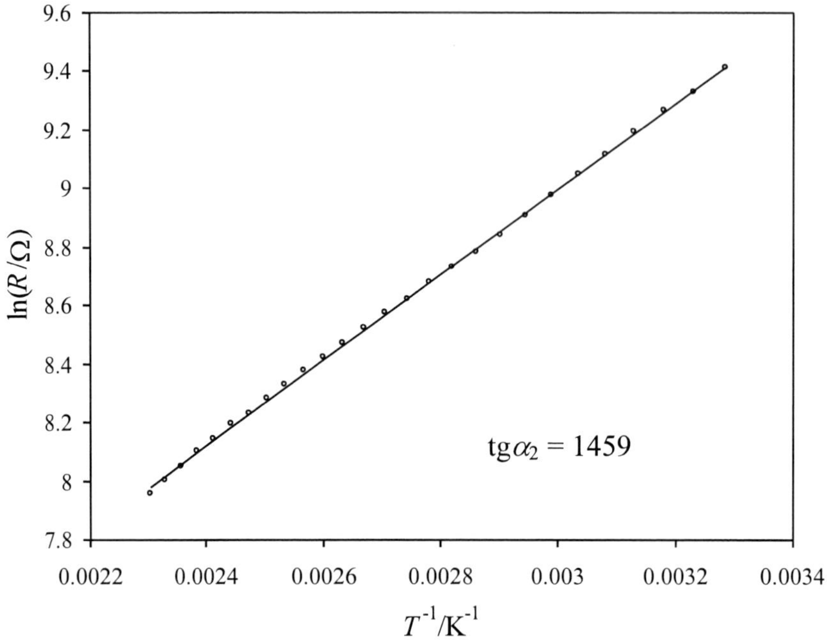

Figure 41. Dependence of ln *R* vs. 1/*T* of Bi_2Se_3 thin films in the range of lower temperatures (from room temperature to 166 °C).

The linear trend which corresponds to higher temperature region (166–210 °C) is due to mechanism of intrinsic conduction, *i.e.* the decrease of the dark electrical resistance with increasing temperature is the result of generation of the free charge carriers (electrons and holes) upon fundamental electronic transitions.

The experimental data of the higher temperature region were fitted with the following equation (Figure 42):

$$\ln R = \frac{E_g}{2k}\frac{1}{T} + \ln R_0 \tag{27}$$

From the linear least-squares fit of the dependence ln $R = f(1/T)$ in the higher temperature region the thermal band gap energy of 0.37 eV was obtained from the slope d(ln *R*)/d(1/*T*). As it has been extensively discussed in our previous publications [117-133], such calculated thermal band gap value is actually an extrapolation to 0 K, *i.e.* it refers to 0 K (in contrast to the band gap energy determined from analysis of e.g. optical absorption spectra, *i.e.* the "optical band gap").

The determined thermal band gap energy value in our studies [5,6] is in excellent agreement with the literature data for this system [162].

The obtained results for the thermal band gap energy and ionization energy of the impurity level in [6] may be regarded as much superior to our previously published data [5], as the latter were obtained in air (non-inert) atmosphere, where chemical changes of the investigated thin films are possible. The measurements of the temperature dependence of the dark electrical resistance are reproducible, as can be seen from Figure 43 where the dependence of ln *R* versus 1/*T* during a single heating–cooling cycle is presented.

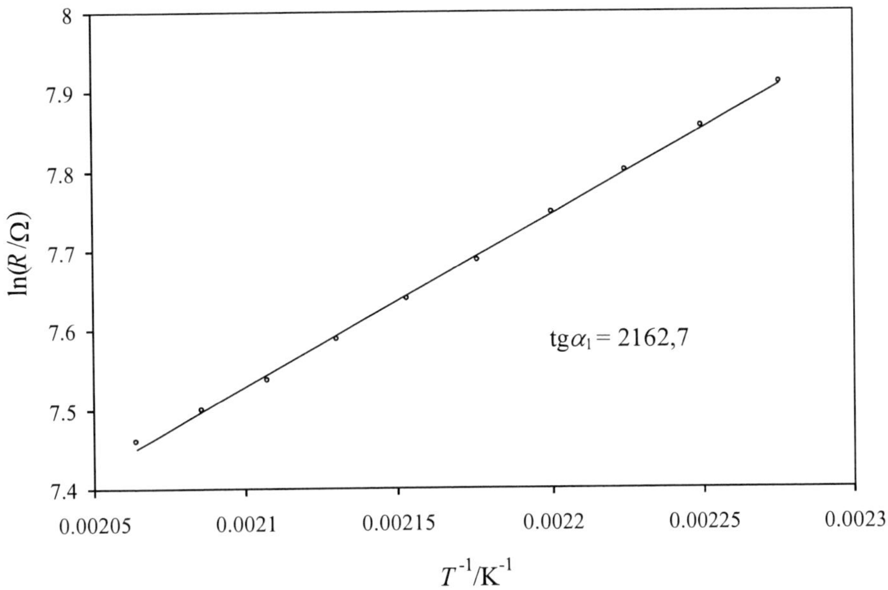

Figure 42. Dependence of ln R vs. $1/T$ of Bi_2Se_3 thin films in the range of higher temperatures (166–210 °C).

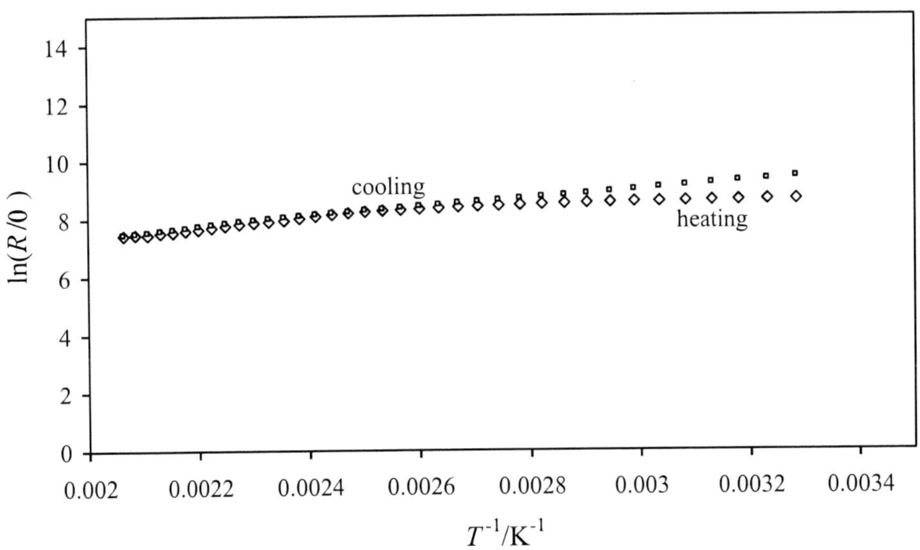

Figure 43. Dependence of ln R vs. $1/T$ during a single heating–cooling cycle for Bi_2Se_3 thin films.

6.4. Temperature Dependence of Resistivity in Iron, Nickel and Cobalt Oxides in Thin Film Form

The variation of resistance of the Fe_2O_3 thin films with temperature was studied in the range from 357 to 588 K. This range corresponds to the intrinsic semiconduction mechanism, for which the resistivity R depends on temperature T according to Arrhenius-like law (first term in equation (25)) [173,174]:

$$R = C \cdot \exp(E_g/2kT) \tag{28}$$

where C is nearly constant (*i.e.* only weekly temperature dependent, $C \propto T^{3/2}$, as implied by Eq. (25)), k is Boltzman constant and T is temperature. The plot of the logarithm of electrical resistivity *vs.* $1/T$ is presented in Figure 44. As can be seen, the dependence is practically linear with slope equal to $E_g/2k$. The estimated thermal band gap energy is 1.74 eV.

The variation of the resistance of NiO thin films with temperature in the range between 323-523 K was studied as well. This range corresponds to the extrinsic semiconduction mechanism, for which the resistivity R depends on temperature T according to the second term in (25) [173,174]:

$$R = C' \cdot \exp(\Delta E_a/kT) \tag{29}$$

where C' is constant, k is the Boltzman constant and T is temperature. The plot of the logarithm of electrical resistivity *vs.* $1/T$ is presented in Figurc 45. As can be seen, the dependence is practically linear with slope equal to $\Delta E/k$. The estimated activation energy is 0.16 eV.

Figure 46 shows the dependence of the logarithm of electrical resistance of cobalt oxide films on the reciprocal temperature in the range of 323-523 K (which, similarly as in the case of NiO, corresponds to the extrinsic conduction mechanism) constructed from the experimental $R = f(T)$ data. The measured resistance fitted the expression of the form (29). The slope of the curve is equal to $\Delta E_a/kT$. On the basis of this plot we have calculated impurity level activation (*i.e.* ionization) energy of 0.90 eV.

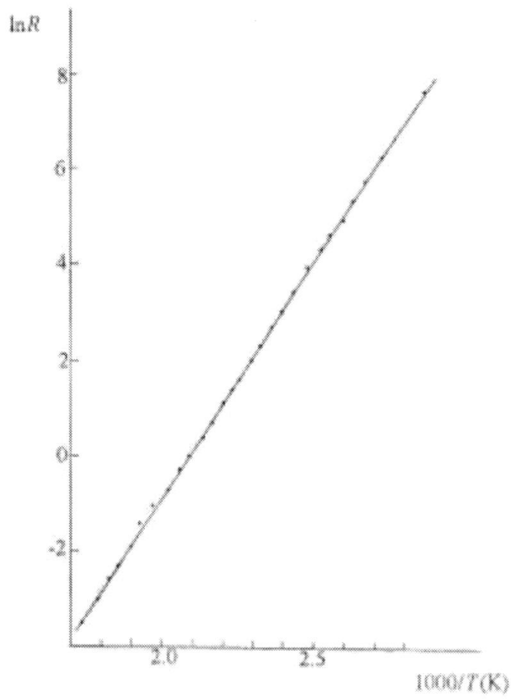

Figure 44. The dependence of $\ln(R/\Omega)$ *vs.* $10^3/T$ for α-Fe_2O_3 film.

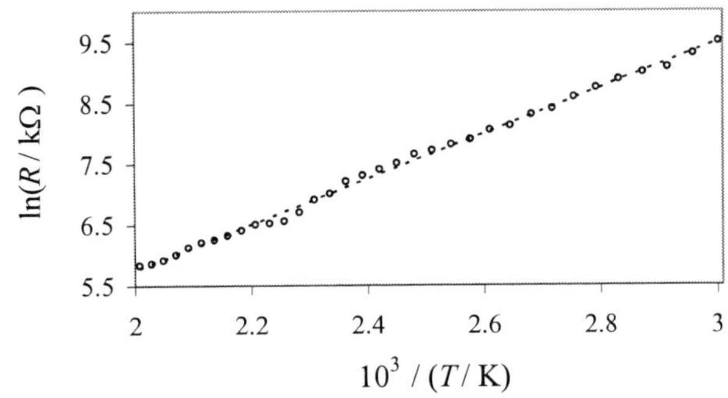

Figure 45. The dependence of ln(R/Ω) *vs*. $10^3/(T/K)$ for NiO thin film.

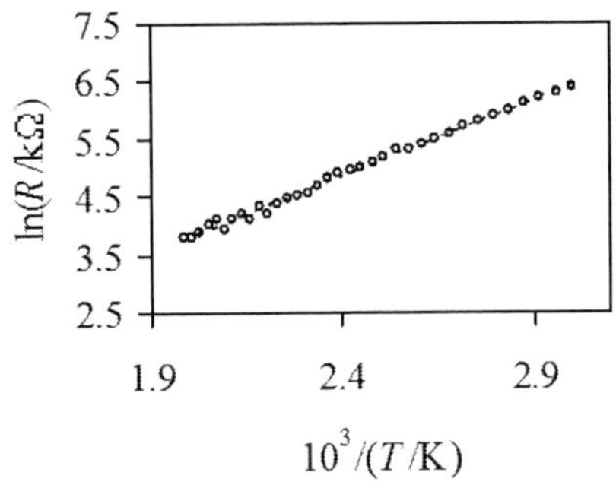

Figure 46. The dependence of ln(R/Ω) *vs*. $10^3/(T/K)$ for Co_3O_4 thin film.

REFERENCES

[1] Sze, S. M. (1985). Semiconductor Devices, Physics and Technology, Wiley, New York.
[2] Dalven, R. (1990). Introduction to Applied Solid State Physics, Plenium Press, New York.
[3] Ohring, M. (1992). The Materials Science of Thin Films, Academic Press, New York.
[4] Pejova, B. & Grozdanov, I. (2001). Appl. Surface Sci., 177, 152.
[5] Pejova, B. & Grozdanov, I. (2002). Thin Solid Films, 408, 6.
[6] Pejova, B., Grozdanov, I. & Tanuševski, A. (2004). Mater. Chem. Phys., 83, 245.
[7] Pejova, B. & Grozdanov, I. (2001). J. Solid State Chem., 158, 49.
[8] Pejova, B., Najdoski, M., Grozdanov, I. & Isahi, A. (2000). J. Mater. Sci.: Materials in Electronics, 11, 405.
[9] Pejova, B., Kocareva, T., Najdoski, M. & Grozdanov, I. (2000). Appl. Surf. Sci., 165, 271.
[10] Pejova, B., Isahi, A., Najdoski, M. & Grozdanov, I. (2001). Mater. Res. Bull., 36, 161.
[11] Tnaniguchi Ichiro, Ohachi Tadashi, Japan. 74 01,387 (Cl. C 23c), 12 Jan 1974, Appl.68 79,451, 31 Oct. 1968.
[12] Buckley, David A. (Xerox Corp.) Ger. Offen. 2,531,408 (Cl. Co1B), 26 Feb 1976, US Appl. 495,242, 06 Aug. 1974; p 7.
[13] Goldstein Irving, S. (Xerox Corp.) U.S. 3,296,762 (Cl. 204-192: C23C, Ho1G, Go3G), 16 Dec 1975, Appl. 508,950, 24 Sep 1974; p 5.

[14] Shinya Ryohei, Miura Shinichi, Aozuka Rikio (Ricoh Co., Ltd.) Brit. UK. Pat. Appl. GB 2,129,018 (Cl. C23C13/00) 10 May 1984, JP Appl. 82/129,802, 30 Aug 1982; p 30.
[15] Greenwood, N. N. & Earnshaw, A. (1997). Chemistry of the Elements, 2nd Edn. Chapter 16, Pergamon Press, Oxford.
[16] Sampath, P. I. (1966). J. Chem. Phys., 45, 3519.
[17] Dessauer, J. H. & Clarc, H. E. (Eds.), (1965). Herography and Related Processes, The Focal Press, London, Chapter 3.
[18] Arimoto Akira, Saito Susumu, Oshima Masahiro, Hirai Tadaaki (hitachi, Ltd.) Jpn. Kokai Tokkyo Koho JP 02,167,579 [90,167,579] (Cl. G03G15/02), 27 Jun 1990, Appl. 88/305,927, 05 Dec 1988; p 6.
[19] Bhattacharya, R. N. & Pramanik, P. (1982). J. Electrochem. Soc., 129, 332.
[20] Bates, C. W. & England, L. (1996). Appl. Phys. Lett., 14, 390.
[21] Hyun, D. B., Hwang, J. S., Oh, T. S., Shim, J. D. & Kolomoets, N. V. (1998). J. Phys. Chem. Solids, 59, 1039.
[22] Pramanik, P., Bhattacharya, R. N. & Mondal, A. (1980). J. Electrochem. Soc., 127, 1857.
[23] Garcia, V. M., Nair, M. T. S., Nair, P. K. & Zingaro, R. A. (1997). Semicond. Sci. Technol., 12, 645.
[24] Sankapal, B. R., Mane, R. S. & Lonkhande, C. D. (2000). Mater. Chem. Phys., 63, 230.
[25] Torane, A. P., Lokhande, C. D., Patil, P. S. & Bhosale, C. H. (1998). Mater. Chem. Phys., 55, 51.
[26] Boyer, A. & Charles, E. (1991). Le Vide Couches Minces, 47, 99.
[27] Loferski, J. J. (1956). J. Appl. Phys., 27, 777.
[28] Okimura, H., Matsumae, T. & Makabe, R. (1980). Thin Solid Films, 71, 53.
[29] Chen, W. S., Stewart, J. M. & Mickelsen, R. A. (1985). Appl. Phys. Lett., 46, 1095.
[30] Mondal, A. & Pramanik, P. (1983). J. Solid State Chem., 47, 81.
[31] Chu, T. L., Chu, S. S., Lin, S. C. & Yue, J. (1984). J. Electrochem. Soc., 131, 2182.
[32] Padam, G. K. (1987). Thin Solid Films, 150, L89.
[33] Haram, S. K., Santhanam, K. S. V., Neumann-Spallart, M. & Lévy-Clément, C. (1992). Mater. Res. Bull., 27, 1185.
[34] Massaccesi, S., Sanchez, S. & Vedel, J. (1993). J. Electrochem. Soc., 140, 2540.
[35] Thouin, L., Rouquette-Sanchez, S. & Vedel, J. (1993). Electrochim. Acta, 38, 2387.
[36] Grozdanov, I. (1994). Synthetic Metals, 63, 213.
[37] Lévy-Clément, C., Neumann-Spallart, M., Haram, S. K. & Santhanam, K. S. V. (1997). Thin Solid Films, 302, 12.
[38] Lakshmi, M., Bindu, K., Bini, S., Vijayakumar, K. P., Kartha, C. S., Abe, T. & Kashiwaba, Y. (2000). Thin Solid Films, 370, 89.
[39] Lee, E. T., Kim, B. J. & Jang, G. E. (1999). Thin Solid Films, 341, 73.
[40] Monanty, S. & Ghose, J. (1991). J. Phys. Chem. Solids, 53, 81.
[41] Sharon, M. & Prasad, B. M. (1983). Solar Energy Materials, 8, 457.
[42] Kanai, H., Mizutani, H., Tanaka, T., Funabiki, T., Yoshida, S. & Takano, M. (1992). J. Mater. Chem., 2, 703.
[43] Kennedy, S. H. & Frese, K. W. (1978). J. Electrochem. Soc., 125, 709.
[44] Vasiliev, A. A. (1992). Sensor Actuators B, 7, 626.
[45] Chai, C. C., Peng, J. & Jan, B. P. (1995). J. Electron. Mater., 24, 912.
[46] Nakatani, Y. & Matsuoka, M. (1982). Jpn. J. Appl. Phys., 21, L 758.

[47] Nakatani, Z., Sakai, N. & Matsuoka, M. (1983). Jpn. J. Appl. Phys., 22, 912.
[48] Young, W., Lee, C. & Lee, D. D. (1991). Thin Solid Films, 200, 329.
[49] Darewards, M. P., Goodenough, J. B., Hamnet, A. & Trevellick, P. R. (1983). J. Chem. Soc. Faraday Trans., I 79, 2027.
[50] Quinn, R. K., Nasby, R. D. & Banghman, R. J. (1976). Mater. Res. Bull., 11, 1011.
[51] Hardee, K. L. & Bard, A. J. (1977). J. Electrochem. Soc., 124, 215.
[52] Gardner, R. F. G., Sweett, F. & Tanner, D. W. (1963). J. Phys. Chem. Solids, 24, 1183.
[53] Lin, J. K., Sivertsen, J. M. & Judy, J. H. (1986). IEEE Trans. Magn. MAG., 22, 50.
[54] Joshi, S., Nawathey, R., Koinkar, V. N., Godbole, V. P., Chaudhari, S. M. & Ogale, S. G. (1988). J. Appl. Phys., 64, 5647.
[55] Langiet, M., Labeau, M. & Joubert, J. C. (1986). IEEE Trans. Magn. MAG., 22, 600.
[56] Qian, Y. T., Niu, C. M., Hannigan, C., Dwight, K. & Wold, A. (1991). J. Solid State Chem., 92, 20.
[57] Kordas, G. & Weeks, R. A. (1985). J. Appl. Phys., 57, 3812.
[58] Abe, M., Tamaura, Y., Oishi, M., Saitoh, T., Itoh, T. & Gomi, M. (1987). IEEE Trans. Magn. MAG., 23, 3432.
[59] Ragan, D. D., Svedlindh, P. & Granqvist, C. G. (1998). Solar Energy Mater. Solar Cells, 54, 247.
[60] Ushio, Y., Ishikawa, A. & Niwa, T. (1996). Thin Solid Films, 280, 233.
[61] Kitao, M., Izawa, K., Urabe, K., Komatsu, T., Kuwano, S. & Yamada, S. (1994). Jpn. J. Appl. Physs., 33, 6656.
[62] Kumagai, H., Matsumoto, M., Toyoda, K. & Obara, M. (996). J. Mater. Sci. Lett., 15, 1081.
[63] Cronin, J. P. et al. (1993). Solar Energy Mater. Solar Cells, 29, 371.
[64] Svensson, J. S. E. M. & Granqvist, C. G. (1987). Solar Energy Mater., 16, 19.
[65] Green, M. & Kang, K. (1988). Displays, 10, 179.
[66] Yasuda, A. & Seto, J. (1992). Solar Energy Mater. Solar Cells, 25, 257.
[67] Baucke, F. G. K. (1987). Solar Energy Mater., 16, 67.
[68] Kocemba, I. (1996). Electron Technol., 4, 29.
[69] Neubecker, A., Pomol, T., Doll, T., Hansch, W. & Eisele, I. (1997). Thin Solid Films, 310, 19.
[70] Hale, J. S., De Vries, M., Dworak, B. & Woollam, J. A. (1998). Thin Solid Films, 313, 205.
[71] Lechner, R. & Thomas, L. K. (1998). Solar Energy Mater. Solar Cells, 54, 139.
[72] Gilmore, G. G. (1995). Handbook of Inorganic Electrochromic Materials, Elsevier, Amsterdam.
[73] Bande, K. & Gamnke, T. (1990). Adv. Mater., 2, 10.
[74] Scrosati, B. (1993). Applications of Electroactive Polymers, Chapman & Hall, London.
[75] Jujii, E., Tomozawa, A., Torii, H. & Takayama, R. (1996). Jpn. J. Appl. Phys., 35, L328.
[76] Šurca, A., Orel, B., Pihlar, B. & Bukovec, P. (1996). J. Electroanal. Chem., 408, 83.
[77] Lampert, C. M., Ourstead, T. R. & Yu, P. C. (1986). Solar Energy Mater., 14, 61.
[78] Varkey, A. J. & Fort, A. F. (1993). Thin Solid Films, 235, 47.
[79] Sato, H., Minami, T., Takata, S. & Jamada, T. (1993). Thin Solid Films, 236, 27.
[80] Scrosati, B. (1994). Electrochemistry of Novel Materials, Wiley VCH, New York.

[81] Rubin, M., Wen, S. J., Richardsonn, T., Kerr, J., Von Rottkay, K. & Slack, J. (1998). Solar Energy Mater. Solar Cells, 54, 59.
[82] Kamath, P. V. & Subbanna, G. N. (1992). J. Appl. Electrochem., 22, 478.
[83] Torresi, R. M., Vazquez, M. V., Gorenstein, A. & Cordoba de Torresi, S. I. (1993). Thin Solid Films, 229, 180.
[84] Fantini, M. & Goresteinm, A. (1987). Solar Energy Mater., 16, 487.
[85] Estrada, W., Andersson, A. M., Granqvist, C. G., Gorestein, A. & Decker, F. (1991). J. Mater. Res., 6, 1715.
[86] Stevensson, J. S. E. M. & Granqvist, C. G. (1986). Appl. Phys. Lett., 49, 1566.
[87] Gorestein, A., Decker, F., Estrada, W., Esteves, S., Andersson, A. M., Passerini, S., Pantaloni, S. & Scrosati, B. (1990). J. Electroanal. Chem., 277, 277.
[88] Nemetz, A., Temminik, A., Bange, K., Gordoba de Torresi, S., Gabrielli, C., Torresi, R. & Hugot-le Goff, A. (1992). Solar Energy Mater. Solar Cells, 25, 93.
[89] Wruck, D. A., Dixon, M. A., Rubin, M. & Bogy, S. N. (1991). J. Vac. Sci. Tech., A 9, 2170.
[90] Cordoba-Torrsi, S. I., Gabrielli, C., Hugot-Le Goff, A. & Tiorresii, R. (1991). J. Electrochem. Soc., 139, 1548.
[91] Decker, F., Pillegi, R., Passerini, S. & Scrosati, B. (1991). J. Electrochem. Soc., 138, 3182.
[92] Agrawal, A., Habibi, H. R., Agrawal, R. K., Cronin, J. P., Popowich, R. C. & Lampert, C. M. (1992). Thin Solid Films, 221, 239.
[93] Nagai, J. (1993). Solar Energy Mater. Solar Cells, 31, 291.
[94] Chigane, M. & Ishikawa, M. (1992). J. Chem. Soc. Faraday Trans., 88, 2203.
[95] Orel, Z. C., Hutchins, M. G. & Mcmeeking, G. (1993). Solar Energy Mater. Solar Cells, 30, 327.
[96] Passerini, S. & Scrosati, B. (1991). Solid State Ionics, 53/56, 520.
[97] Scarmino, J., Estrada, W., Andersson, A., Gorenstein, A. & Decker, F. (1992). J. Electrochem. Soc., 139, 1236.
[98] Faria, I. C., Torresi, R. & Gorenstein, A. (1993). Electrochim. Acta, 38, 2765.
[99] Wruck, D. A. & Rubin, M. (1993). J. Electrochem. Soc., 140, 1097.
[100] Conell, R. S., Korrigan, D. A. & Powell, B. R. (1992). Solar Energy Mater., 25, 301.
[101] Maruyama, T. & Nakai, T. (1991). Solar Energy Mater., 23, 25.
[102] Burke, L. D., Lyons, M. E. & Murphy, O. J. (1982). J. Electroanal. Chem., 132, 247.
[103] Monk, P. M. S. & Farooq, N. (1995). J. Mater. Sci.: Materials in Electronics, 6, 389.
[104] Cotton, F. A. & Wilkinson, G. (1980). Advanced Inorganic Chemistry, 4th Edn., Wiley, Chicester, 767.
[105] Greenwood, N. N. & Earnshaw, A. (1984). Chemistry of the Elements, 4th Edn., Pergamon Press, Oxford, 1297.
[106] Gennero de Chialvo, M. R. & Chialvo, A. C. (1990). Electochim. Acta, 35, 427.
[107] Bendert, M. & Corrigan, D. A. (1989). J. Electrochem. Soc., 136, 1369.
[108] Monk, P. M. S., Chester, S. L., Higham, D. S. & Partridge, R. D. (1994). Electochim. Acta, 39, 2277.
[109] Ragan, D. D., Svedlindh, P. & Granqvist, C. G. (1998). Solar Energy Mater. Solar Cells, 54, 247.
[110] Cook, J. G. & Van der Meer, M. P. (1986). Thin Solid Films, 144, 165.
[111] Chidambaram, K., Malhotra, L. K. & Chopra, K. L. (1982). Thin Solid Films, 87, 365.

[112] Cathro, K. J. (1984). Solar Energy Mater., 9, 433.
[113] Kruidhof, W. & van der Liij, M. (1979). Solar Energy Mater., 2, 69.
[114] Smith, G. B., Ignatiev, A. & Zajac, G. (1980). J. Appl. Phys., 51, 4186.
[115] Hutchins, M. G., Wright, P. J. & Grebeniik, P. D. (1987). Solar Energy Mater., 16, 113.
[116] Varkey, A. J. & Fort, A. F. (1998). Solar Energy Mater. Solar Cells, 31, 247.
[117] Pejova, B., Najdoski, M., Grozdanov, I. & Dey, S. K. (1999). J. Mater. Chem., 9, 2889.
[118] Pejova, B., Najdoski, M., Grozdanov, I. & Dey, S. K. (2000). Mater. Lett., 45, 2694.
[119] Pejova, B., Tanuševski, A. & Grozdanov, I. (2003). J. Solid State Chem., 172, 381.
[120] Pejova, B., Tanuševski, A. & Grozdanov, I. (2003). J. Solid State Chem., 174, 276.
[121] Pejova, B. & Grozdanov, I. (2004). Mater. Lett., 58, 666.
[122] Pejova, B., Tanuševski, A. & Grozdanov, I. (2004). J. Solid State Chem., 177, 4785.
[123] Pejova, B. & Grozdanov, I. (2005). Mater. Chem. Phys., 90, 35.
[124] Pejova, B., Tanuševski, A. & Grozdanov, I. (2006). J.Solid State Chem., 178, 1786.
[125] Pejova, B. & Grozdanov, I. (2006). Czech. J. Phys., 56, 75.
[126] Pejova, B. & Grozdanov, I. (2006). Mater. Chem. Phys., 99, 39.
[127] Pejova, B. & Grozdanov, I. (2007). Thin Solid Films, 515, 5203.
[128] Pejova, B. (2008). Mater. Res. Bull., in press, doi:10.1016/ j.materresbull.2007.12.010.
[129] Pejova, B. & Tanuševski, A. (2008). J. Phys. Chem. C, 112, 3525.
[130] Pejova, B., Grozdanov, I., Nesheva, D. & Petrova, A. (2008). Chem. Mater., 20, 2551.
[131] Pejova, B. (2008). J. Solid State Chem., in press, doi:10.1016/j.jssc.2008.03.038.
[132] Pejova, B. (2007). In: Progress in Solid State Chemistry Research, (Ronald W. Buckley, editor), Nova Science Publishers, New York, 55.
[133] Pejova, B. In: Quantum Dots: Research, Technology and Applications, Nova Science Publishers, New York, in press.
[134] Handbook of Chemistry and Physics, CRC Press, 64th edition, 1983-1984.
[135] Doña, J. M. & Herreo, J. (1995). J. Electrochem. Soc., 142, 764.
[136] Gorer, S., Albu-Yaron, A. & Hodes, G. (1995). Chem. Mater., 7, 1243.
[137] Froment, M. & Lincot, D. (1995). Electrochim. Acta, 40, 1293.
[138] Chen, Q. W., Qian, Y. T., Qian, H., Chen, Z. Y., Wu, W. B. & Zhang, Y. H. (1995). Mater. Res. Bull., 30, 443.
[139] Greenwood, N. N. & Earnshaw, A. (1997). Chemistry of the Elements, Butteerworth Heinnemann, Oxford, 1080.
[140] Rane, K. S., Verenkar, V. M. S., Pednekar, R. M. & Sawant, P. Y. (1999). J. Mater. Sci.: Materials in Electronics, 10, 121.
[141] Gorer, S. & Hodes, G. (1994). J. Phys. Chem., 98, 5338.
[142] Abramowitz, M. & Stegun, I. A. (Eds.) (1972). Handbook of Mathematical Functions, Dover, New York.
[143] Heavens, O. S. (1955). Optical Properties of Thin Solid Films, Academic Press Inc., New York.
[144] Basavalingu, B., Tareen, J. A. K. & Bhandage, G. T. (1986). J. Mater. Sci. Lett., 5, 1227.
[145] Mehandjiev, D. & Nikalova-Zheeheva, E. (1980). Thermochim. Acta, 37, 45.
[146] JCPDS - International Centre for Diffraction Data set no. 06-0362.
[147] Weller, M. T. (1994). Inorganic Materials Chemistry, Oxford Science, Oxford, 25.
[148] JCPDS - International Centre for Diffraction Data set no. 33-0214.
[149] JCPDS – International Centre for Diffraction Date set no. 08-0279.

[150] JCPDS – International Centre for Diffraction Date set no. 34-0171.
[151] Inorganic Index to the Powder diffraction file, Joint Committee on Powder Diffraction Standards, 1972.
[152] Hotovy, I., Buc, D. & Nennewitz, O. (1998). Vacuum, 50, 41.
[153] Azaroff, L. V. & Buerger, M. J. (1958). The Powder Method, Mc Graw-Hill Book Company, New York.
[154] Peiser, H. S., Rooksby, H. P. & Wilson, A. J. C. (Eds.) (1960). X-ray Diffraction by Polycrystalline Materials, Champan & Hall, London.
[155] Ross, S. D. (1972). In: Inorganic Infrared and Raman Spectra, McGraw-Hill Book Company, Maidenhead, 106.
[156] Lawson, K. E., Corporation, S., Base, S. & Albuquerque, N. M. (1961). In: Infrared Absorption of Inorganic Materials, Reinhold Publishing Corporation, New York, 62.
[157] Inorganic library of FTIR spectra, 2nd ed., Nicodom 1998, p. 115, 123.
[158] Nakamoto, K. (1970). Infrared Spectra of Inorganic and Coordination Compounds, 2nd Ed. Wiley- Interscience, New York.
[159] MICROSOFT EXCEL, 1985-2003 Microsoft Corporation.
[160] Klingshirn, C. F. (1997). Semiconductor Optics, Springer, Berlin.
[161] Yu, P. Y. & Cardona, M. (1999). Fundamentals of Semiconductors, Second Edition, Springer-Verlag, Berlin.
[162] Novoselova, A. B. (Ed.) (1978). Physical and Chemical Properties of Semiconductors, Handbook, Moscow.
[163] Mishara, S. K., Satpathy, S. & Jepsen, O. (1997). J. Phys.: Condens. Mat., 9, 461.
[164] Mastai, Y. & Hodes, G. (1997). J. Phys. Chem. B., 101, 2685.
[165] Kunz, A. B. (1981). J. Phys. C, 14L, 445.
[166] Adler, D. & Feinleib, J. (1970). Phys Rev., B 2, 3112.
[167] Rifkin, S. M. (1963). Fotoelektrichesskiye Yavleniya v Poluprovodnikhah, GI Fiz-mat. literaturi, Moskva.
[168] Seeger, K. (1997). Semiconductor Physics, Springer-Verlag, New York.
[169] Green, M. & Miles, R. E. (1973). J. Phys. D: Appl. Phys., 6, L45.
[170] Hiraku, K. (1988). Oe Bucuri, 57, 1571.
[171] Handbook of Chemistry and Physics, 64th Ed., CRC Press Inc., Boca Raton, Florida, 1984, p. E-42.
[172] Antolini, E. (1992). J. Mater. Sci., 27, 3335.
[173] Hutchison, T. S. & Baird, D. C. (1968). In: The Physics of Engineering Solids, Wiley, New York, Sect. 14.3.
[174] Bromberg, J. P. (1984). In: Physical Chemistry, Allyn, Boston, Sect. 36.6.

In: Bulk Materials: Research, Technology and Applications ISBN: 978-1-60692-963-6
Editors: T. Frias, V. Maestas, pp. 245-266

Chapter 5

COMBINATORIAL FRICTION MATERIALS RESEARCH

Yafei Lu**, *Yuxiong Liu and Yanli Fan
The Key Laboratory of Beijing City on Preparation and Processing of Novel Polymer Materials (Box 82), Beijing University of Chemical Technology, Beijing 100029, P. R. China
Choong-Fong Tang
Center for Advanced Friction Studies, Southern Illinois University Carbondale, IL 62901-4343, USA

ABSTRACT

Combinatorial friction materials are novel brake friction composites defined as the raw materials used were optionally selected according to scientific criteria using a combinatorial approach and the multi-component friction material formulations were optimized by a optimization technique including design of friction material formulations by a Golden Section approach, relational grade analysis of the relationships between friction performance and compositions, and comprehensive evaluation of friction performance. The combinatorial friction materials research (CFMR) was proposed and performed to solve the critical problems related to automotive friction materials to meet the requirements of brake industry including mainly (1) to screen raw materials with good friction performance and find out the interaction effects among ingredients by a combinatorial approach; (2) to design friction material formulation by a Golden Section approach; (3) to optimize friction material formulations by statistical analysis and multi-objective optimization method; (4) to characterize the structural development of friction layer formed on the friction surface and explain the friction mechanisms of friction materials during the brake process; and (5) to minimize the environment impact and development of eco-friendly friction materials. The objective of the CFMR is to shift the friction materials research from experience to science.

Keywords: Friction materials, combinatorial approach, optimization, design of experiments (DOE), friction performance, friction mechanisms.

---

* Corresponding author: Email: ylu623@gmail.com

INTRODUCTION

Friction materials are widely applied as linings in automotive brakes (Figure 1a). The role of brakes is to protest the safety of human's life. Braking that stops automobiles and converts mechanical work into thermal energy happens at the interface between disk (grey cast iron) and friction materials (Figure 1b). Therefore, braking process is complex and during braking, friction heat, wear debris (third body), noise, vibration, tribochemistry, and formation of new phases have to be produced.

Currently two kinds of phenolic bound friction materials, semi-metallic and non-asbestos organic (NAO), are popularly used in vehicles [1,2]. Large amounts of steel wool and iron particulates are added to the semi-metallic friction materials. The NAO friction materials can be divided into three sub-types: (1) low-metallic, large amount of organic or inorganic fibers are used as main reinforcements coupled with small amount of steel wool; (2) non-metallic, Twaron (aramid pulp) is used as main reinforcement and no any metal fibers are used; and (3) ceramic, potassium titanate is used as main reinforcement coupled with aramid pulp, brass, or copper chips. Thousands of ingredients have been used as raw materials in friction formulations till now, but only 7-20 ingredients are contained in each industrial brake friction composite. The compositions of brake friction materials including fibers, fillers, abrasives (lubricants, and others), and binders are shown in Figure 2.

In order to be effective to the safety of driving a vehicle, it requires that friction materials must exhibit optimized friction performance and low cost. However, although friction materials have been developed for a century, most commercially available friction formulations were achieved through trial and error coupled with prior experience and testing expertise. The main problems related to brake friction materials are how to select raw materials, how to find out the interactions among the ingredients, and how to optimize the formulations. To solve these critical requirements, combinatorial friction materials research (CFMR) methodology was proposed. CFMR including a combinatorial approach to screening of raw materials, a Golden Section approach to design friction formulations, and a multi-objective optimization method coupled with statistical analysis to optimizing of friction performance was conducted. The combinatorial friction materials so-called are defined as the raw materials used were optionally selected according to scientific criteria and the multi-component formulations were optimized by a multi-objective optimization technique. The framework of CFMR is shown in Figure 3 and the purpose of CFMR is to establish scientific standards for the choice of raw materials, to understand the role of each ingredient and the interactions among the multi-components, to optimize the friction performance of friction materials with good friction performance and low cost, to understand the friction mechanisms, and to realize the friction materials research from experience to science.

In this work the CFMR including combinatorial screening of raw materials, rational design of friction material formulations, optimization of friction performance, and friction and wear mechanisms is summarized. According to CFMR the new selection standards for raw materials, the roles of raw materials, the matching rules of raw materials, and friction performance spectra of raw materials were proposed. Both drag types of Friction Assessment and Screening Test (FAST, Figure 4a) and Tribotester with Constant Speed (TCS, Figure 4b) were utilized to perform the CFMR. One piece of sample with the size of 13.3 mm × 12.6 mm × 4.7 mm is required for FAST. The curve of friction coefficient (μ) varies with time (90 min) and an averageμ as well as a wear rate expressed as either wt.% or thickness % can be obtained by performing FAST. Two pieces of samples with the size of 25 mm × 25 mm × 5 mm are needed for TCS tests. Multiple μ values at different temperature (100, 150, 200, 250, 300, 350, 300, 250, 200, 150, and 100°C) and six wear rate at 100, 150, 200, 250, 300, and 350°C can be measured by utilizing TCS.

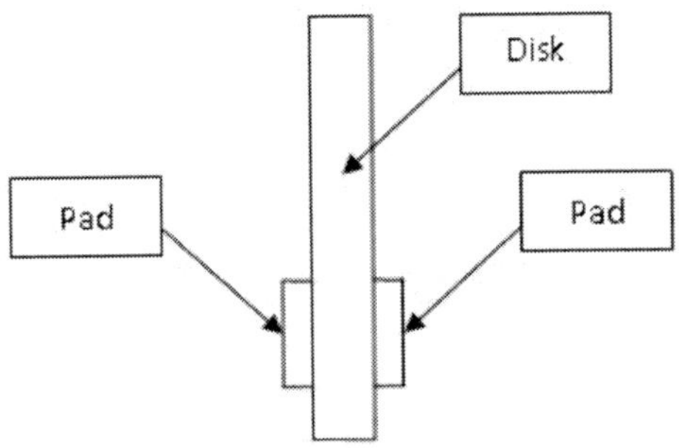

(a)

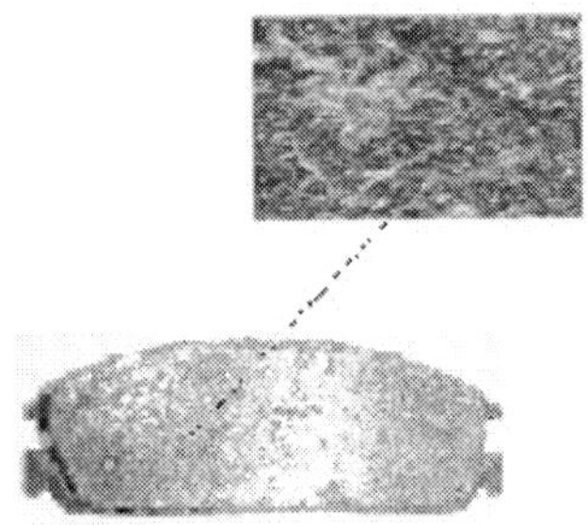

(b)

Figure 1. (a) Schematic of Disk Brake and (b) Brake Pad.

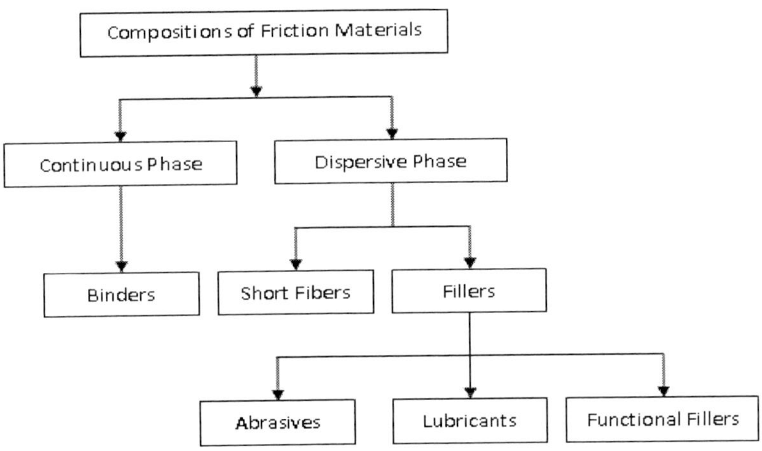

Figure 2. Compositions of Brake Friction Materials.

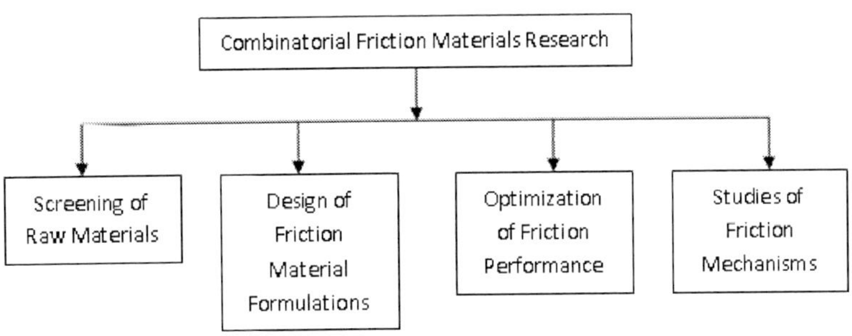

Figure 3. Framework of CFMR.

(a) (b)

Figure 4 (a) FAST and (b) TCS machines.

Table 1. Synergism between Ingredients in Group IIA and in Group I

Twaron	Steel Wool	Alumina	Graphite
$BaSO_4$	$CaCO_3$	Brass chips	Brass chips
FKF fiber	$Ca(OH)_2$	Copper chips	Copper chips
Ultrafibe (Wollastonite)	MgO		Potassium titanate
Potassium titanate	OF		
Vermiculite	Sb_2S_3		
	Wollastonite		

Combinatorial Approach to Screening of Raw Materials

Combinatorial Approach and Classification of Raw Materials

A combinatorial approach was developed to systematical screening of ingredients used in automotive friction materials. The approach, as shown in Figure 5, includes mainly (1) two-component (A + B) systems where A is a fiber or a filler and B is a binder and (2) three-component (A + B + C) systems where A and C are two fibers, two fillers, or a fiber and a filler, B is a binder.

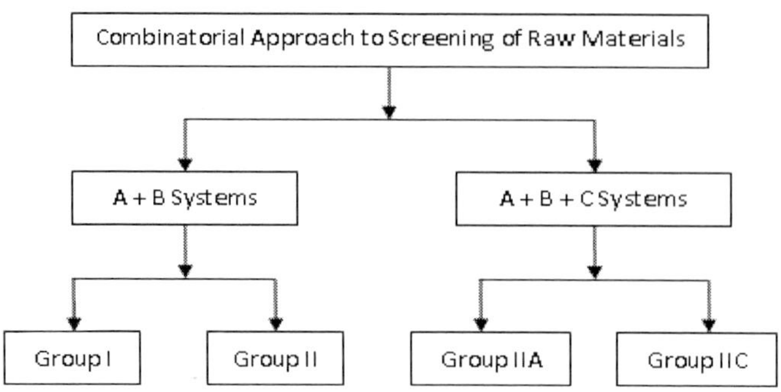

Figure 5. Combinatorial Approach to Screening of Raw Materials.

A + B system was utilized to classify the raw materials into a good wear resistant ingredient and a poor wear resistant ingredient and to find out the adhesion between filler or fiber and binder [3]. All raw materials can be divided into two groups according to their wear resistant performance: group I, which has good wear resistance, includes alumina (Al_2O_3), Twaron (aramid pulp), graphite and coke, MoS_2, steel wool, rubbers including nitrile rubber (NBR) and styrene butadiene rubber (SBR), and organic binders (phenolic, boron-phenolic, phenolic triazine, and benzoxazine) and group II, which has poor wear resistance, includes baryte ($BaSO_4$), BN, B_2O_3, brass chips, $CaCO_3$, $Ca(OH)_2$, cashew, copper chips, CuS, Cu_2S, FKF fiber (mineral fiber), H_3BO_3, iron powder, MgO, oxidized polyacrylonitrile fiber (OF), PMF (slag fiber, SiO_2+CaO), potassium titanate (K_2O6TiO_2), stibnite (Sb_2S_3), Ultrafibe (wallastonite, $CaSiO_3$), zircon ($ZrSiO_4$), and silicon carbide (SiC). Therefore, the raw materials belonging to group I should be primarily chosen in friction formulations due to their good friction and wear performance. This is important because if the raw materials belonging to the group I are used in the friction composites, good wear performance can be easily obtained. However, the ingredients with good friction performance can also be obtained from the combinations of the raw materials belonging to group II with steel wool for developing semi-metallic or with Twaron for developing NAO friction materials. Therefore, the components in group II can be also chosen as raw materials due to their low cost. A + B + C systems where both A and C are ingredients of group I were performed to find the combined effects of A and C on friction performance. The interaction effects of morphological (fiber-fiber, fiber-filler, and filler-filler) and material nature (abrasive-abrasive, abrasive-lubricant, and lubricant-lubricant) combinations on friction performance were experimentally determined [4]. For improving wear, the fiber-filler combinations are better than fiber-fiber and filler-filler combinations and the lubricant-lubricant combinations are better than abrasive-abrasive and abrasive-lubricant combinations.

A + B + C systems where A is an ingredient in group one raw materials like steel wool, aramid pulp, alumina, or graphite and C is an ingredient belonging to the group II have been utilized to screening of group II ingredients for semi-metallic and NAO friction materials [5]. The screening results for ingredients (group IIA) with good wear resistance when they combination with group I are given in Table 1. According to the wear of the combinations, the second raw materials selection criterion was proposed that those showing good wear resistant ingredients combined with group I should be also preferably chosen in friction formulations.

Raw Materials Selection Standards

Five selection standards for raw materials used in friction materials have been proposed. They are: ① ingredients of group I should be preferred selected; ② ingredients of group IIA should be preferred selected; ③ in order to keep the higher and stable friction coefficient, abrasives have to be selected and added into the formulations; ④ in order to meet the special requirements, for instance, to reduce the vibration and noise, the ingredients with special function should be selected; and ⑤ in order to be eco-friendly, natural plant fibers and regenerated fibers, natural mineral fibers and fillers, and binders should be preferred selected. The last standard ⑤ is a novel direction of the friction materials in the near future and the eco-friendly friction materials for brake applications using CFMR are investigating in our laboratory. According to the selection criteria, four types of phenolic based friction materials, semi-metallic, NAO (low-metallic), NAO (non-metallic), and NAO (ceramic), were designed and their compositions are shown in Table 2 [6, 7].

Table 2. Selected Compositions for Friction Materials*

Semi-metallic	NAO (Low-metallic)	NAO (Non-metallic)	NAO (Ceramic)
Steel wool ①*	Steel wool①	Twaron ①	Twaron ①
OF ②*	OF②	FKF fiber ②	Potassium titanate ②
Wollastonite ②	Wollastonite ②	Wollastonite ②	Wollastonite ②
$CaCO_3$ ②	$BaSO_4$ ④*	$BaSO_4$②	Copper chips ②
$Ca(OH)_2$②	Graphite①	Graphite ①	Graphite①
MgO ②	Al_2O_3 ①	Rubber ①	MgO ④
Sb_2S_3 ②	Rubber ①	Phenolic resin ①	Sb_2S_3 ④
$ZrSiO_4$ ③*	Phenolic resin ①	$ZrSiO_4$ ③	$ZrSiO_4$ ③
Coke ①			Coke ①
Rubber ①			Rubber ①
Phenolic resin ①			Phenolic resin ①
			Mica ④
			Vermiculite ②

* ①, ②, ③, and ④ represent the selection standards.

Wear Mechanisms and Friction Fractography of Raw Materials

CFMR provides a new route to study friction and wear mechanisms and the effects of raw materials used on friction performance. The wear of friction materials can be improved by addition of ingredients in group I and they represent five different wear improvement approaches: (1) abrasive mechanism (Al_2O_3); (2) lubrication mechanism (graphite, coke, and MoS_2); (3) reinforcing mechanism (aramid pulp); (4) friction layer protesting mechanism (steel wool); and (5) toughen mechanism (rubbers).

Wear is a surface failure phenomenon. In order to understand the wear behavior of the ingredients and friction mechanisms, the friction fractography and wear debris of the friction materials based on A + B and A + B + C systems were observed by SEM which is a useful tool for failure analysis. Fractography has been developed to describe the fracture origins and failure mechanisms through observation of fracture surface. Therefore, the friction fractography here is defined as the observation of fractured surface caused by friction.

Twaron (aramid pulp) is an excellent reinforcement for improving the wear rate of NAO friction materials because it has higher mechanical strength and chemical stability. Twaron has very fine diameter (< 1 μm), as shown in Figure 6a and there is no obvious fracture of Twaron observed from SEM micrograph. OF and wollastonite are not wear resistant fibers. The fracture of them caused by friction is obviously shown in Figures 6b and 6c, respectively.

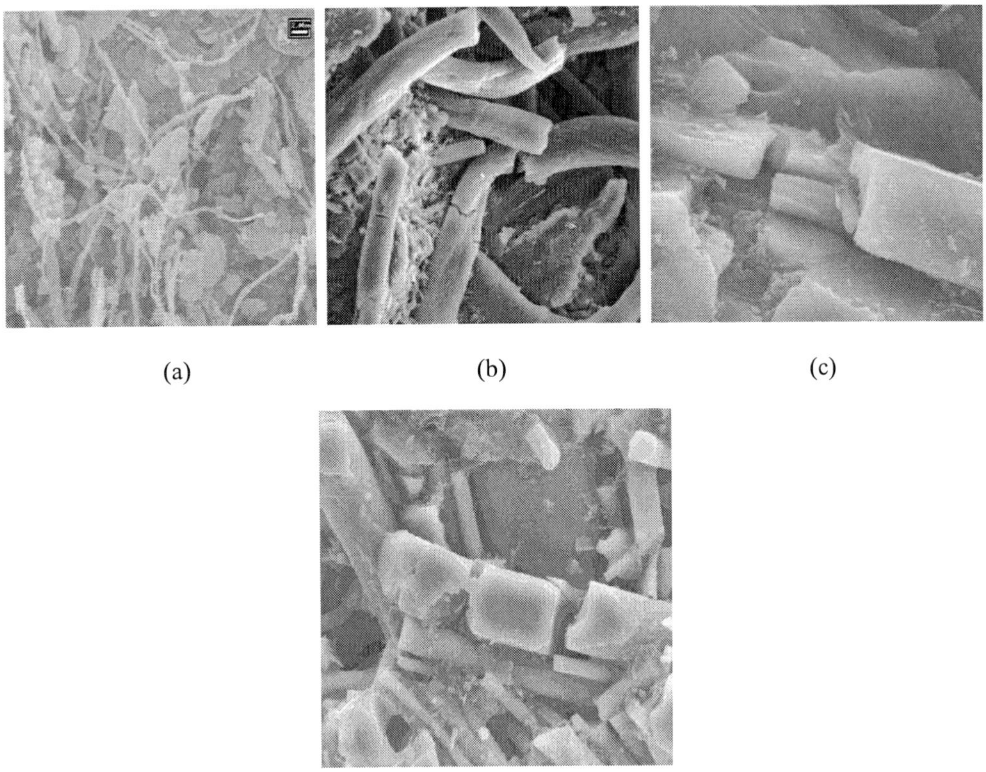

(a) (b) (c)

(d)

Figure 6. Friction Fractography of (a) Twaron; (b) OF; (c) Wollastonite, and (d) Slag Fiber.

Steel wool has good wear resistant performance and it is a main reinforcement in currently used semi-metallic friction materials. The friction fractography of steel wool is shown in Figure 7a. Brass and copper chips have relative poor wear resistance. Their friction fractography is shown in Figure 7b and 7c, respectively. It is obvious that the fracture-resistant behavior of steel wool is better than brass and copper chips.

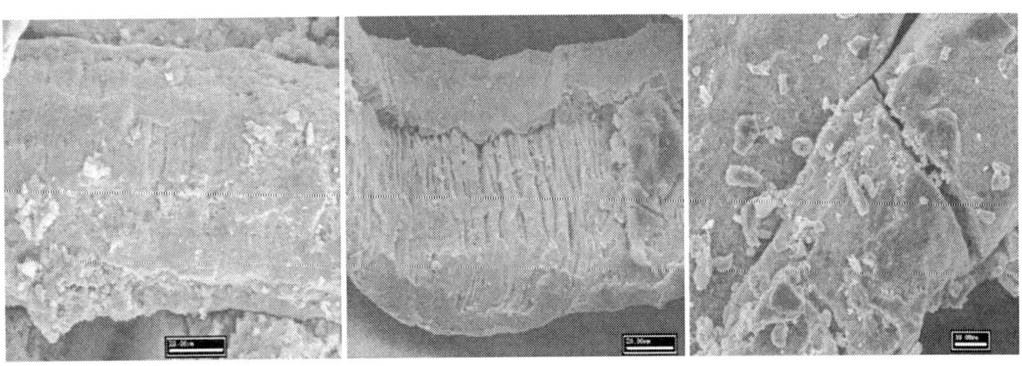

(a) (b) (c)

Figure 7. Friction Fractography of (a) Steel Wool; (b) Brass Chips; and (c) Copper Chips.

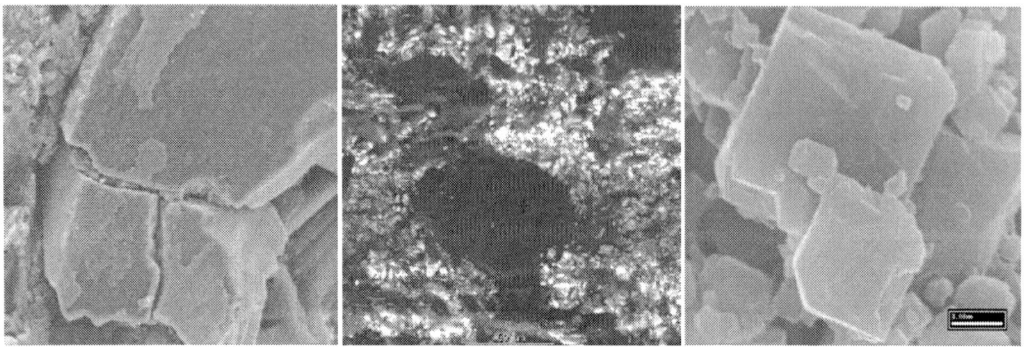

(a) (b) (c)

Figure 8. Friction Fractography of (a) Al_2O_3; (b) Graphite; and (c) $CaCO_3$.

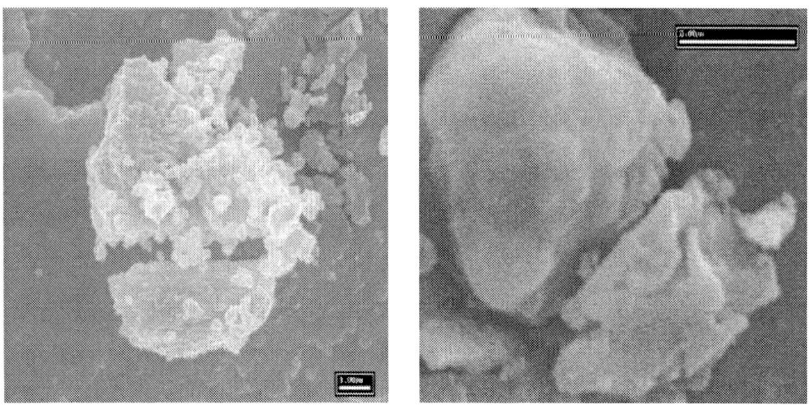

(a) (b)

Figure 9. Wear Debris of (a) CuS and (b) Cu_2S.

Alumina (Al_2O_3) and graphite are abrasive and lubricant, respectively, and they have good wear resistant performance. The friction fractography of alumina and graphite is shown in Figure 8a and 8b. The fracture of alumina was caused by compression force (load) during the friction. The particle size of alumina varied from big to small is favorable to the wear process. There is no fracture of graphite observed from SEM micrograph and this is different with the fracture of abrasives. The friction fractography of $CaCO_3$, the ingredient in group II, is shown in Figure 8c. It's poorly wear resistance and it's easy to be fractured during the braking. The wear debris of CuS and Cu_2S, both are the ingredients in group II, are shown in Figure 9a and 9b, respectively.

The friction fractography of benzoxazine (a ring-opening polymerized phenolic resin) is shown in Figure 10. It's interested that the fractured zone perpendicular to the friction direction and its shape is like a dune (Figure 10a). The fine structure inside the fractured zone is shown in Figure 10b and 10c. The fracture is obviously caused by thermal rolling which are the combined effects of thermal oxidation (the color of benzoxazine during friction changes from light yellow, through red, Figure 10a, to black finally) and shear force (Figure 10b and 10c) during the friction.

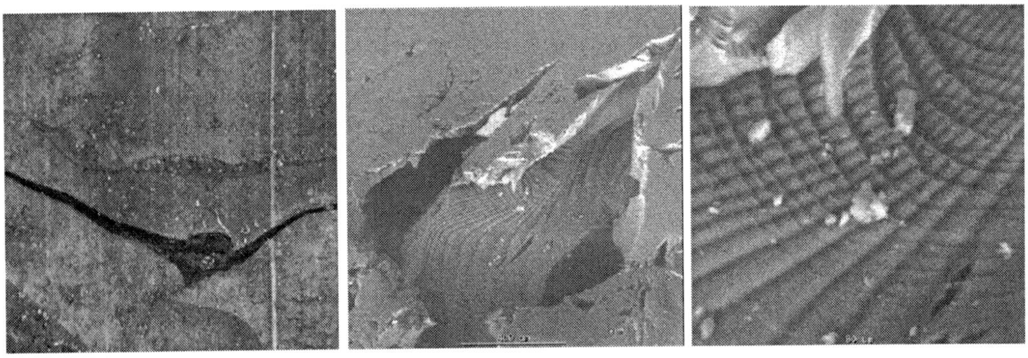

(a) (b) (c)

Figure 10. Friction Fractography of Benzoxazine (a) LM Photograph, (b) and (c) SEM Micrographs.

Roles and Friction Performance Spectrum of Ingredients

The friction performance spectrum of raw material was drawn to correlate the friction performance, transfer matter from friction materials to disc made of cast iron or from disc to friction materials, and material nature, as shown in Figure 11. Raw materials are divided into lubricant, abrasive, and transition zone. The ingredients in lubricant zone are graphite, coke, and MoS_2 and they show lower μ and good wear resistance. The characteristics of the lubricants are lower hardness and adhesion effect on interaction with disk. The main role of lubricants in friction materials is to improve wear. If there are lubricants in friction materials, the wear debris formed from the lubricants can either transfer to the surface of disk or spread on the surface of friction materials. The ingredients in abrasive zone like alumina show higher hardness, intermediate μ and better wear rate. The roles of the abrasives are enhancing μ and abrasion effect in interaction with disk. If there are abrasives in friction materials, the iron debris from disk abraded by abrasives will transfer to the surface of friction materials. Meanwhile, the abrasives can be fractured when they subject to braking force and the wear debris of abrasives can be dispersed on the surface of friction materials. We have observed that the larger size of abrasive, the more abrasion effect on friction performance. Most ingredients are located in transition zone. They show either higher or lower μ. Steel wool is located in the boundary between abrasive and transition zone. Aramid pulp is located in the boundary between adhesive and transition zone. Phenolic resin acts as binder in friction materials [8]. The role of the phenolic resin is to bind all other ingredients together. There is a critical composition of phenolic resin depending upon the size of the ingredients. The larger size of the ingredients uses, the more amount of phenolic resin is needed.

The formation mechanisms of friction layers during braking are an important issue in CFMR. Friction layers were formed on both the surface of friction materials and disk. The composition and morphology of friction layers differ with the bulk of friction materials. The formation and destruction processes of friction layers, that are the development of the structure of friction layers, are the bridge to connect the relationships between the composition of friction materials and friction performance. Three main types of mechanisms related to the formation of friction layers were observed. The friction layer formed by accumulation and compaction of wear debris of baryte on the friction surface of non-metallic friction materials is shown in Figure 12. Two types of iron films were observed on the surface of the semi-metallic friction materials. Steel wool itself is easy to form friction layer (iron film I) to cover other ingredients during braking, as shown in Figure 13a. The selective transfer of matters caused by abrasives such as alumina, zircon, and silicon carbide has higher hardness than that of cast iron disk and has an ability to transfer iron from the disk to the surface of the friction materials (iron film II) is shown in Figure 13b.

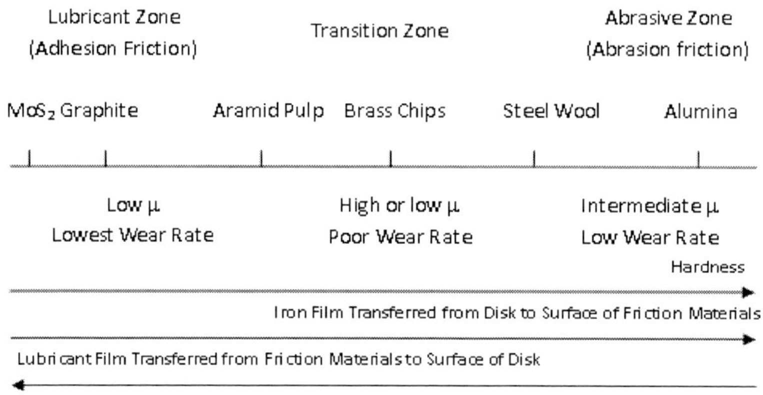

Figure 11. Friction Performance Spectrum of Raw Materials.

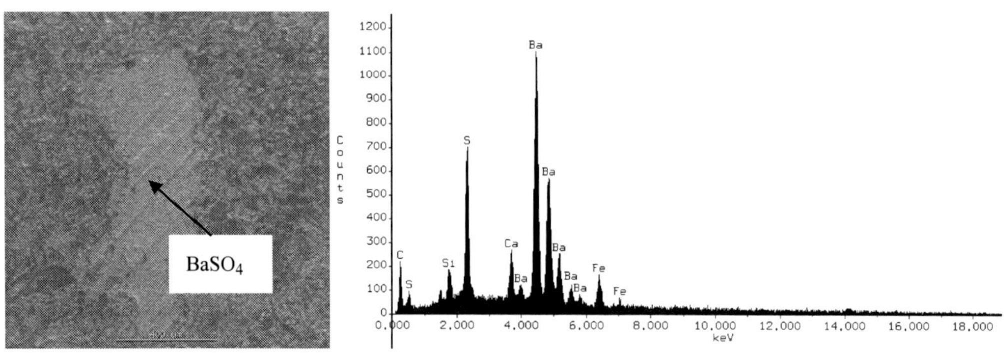

(a) (b)

Figure 12. Friction Layer Formed by Accumulation and Compaction of Baryte Debris on Surface of Non-metallic Friction Materials (a) SEM Micrograph and (b) EDX Spectrum.

Friction layers have following characteristics: (1) Friction layer has compositional dependence that the raw materials used in friction formulation determine the structure of friction layers; (2) New phases in friction layers can be formed due to tribochemical reactions; (3) Friction layers are dynamic that formation, destroy, reformation, and redestroy occur; (4) Friction layers are changeable that depends on friction conditions such as temperature, pressure, speed, and environment; and (5) Friction layers are controllable that high performance of friction materials with stable friction coefficient, low wear, and low noise can be received through life cycle assessment for debris and controlling the structure of the friction layer.

(a) (b)

Figure 13. (a) Friction Layer (Iron Film I) Formed by Steel Wool and (b) Friction Layer (Iron Film II) Formed by Transferring of Iron Debris from Disk to the Surface of NAO Friction Materials Containing Abrasive.

DESIGN OF FRICTION MATERIAL FORMULATIONS AND OPTIMIZATION OF FRICTION PERFORMANCE

Optimization of friction performance for brake friction materials is a multi-objective problem, for instance, it needs to reach a moderate friction coefficient, to minimize the stability of friction coefficient, to minimize the wear rate, and to minimize the cost [1,7]. The statistical analysis [9], relational grade analysis (RGA), and comprehensive evaluation [7,10] for friction performance evaluation are utilized. For using the drag modes testing machines (FAST and TCS), the optimization technique developed is shown in Figure 14.

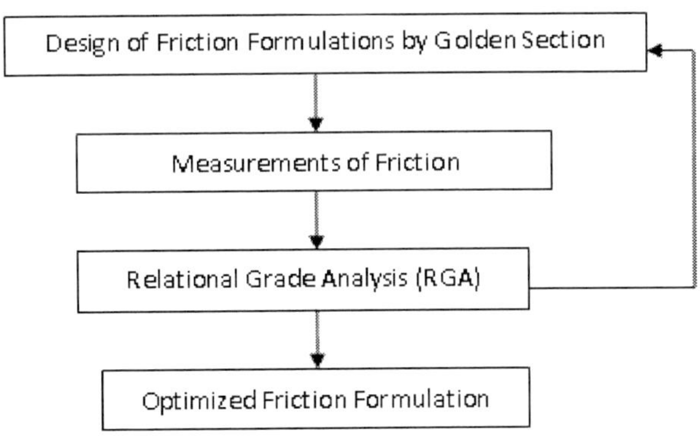

Figure 14. Optimization of Friction Performance.

Golden Section Approach to Design of Friction Material Formulations

Golden Section is an effective search algorithm for multivariate functions. The Golden ratio:

$$r = \frac{1}{2}(\sqrt{5} - 1) = 0.618 \tag{1}$$

satisfies the following equation:

$$r^2 = 1 - r = 0.382 \tag{2}$$

If in an interval [a, b] it contains the minimum point x^*, then the search in one-dimensional space can be done by:

$$\begin{aligned} x &= a + r(b - a) \\ y &= a + r^2(b - a) \end{aligned} \tag{3}$$

If the interval is [0, 1] (a = 0 and b = 1), equation (3) becomes:

$$\begin{aligned} x &= r \\ y &= r^2 \end{aligned} \tag{4}$$

However, automotive friction materials contain multiple components. Following equations can be derived:

$$\begin{aligned} r^3 &= r(1 - r) = 0.236 \\ r^4 &= r^2(1 - r) = 0.146 \\ r^n &= r^{n-2}(1 - r) \end{aligned} \tag{5}$$

A formulation design method based on Golden Section was developed [11-15]. The ingredients were divided into three groups: fibers (F), fillers (f), and binder (b) for NAO friction materials, and metal (m), non-metal (nm), and binder (b) for semi-metallic friction materials. Then permutation of fibers or metal group (V_F or V_m = 0.382), fillers group (V_f or V_{nm} = 0.382), and binder group (V_b = 0.236) according to the 0.618 principle, all possible formulations can be calculated so that the friction material formulations can be automatically formed. The relationship between volume fraction of each ingredient (V_j) within the friction formulations and Golden Section ratio (r=0.618) can be correlated by:

$$V_j = a + r^n(b - a) \tag{6}$$

When a = 0 and b = V_i where i represents ingredients, the equation 1 is:

$$V_j = V_i r^n \tag{7}$$

Three phases were designed by means of Golden Section principle to optimize the compositions of friction materials. In phase 1, the optimal proportion among the fibers and fillers used within their given V_F and V_f regions was evaluated. In phase 2, the optimal proportion between V_F and V_f was searched and in phase 3, the optimal proportion between V_b and $V_{(VF+Vf)}$ was obtained. Usually, about 20 formulation experiments are needed to receive an optimized one. As an example, the semi-metallic friction material formulations were designed and optimized by the Golden Section approach coupled with relational grade analysis. The friction performance was tested by FAST.

Phase 1

Ten ingredients were selected for the design and optimization of semi-metallic friction formulations according to the criteria of selection of raw materials listed in Tables 1 and 2. Ten formulations were designed to perform experiments to obtain μ and wear loss, as shown in Table 3 where $V_F = V_f = 0.382$ and $V_b = 0.236$, as shown in Figure 15. Among the ten formulations S-1 has relational low wear loss and suitable value of μ.

Table 3. Ten Formulations Were Selected as Phase 1

	S-1	S-2	S-3	S-4	S-5	S-6	S-7	S-8	S-9	S-10
OF	0.09	0.056	0.09	0.146	0.056	0.09	0.146	0.236	0.236	0.146
Wallastonite	0.056	0.09	0.146	0.09	0.236	0.236	0.146	0.056	0.09	0.146
Steel Wool	0.236	0.236	0.146	0.146	0.09	0.056	0.09	0.09	0.056	0.09
Sb_2S_3	0.056	0.021	0.056	0.09	0.056	0.021	0.146	0.034	0.034	0.034
Coke	0.034	0.034	0.034	0.146	0.034	0.034	0.09	0.056	0.021	0.056
$Ca(OH)_2$	0.008	0.034	0.009	0.021	0.008	0.146	0.021	0.008	0.034	0.009
MgO	0.013	0.146	0.034	0.056	0.013	0.034	0.056	0.013	0.056	0.034
$CaCO_3$	0.034	0.056	0.021	0.013	0.034	0.056	0.013	0.034	0.146	0.021
SBR	0.236	0.09	0.146	0.056	0.236	0.09	0.056	0.236	0.09	0.146
Benzoxazine	0.236	0.236	0.236	0.236	0.236	0.236	0.236	0.236	0.236	0.236
Wear, Wt. %	14.12	88.94	35.29	25.47	39.65	46.95	43.02	27.57	43.28	32.81
Average μ	0.33	0.34	0.34	0.31	0.28	0.35	0.31	0.32	0.33	0.34

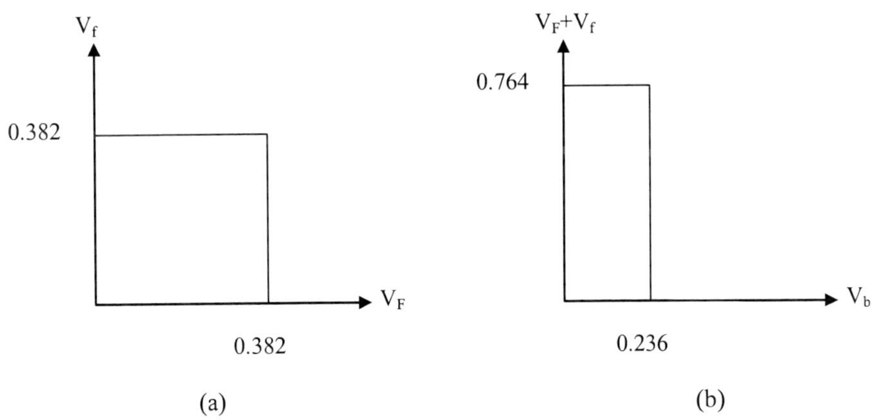

Figure 15. (a) V_F, V_f, and (b) V_b in Phase 1.

Relational grade analysis was utilized to establish the relationship between the ingredient (and its amount) and friction performance [16-18]. The analysis is performed by following steps.

Set up experimental data matrix X' (m = numbers of the ingredients, n =numbers of the formulations) and wear or friction coefficient data matrix Y':

$$X' = \begin{pmatrix} x_{11} & x_{12} & \cdots & x_{1n} \\ x_{21} & x_{22} & \cdots & x_{2n} \\ \cdots & & & \\ x_{m1} & x_{m2} & \cdots & x_{mn} \end{pmatrix}$$

$$Y' = \begin{pmatrix} y_{11} & y_{12} & \cdots & y_{1n} \\ y_{21} & y_{22} & \cdots & y_{2n} \\ \cdots & & & \\ y_{m1} & y_{m2} & \cdots & y_{mn} \end{pmatrix}$$

Normalization of X' and Y':

$$X = \begin{pmatrix} x_{11}/x_{11} & x_{12}/x_{11} & \cdots & x_{1n}/x_{11} \\ x_{21}/x_{21} & x_{22}/x_{21} & \cdots & x_{2n}/x_{21} \\ \cdots & & & \\ x_{m1}/x_{m1} & x_{m2}/x_{m1} & \cdots & x_{mn}/x_{m1} \end{pmatrix}$$

$$Y = \begin{pmatrix} y_{11}/y_{11} & y_{12}/y_{11} & \cdots & y_{1n}/y_{11} \\ y_{21}/y_{21} & y_{22}/y_{21} & \cdots & y_{2n}/y_{21} \\ \cdots & & & \\ y_{m1}/y_{m1} & y_{m2}/y_{m1} & \cdots & y_{mn}/y_{m1} \end{pmatrix}$$

Calculation of relational coefficient:

$$\xi_i(k) = \frac{\min\min|y_i(k) - x_i(k)| + 0.5\max\max|y_i(k) - x_i(k)|}{|y_i(k) - x_i(k)| + 0.5\max\max|y_i(k) - x_i(k)|} \tag{8}$$

where $|y_i(k) - x_i(k)|$ means the distance between y(k) and x(k). Due to min min $|y_i(k) - x_i(k)| = 0$ in the normalized case and 0.5 max max $|y_i(k) - x_i(k)|$ is a constant so that $\xi_i(k) \propto 1/|y_i(k) - x_i(k)|$. The smaller of $|y_i(k) - x_i(k)|$, the larger of $\xi_i(k)$.

Calculation of relational grade:

$$\gamma_i = \frac{1}{n}\sum \xi(k) \tag{9}$$

The calculated results for both μ and wear are shown in Table 4. The rank of ingredients for μ is as follow:

SBR > Steel Wool > Sb_2S_3 > OF > $CaCO_3$ > Coke > Wallastonite > MgO > $Ca(OH)_2$
4 3 2 1 0 -1 -2 -3 -4

And the rank of ingredients for wear is as follow:

$Ca(OH)_2$ > Steel Wool > SBR > $CaCO_3$ > Sb_2S_3 > Coke > OF > MgO > Wallastonite
4 3 2 1 0 -1 -2 -3 -4

Therefore, the content of OF (1, -2) and wallastonite (-2, -4) has to be decreased and the content of steel wool (3, 3) has to be increased. The formulations designed according to γ_i and based S-1 are shown in Table 5. Compared with S-1, S-11 has lower wear loss.

Table 4. Relational Grade (γ_i) for μ and Wear

	γ_i for μ	γ_i for Wear
OF	0.9386	0.8528
Wallastonite	0.8773	0.8940
Steel wool	0.9499	0.7789
Sb_2S_3	0.9468	0.8242
Coke	0.9339	0.8314
$Ca(OH)_2$	0.7351	0.7435
MgO	0.8130	0.8615
$CaCO_3$	0.9357	0.8172
SBR	0.9526	0.7867
Benzoxazine	0.9947	0.8191

Phase 2

The purpose of phase 2 is to search the optimized compositions between V_F and V_f, as shown in Figure 16. Based on S-11, four formulations were designed by increasing the volume fraction of fibers group and instantaneous decreasing the volume fraction fillers group by 0.056 (S-13) and 0.090 (S-14) as well as by decreasing the volume fraction of fibers group and instantaneous increasing the volume fraction fillers group by 0.056 (S-15) and 0.090 (S-16), as shown in Table 6. After testing their friction performance S-11 is still a better formulation compared with S-13 – S-16.

Table 5. Formulations Designed According to Relational Grade

	S-11	S-12
OF	0.056	0.034
Wallastonite	0.022	0.022
Steel wool	0.304	0.326
Sb_2S_3	0.056	0.056
Coke	0.034	0.034
$Ca(OH)_2$	0.008	0.008
MgO	0.013	0.013
$CaCO_3$	0.034	0.034
SBR	0.236	0.236
Benzoxazine	0.236	0.236
Wear, Wt. %	10.73	45.97
Avcrage μ	0.30	0.33

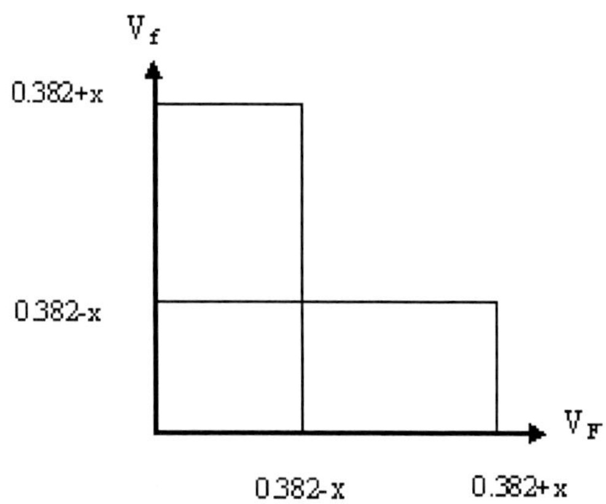

Figure 16. V_F and V_f inPhase 2.

Table 6. Formulations Designed as Phase 2

	S-13	S-14	S-15	S-16
OF	0.064	0.069	0.048	0.043
Wallastonite	0.025	0.027	0.018	0.017
Steel wool	0.349	0.376	0.259	0.232
Sb_2S_3	0.047	0.042	0.064	0.069
Coke	0.029	0.026	0.039	0.042
$Ca(OH)_2$	0.007	0.006	0.009	0.01
MgO	0.011	0.01	0.015	0.016
$CaCO_3$	0.029	0.026	0.039	0.042
SBR	0.201	0.180	0.271	0.292
Benzoxazine	0.236	0.236	0.236	0.236
Wear, wt. %	16.93	11.30	15.68	12.00
Average μ	0.34	0.36	0.35	0.29

Phase 3

The purpose of phase 3 is to search the optimized compositions between V_b and $V_F + V_f$, as shown in Figure 17. Based on S-11 again, five formulations were calculated by decreasing the volume fraction of binder group and instantaneous increasing the volume fraction of the fibers + fillers group by 0.056 (S-17), 0.090 (S-18), and 0.146 (S-19) and by increasing the volume fraction of binder group and instantaneous decreasing the volume fraction of the fibers + fillers group by 0.056 (S-20) and 0.090 (S-21). The effect of binder content on wear shown in Figure 18 is like a wear window [19]. The critical volume fraction of binder ($V_{b,cr}$) is around 0.09. Below $V_{b,cr}$, the wear of friction materials is poor. But the wear of friction materials is poor too when V_b is over 0.30.

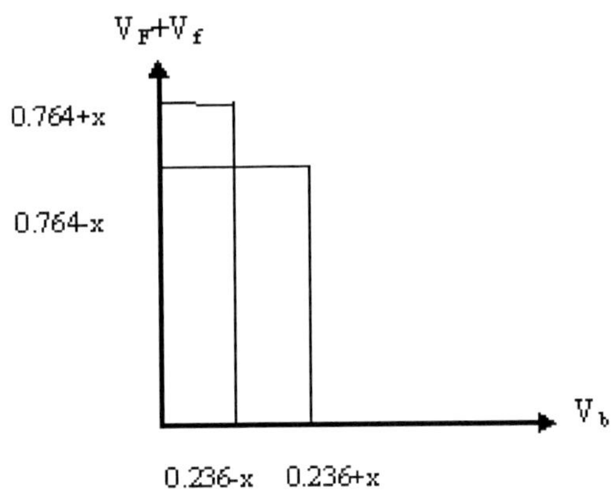

Figure 17. V_b in Phase 3.

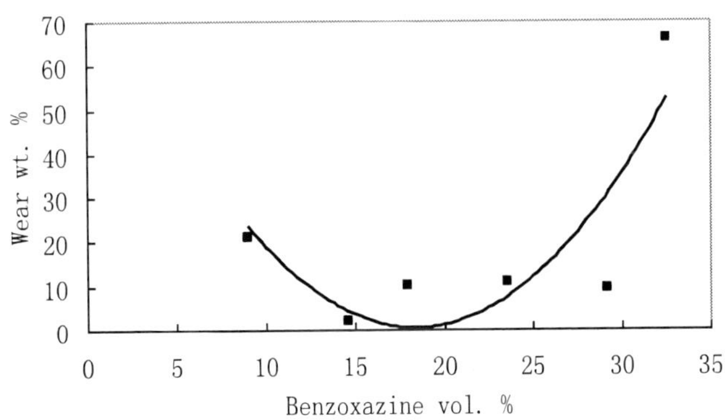

Figure 18. Effect of Binder Content on Wear.

Table 7 shows the results of friction coefficient and wear loss for each formulation. The best result was obtained in S-18 with lowest wear loss of 1.96 wt % and average μ of 0.42.

Table 7. Formulations Designed as Phase 3

	S-17	S-18	S-19	S-20	S-21
OF	0.074	0.077	0.082	0.064	0.061
Wallastonite	0.029	0.030	0.032	0.025	0.024
Steel wool	0.403	0.420	0.447	0.348	0.331
Sb_2S_3	0.046	0.048	0.051	0.039	0.037
Coke	0.028	0.030	0.031	0.024	0.023
$Ca(OH)_2$	0.007	0.007	0.008	0.006	0.006
MgO	0.011	0.011	0.012	0.009	0.009
$CaCO_3$	0.028	0.030	0.032	0.025	0.023
SBR	0.194	0.202	0.215	0.167	0.159
Benzoxazine	0.180	0.146	0.09	0.292	0.326
Wear, wt. %	9.78	1.96	21.06	9.13	65.65
Average μ	0.31	0.42	0.32	0.43	0.27

FUZZY COMPREHENSIVE EVALUATION OF FRICTION PERFORMANCE

Friction performance of friction materials as functions of temperature, speed, load, and environments is non-linear and complex. To evaluate friction performance of different friction material formulations tested by TCS, a fuzzy comprehensive evaluation method was developed [10].

Friction Stability

The friction stability ($\Delta\mu^*$), friction coefficient varies as a function of temperature, can be quantitatively calculated by:

$$\Delta\mu^* = \frac{\sum_{i=1}^{5} \mathrm{N}_i \times \frac{\left|\mu_f - \mu_r\right|_{\max} - \left|\mu_f - \mu_r\right|_i}{\left|\mu_f - \mu_r\right|_{\max} - \left|\mu_f - \mu_r\right|_{\min}} + \mathrm{N}_6}{n} \tag{10}$$

$$\mathrm{N} = \begin{cases} 1, & \mu \in [0.4,0.5] \\ 0.75, & \mu \in [0.35,0.4) \cup (0.5,0.55] \\ 0.5, & \mu \in [0.30,0.35) \cup (0.55,0.60] \\ 0.25, & \mu \in [0.25,0.30) \cup (0.60,0.70] \\ 0, & \mu \in [0,0.25) \cup (0.7,\ \mathrm{X}] \end{cases} \tag{11}$$

where N_i are weight factors for values of μ measured at i = 1 (100°C), 2 (150°C), 3 (200°C), 4 (250°C), and 5 (300°C), $N_6 = \mu_f$ at 350°C, μ_f and μ_r are friction coefficient of fade and recovery at different temperature, respectively. The up and bottom limitations of both μ_f and μ_r at the certain temperature allowed by GB5763-1998 procedure are shown in Figure 19a.

Friction materials can not be used if the values of μ_f and μ_r are over the ranges at the certain testing temperature. The larger of $\Delta\mu^*$ is, the better of friction stability is.

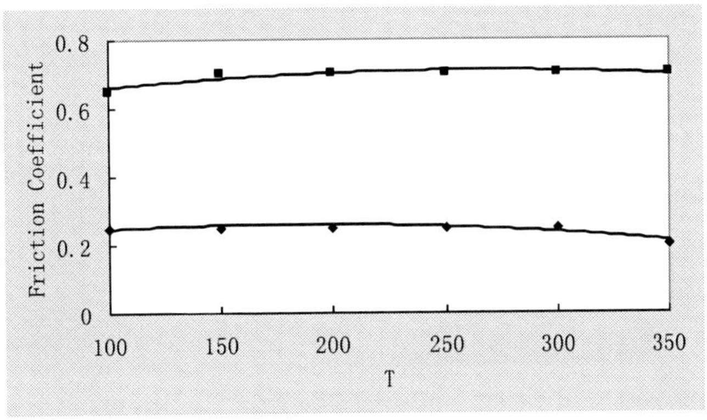

(a)

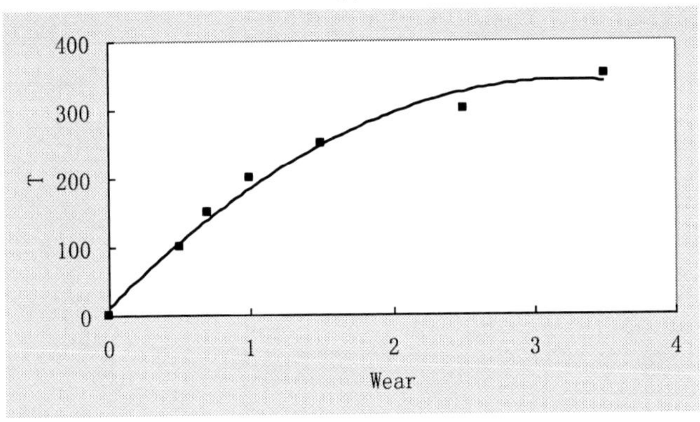

(b)

Figure 19. (a) Up and Bottom Limitations of μ Values and (b) Up Limitations of Wear ($\times 10^{-7}$cm^3/Nm) at Different Temperature (°C).

Overall Wear Rate

The overall wear rate (M_A), wear changes as a function of temperature, can be also quantitatively described by:

$$M_A = \frac{\sum_{j=1}^{6}\left(\frac{\omega(T)}{S(T)}\right)_j}{n} \tag{12}$$

where $\omega(T_j)$ and $S(T_j)$ are the determined wear values and the maximum values allowed by GB5763-1998 at jth temperature (j = 1 (100°C), 2 (150°C), 3 (200°C), 4 (250°C), 5 (300°C), and 6 (350°C), respectively), as shown in Figure 19b. Friction materials can not be applied if the wear value at the certain testing temperature is over the up limitation allowed. The smaller of M_A is, the lower of wear rate is.

Comprehensive Evaluation of Friction Performance

To evaluate the wear of different types of friction materials, a new parameter M^*_A was proposed:

$$M_A^* = 1 - \frac{M_A - M_{A\min}}{M_{A\max} - M_{A\min}} = \frac{M_{A\max} - M_A}{M_{A\max} - M_{A\min}} \tag{13}$$

Table 8. Formulations and P* of Semi-metallic Friction Materials Containing Zircon

	Semi-Zr-0	Semi-Zr-3.4	Semi-Zr-5.6	Semi-Zr-9.0	Semi-Zr-14.6
OF	0.077	0.0739	0.0719	0.0689	0.0638
Wallastonite	0.030	0.0288	0.028	0.0268	0.0248
Steel wool	0.420	0.4033	0.3924	0.3757	0.3480
Sb_2S_3	0.048	0.0461	0.0448	0.0429	0.0398
Coke	0.030	0.0288	0.028	0.0268	0.0248
$Ca(OH)_2$	0.007	0.0067	0.0065	0.0062	0.0058
MgO	0.011	0.0105	0.0103	0.0098	0.0091
$CaCO_3$	0.030	0.0288	0.028	0.0268	0.0248
NBR	0.202	0.1940	0.1887	0.1807	0.1670
PF	0.146	0.1460	0.1460	0.1460	0.1460
$ZrSiO_4$	0	0.0340	0.0560	0.0900	0.1460
P*	0.22	0.56	0.62	No*	No

* No means the μ_f, μ_r, or$\omega(T_j)$ is over the allowed range and the formulation can not be used.

where $M_{A\max}$ and $M_{A\min}$ are the maximum and minimum wear rate of the friction materials evaluated, respectively. The total friction performance (P*) of the different friction materials can be compared by:

$$P^* = A \cdot \Delta\mu^* + B \cdot M_A^* \tag{14}$$

where A (= 0.7) and B (= 0.3) are weighting factors for both $\Delta\mu^*$ and M_A, respectively. Therefore, the larger of P^* is, the better of friction performance is.

Case Study

The friction performance of the semi- metallic friction materials containing different amount of zircon ($ZrSiO_4$) tested by TCS is comprehensively evaluated. The formulations designed by proportional increasing of zircon and instantaneous decreasing of all other ingredients are shown in Table 8 [6] and the Semi-Zr-0 is similar to S-18 as shown in Table 7. Addition of zircon is to enhance and stabilize μ and the role of zircon is already shown in Table 2. The evaluated results are also shown in Table 8. It indicates that the best friction performance is the formulation containing 5.6 vol.% of $ZrSiO_4$ (Semi-Zr-5.6) which shows the highest P*.

CONCLUSION

The combinatorial friction materials research (CFMR) currently conducted in our laboratory was summaried. The CFMR offers several advantages over the traditional research on friction materials that CFMR decomposes the multi-component friction materials into two- and three-component model composites using a combinatorial approach and therefore, CFMR is a solution to discover the essential of friction materials from ingredients. CFMR provides a new methodology to promote the evolution of friction materials, especially, to select raw materials with required performance, to find out the interaction effects among the ingredients and to decide their amounts, to optimize friction formulations and to establish the relationships among compositions, friction layer, and friction performance. Five criteria were suggested to select the ingredients with good friction performance. An attempt to understand why the wear of raw materials is different and to correlate the relationship between wear mechanism and ingredients was performed by utilizing the friction fractography. The friction performance spectrum of raw materials were proposed to explain the roles of abrasives, lubricants, and binders. Golden Section approach coupled with relational grade analysis was utilized to design friction mmaterial formulations, optimize friction performance, and correlate the relationships between ingredients and friction performance. A fuzzy comprehansive evaluation method on friction stability ($\Delta\mu^*$), overall wear rate (M_A), and total friction performance (P*) was developed to describe quantitatively non-linear friction coefficient or wear rate as functions of compositions and friction temperature.

ACKNOWLEDGMENTS

The combinatorial friction materials research (CFMR) was financially supported by National Natural Science Foundation of China (50373002, 50673012), Center for Advanced Friction Studies, Southern Illinois University Carbondale, USA (from 1998 through 2002), and Twaron Research Fund (Teijin Twaron GmbH, Germany, 2007).

REFERENCES

[1] Fan, Y., Matejka, V., Kratosova, G. & Lu, Y. (2008). Roles of Al_2O_3 in semi-metallic friction materials and its effects on friction performance, *Tribology Transactions*, *51*, 771-778.

[2] Lu, Y., Wright, M. A. & Policandriotes, T. (2002). Modeling wear traces of automotive friction materials by Cantor set, *Tribology Transactions*, *45*, 258-262.

[3] Lu, Y. (2006). A combinatorial approach for automotive friction materials: Effects of ingredients on friction performance, *Composites Science and Technology*, *66*, 591-598.

[4] Tang, C. F. & Lu, Y. (2004). Combinatorial screening of ingredients for steel wool based semi-metallic and aramid pulp based non-metallic friction materials, *Journal of Reinforced Plastics and Composites*, *23*, 51-63.

[5] Lu, Y. (2002). A combinatorial approach for automotive friction materials: Combined effects of ingredients on friction performance, *Polymer Composites*, *23*, 814-822.

[6] Matejka, V., Martynkova, G. S., Ma, Y. & Lu, Y. (2009). Semimetallic brake friction materials containing $ZrSiO_4$: friction performance and friction layers evaluation, *Journal of Composite Materials*, *43*, 1421-1434.

[7] Ma, Y., Martynkova, G. S., Valaskova, M., Matejka, V. & Lu, Y. (2008). Effects of $ZrSiO_4$ in non-metallic brake friction materials on friction performance, *Tribology International*, *41*, 166-174.

[8] Lu, Y. (2002). Friction performance of phenolic binders, *International SAMPE Symposium and Exhibition Proceedings*, *47*, 236-244.

[9] Zhao, Y., Lu, Y. & Wright, M. A. (2006). Sensitivity series and friction surface analysis of non-metallic friction materials, *Materials & Design*, *27*, 833-838.

[10] Tomasek, V., Kratosova, G., Yun, R., Fan, Y. & Lu, Y. (2009). Effects of alumina in nonmetallic brake friction materials on friction performance, *Journal of Materials Science*, *44*, 266-273.

[11] Lu, Y. Tang, C. F. & Wright, M. A. (2002). Optimization of a commercial brake pad formulation, *Journal of Applled Polymer Science*, *84*, 2498-2503.

[12] Lu, Y. (2003). A Golden Section approach to optimization of automotive friction materials, *Journal of Materials Science*, *38*, 1081-1085.

[13] Tang, C. F. & Lu, Y. (2003). Optimization of an industrial friction material formulation, *Tribology Transactions*, *46*, 19-23.

[14] Lu, Y., Tang, C. F., Zhao, Y. & Wright, M. A. (2004). Optimization of a semimetallic friction material formulation, *Journal of Reinforced Plastics & Composites*, *23*, 1537-1545.

[15] Han, L., Huang L., Zhang, J. & Lu, Y. (2006). Optimization of ceramic friction materials, *Composites Science and Technology*, *66*, 2895-2906.

[16] Wu, H. (1998). Effects of carbon fibers on friction performance of NAO friction materials, *M.S. Thesis*, Beijing University of Chemical Technology [in Chinese].

[17] Lu, Y. & Wright, M. A. (2000). Optimizing friction formulation technique with emphasis of relational grade analysis, *International SAMPE Symposium and Exhibition Proceedings*, *45(2)*, 1972-1978.

[18] Han, L., Liu, X., Lu, Y., Zhao, Y., Li, W. & Zhang, Z. (2006). Combinatorial tribological materials research: five-component composite systems, *International SAMPE Symposium and Exhibition Proceedings* on CD.

[19] Lu, Y. & Zhao, Y. (2002). Experimental and genetic algorithm simulation of wear of two phase model friction materials, *International SAMPE Symposium and Exhibition Proceedings*, *47*, 245-252.

In: Bulk Materials: Research, Technology and Applications ISBN: 978-1-60692-963-6
Editors: T. Frias, V. Maestas, pp. 267-305

Chapter 6

GRAIN-SIZE DEPENDENT THERMAL, ELECTRICAL AND MECHANICAL PROPERTIES OF BULK NANOCRYSTALLINE MATERIALS

Linli Zhu and Xiaojing Zheng*
Key Laboratory of Mechanics on Western Disaster and Environment,
Ministry of Education, Department of Mechanics and Engineering Science,
College of Civil Engineering and Mechanics, Lanzhou University,
Lanzhou, Gansu 730000, P.R. China

ABSTRACT

For the bulk polycrystalline materials, the size of microstructures such as the grain size shrinking to the nanometer scale will result in an apparent change of the physical and mechanical performance of materials, which are distinguished remarkably with the ones of coarse-grain polycrystalline materials. In order to understand the influence of grain size on the properties of polycrystalline materials completely, the theoretical models with the point of view of the microstructures are presented to analyze grain size effects on the phonon thermal conductivity, electrical conductivity and the elastic modulus. For the phonon thermal conductivity, the phonon transport in both the grain interior and grain-boundary are discussed by using the phonon radiative transfer equation, and the analytical expression of conductivity is addressed to describe the influence of grain size on the thermal properties of polycrystalline materials. The electron grain-boundary scattering in the polycrystalline materials is taken into account as the boundary conditions of electron transport in the grain interior, to investigate the grain size effect on electrical conductivity. The predictions of proposed models for the phonon thermal conductivity and the electrical conductivity can agree well with experimental results. For elastic properties of nanocrystalline materials, a framework based on the continuum theory of mechanics is developed by considering the interface energy effects and grain-boundary effects to describe grain-size effects on elastic modulus. Base on the numerical results, it is illustrated that the effects of interface energy and grain-boundary can not be considered for large grain sizc. When the grain size is under several tens nanometers, the grain-size

* Corresponding author: Email:xjzheng@lzu.edu.cn.

effects are significant and the effects of interface energy and grain-boundary must be taken into account.

Keywords: Bulk polycrystalline materials; grain size effect; grain boundary scattering; surface energy; phonon thermal conductivity; electrical conductivity; effective elastic modulus

1. INTRODUCTION

With the rapid development of modern industry, nano-science and nano-technology have simulated a great deal of interest in recent two decades, and have become focus and front subject of every discipline. For the structures including microstructures with the size lower down to 100 nanometers, the resulting surface/interface and quantum confinement effects are expected to leading to novel or enhanced physical, mechanical and chemical performance with respect to the corresponding ones of macroscopic structures. For the polycrystalline materials with the grain size down to 100nm in spite of the materials still in the macroscopic scale, the electric, magnetic, optical and mechanical properties are distinguished from the ones with a large grain size but the same chemical compositions, in the presence of the obvious grain size effect due to the large surface-to-volume ratio available at the nanoscale. As a result that there are great existing or potential applications including easy shaping of intermetallics, damage tolerant ceramic composites for brake discs, wear resistant coatings, hydrogen storage for the zeroemission vehicle, the magnetic recording systems[1-3]. It has been observed in experiments that the electric, thermal and elastic properties of bulk polycrystalline materials are sensitive to the grain size which descends to the nanometer scale [4-9]. In this work, we will propose to develop the theoretical models to describe the grain size effect on the phonon thermal conductivity, electrical conductivity and elastic modulus of bulk polycrystalline material with the point of view of the microstructures.

Phonon thermal conductivity which plays a vital role in applications of the semiconductors has attracted much attention especially for the nanostructural materials. There are numerous works focusing on the phonon thermal conductivity of low-dimensional structures, such as nanowires, nanotubes, nanofilms and superlattice structures [10-14], as well as many studies to discuss the thermal conductivity of bulk materials [15-23]. Because of the great potential and extensive application of the bulk thermoelectric materials in the system with large heat load such as a refrigerator [24], it is necessary to find an efficient method to decrease the phonon thermal conductivity in order to enhance the thermoelectric performance of the materials. The experimental investigations demonstrated that, for the bulk polycrystalline material, the phonon thermal conductivity can be reduced remarkably with the aid of decreasing the grain size. It is due to the fact that the grain-boundary scattering of phonons becomes dominant with the grain size down to nanometer scale, leading to the modification in the phonon thermal conductivity of bulk polycrystalline materials. Here, we will investigate the phonon transport in the microstructures such as grains and grain boundary based on the phonon transport equation to establish a theoretical model (in section 1), which is expected to describe the grain-size dependent phonon thermal conductivity of bulk polycrystalline materials.

Similarly, the electrical conductivity as one of the most important material parameters has been widely studied in the perfect or semi-conducting nanostructural materials [25-28]. In general, the electrical conductivity of materials is associated with different electron scattering from the impurities, lattice defects, surfaces, and grain boundaries. The surface and grain boundary scattering of electrons has been discussed in theory [29,30] since the 50's of last century to predict size-dependent electrical conductivity for the wires and thin films, in which electrical conductivity are dependent on the structural size such as the diameter of wires and thickness of films. When the structural size falls down further to the nanometer scale, the size effect on the electrical conductivity becomes more significantly. Due to the increment of the scattering centers for conduction electrons from grain boundaries, the nanocrystalline materials exhibit lower electrical conductivity which is grain-size dependent remarkably. We will study the electron transport in the microstructures of bulk polycrystalline materials from the transport theory to present the analytical expression of the electrical conductivity (in section 2). The proposed model can not only describe the grain size dependent electrical conductivity but overcome the shortcoming of the previous description for grain boundary scattering of electrons.

The investigation of the mechanical properties is also very important in the researches of nanostructural materials, sine the mechanical properties are indeed relevant to the fabrication of the nanoelectromechanical systems, and the stability and performance of nanodevices are strongly sensitive to the mechanical reliability of nanostructures. On account of the distinct mechanical properties for the surface compared to the bulk ones, the surface effect become more important on the mechanical performance of the structures, resulting in the size-dependent elastic phenomena in the length scale range of 1-10nm due to the large surface-to-volume ratios of nanostructures [31-34]. For the bulk nanocrystalline materials, the interfaces can change the elastic properties notably as well as the thicker grain boundary induced by the smaller grain size and the vacuity. Here, the nanocrystalline materials are considered into the two phase: grain one and the grain-boundary one. With the framework of the continuum mechanics theory, the analytical expression of the effective elastic modulus of the bulk polycrystalline materials will be addressed (in section 3) by considering the surface effect of the grains and the grain boundary effect, to predict the grain-size dependent Young's modulus of nanocrystalline materials in experiments.

2. Phonon Thermal Conductivity of Bulk Thermoelectric Materials

2.1. Introduction

In the past decade, a tremendous effort has been witnessed in searching for new thermoelectric materials of improved performance [35]. The performance is measured by the value of the thermoelectric figure of merit $ZT = \alpha^2 \sigma T / \kappa$, where α, σ, κ, T are the Seebeck coefficient, electrical conductivity, thermal conductivity and temperature. To increase *ZT*, many material exploration and optimization investigations have been undertaken. Several state-of-the-art thermoelectric materials were discovered such as complex crystal structure systems[36], strong correlation system as Kondo insulators [37,38], layered structure materials and mesoscopic structure systems like superlattices [39,40].

At present, extensive researches focus on the thin films or nanowires with low-dimension structures to get good thermoelectric materials with high *ZT* value. Besides, one still pays attention to bulk thermoelectric materials [15-23] because there exist significantly important and extensive applications for bulk materials in systems with large heat loads such as a home refrigerator [24]. Since the thermal conductivity of the bulk materials is too high to obtain a figure of merit suitable for applications, many attempts have been undertaken to lower the thermal conductivity. Filling the voids in the skutterudite structure with, e.g., rare earth elements such as Ce, La, and Yb, has been explored with the hope of reducing the phonon thermal conductivity [15,20]. Substitutional doping has also been performed, on either the cobalt or antimony site, resulting in the reduction of the thermal conductivity with different scattering mechanism [17]. In addition, the expected enhancement in point-defect or large-defect scattering suggests an alternative approach to follow in reducing the phonon thermal conductivity [41,42].

In addition, an alternative approach by controlling the microstructures of the bulk materials has been expected to lower the phonon thermal conductivity for improving the performance of the bulk thermoelectric materials [6, 43]. They attributed this mainly to the different carrier scattering mechanism dependent of the grain size. It has been noted that the microstructure of bulk thermoelectric materials can also play an important role in the transport properties of the materials. Nevertheless, the current works haven't quantitatively clarified the relationship between the grain size and the transport properties of polycrystalline, and only investigated with large grain size samples but not with even smaller grain sizes such as several tens or several hundreds nanometers. Otherwise, although there are a great number of works denoted to study the transport properties of nanoscale materials or structures in experiment or theory, there are few detailed studies on grain size effect (or grain boundary effect) on the transport properties of bulk materials and particularly in theory there are scarce theoretical models applicable to describe such size effects. To improve the thermoelectric performance, it is important to understand the relationship between microstructures and transport properties in detail.

It is the purpose in this paper to develop the models for the grain size effects on the thermal conductivity of bulk thermoelectric materials. Taken into account the influence of grain-boundary scattering, the phonons transports in polycrystalline materials are investigated. The analytical expressions indicate that the thermal conductivity is the function of the grain size. Comparing with the experimental data, the predictions are in good agreement with the measurement.

2.2. Model

As we known that the phonon thermal conductivity of a bulk material is usually estimated from a simple kinetic formula $\kappa_{\mathrm{p}} = C_V v_{\mathrm{p}} \Lambda_{\mathrm{pT}} / 3$, where C_V is the specific heat per unit volume which usually is a constant, v_{p} is the group velocity of phonons and Λ_{pT} is the total phonon mean free path. We can find from this formula that there is no an explicit relation between the phonon thermal conductivity and the grain-sizes of a bulk thermoelectric material even the total phonon mean free path Λ_{pT} should be dependent on the grain size. Thus, a new formula should be developed to give a correct prediction and explanation on our experiment results. Here, a theoretical model will focus on the phonon thermal conductivity since it is a dominator in the thermal conductivity. In order to develop an effective and simple theoretical model for the phonon transport in the bulk CoSb3, the following assumptions are employed: (1) all grains are assumed to be uniform spheres with the diameter, *D*, and the minimum span distance, *H*, between two grain surfaces (see figure 1); (2) the phonon transport along the heat flowing direction, denoted by *z*-axis, is considered only so that the phonon intensity I_m is a function of the polar angle θ_m of the phonon velocity and the position coordinate z_m, in which the subscript $m(=\mathrm{g},\mathrm{b})$ denotes the grain g and the grain-boundary b , respectively; (3) from grain to grain in the bulk, the energy transmissivity T_{mn} and reflectivity R_{mn} at the interface between a grain and the grain-boundary are independent on the grain size and the polar angle of phonon velocity; and (4) the intensity gradient of equilibrium phonon in the grains is a constant. These assumptions are acceptable because of small grain sizes.

According to the above assumptions and the steady-state Boltzmann transfer equation under the relaxation time approximation [44], the phonon intensity function I_m in a grain and in a grain-boundary of a bulk material can be described by a following equation

$$sin\theta_m \frac{\partial I_m}{\partial z_m} = -\frac{I_m}{\Lambda_{\mathrm{pb}}} + \frac{I_0}{\Lambda_{\mathrm{pb}}}\delta_{m\mathrm{g}}, \qquad (2\text{-}1)$$

where $\delta_{m\mathrm{g}}$ means the Kronecker δ and it is taken as 1 for $m=\mathrm{g}$ and 0 for $m\neq\mathrm{g}$, respectively, Λ_{pb} is the phonon background mean free path independent of the grain-sizes. Having taken $\mu_m = \cos\theta_m$, $\eta_m = z_m/\Lambda_{\mathrm{pb}}$, $\xi_{\mathrm{g}} = D/\Lambda_{\mathrm{pb}}$ and $\xi_{\mathrm{b}} = H/\Lambda_{\mathrm{pb}}$, and considered the energy balance at the interfaces [45], the diffuse interface-scattering boundary conditions of Eq. (1) for the grain and for the grain-boundary are respectively established on account of the diffuse interface scattering occurring at most interfaces as the key factor in the interface scattering

$$\begin{cases} I_m^+(0,\ \ \mu_m) = 2R_{mn}\int_{2\pi} I_m^-(0,-\mu_m)\mu_m d\mu_m + 2T_{nm}\int_{2\pi} {}^{\mathrm{P}}I_n^+(\xi_n,\mu_n)\mu_n d\mu_n \\ I_m^-(\xi_m,-\mu_m) = 2R_{mn}\int_{2\pi} I_m^+(\xi_m,\mu_m)\mu_m d\mu_m + 2T_{nm}\int_{2\pi} {}^{\mathrm{N}}I_n^-(0,-\mu_n)\mu_n d\mu_n \end{cases}, \qquad (2\text{-}2)$$

in which $m,n=\mathrm{g},\mathrm{b},m\neq n$, "+" for $0<\mu_m<1$ and "-" for $-1<\mu_m<0$, P and N represent the preceding and next grain unit respectively. In fact, the relation of ${}^{\mathrm{P}}I_{\mathrm{g}} \approx I_{\mathrm{g}} \approx {}^{\mathrm{N}}I_{\mathrm{g}}$ can be acceptable for the small grain diameter with the small variation of local temperature. Having solved Eqs. (1) and (2) to achieve the phonon intensity function I_m and used the formula of the local heat flux in a grain and in a grain-boundary as follows

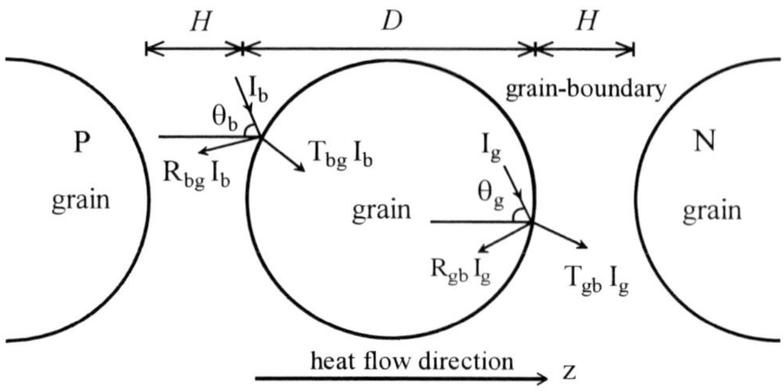

Figure 1. Schematic diagram of a phonon reflection and transmission at the interface between a grain and a grain-boundary.

$$q_m(\eta_m) = \pi \int_{4\pi} I_m \sin 2\theta_m d\theta_m \quad (m = \mathrm{g}, \mathrm{b}), \tag{2-3}$$

the phonon thermal conductivity, respectively for the grain and the grain-boundary, can be expressed as follows

$$\begin{aligned}\kappa_{\mathrm{pg}}(z_{\mathrm{g}}) &= C_V v_{\mathrm{p}} \Lambda_{\mathrm{pb}} \int_0^1 \{ (c_1 + \mu_{\mathrm{g}})[e^{-z_{\mathrm{g}}/(\Lambda_{\mathrm{p}}\mu_{\mathrm{g}})} + e^{-(D-z_{\mathrm{g}})/(\Lambda_{\mathrm{p}}\mu_{\mathrm{g}})}] - 2\mu_{\mathrm{g}} \\ &\quad - Dc_2[e^{-z_{\mathrm{g}}/(\Lambda_{\mathrm{p}}\mu_{\mathrm{g}})} - e^{-(D-z_{\mathrm{g}})/(\Lambda_{\mathrm{p}}\mu_{\mathrm{g}})}] \} \mu_{\mathrm{g}} d\mu_{\mathrm{g}} / 2 \\ &= C_V v_{\mathrm{p}} \Lambda_{\mathrm{pb}} G_{\mathrm{g}}(z_{\mathrm{g}}, D, H) / 2\end{aligned} \tag{2-4}$$

$$\begin{aligned}\kappa_{\mathrm{pb}}(z_{\mathrm{b}}) &= C_V v_{\mathrm{p}} \Lambda_{\mathrm{pb}} \int_0^1 \{ Dc_3[e^{-z_{\mathrm{b}}/(\Lambda_{\mathrm{p}}\mu_{\mathrm{b}})} - e^{-(H-z_{\mathrm{b}})/(\Lambda_{\mathrm{p}}\mu_{\mathrm{b}})}] \\ &\quad - c_4[e^{-z_{\mathrm{b}}/(\Lambda_{\mathrm{p}}\mu_{\mathrm{b}})} + e^{-(H-z_{\mathrm{b}})/(\Lambda_{\mathrm{p}}\mu_{\mathrm{b}})}] \} \mu_{\mathrm{b}} d\mu_{\mathrm{b}} / 2 \\ &= C_V v_{\mathrm{p}} \Lambda_{\mathrm{pb}} G_{\mathrm{b}}(z_{\mathrm{b}}, D, H) / 2\end{aligned} \tag{2-5}$$

where $c_i (i = 1, 2, 3, 4)$ given in appendix A is related to the diameter D and the thickness H as well as the energy transmissivity T_{mn} and reflectivity R_{mn} which can all be determined owing to the relation $T_{mn} = R_{nm} = 1 - T_{mn}$, $(m, n = \mathrm{g}, \mathrm{b}, m \neq n)$ [46] and also be independent of the grain-sizes according to above assumptions. Moreover, the functions $G_{\mathrm{g}}(z_{\mathrm{g}}, D, H)$ and $G_{\mathrm{b}}(z_{\mathrm{b}}, D, H)$ given in appendix are only related to the geometrical parameters D and H of the grain and the boundary, in which $z_{\mathrm{g}} \in [0, D]$ and $z_{\mathrm{b}} \in [0, H]$, so that they are actually a kind of scale functions. Compared the expressions (4) and (5) with the existing formula $\kappa_{\mathrm{p}} = C_V v_{\mathrm{p}} \Lambda_{\mathrm{pT}} / 3$, it can be found that Λ_{pT} in the existing formula is respectively replaced by $\Lambda_{\mathrm{pb}} G_{\mathrm{g}}(z_{\mathrm{g}}, D, H)$ and $\Lambda_{\mathrm{pb}} G_{\mathrm{b}}(z_{\mathrm{b}}, D, H)$. Since C_V, v_{p} and Λ_{pb} are usually considered as independence of the grain-sizes, the effect of grain size on the phonon thermal conductivity, respectively for the grain and the grain-boundary, are actually exhibited in explicit form by the scaling functions $G_g(z_g, D, H)$ and $G_b(z_b, D, H)$. With the aid of the volume percentage $p_b = 3D_b/(D_g + D_b)$ for the grain-boundary region in a bulk material, the phonon thermal conductivity of the bulk material can be written as $\kappa_p = (1-p_b)\kappa_{pg} + p_b\kappa_{pb}$ and the total scaling function G_B of the phonon thermal conductivity can be expressed as $G_B = (1-p_b)G_g + p_bG_b$. So, the theoretical model suggested in this paper can clearly reveal the effect of the grain size on the phonon thermal conductivity of the bulk materials.

2.3. Comparison and Discussion

In order to confirm the theoretical models suggested above for describing the grain size effects on the phonon thermal conductivity of the bulk polycrystalline in detail, the theoretical results of the thermal conductivity is calculated to fit with the experimental data as shown in figure 2, respectively. It can be found from figure 2 that the fits for the phonon thermal conductivity show quite good agreement both at 300K and 600K with phonon background mean free path Λ_{pb} as 300nm and 80nm respectively. The curve has the correct slope and passes through these experimental data almost entirely. It should be noted that the given parameters of phonon background mean free path are reasonable for the thermoelectric materials [47]. Consequently, it is concluded that the proposed models of the phonon thermal conductivity can successfully predict the experimental results which exhibit the influence of grain size on the phonon thermal conductivity of the bulk material.

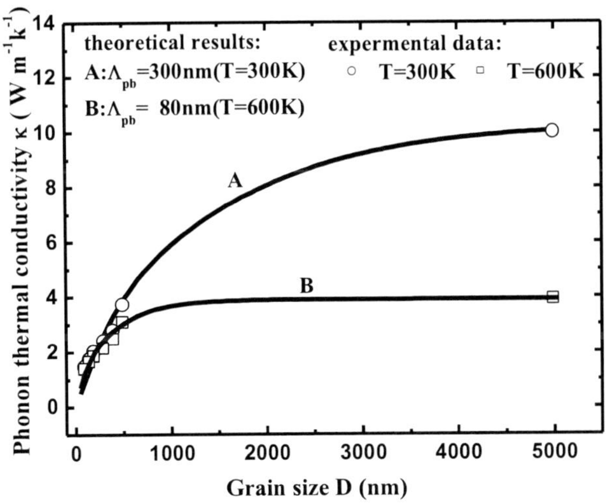

Figure 2. Fit of the experimental data on the bulk CoSb3 to the proposed model respectively at temperatures of 300K and 600K in κ_p vs. grain size D in which $\eta_g = \eta_b = 0.5$, $T_{gb} = 0.5$, $H = 2\text{nm}$ and $\Lambda_{pb} = 300\text{nm}$ for 300K and $\Lambda_{pb} = 80\text{nm}$ for 600K.(Symbols: experimental data, lines: calculations with the proposed model).

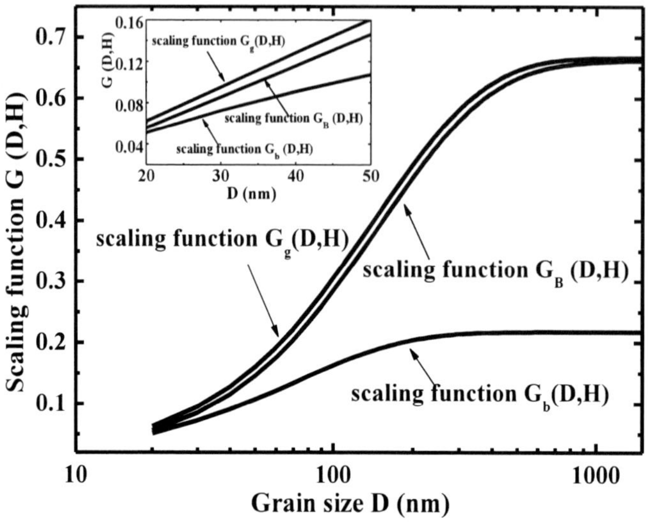

Figure 3. Comparison among the scaling functions G_g, G_b and G_B vs. the grain size D in the regions of 20~1500nm when $\eta_g = \eta_b = 0.5$, $T_{gb} = 0.5$, $\Lambda_{pb} = 100\text{nm}$, $H = 2\text{nm}$.

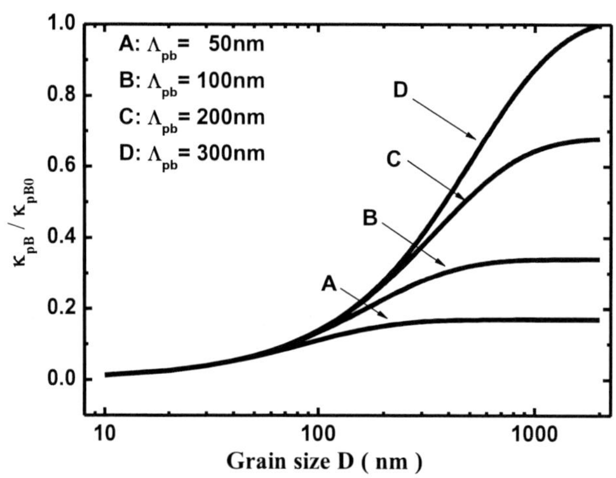

Figure 4. Theoretically calculated ratio $\kappa_{pB} / \kappa_{pB0}$ vs. grain size D with different background mean free path Λ_{pb} for the parameters $\eta_g = \eta_b = 0.5$, $T_{gb} = 0.5$, $H = 2\text{nm}$.

For the sake of further discussion of the grain size effects predicted by the proposed models, the curves of the scaling functions $G_g(z_g, D, H)$ for the grain, $G_b(z_b, D, H)$ for the grain boundary and $G_B(z_g, z_b, D, H)$ for the bulk varying with the grain size D are plotted in Figure 3. It can be found from Figure 3 that all of them declined with the reducing the grain size D and the scaling functions $G_g(z_g, D, H)$ and $G_b(z_b, D, H)$ have different contributions. When the grain size is D>50nm, the scaling function $G_B(z_g, z_b, D, H)$ for the bulk is more and more close to the scaling function $G_g(z_g, D, H)$ for the grain. However, when the grain size D is smaller, such as D<50nm, the scaling function $G_B(z_g, z_b, D, H)$ for the bulk is more and more close to the function $G_b(z_b, D, H)$ for the grain boundary with the reducing the grain size D, which means the scaling function $G_b(z_b, D, H)$ for the grain boundary is predominated and important for the smaller grains.

The ratio of the phonon thermal conductivity, i.e. $\kappa_{\mathrm{pB}} / \kappa_{\mathrm{pB0}}$, varying with the grain size D is plotted in Figs. 4 with different phonon background mean free path Λ_{pb}. Here κ_{pB0} is the reference value of the bulk phonon thermal conductivity κ_{pB} with large grain size such as $D = 5\mu m$ and the phonon background mean free path as 300nm, at which the effect of the grain size is very small. It is found from Figure 4 that the ratio $\kappa_{\mathrm{pB}} / \kappa_{\mathrm{pB0}}$ falls down with the reducing of the grain size D and the phonon background mean free path has significant influence on the ratio $\kappa_{\mathrm{pB}} / \kappa_{\mathrm{pB0}}$, especially when the grain size D>100nm. The falling rates of the ratio $\kappa_{\mathrm{pB}} / \kappa_{\mathrm{pB0}}$ are related to both the grain size D and the phonon background mean free path Λ_{pb}. The effect of the grain size D on the ratio $\kappa_{\mathrm{pB}} / \kappa_{\mathrm{pB0}}$ is obvious only for the phonon background mean free path $\Lambda_{\mathrm{pb}} \geq 100$ nm and the falling rates of the ratio $\kappa_{\mathrm{pB}} / \kappa_{\mathrm{pB0}}$ for the grain size D in the region between 100nm to 1000nm are greater than those for the grain size in other regions. This means that the effect of the grain size D on the bulk phonon thermal conductivity κ_{pB} is notable only when the grain size D are in the region that mentioned above and the phonon background mean free path Λ_{pb} >100nm. Moreover, the effect of the phonon background mean free path Λ_{pb} on the bulk phonon thermal conductivity κ_{pB} is also significant, especially when the grain size D>500nm. The greater the grain size is, the more obvious the effect. When the grain size is D<100nm, the phonon background mean free path Λ_{pb} has no influence on the bulk phonon thermal conductivity κ_{pB}. Besides that, the bulk phonon thermal conductivity κ_{pB} increases with the increment of the phonon background mean free path Λ_{pb}. Since the phonon background mean free path Λ_{pb} is related to the temperatures for a given material, the lower the temperature is, the larger the phonon background mean free path. The effect of the phonon background mean free path on the bulk phonon thermal conductivity κ_{pB} is actually a kind of the effect of the temperature. So, these predicted results have good agreement in quality with the experimental results, which also exhibited that the effect of the temperature on the thermal conductivity where they are obvious only for the greater grain sizes and the temperature makes the thermal conductivity increase.

2.4. Summary

In this section, we developed the theoretical models that take into account the grain boundary scattering for the phonons to describe the grain size effects on the thermal conductivity of bulk thermoelectric materials. Excellent fits to experimental data of the phonon thermal conductivity and the electrical conductivity confirm the validity of the model. The numerical results from the proposed models show that the phonon background mean free path makes the thermal conductivity increase. The results in this paper may provide a useful guidance for future improvements in the thermoelectric figure of merit of bulk thermoelectric materials.

3. Electrical Conductivity of Bulk Polycrystalline Materials

3.1. Introduction

The size effects on the electrical conductivity of low-dimensional structures such as the thin films and the wires have been the subject of intense research for decades. It is well known that the size effects of low-dimensional structures can be simply interpreted as geometrical-scale variation of structures in certain dimensions, and FS theory [29] and MS theory [30] are often applied to describe the surface scattering and grain boundary scattering of the electrical conductivity respectively [25-28]. However, there are few investigations indeed probing into the size effects of the electrical conductivity with the viewpoint of the grains in materials [6], and it is very scarce in theories to study the grain- size effects of the polycrystalline systems in electrical conductivity. Despite FS theory and MS theory availably describing the size effects on the electrical conductivity, establishing the relative magnitudes of the different scattering mechanisms in polycrystalline systems was found to be a difficult task [5,48].

The MS theory applying in the bulk materials is controversial to describe the size effects stemmed from the grain-boundary scattering, because the MS theory is quasi one dimensional neglecting lateral percolation of conducting electrons through neighboring grains [5]. The another problem is that the assumption of the parallel plane in the MS theory is questionable in the further rigorous investigation owing to arbitrary shape and size of the grain boundaries or the grains in bulk materials. Furthermore, for simultaneous different origins of electron scattering, Matthiessen's rule will be not satisfied to define a characteristic mean free path self-consistently [48,49].

This issue of focusing on the grain-boundary scattering effects in the electrical conductivity of the polycrystalline systems is the first time to investigate the grain-size effects of electrical conductivity with the aide of discussing the conduction electrons distribution in the grains. Obviously, the problem of computing the electron distribution presents enormous difficulties for grain-boundary scattering considering in scattering terms as well as isotropic background scattering in Boltzmann transport equation on account of arbitrary shape, size, and orientation of the grain boundary or the grains in polycrystalline materials. However, with alternative approach and assumptions for the grain boundary scattering, the problem can be made more tractable and simplified. Usually, the scattering effects of the surface and the grain boundary to the electrical conductivity can be treated in Boltzmann equation by two manners, namely the scattering items or the boundary conditions. The treatment on the grain-boundary scattering as the boundary conditions chosen in this letter can not only consider the various grain shapes conveniently in detail but also avoid the invalidation of the Matthiessen's rule. As the polycrystalline materials, the influence of grain-boundary scattering is dominant resulting in that the surface scattering can be safely neglected [50].

3.2. Model

The Boltzmann equation for this problem is solved to find electrical conductivity considering the grain-boundary scattering as the boundary conditions in grains. For the sake of the simplification, it is first taken into account the isotropic approximation in the polycrystalline materials. The electrical conductivity in the single grain unit is represented as the total one on the assumption that the polycrystalline materials are consistent of the perfect grain units with periodic boundaries. When the external electric field E is in the direction of the x coordinate, the electron distribution function of the periodic unit can be considered as the function of the coordinate x due to the grain-boundary scattering. Therefore, the Boltzmann equation reduces to a differential equation dependent on the coordinate x in grains,

$$\frac{\partial f_1}{\partial x}+\frac{f_1}{\tau v_x}=\frac{eE_x}{mv_x}\frac{\partial f_0}{\partial v_x}. \tag{3-1}$$

Here f_0 is the equilibrium distribution function and f_1 is the correction to the distribution function.; τ is a relaxation time to describe the scattering from other sources (point defects and phonons); e is the electron charge, and m and v_x are the electron mass and x component of velocity, respectively. So, the electron distribution function in grains can be written as $f=f_0+f_1(v,x)$.

For the boundary conditions in grains, the parameter Ts for the grain boundary effect is introduced as the transmisitivity of the conduction electrons with tunneling the grain boundary barrier from a grain to the next one [51]. One simple assumption is to suppose that the scattering is entirely diffuse so that the distribution function of the electrons leaving each boundary must then be independent of direction. Based on the electrons balance at the boundaries between two grains, the boundary conditions in the periodic unit can be written as

$$\begin{cases}\int_s f_1^+(0,v_x)ds=(1-\mathrm{Ts})\int_s f_1^-(0,-v_x)ds+\mathrm{Ts}\int_s f_\mathrm{P}^+(d,v_x)ds\\ \int_s f_1^-(d,v_x)ds=(1-\mathrm{Ts})\int_s f_1^+(d,-v_x)ds+\mathrm{Ts}\int_s f_\mathrm{N}^-(0,v_x)ds\end{cases}. \tag{3-2}$$

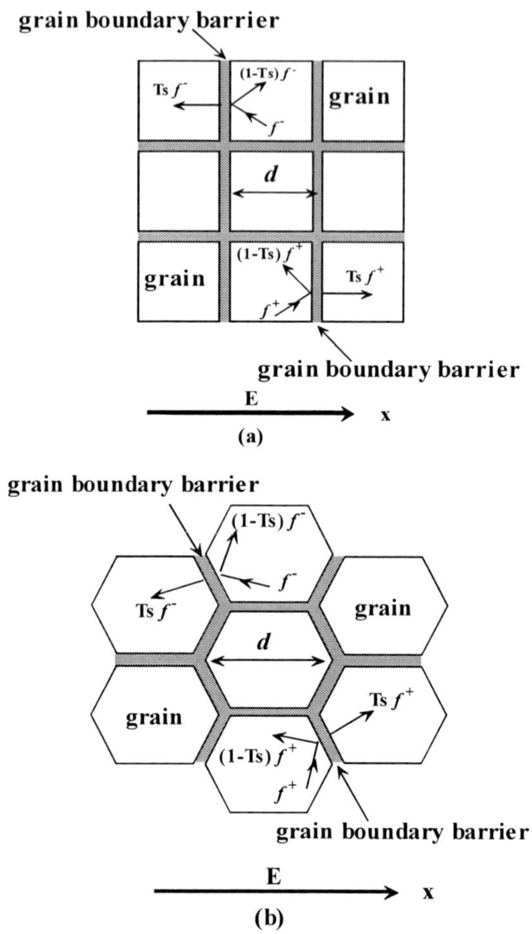

Figure 5. Two-dimensional schematic illustration of conduction electrons tunneling the grain boundary barrier between two neighboring with the assumption of the square grain shape (a) and the hexagonal grain shape (b) .

Here s represents the boundary areas of the grains varied with the geometrical shapes for the grains; d is the grain unit size; P represents the preceding unit, and N represents the nest unit in the x direction; and there are two distribution functions, f_1^+ for electrons with $v_x > 0$ and f_1^- for electrons with $v_x < 0$.

We give the square grain shape and the hexagonal grain shape (shown in figure 5), respectively, as an example to obtain the distribution function f_1 as follows

$$f_1(\boldsymbol{v},x) = \begin{cases} \dfrac{e\tau E_x}{m}\dfrac{\partial f_0}{\partial v_x}\left[1 - C_k^+(v_x)\exp(-\dfrac{x}{\tau v_x})\right], v_x > 0 \\ \dfrac{e\tau E_x}{m}\dfrac{\partial f_0}{\partial v_x}\left[1 - C_k^-(v_x)\ \exp(\dfrac{d-x}{\tau v_x})\right], v_x < 0 \end{cases} \quad (k = \mathrm{q,\ h}). \qquad (3\text{-}3)$$

where $C_k^{\pm}(\boldsymbol{v})$ are given for $k = \mathrm{q}$ as the square shape of the grain and for $k = \mathrm{h}$ as the hexagonal shape of the grain as follows

$$C_{\mathrm{q}}^{\pm}(v_x) = -\frac{2(1-Ts)}{Ts(1-\mathrm{e}^{\mp d/\tau v_x})}$$
$$C_{\mathrm{h}}^{\pm}(v_x) = -d(1-Ts)[\mathrm{e}^{\frac{\mp 3d}{4\tau v_x}} - \mathrm{e}^{\frac{\mp d}{\tau v_x}} - (1-\mathrm{e}^{\frac{\mp d}{4\tau v_x}})](2\tau v_x)^{-1}\{(\mathrm{e}^{\frac{\mp 3d}{4\tau v_x}} - \mathrm{e}^{\frac{\mp d}{\tau v_x}})[T_s(1-\mathrm{e}^{\frac{\mp d}{4\tau v_x}})\cdot \quad (3\text{-}4)$$
$$+(1-2T_s)(\mathrm{e}^{\frac{\mp 3d}{4\tau v_x}} - \mathrm{e}^{\frac{\mp d}{\tau v_x}})] - (1-\mathrm{e}^{\frac{\mp d}{4\tau v_x}})[1-\mathrm{e}^{\frac{\mp d}{4\tau v_x}} - T_s(\mathrm{e}^{\frac{\mp 3d}{4\tau v_x}} - \mathrm{e}^{\frac{\mp d}{\tau v_x}})]\}^{-1}.$$

Having found the various f_1's by means of the usual formula of the current density, we can find the effective electrical conductivity of the grain

$$\sigma = \frac{1}{E_x V}\iiint_V (-4)e^2\tau E_x \frac{1}{h^3}(\frac{m}{2})^{1/2}\int_0^\infty \varepsilon^{3/2}\frac{\partial f_0}{\partial \varepsilon}d\varepsilon\{\ \int_0^\pi d\theta(\int_0^{\frac{\pi}{2}} d\varphi + \int_{\frac{3\pi}{2}}^{2\pi} d\varphi)\sin^3\theta\cos^2\varphi ,$$
$$\times[1-C_k^+\exp(-\frac{x}{\Lambda\sin\theta\cos\varphi})] + \int_0^\pi d\theta\int_{\frac{\pi}{2}}^{\frac{3\pi}{2}} d\varphi\sin^3\theta\cos^2\varphi[1-C_k^-\exp(\frac{d-x}{\Lambda\sin\theta\cos\varphi})]\ \}dV$$
$$(3\text{-}5)$$

where $\Lambda = \tau v, v = |\boldsymbol{v}|$ is the background free path of the electrons, V is the volume of the unit. Since the coefficients $C_k^{\pm}$ in Eq. (4) are subjected to the grain size d, the theoretical model suggested here can demonstrate the effect of the grain size on the effective electrical conductivity shown in Eq.(5). It can be noted from this model that there exist two parameters including the background mean free path Λ and the transmisitivity Ts. Both of them are not the phenomenological parameters and can be confirmed directly or indirectly by theoretical derivation or experimental measure.

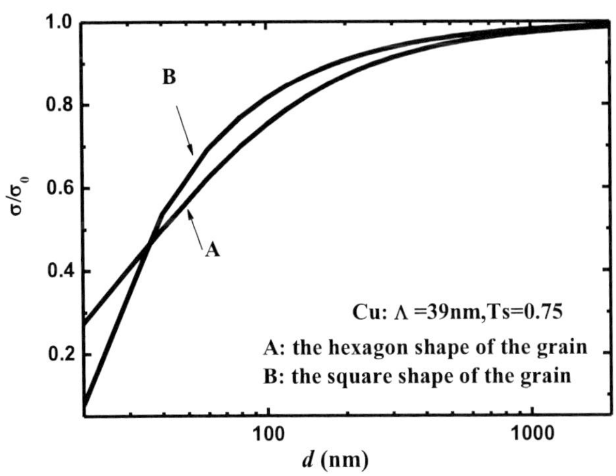

Figure 6. Theoretical results in electrical conductivity vs. grain size d with the square and hexagonal grain shapes for the proposed model given polycrystalline Cu an example with the parameters $\Lambda = 39\mathrm{nm}$ and $\mathrm{Ts} = 0.75$.

3.3. Numerical Results and Discussion

The electrical conductivities with different grain sizes have been calculated in the cases of the square grain shape and the hexagonal grain shape respectively by applying proposed model. In order to perform the influence of the grain shape on the electrical conductivity qualitatively, the curves of the electrical conductivity ratio σ / σ_0 verse the grain size d are shown in Figure 6 for the Cu polycrystalline materials. Here, σ_0 is the reference value of the bulk electrical conductivity with large grain size or single crystalline. It can be observed from Figure 6 that the influence of grain shape is weak on the electrical conductivity when the gain size is large such as $d > 500\text{nm}$, yet it is gradually marked along with the reduction of the grain size: the electrical conductivity with the square shape is 8.5% larger than the one with the hexagonal shape when the grain size $d = 100\text{nm}$, and the conductivity with square shape is 30.4% lower than the one with the hexagonal shape when the grain size $d = 25\text{nm}$.

In order to evaluate the proposed model and the MS theory which one is more appropriate, the experimental data of Cu polycrystalline nanowire [52] and Au polycrystalline nanofilm [53] are analyzed in terms of the proposed model with the hexagonal grain shape and the MS theory, and the comparisons are respectively shown in Figs. 7(a) and 7(b). It is worth to note that the MS theory is generally suitable to describe the size effect of the low-dimensional polycrystalline systems. Meanwhile, due to the isotropic assumption of the proposed model to discuss the electrical conductivity in one direction, it can be approximately applied to predict the size effects of the electrical conductivity for low-dimensional polycrystalline systems directly in which the surface scattering can be safely neglected. It can be noticed from Figs. 7(a) and 7(b) that the proposed model can predict size effects on the electrical conductivity in presence of well agreement with the experimental results of Cu nanowire and Au nanofilm in all investigated region of grain size, and the grain-size dependence of the electrical conductivity for Cu nanowire and Au nanofilm with all region of grain sizes could be modeled by the proposed model fairly better than by the MS theory. Moreover, the theoretical predictions from the proposed model and the MS theory are consistent basically when the grain size $d > 100\text{nm}$, whereas the prediction of the MS theory is greater than the proposed model's and the difference gradually becomes significant when the grain size $d < 100\text{nm}$. This is due to the fact that the parallel plane assumption of the grains is more reasonable for the large grain size than for the small grain size in polycrystalline systems, yet the grain shape considered as the hexagon is independent on the grain size.

The proposed model can be applied to predict the electrical conductivity of bulk materials on account of the isotropic simplification reasonable to discuss the electrical conductivity of the polycrystalline bulk materials in general. Figure 8 plots the experimental data of bulk CoSb3[54] and the calculations of the proposed model and the MS theory. It can be found from Figure 8 that the theoretical results of two models are very close to each other and in good agreement with the experiment when the grain size $d > 200\text{nm}$, though the results of the MS model is higher than the ones of the proposed model when $d < 200\text{nm}$ and the discrepancy between two models increases with the decreasing grain size. It is found from the comparison with the experimental data that the results of the proposed model agree with the experiment better while the prediction of the MS model is 57% larger than the experimental data when $d = 100\text{nm}$. This analysis indicates that the MS model based on the assumption of parallel planes cannot be suitable to predict the electrical conductivity of the bulk polycrystalline materials, and the proposed model has more extensive applicability rather than the MS model.

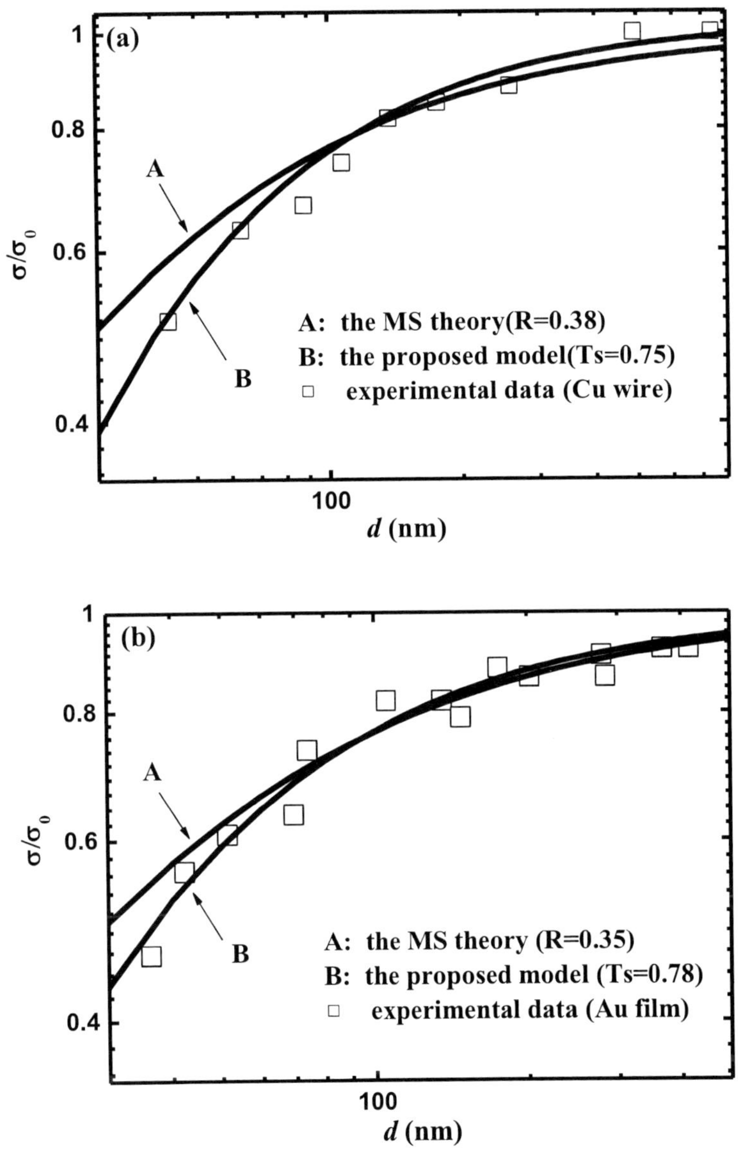

Figure 7. The comparison of the proposed model with the hexagonal grain shape and the MS theory in σ/σ_0 vs. grain size d for the experimental data of Cu nanowire with the background mean free path $\Lambda = 39\text{nm}$, $\text{R} = 0.38$ for the MS theory and $\text{Ts} = 0.75$ for the proposed model(a), and for the experimental data of Au nanofilm with the background mean free path $\Lambda = 40\text{nm}$, the parameter $\text{R} = 0.35$ for the MS theory and the parameter $\text{Ts} = 0.78$ for the proposed model(b).

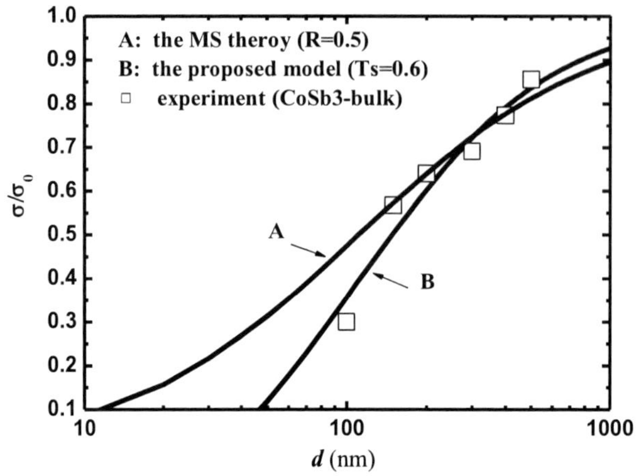

Figure 8. The comparison of the proposed model with the hexagonal grain shape and the MS theory in σ / σ_0 vs. grain size d for the experimental data of CoSb3 bulk material with the background mean free path $\Lambda = 85\text{nm}$, $\text{R} = 0.5$ for the MS theory and $\text{Ts} = 0.6$ for the proposed model.

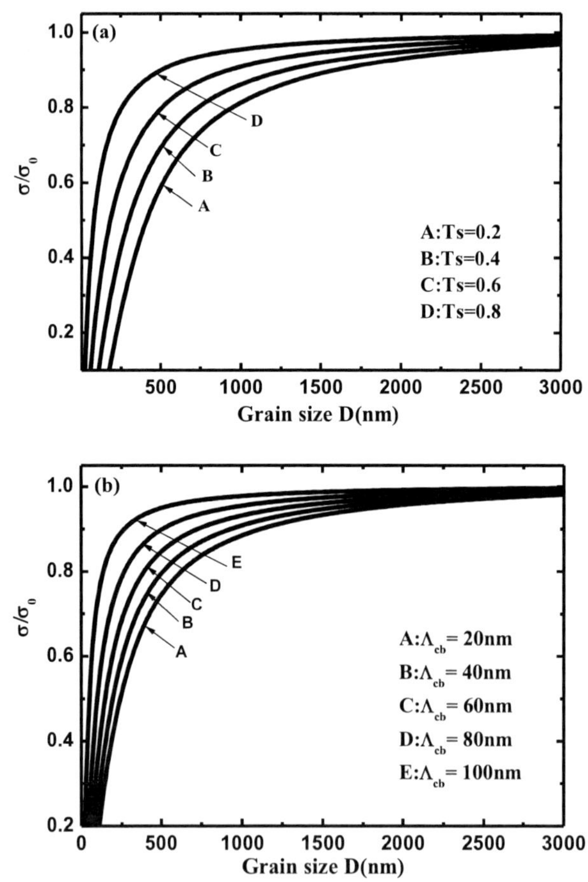

Figure 9. Theoretically calculated σ / σ_0 vs. grain size D with different transmisitivity Ts (a) and different background mean free path Λ_{cb} (b).

The effect of the grain size on the ratio σ / σ_0 of the electrical conductivity is plotted in figure 9 for several given background mean free path Λ_{cb} and transmisitivity Ts. Here, σ_0 is the reference value of the bulk electrical conductivity with independence of the grain size . It can be found from figure 9 that the electrical conductivity has a fast falling down after the grain size D<500nm. The smaller the grain size is, the faster the falling rate of the electric conductivity. Moreover, the figure 9(a) shows that the transmisitivity Ts makes the electrical conductivity increase. It is due to the fact that the larger transmisitivity means that there are more electrons across the barrier. The figure 9(b) shows that the conductivity falls down with decreasing of the background mean free path Λ_{cb}, which means that the smaller background mean free path stems from the more significant electrons scattering from point defects and phonons resulting in the reduction of the electrical conductivity.

3.4. Summary

In this section, taking into account the grain-boundary scattering with the point of view of dealing with the scattering as the boundary conditions of grains, we have derived the analytical formula of the electrical conductivity applicable to 1-3 dimensional materials and different grain shapes resulting in regarding the influence of grain boundary shape indirectly. It also can avoid the invalidation of the Matthiessen's rule on the characteristic mean free path. Compared with the MS model subjected to the assumption of parallel planes for the grain boundary scattering, the experimental results lead us to conclude that the proposed model can agree well with the experiment in all investigated region of the grain size while the MS model can only be appropriate to the regions of large grain size due to the assumption of parallel planes. Based on this novel model, it is proposed that the relative magnitudes of different scattering mechanisms on the electrical conductivity can be established in polycrystalline systems.

4. Effective Young's Modulus of Bulk Nanocrystalline Materials

4.1. Introduction

Generally nanomaterials can be classified into two categories, namely nano-sized structural elements and nano-structured materials. The first category can be defined that at least one of the overall dimensions of a structural element is in the nanometer range including nanoparticles, nanowires, nanotubes, nanobelts, nanospring, etc. The second one is that the characteristic length of the microstructure is in the nanometer range, such as the grain size of a polycrystalline and the void size of porous material. In the past decades, many researches of nanomaterials have been done to investigate the properties of nanotubes, nanowires, nanofilms, nanosprings, nanocrystalline, etc, while the study of the mechanical properties is one of the most important researches for nanomaterials [55] because the mechanical behavior of materials at the nanoscale is different from that at the macroscopic scale thanks to the increasing ratio of the surface/interface to the volume. For the grain size of polycrystalline decreasing into nanometer range, the elastic properties of the materials will be modified significantly, that is the main objective in this work.

It is well-known that the surface/interface ratio to the volume increase for the materials in the nanometer range resulting in the effect of surface/interface may become quite remarkable on the elastic properties of the materials because of the increased contribution to the total energy from the surface/interface. In the previous works for elastic properties of nanometerials, most of researchers focus on the nanowires/beams, nanofilms/plates, nanosprings, etc to take into account of the surface effects on the elastic properties of the materials. It has been found [56-58] that the surface free energy can enhance the axial modulus of thin film when the thickness of the film is in the range of several nanometers; and Streitz et al. predicted the modulus enhancement/reduction of 20% for the thickness of thin film below 5nm [59]. Recently, Miller and Shenoy developed a simple model to incorporated surface stress effects in examining unidirectional tension and pure bending of nano-scale bars and plates [32]. They concluded that the effect of surface/interface stress can be used to interpret the size dependence of the elastic modulus of nano-sized elements very well. Sun and Zhang presented a semi-continuum model to investigate size- dependent elastic properties of platelike nanomaterials [60]. Their theoretical results indicated that the Young's modulus of nanomaterials may be remarkable lower than its bulk counterpart. Diao et.al. studied the surface stress effect on the elastic properties of the Cu nanowire by atomistic simulation[61]; Zhou and Huang also used molecular dynamic simulations to show the effective elastic modulus of thin films increasing or decreasing as the film thickness decreases due to surface effect [62]. Yang obtained the size-dependent effective modulus of and elastic composite materials containing spherical nano-cavities taken account of the surface energy effect [63]. Cuenot et al. analyzed the surface effect on the elastic modulus of silver and lead nanowires, attributing the size dependence of elastic modulus to the surface tension [64]. Dingreviller et. al. incorporated the surface free energy into the continuum theory of mechanics to demonstrate the size-dependence of the elastic behavior of structural elements [65]. Guo and Zhao developed a three-dimensional model with the relaxation on the surface of nanofilm taken into consideration to investigate the size-dependent elastic constants [66]. Jing et al. showed in experimental that the apparent Young's modulus increases remarkably with the decrease of the diameter of silver nanowires, and presented a theoretical analysis through considering the surface effect [67].

Besides of surface effect, a great deal of research has been carried out to discuss the effects of the interface energy or interface stress on elastic properties of nanomaterials. There are two different interface introduced in the literature: one is defined as separating two homogeneous phases in which the tangential strains are equal on both sides of the phases, the other is associated with heterogeneous phases in which atomic bonds may be broken with different tangential strains occurring at the interface. Development of a generic and mathematical exposition on interface elasticity has been pioneered by Gurtin and his co-workers [31,68]. As to nano-scale inclusions, related studies include Sharma and his co-workers who employed analytical expression for the size-dependent strain stated caused by nano- inhomogeneities including nano-inclusions and spherical quantum dots and pores with the aid of taking into account of interface energies [69-71]. Lim et al. obtained an analytical solution for the elastic state of nano-scale spherical inclusion leading to size dependent elastic field due to the effect of interface stress[72]. Duan et al. generalized several micromechanical averaging procedures to take into account of interface stress effect on the effective elastic constants of isotropic heterogeneous solid containing nano-inhomogenerities [73]. Chen et al. assessed the overall thermoelastic behavior of solids containing spherical inclusion with interface stresses to obtain the effective elastic moduli and thermal expansion coefficient [74]. On the other hand, the microstructure component of the polycrystalline can be modified notably with the decrement of grain size, namely that the ratio of the interface to the volume will be enhanced, so will the ratio of the grain-boundary to the volume, while the interface energies and grain-boundaries obviously influence the elastic properties leading to grain size effect on elastic modulus of polycrystalline materials. Wolf et al. [75,76] and Adams et al.[77] studied the superlattices of twist boundaries and the twist boundaries of a thin film of copper to find the elastic moduli of the grain boundary domain different from those in the bulk. Bassan and co-wokers used the molecular dynamic simulations to determine local elastic modulus tensor near grain boundaries of fcc metals [78]. Schiotz et al. calculated the Young's modulus of nanocrystalline metal though atomistic simulation, and found that the Young's modulus decreased significantly when the grain size is reduced to 5nm [79]. Nan et al. [80] and Sharma and Ganti [81] presented the relationship between grain size and the elastic modulus of nanocrystalline materials and obtained the Young's modulus smaller than those of its coarse-sized counterpart. Wang et al. investigated the effect of the interface phase in nanocrystalline materials on the effective elastic property and suggested that the distorted atomic structures and the increased atomic spacing in interface regions as two micromechanisms influence the effective stiffness of the materials [82].

Nevertheless, scarce works in present researches discuss the grain-size effect on nancrystalline materials from the point of the view of the relationship between elastic property and interface energy stemmed from the grain surfaces. There are also no attempts to analyze the relationship among the grains, interfaces and grain-boundaries, and the elastic property in the framework of continuum theory of mechanics. Therefore, the objective of this paper is to investigate the grain-size effect on the Young's modulus of nanocrystalline materials taken into consideration of the interface energy effect from the grain surface and the grain-boundary effect based on the framework of continuum mechanics. We suppose the nanocrystalline materials composed of the grains region and the grain-boundaries region to obtain the effective Young's modulus of nanocrystalline materials. Here, we take into account of interface energy effect on the elastic energy of the grain to estimate the effective Young's modulus in the region of grains. The local theory in the classical elasticity is invalid as the thickness of the grain-boundary is in the several nanometers scale, so the nonlocal interaction should be considered for the grain-boundaries region. The nonlocal interaction in the nanometer scale based on the continuum mechanics can be regarded by applying the strain gradient elasticity approximately [82-84]. The layout of the paper is as follows: In Section 2, we derive the free energies in the regions of the grains and the grain-boundaries to obtain the corresponding effective elastic modulus based on the framework of the continuum mechanics. Section 3 shows the simplification of the isotropy for the cubic grains to generate the effective Young's modulus of the grain and the grain-boundary respectively. Numerical calculations in section 4 demonstrate the effects of the interface energy and grain-boundary on the Young's modulus of nanocrystalline materials.

4.2. Theoretical Description

4.2.1. Interface effect in the grains region

For the grain surface as the interfaces between the grains region and the grain-boundaries region, a few atomic layers along the interior of the grains differ from those of the grain interiors. The atomic space of the layers for interfaces must be larger than that in the grains, resulting in the elastic property of interfaces different from that of the interior of the grains. Hence, we attempt to incorporate the interface energy into free energy of the grains in order to analyze the effect of the interfaces on elastic property of the grains. As there exist distinct stresses and strains for the interfaces, the total free energy of grains with the regard of interface energy can be addressed to derive the effective elastic modulus of the grains. The interface energy density can be expanded into power a series of interface strain,

$$\Gamma^{Ig} = \Gamma_0^{Ig} + \Gamma_{\alpha\beta}^{Ig(1)}\varepsilon_{\alpha\beta}^{Ig} + \frac{1}{2}\Gamma_{\alpha\beta\kappa\lambda}^{Ig(2)}\varepsilon_{\alpha\beta}^{Ig}\varepsilon_{\kappa\lambda}^{Ig} + \frac{1}{6}\Gamma_{\alpha\beta\kappa\lambda\gamma\eta}^{Ig(3)}\varepsilon_{\alpha\beta}^{Ig}\varepsilon_{\kappa\lambda}^{Ig}\varepsilon_{\gamma\eta}^{Ig} + \dots, \tag{4-1}$$

where $\varepsilon_{\alpha\beta}^{Ig}$ is the (2×2) interface strain tensor, $\Gamma_{\alpha\beta}^{Ig(1)}, \Gamma_{\alpha\beta\kappa\lambda}^{Ig(2)}, \Gamma_{\alpha\beta\kappa\lambda\gamma\eta}^{Ig(3)}$ are interface elasticity tensor, dependent on the material and interface. So, the interface stress associated with the interface strain can be written as

$$\sigma_{\alpha\beta}^{Ig} = \frac{d\Gamma^{Ig}}{d\varepsilon_{\alpha\beta}^{Ig}} = \Gamma_{\alpha\beta}^{Ig(1)} + \Gamma_{\alpha\beta\kappa\lambda}^{Ig(2)}\varepsilon_{\kappa\lambda}^{Ig} + \frac{1}{2}\Gamma_{\alpha\beta\kappa\lambda\gamma\eta}^{Ig(3)}\varepsilon_{\kappa\lambda}^{Ig}\varepsilon_{\gamma\eta}^{Ig} + \dots. \tag{4-2}$$

Conventional summation rules apply unless otherwise stated. It then follows that the strain energy stored in the deformed interface is given by

$$U_{Interface}^{g} = \int_S [\int_0^{\varepsilon_{\alpha\beta}^s} \sigma_{\alpha\beta}^{Ig}(e_{\alpha\beta}^{Ig})de_{\alpha\beta}^{Ig}]ds = \int_S [\int_0^{\varepsilon_{\alpha\beta}^s} \frac{d\Gamma^{Ig}}{de_{\alpha\beta}^{Ig}} de_{\alpha\beta}^{Ig}]ds. \tag{4-3}$$

The interface strain is related to the absolute strain $\varepsilon_{\alpha\beta}^{g}$ within the grain through a coordinate transformation $\varepsilon_{\alpha\beta}^{Ig} = z_{i\alpha}z_{j\beta}\varepsilon_{\alpha\beta}^{g}|_s$, in which $z_{i\alpha}$ is the coordinate transformation tensor. Hence, the interface energy can be rewritten with the interior strain of the grain as

$$U_{Interface}^{g} = \frac{2V_0}{d}\mathrm{T}_{ij}\varepsilon_{ij}^{g} + \frac{V_0}{d}Q_{ijkl}\varepsilon_{ij}^{g}\varepsilon_{kl}^{g} + \frac{V_0}{3d}P_{\alpha\beta\kappa\lambda\gamma\eta}^{Ig(3)}\varepsilon_{\alpha\beta}^{g}\varepsilon_{\kappa\lambda}^{g}\varepsilon_{\gamma\eta}^{g}, \tag{4-4}$$

where d is the grain size here and

$$
\begin{aligned}
&\mathrm{T}_{ij} = \frac{d}{2V_0}\int_{S_0}\Gamma^{Ig(1)}_{\alpha\beta} z_{\alpha i} z_{\beta j} dS_0, Q_{ijkl} = \frac{d}{2V_0}\int_{S_0}\Gamma^{Ig(2)}_{\alpha\beta\kappa\lambda} z_{\alpha i} z_{\beta j} z_{\kappa k} z_{\lambda l} dS_0 \\
&P_{ijklmn} = \frac{d}{2V_0}\int_{S_0}\Gamma^{Ig(3)}_{\alpha\beta\kappa\lambda\gamma\eta} z_{\alpha i} z_{\beta j} z_{\kappa k} z_{\lambda l} z_{\gamma m} z_{\eta n} dS_0
\end{aligned}
\quad . \qquad (4\text{-}5)
$$

For the interior of the grain, the strain energy is expressed as

$$
U^g = \int_{V_0}[\int_0^{\varepsilon_0} \frac{d\Phi^g}{de^g_{ij}} de^g_{ij}]dV_0, \qquad (4\text{-}6)
$$

where Φ^g is the bulk elastic potential, which can be expanded into a series of the bulk strain tensor

$$
\Phi^g = \frac{1}{2}C^g_{ijkl}\varepsilon^g_{ij}\varepsilon^g_{kl} + \frac{1}{6}C^{g(3)}_{ijklmn}\varepsilon^g_{ij}\varepsilon^g_{kl}\varepsilon^g_{mn} + \cdots . \qquad (4\text{-}7)
$$

Here, C^g_{ijkl} $C^{g(3)}_{ijklmn}$ are tensors of second- and third-order elastic constants of the perfect crystal lattice respectively. The total strain energy of the grain corresponding to $\varepsilon^g_{\alpha\beta}$ can be then written as,

$$
\begin{aligned}
U &= U^g + U^g_{Interface} \\
&= \frac{2V_0}{d}\mathrm{T}_{ij}\varepsilon^g_{ij} + \frac{V_0}{2}(C^g_{ijkl} + \frac{2}{d}Q_{ijkl})\varepsilon^g_{ij}\varepsilon^g_{kl} + \frac{V_0}{6}(C^{g(3)}_{ijklmn} + \frac{2}{d}P_{ijklmn})\varepsilon^g_{ij}\varepsilon^g_{kl}\varepsilon^g_{mn}
\end{aligned}
\quad . \qquad (4\text{-}8)
$$

Having minimized the total strain energy of the grain, that is

$$
\frac{\partial U}{V_0 \partial \varepsilon_{ij}}|_{\varepsilon_{ij}=\tilde{\varepsilon}_{ij}} = 0, \qquad (4\text{-}9)
$$

the strain tensor $\tilde{\boldsymbol{\varepsilon}}^g$ of the grain at the self-equilibrium state can be obtained. Then the effective elastic modulus tensor of the grain at the state of self-equilibrium can be defined as,

$$
\bar{C}^g_{ijkl} = \frac{\partial^2 U}{V_0 \partial\varepsilon^g_{ij}\partial\varepsilon^g_{kl}}|_{\varepsilon^g_{ij}=\tilde{\varepsilon}^g_{ij}} = C^g_{ijkl} + \frac{2}{d}Q_{ijkl} + (C^{g(3)}_{ijklmn} + \frac{2}{d}P_{ijklmn})\tilde{\varepsilon}^g_{mn} . \qquad (4\text{-}10)
$$

For $\tilde{\varepsilon}^g_{ij} \ll 1$, the quadratic term in Eq.(4-9) can be neglected. Having considered the assumption of $C'^g_{ijkl} \gg 2Q_{ijkl}/d$, $C^{g(3)}_{ijklmn} \gg 2P_{ijklmn}/d$, the explicit expression of the self-equilibrium and the effective elastic modulus tensor of the grain can be given as

$$\tilde{\varepsilon}_{kl}^{g} \approx \frac{2}{d} C_{ijkl}^{g\ -1} \mathrm{T}_{ij} = \frac{2}{d} M_{ijkl}^{g} \mathrm{T}_{ij}, \tag{4-11}$$

$$\bar{C}_{ijkl}^{g} = C_{ijkl}^{g} + \frac{2}{d}(Q_{ijkl} + C_{ijklmn}^{g(3)} M_{mnpq}^{g} \mathrm{T}_{pq}). \tag{4-12}$$

Here, M_{mnpq}^{g} is the compliance tensor of the bulk crystal, and $M_{ijkl}^{g}, C_{ijklmn}^{g(3)}, Q_{ijkl}, \mathrm{T}_{pq}$ are all independent of the grain size d. It can be found from Eq.(4-12) that the first term is the bulk elastic modulus tensor of the grain and the second one is relative to the interfaces which is inversely proportional to the grain size.

4.2.2. The strain gradient elasticity in the grain-boundary region

For the thickness of grain-boundary in several nanometers scale, stress-strain at a point at the nanometer scale must be dependent of the elastic state of all surrounding neighborhood points, violating the coarse-grained classic elastic assumption of locality. At such small length scales one may use approximately classical continuum mechanics framework to capture nonlocal interaction, the algebraic constitutive equations are replaced by integral equations whereby the stress depends not only on the strain but also the strain gradient, namely the so-called strain gradient elasticity where strain gradients with suitable coupling constants are added to the classical elastic Lagrangian. So the strain energy density subjected to the strain gradient can be represented as

$$\begin{aligned} w^{b} &= C_{0}^{b} + C_{ij}^{b}\varepsilon_{ij}^{b} + \frac{1}{2}C_{ijkl}^{b}\varepsilon_{ij}^{b}\varepsilon_{kl}^{b} + \frac{1}{6}C_{ijklmn}^{b}\varepsilon_{ij}^{b}\varepsilon_{kl}^{b}\varepsilon_{mn}^{b} \\ &+ D_{1}^{b}\xi_{ijk}^{b}\xi_{ijk}^{b} + D_{2}^{b}\xi_{ijk}^{b}\xi_{jki}^{b} + D_{3}^{b}\xi_{jji}^{b}\xi_{kki}^{b} + D_{4}^{b}\xi_{iik}^{b}\xi_{kkj}^{b} + D_{5}^{b}\xi_{kki}^{b}\xi_{kjj}^{b} \end{aligned}, \tag{4-13}$$

where $\varepsilon_{ij}^{b} = (u_{i,j}^{b} + u_{j,i}^{b})/2$, $\xi_{ijk}^{b} = u_{k,ij}^{b} = \varepsilon_{ki,j}^{b}$. And the total strain energy can be written by

$$W^{b} = \int_{V} w^{b}(\varepsilon_{ij}^{b}, \xi_{ijk}^{b}) dV. \tag{4-14}$$

For an arbitrary variation of the displacement field, the energy increment is expressed as

$$\delta W^{b} = \int_{V} (\sigma_{ij}^{b}\delta\varepsilon_{ij}^{b} + \tau_{ijk}\delta\xi_{ijk}^{b}) dV. \tag{4-15}$$

where the conventional Cauchy stress tensor σ_{ij}^{b} and the higher stress tensor τ_{ijk} are their work-conjugate variables respectively, defined as

$$\sigma_{ij}^{b} = \frac{\partial w^{b}}{\partial \varepsilon_{ij}^{b}}, \ \tau_{ijk} = \frac{\partial w^{b}}{\partial \xi_{ijk}^{b}}. \tag{4-16}$$

To identity the independent traction boundary conditions, the deformation gradient $\delta u_{i,j}^{b}$ are decomposed into a surface gradient $D_j \delta u_i^b$ and a normal gradient $n_j D\delta u_i^b$, namely [85]

$$\delta u_{i,j}^{b} = D_j \delta u_i^b + n_j D \delta u_i^b , \tag{4-17}$$

where the calculation operators D_i are D defined as

$$D_i = (\delta_{jp} - n_j n_p)\frac{\partial}{\partial x_p}, D = n_p \frac{\partial}{\partial x_p}, \tag{4-18}$$

Denote the surface traction and the double stress on the surface as t_i^b and r_i^b, based on the principle of virtual work, the energy increment can be given as

$$\delta W^b = \int_S (t_i^b \delta u_i^b + r_i D\delta u_i^b) dS . \tag{4-19}$$

Combining Eqs.(4-15) and (4-19) and using the divergence theorem yield the equilibrium equation

$$\sigma_{ij,j}^{b} - \tau_{ijk,jk} = 0 , \tag{4-20}$$

and the traction boundary conditions

$$\begin{aligned} t_i^b &= n_j(\sigma_{ij}^b - \tau_{ijk,k}) + n_j n_k \tau_{ijk}(D_p n_p) - D_j(n_k \tau_{ijk}) \\ r_i^b &= n_j n_k \tau_{ijk} \end{aligned} . \tag{4-21}$$

4.3. Effective Modulus of Cubic Grain

For the nanocrystalline materials with cubic grains under the isotropic assumption (Figure 10), the nonzero components of elastic constants tensors C_{ijkl}^{g}, M_{ijkl}^{g}, $C_{ijklmn}^{g(3)}$ satisfied the symmetric conditions as shown in Appendix B, therefore, the independent components of the elastic constants tensors, respectively, are C_{11}^{g}, C_{12}^{g}, C_{44}^{g}, M_{11}^{g}, M_{12}^{g}, M_{44}^{g} and $C_{111}^{g}, C_{144}^{g}, C_{112}^{g}, C_{155}^{g}, C_{123}^{g}, C_{456}^{g}$ where $(11) \to (1)$, $(22) \to (2)$, $(33) \to (3)$, $(12) \to (6)$, $(13) \to (5)$, $(23) \to (4)$. The elastic constants tensors related to the interfaces can be expressed as following with the aid of the isotropic assumptions and symmetric conditions,

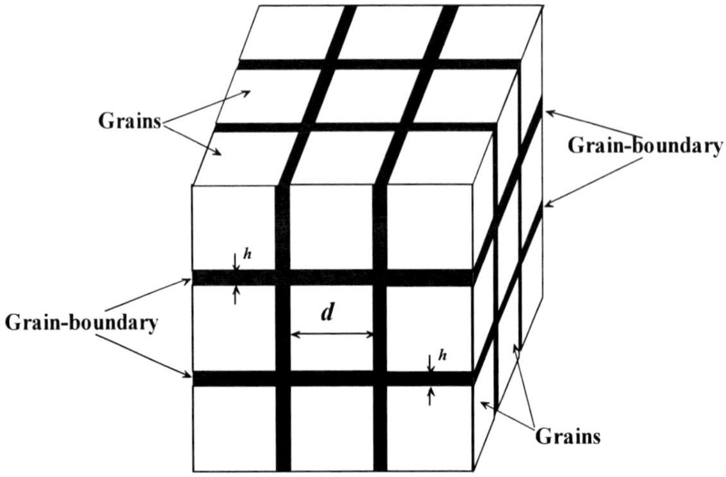

Figure 10. Three-dimensional schematic of nanocrystalline materials with cubic grains.

$$\begin{aligned} & \mathrm{T}_{11} = \mathrm{T}_{22} = \mathrm{T}_{33} = \mathrm{T}_0 \\ & Q_{1111} = Q_{2222} = Q_{3333} = \mathrm{K}^I + \Upsilon^I \\ & Q_{1122} = Q_{2233} = Q_{1133} = \mathrm{K}^I - \Upsilon^I \\ & Q_{1212} = Q_{2323} = Q_{1313} = \Upsilon^I \end{aligned} \quad . \tag{4-22}$$

Define $\mathrm{N}_{ijkl} = C^{g(3)}_{ijklmn} M^g_{mnpq} \mathrm{T}_{pq}$, the nonzero components of effective elastic tensor of the grain $\bar{C}^g_{ijkl}$ can be obtained from Eq.(4-12), that is

$$\begin{aligned} & \bar{C}^g_{11} = \bar{C}^g_{22} = \bar{C}^g_{33} = C^g_{11} + \frac{2}{d}(Q_{1111} - \mathrm{N}_{1111}) \\ & \bar{C}^g_{12} = \bar{C}^g_{13} = \bar{C}^g_{12} = C^g_{12} + \frac{2}{d}(Q_{1212} - \mathrm{N}_{1212}) \\ & \bar{C}^g_{44} = \bar{C}^g_{55} = \bar{C}^g_{66} = C^g_{44} + \frac{2}{d}(Q_{4444} - \mathrm{N}_{4444}) \end{aligned} \quad , \tag{4-23}$$

where d is the size of the grain. According to the definition of isotropic Young's modulus, the effective Young's modulus of the cubic grain can be presented as

$$\bar{E}^g = \bar{C}^g_{11} - \frac{2\bar{C}^g_{12}}{(\bar{C}^g_{11} + \bar{C}^g_{12})} = E^g + \frac{1}{d}\left(\frac{B_1\mathrm{K}^I + B_2\Upsilon^I + B_3\mathrm{T}_0}{A_1 + A_2\mathrm{K}^I/d + A_3\mathrm{T}_0/d}\right), \tag{4-24}$$

Where the parameters $A_1, A_2, A_3, B_1, B_2, B_3$ are associated with the elastic tensor C^g_{ijkl}, as shown in following

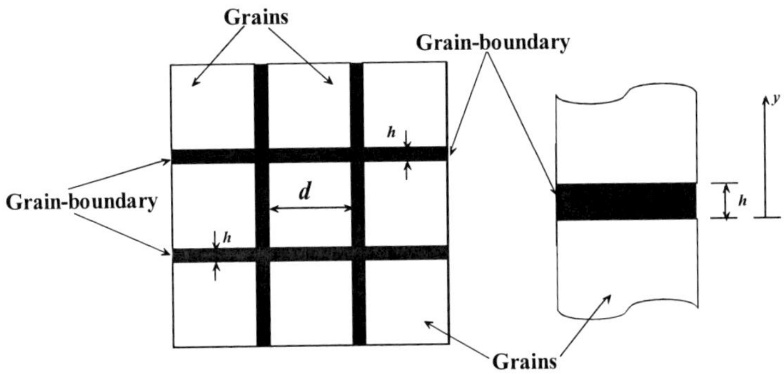

Figure 11.Two-dimensional schematic of grain-boundary in nanocrystalline materials with cubic grains.

$$\begin{aligned} &A_1=(C_{11}^g+C_{12}^g)^2, A_2=4(C_{11}^g+C_{12}^g), A_3=-2\frac{(C_{11}^g+C_{12}^g)}{(C_{11}^g+2C_{12}^g)}(C_{111}^g+C_{123}^g+4C_{112}^g),\\ &B_1=2(C_{11}^g-C_{12}^g)^2, B_2=2(C_{11}^g+5C_{12}^g)(C_{11}^g+C_{12}^g),\\ &B_3=4\frac{(C_{11}^g-C_{12}^g)(C_{11}^g+C_{12}^g)}{(C_{11}^g+2C_{12}^g)}C_{112}^g+2\frac{(C_{11}^{g2}+2C_{12}^gC_{11}^g+C_{12}^{g2})}{(C_{11}^g+2C_{12}^g)}C_{111}^g+4\frac{C_{12}^g(C_{11}^g+2C_{12}^g)}{(C_{11}^g+2C_{12}^g)}C_{123}^g \end{aligned} \tag{4-25}$$

4.4. Effective Modulus of the Grain-Boundary

For convenience to investigate the effective Young's modulus of the grain-boundary, we suppose that the upper and bottom boundaries of the grain-boundary are subjected to a normal stress σ_0 in the direction of y axial shown in Figure 11. From Eq.(4-20), the equilibrium equation can be addressed by the strain ε_{ij}^b,

$$E^b\varepsilon_{22,2}-2(D_1+D_2+D_3+D_4+D_5)\varepsilon_{22,222}=0. \tag{4-26}$$

Let $E^b\gamma^2=(D_1+D_2+D_3+D_4+D_5)$, in which γ is the intrinsic materials lengths of the grain-boundary, Eq.(4-26) can be replaced by

$$\varepsilon_{22,2}^b-2\gamma^2\varepsilon_{22,222}^b=0. \tag{4-27}$$

The general solution of Eq.(4-27) is

$$\varepsilon_{22}^b=A^b+B^b\exp(-\frac{1}{\sqrt{2}\gamma}y)+C^b\exp(\frac{1}{\sqrt{2}\gamma}y), \tag{4-28}$$

where the parameters A^b, B^b, C^b can be determined by boundary conditions. On the boundaries between the grain and the grain-boundary, it is required not only for the continuity of the displacements and the tractions, but also for the continuity of the normal displacement gradient and the high order tractions. So, the boundary conditions can be presented as

$$t_2^b = \sigma_0, \quad \varepsilon_{22}^b = \varepsilon^{Ig}, \qquad at\ y=0, h. \tag{4-29}$$

Here, the strain ε^{Ig} is approximated as $\varepsilon^{Ig} \approx \bar{\varepsilon}^g = \sigma_0 / \bar{E}^g$. The effective normal strain of the grain-boundary then can be solved through carrying out the integration over y in the grain-boundary region, that is

$$\begin{aligned}\bar{\varepsilon}_{22}^b = \frac{1}{h}\{\frac{\sigma_0}{E^b}h + (\frac{1}{\bar{E}^g} - \frac{1}{E^b})\frac{\sqrt{2}\gamma}{h}\frac{[\exp(h/\sqrt{2}\gamma)-1]}{[\exp(h/\sqrt{2}\gamma)+1]} \\ + (\frac{1}{\bar{E}^g} - \frac{1}{E^b})\frac{\sqrt{2}\gamma}{h}\frac{[\exp(-h/\sqrt{2}\gamma)-1]}{[\exp(-h/\sqrt{2}\gamma)+1]}\},\end{aligned} \tag{4-30}$$

where h is the thickness of the grain-boundary. Accordingly, the effective Young's modulus of the grain-boundary can be expressed as

$$\begin{aligned}\bar{E}^b = \frac{\sigma_0}{\bar{\varepsilon}_{22}^b} = \{\frac{1}{E^b} + (\frac{1}{\bar{E}^g} - \frac{1}{E^b})\frac{\sqrt{2}\gamma}{h}\frac{[\exp(h/\sqrt{2}\gamma)-1]}{[\exp(h/\sqrt{2}\gamma)+1]} \\ + (\frac{1}{\bar{E}^g} - \frac{1}{E^b})\frac{\sqrt{2}\gamma}{h}\frac{[\exp(-h/\sqrt{2}\gamma)-1]}{[\exp(-h/\sqrt{2}\gamma)+1]}\}^{-1},\end{aligned} \tag{4-31}$$

Finally, with the basic of the effective Young's modulus of the grain and the grain-boundary obtained above, one may suggest the effective Young's modulus of the nanocrystalline materials according to the grain-to-volume ratio $p = 2d/(2d+h)$ and the ratio of the grain-boundary to the volume $(1-p)$, that is

$$\begin{aligned}\bar{E}^B &= p\bar{E}^g + (1-p)\bar{E}^b \\ &= p[E^g + \frac{1}{d}(\frac{B_1\mathrm{K}^I + B_2\Upsilon^I + B_3\mathrm{T}_0}{A_1 + A_2\mathrm{K}^I/d + A_3\mathrm{T}_0/d})] + (1-p)\{\frac{1}{E^b} + \\ &[\frac{1}{E^g + \frac{1}{d}(\frac{B_1\mathrm{K}^I + B_2\Upsilon^I + B_3\mathrm{T}_0}{A_1 + A_2\mathrm{K}^I/d + A_3\mathrm{T}_0/d})} - \frac{1}{E^b}]\frac{\sqrt{2}\gamma}{h}\frac{[\exp(h/\sqrt{2}\gamma)-1]}{[\exp(h/\sqrt{2}\gamma)+1]} + \\ &[\frac{1}{E^g + \frac{1}{d}(\frac{B_1\mathrm{K}^I + B_2\Upsilon^I + B_3\mathrm{T}_0}{A_1 + A_2\mathrm{K}^I/d + A_3\mathrm{T}_0/d})} - \frac{1}{E^b}]\frac{\sqrt{2}\gamma}{h}\frac{[\exp(-h/\sqrt{2}\gamma)-1]}{[\exp(-h/\sqrt{2}\gamma)+1]}\}^{-1}\end{aligned} \tag{4-32}$$

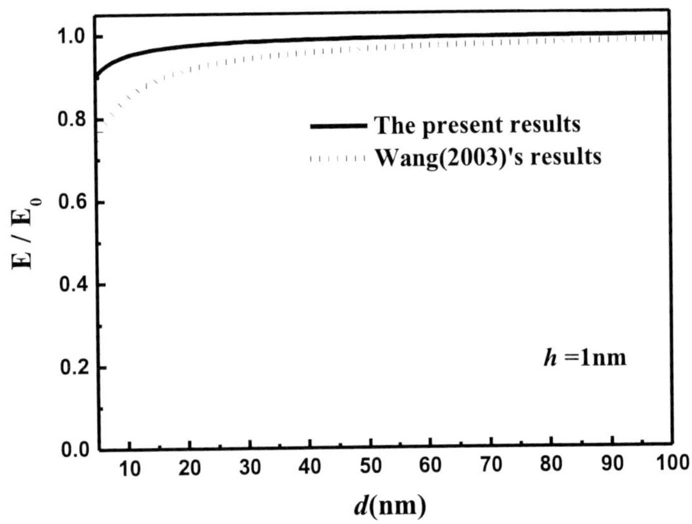

Figure 12.Young's modulus of nanocrystalline materials as a function of grain size for different models.

4.5. Numerical Results and Discussion

Having the expression for the effective Young's modulus in Eq.(4-31), it is possible to discuss some natures about the effects of interface energy and grain-boundary on the effective elastic modulus of nanocrystalline materials. Here, we consider the copper nanocrystalline material with the parameters used in the calculation as follows: $C_{11}^{g}=167.38\text{GPa}$, $C_{12}^{g}=124.11\text{GPa}$, $C_{111}^{g}=832.02\text{GPa}$, $C_{112}^{g}=-621.92\text{GPa}$, $C_{123}^{g}=15.29\text{GPa}$, $K^{s}=2.6755\text{Jm}^{-2}$, $\mu^{s}=-3.5524\text{Jm}^{-2}$ and $\text{T}^{0}=-1.3961\text{Jm}^{-2}$, respectively [65]. As the intrinsic materials lengths γ is to the order of the atomic spacing, we take $\gamma=0.25nm$ approximately [84]. The numerical results show that the Young's modulus of nanocrystalline materials with cubic grain is independent on the grain size for the size larger than several tens nanometers. When the grain size is reduced to lower 20nm, the modulus varies with the grain size smoothly. The reduction of the modulus is only less than 10% for the grain size $d=10\text{nm}$ as shown in Figure 12. The present experimental measurments illustrated that there is a less reduction in the modulus of the nanocrystalline materials with high density. While we take 1nm thickness of the grain-boundary to make the material have high density, so the theoretical results are agreeable with the experimental observation. We also compare the results with those of Wang [82] in Figure 12. The results from Wang's paper in which the gradient elasticity is applied in the region of grains and grain-boundaries are remarkably lower than proposed results when the grain size is lower than 50nm. That means the gradient elasticity applied in the region of the grain may overestimate the grain-size effect on the Young's modulus of the nanocrystalline materials while the interface energy effect considered in the grain region can character the influence of the grain size on the effective of elastic modulus more reasonably for the high density from the point of the view of the experimental observation.

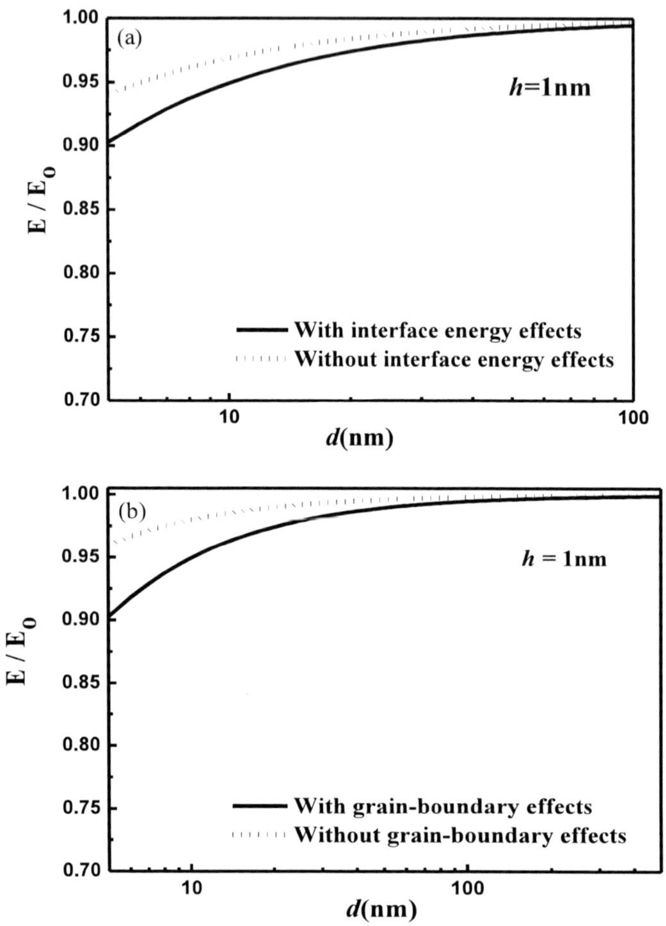

Figure 13. Young's modulus of nanocrystalline materials varied with the grain size considering the interface energy effect or not(a) and grain-boundary effect or not(b).

To analyze the influence of the interface between the grain and the grain- boundary on the elastic modulus, the Young's modulus of the nanocrystalline materials with and without regard of interface energy effect as functions of the grain size are plots in Figure 13a. Significant departure for the results of considering and unconsidering interface energy effect can be noted at small grain size. The interface energy effect can be neglected for the larger grain sizes such as $d > 80\text{nm}$. With the reduction of the grain size, the influence of the interface energy becomes more notable. Especially when the grain size is reduced to lower 20nm, the interface energy effect on the Young's modulus must be regarded. Figure 13b shows the grain-boundary effect on the elastic modulus of the nanocrystalline materials as functions of the grain size. It can be found from Figure 13b that the influence of the grain-boundary must be taken into account when the grain size is lower 100nm. The smaller the grain size is, the more significant the influence is. When the grain size is 10nm, the reduction of the Young's modulus is less than 5% without regard of the grain-boundary effect while almost 10% for considering the influence of the grain-boundary on the Young's modulus.

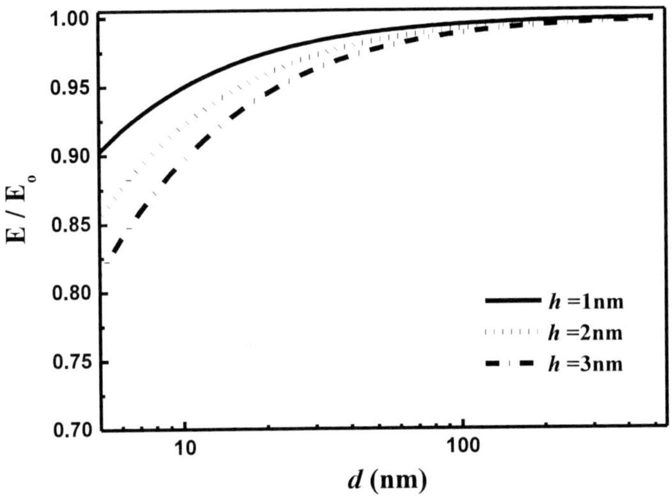

Figure 14. Grain-size dependent Young's modulus of nanocrystalline materials for different thickness of grain-boundary.

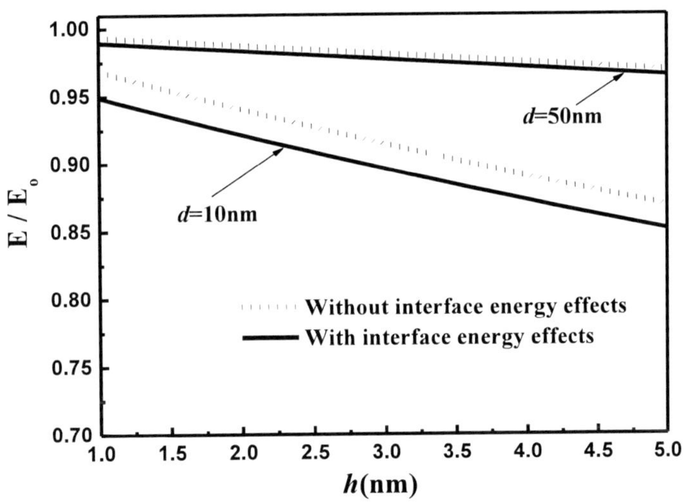

Figure 15.Young's modulus of nanocrystalline materials varied with thickness of grain-boundary for different grain size.

The density of the polycrystalline plays an important role in the elastic modulus of the materials, especially in the region of the grain size less than 100nm, while the thickness of the grain-boundary is one of key factor for the density of the nanocrystalline materials. The Young's modulus of the nanocrystalline materials for different thickness of the grain-boundary as functions of the grain size are plotted in Figure 14. It can be seem for this figure that the thickness of the grain-boundary influence on the Young's modulus is gradually enhanced with the reduction of the grain size. For the larger grain size, the Young's modulus is independent on the thickness of the grain-boundary. When the grain size is reduced to less than 100nm, the influence of the thickness of grain-boundary must be considered. A further point to be noted that the larger the thickness is, the less the modulus is, and the more notable the dependence of the grain size is for the modulus. That is due to the density is smaller for the thicker grain-boundary resulting in the decrement of the modulus of the materials which is identical to the experimental results. Figure 15 shows the Young's modulus of the nanocrystalline materials for different grain size as function of the thickness of grain-boundary in the present of the interface energy effect or not. It can be noted for Figure 15 that the modulus reduces linearly with the increase of the thickness and more rapidly for the grain size of 10nm than for 50nm as the same of the results in Figure 14. It also can be found that the interface energy effect represents more obviously for smaller grain size, similar to the results of Figure 13

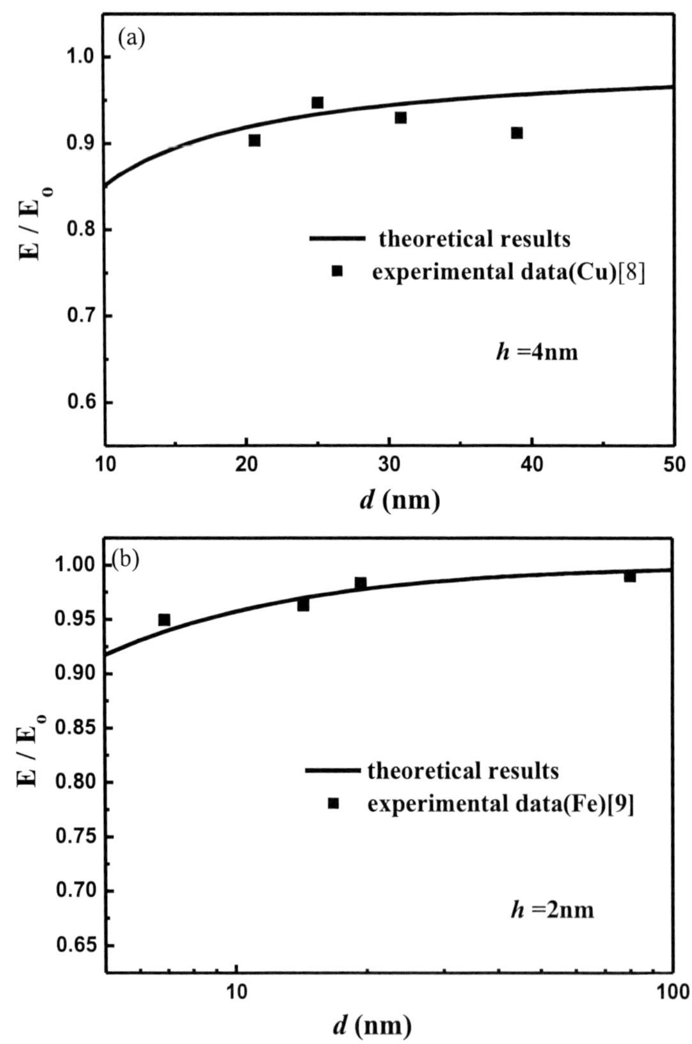

Figure 16. Comparison of grain-size dependent Young's modulus of Cu(a) and Fe(b) nanocrystalline materials for the present model with experimental measurement. Figure

Furthermore, the predicted Young's modulus of the Cu nanocrystalline material with the thickness of grain-boundary of 4nm is compared with experimental measurements [8] as shown in Figure 16a. It is clear from Figure 16a that the Young's modulus of Cu nanocrystalline material is reduced about 10% with the grain size in the region of 20nm to 50nm. That is because the lower density resulting in the biggish reduction of the Young's modulus of the material. And the modeled woks well in describing the Young's modulus dependent on the grain size of Cu nanocrystalline material. Figure 16b plots the comparison between the experimental data and the calculations for Fe nanocrystalline materials [9]. Here, we take the thickness of the grain-boundary in calculation as 2nm. The Young's modulus of Fe nanocrystalline material lower 8% when the grain size is reduced to 6nm due to the higher density, and the proposed numerical results can agree well with the experiments.

4.6. Summary

The effects of interface energy and grain-boundary on the Young's modulus of nanocrystalline materials are investigated in the framework of continuum elasticity. The effective elastic modulus is addressed explicitly to describe the grain-size dependence of the Young's modulus of nanocrystalline material through discussing the Young's modulus of the grains region and the grain-boundaries region respectively. The numerical results indicate that the proposed analysis can describe the grain-size effect on the Young's modulus of nancrystalline materials qualitatively and quantitatively, and identical with the experimental observation. In addition, the influence of the density is also discussed through analyzing the Young's modulus associated with the thickness of grain-boundary. Finally, the experimental data are supported to compare with the numerical results, and the proposed model is in good agreement with the experimental results.

5. Conclusion and Outlook

To conclude, the effect of the grain size on thermal conductivity, electrical conductivity and the elastic modulus of bulk polycrystalline materials have been investigated theoretically with the aid of the point of view of the microstructures. Because of the decrement of the grain size to nanometer scale, the increasing surface/interface-volume ratio leads to the modification of the physical and mechanical properties of bulk polycrystalline materials. With increasing the scattering of phonons and electrons from grain boundaries, the transports properties in polycrystalline materials are changed significantly, demonstrating in the corresponding models in this chapter that the change of the electrons and phonons in microstructures results in the thermal and electric properties related to the grain size which is in nanometer scale. For the elastic performance of the bulk polycrystalline materials, the interface energy and grain boundaries will affect the mechanical properties remarkably due to that the surface/interface mechanical behavior is quite different from the one in the prefect crystalline. Through considering the polycrystalline made up of the grain phase and the grain boundary phase, the effective Young's modulus of the nanocrystalline materials can be derived based on the continuum mechanics theory. With the comparison to the present experimental results, the predictions from the proposed models can agree well with the measurements. It is believed that the present results will support a comprehensive theoretical description of the grain size effect on the thermal, electric and elastic properties to completely understand the novel physical and mechanical performance in bulk polycrystalline materials with the nanometer-sized grains.

For the future, the further research of nanocrystalline materials should also be focused on the optimization of the functional properties for the corresponding applications by controlling the size of the microstructures and applied field such as electric fields. That is due to the fact that the atomistic and electronic processes at surfaces/interfaces play a vital role in the physical and chemical properties, leading to that the changes in the electronic structures due to polarization in nanocrystalline materials may be exploited to create novel functional nanomaterials with tunable properties as pointed out by Gleiter [86]. Moreover, this issue will give rise to the resulting multi-fields effect on the functional properties of nanocrystalline materials, which has become crucial and challenging topic in the applications of nanostructural materials.

6. ACKNOWLEDGMENT

This work is supported in part by the National Science Foundation of China (No. 90405005) and the Ph.D Fund of the Ministry of Education of China (No.20050729016). It is also supported by the National Basic Research Program of China (2007CB607506) and the Fund of the State Key Laboratory of Advanced Technology for Materials Synthesis and Processing (Wuhan University of Technology) (No. WUT2005Z04).

APPENDIX A

$$c_1 = \left(\frac{2R_{\mathrm{gb}} + 2T_{\mathrm{bg}}E_3^{\mathrm{b}}\dfrac{4R_{\mathrm{bg}}T_{\mathrm{gb}}E_3^{\mathrm{b}}}{1-4R_{\mathrm{bg}}{}^2E_3^{\mathrm{b}2}} - 2T_{\mathrm{bg}}E_3^{\mathrm{b}}\dfrac{2T_{\mathrm{gb}}}{1-4R_{\mathrm{bg}}{}^2E_3^{\mathrm{b}2}}}{1-\dfrac{4T_{\mathrm{gb}}T_{\mathrm{bg}}E_3^{\mathrm{b}}E_3^{\mathrm{g}}}{1-4R_{\mathrm{bg}}{}^2E_3^{\mathrm{b}2}} + 2R_{\mathrm{gb}}E_3^{\mathrm{g}} + 8T_{\mathrm{bg}}E_3^{\mathrm{b}}\dfrac{T_{\mathrm{gb}}R_{\mathrm{bg}}E_3^{\mathrm{g}}E_3^{\mathrm{b}}}{1-4R_{\mathrm{bg}}{}^2E_3^{\mathrm{b}2}}}\right)\left(\frac{1}{3} - E_4^{\mathrm{g}}\right)$$

$$c_2 = \frac{1-R_{\mathrm{gb}} - 4T_{\mathrm{bg}}E_3^{\mathrm{b}}\dfrac{R_{\mathrm{bg}}T_{\mathrm{gb}}E_3^{\mathrm{b}}}{1-4R_{\mathrm{bg}}{}^2E_3^{\mathrm{b}2}} - 4T_{\mathrm{bg}}E_3^{\mathrm{b}}\dfrac{T_{\mathrm{gb}}}{1-4R_{\mathrm{bg}}{}^2E_3^{\mathrm{b}2}}}{1-\dfrac{4T_{\mathrm{gb}}T_{\mathrm{bg}}E_3^{\mathrm{b}}E_3^{\mathrm{g}}}{1-4R_{\mathrm{bg}}{}^2E_3^{\mathrm{b}2}} + 2R_{\mathrm{gb}}E_3^{\mathrm{g}} + 8T_{\mathrm{bg}}E_3^{\mathrm{b}}\dfrac{T_{\mathrm{gb}}R_{\mathrm{bg}}E_3^{\mathrm{g}}E_3^{\mathrm{b}}}{1-4R_{\mathrm{bg}}{}^2E_3^{\mathrm{b}2}}}$$

$$c_3 = [1 - 2E_3^{\mathrm{g}}(\frac{1 - R_{\mathrm{gb}} - 4T_{\mathrm{bg}}E_3^{\mathrm{b}}\dfrac{R_{\mathrm{bg}}T_{\mathrm{gb}}E_3^{\mathrm{b}}}{1-4R_{\mathrm{bg}}{}^2E_3^{\mathrm{b2}}} - 4T_{\mathrm{bg}}E_3^{\mathrm{b}}\dfrac{T_{\mathrm{gb}}}{1-4R_{\mathrm{bg}}{}^2E_3^{\mathrm{b2}}}}{1 - \dfrac{4T_{\mathrm{gb}}T_{\mathrm{bg}}E_3^{\mathrm{b}}E_3^{\mathrm{g}}}{1-4R_{\mathrm{bg}}{}^2E_3^{\mathrm{b2}}} + 2R_{\mathrm{gb}}E_3^{\mathrm{g}} + 8T_{\mathrm{bg}}E_3^{\mathrm{b}}\dfrac{T_{\mathrm{gb}}R_{\mathrm{bg}}E_3^{\mathrm{g}}E_3^{\mathrm{b}}}{1-4R_{\mathrm{bg}}{}^2E_3^{\mathrm{b2}}}})]$$

$$\cdot(1+2R_{\mathrm{bg}}E_3^{\mathrm{b}})/(1-4R_{\mathrm{bg}}{}^2E_3^{\mathrm{b2}})$$

$$c_4 = [1 - E_3^{\mathrm{g}}\frac{(2R_{\mathrm{gb}} + 2T_{\mathrm{bg}}E_3^{\mathrm{b}}\dfrac{4R_{\mathrm{bg}}T_{\mathrm{gb}}E_3^{\mathrm{b}}}{1-4R_{\mathrm{bg}}{}^2E_3^{\mathrm{b2}}} - 2T_{\mathrm{bg}}E_3^{\mathrm{b}}\dfrac{2T_{\mathrm{gb}}}{1-4R_{\mathrm{bg}}{}^2E_3^{\mathrm{b2}}})}{(1 - \dfrac{4T_{\mathrm{gb}}T_{\mathrm{bg}}E_3^{\mathrm{b}}E_3^{\mathrm{g}}}{1-4R_{\mathrm{bg}}{}^2E_3^{\mathrm{b2}}} + 2R_{\mathrm{gb}}E_3^{\mathrm{g}} + 8T_{\mathrm{bg}}E_3^{\mathrm{b}}\dfrac{T_{\mathrm{gb}}R_{\mathrm{bg}}E_3^{\mathrm{g}}E_3^{\mathrm{b}}}{1-4R_{\mathrm{bg}}{}^2E_3^{\mathrm{b2}}})}]$$

$$\cdot(\frac{1}{3} - E_4^{\mathrm{g}})(1-2R_{\mathrm{bg}}E_3^{\mathrm{b}})/(1-4R_{\mathrm{bg}}{}^2E_3^{\mathrm{b2}})$$

$$G_{\mathrm{g}}(z_{\mathrm{g}}, D, H) = \int_0^1 \{\ (c_1 + \mu_{\mathrm{g}})[e^{-z_{\mathrm{g}}/(\Lambda_{\mathrm{pb}}\mu_{\mathrm{g}})} + e^{-(D-z_{\mathrm{g}})/(\Lambda_{\mathrm{pb}}\mu_{\mathrm{g}})}] - 2\mu_{\mathrm{g}}$$
$$- Dc_2[e^{-z_{\mathrm{g}}/(\Lambda_{\mathrm{pb}}\mu_{\mathrm{g}})} - e^{-(D-z_{\mathrm{g}})/(\Lambda_{\mathrm{pb}}\mu_{\mathrm{g}})}]\ \}\mu_{\mathrm{g}}d\mu_{\mathrm{g}}$$

$$G_{\mathrm{b}}(z_{\mathrm{b}}, D, H) = \int_0^1 \{\ Dc_3[e^{-z_{\mathrm{b}}/(\Lambda_{\mathrm{pb}}\mu_{\mathrm{b}})} - e^{-(H-z_{\mathrm{b}})/(\Lambda_{\mathrm{pb}}\mu_{\mathrm{b}})}]$$
$$- c_4[e^{-z_{\mathrm{b}}/(\Lambda_{\mathrm{pb}}\mu_{\mathrm{b}})} + e^{-(H-z_{\mathrm{b}})/(\Lambda_{\mathrm{pb}}\mu_{\mathrm{b}})}]\ \}\mu_{\mathrm{b}}d\mu_{\mathrm{b}}$$

APPENDIX B

$$C_{11}^g = C_{22}^g = C_{33}^g,\ C_{12}^g = C_{13}^g = C_{23}^g,\ C_{44}^g = C_{55}^g = C_{66}^g$$

$$M_{11}^g = M_{22}^g = M_{33}^g,\ M_{12}^g = M_{13}^g = M_{23}^g,\ M_{44}^g = M_{55}^g = M_{66}^g$$

$$C_{111}^g = C_{222}^g = C_{333}^g = L + 6M + 8N$$

$$C_{144}^g = C_{255}^g = C_{366}^g = M$$

$$C_{112}^g = C_{113}^g = C_{122}^g = C_{133}^g = C_{223}^g = C_{233}^g = L + 2M$$

$$C_{155}^g = C_{166}^g = C_{244}^g = C_{266}^g = C_{344}^g = C_{355}^g = M + 2N$$

$$C^g_{123} = L\,,\ C^g_{456} = N$$
$$\mathrm{T}_{11} = \mathrm{T}_{22} = \mathrm{T}_{33} = \mathrm{T}_0$$

$$Q_{1111} = Q_{2222} = Q_{3333} = \mathrm{K}^I + \Upsilon^I$$

$$Q_{1122} = Q_{2233} = Q_{1133} = \mathrm{K}^I - \Upsilon^I$$

$$Q_{1212} = Q_{2323} = Q_{1313} = \Upsilon^I$$

$$\mathrm{N}_{ijkl} = C^{g(3)}_{ijklmn} M^g_{mnpq} \mathrm{T}_{pq}$$

$$\mathrm{N}_{1111} = \mathrm{N}_{2222} = \mathrm{N}_{3333} = \mathrm{T}_0 (M_{11} + M_{12} + M_{23})(C_{111} + 2C_{112})$$

$$\mathrm{N}_{1122} = \mathrm{N}_{2233} = \mathrm{N}_{1133} = \mathrm{T}_0 (M_{11} + M_{12} + M_{23})(C_{123} + 2C_{112})$$

$$\mathrm{N}_{1212} = \mathrm{N}_{2323} = \mathrm{N}_{1313} = \mathrm{T}_0 (M_{11} + M_{12} + M_{23})(C_{144} + 2C_{155}),$$

in which, $M_{11} = \dfrac{(C_{11} + C_{12})}{(C_{11} - C_{12})(C_{11} + 2C_{12})}$, $M_{12} = M_{23} = -\dfrac{C_{12}}{(C_{11} - C_{12})(C_{11} + 2C_{12})}$.

7. References

[1] Schulz a, R., Huot, J., Liang, G., Boily, S., Lalande, G., Denis, M. C. & Dodelet, J. P. (1999). Recent developments in the applications of nanocrystalline materials to hydrogen technologies. *Mater. Sci. Eng. A*, *267*, 240-245.

[2] Suryanarayana, C. & Koch, C. C. (2000). Nanocrystalline materials – Current research and future directions. *Hyperfine Interactions*, *130*, 5-44.

[3] Herzera, G., Vazquezb, M., Knobelc,M., Zhukovd, A., Reininger, T., Grössingerg, R. & Sanchez, J. L. (2005). Round table discussion: Present and future applications of nanocrystalline magnetic materials. *Journal of Magnetism and Magnetic Materials*, *294*, 252-266.

[4] Botcharova, E., Freudenberger, J. & Schultz, L. (2006). Mechanical and electrical properties of mechanically alloyed nanocrystalline Cu–Nb alloys. *Acta Mater.*, *54(12)*, 3333-3341.

[5] Bietsch, A. & Michel, B. (2002). Size and grain-boundary effects of a gold nanowire measured by conducting atomic force microscopy. *Appl. Phys. Lett.*, *80(18)*, 3346-3348.

[6] Anno, H., Hatada, K., Shimizy, H., Matsubara, K., Motohara, Y., Sakakibara, T., Tashiro, H. & Motoya, K. (1998). Structural and electronic

transport properties of polycrystalline p-type $CoSb_3$. *J. Appl. Phys.*, *83(10)*, 5270-5276.

[7] Zhao, Z., Buscaglia, V. , Viviani, M., Buscaglia, M. T., Mitoseriu, L., Testino, A., Nygren, M., Johnsson, M. & Nanni, P. (2004). Grain-size effects on the ferroelectric behavior of dense nanocrystalline $BaTiO_3$ ceramics. *Phys. Rev. B*, *70(2)*, 024107(1-8).

[8] Sanders, P. G., Eastman, J. A. & Weertman, J. R. (1996). Tensile Behavior of Nanocrystalline Copper. *In Processing and Properties of Nanocrystalline Materials*, edited by C. Suryanarayana, J. Singh, & F. H. Froes (The Minerals, Meals and Materials Society, Warrendale, PA). 379-386.

[9] Shen, T. D., Koch, C. C., Tsui, T. Y. & Pharr, G. M. (1995). On the elastic-moduli of nanocrystalline Fe, Cu, Ni, and Cu-Ni alloys prepared by mechanical milling/alloying. *J. Mater. Res.*, *10(11)*, 2892- 2896.

[10] Simkin, M. V. & Mahan, G. D.(2000). Minimum Thermal Conductivity of Superlattices. *Phys. Rev. Lett.*, *84(5)*, 927-930.

[11] Broido, D. A. & Reinecke, T. L.(2004). Lattice thermal conductivity of superlattice structures. *Phys. Rev. B*, *70(8)*, 081310(1-4).

[12] Mingo, N. (2003). Calculation of Si nanowire thermal conductivity using complete phonon dispersion relations. *Phys. Rev. B*, *68(11)*, 113308(1-4).

[13] Dames, C. & Chen, G.(2004). Theoretical phonon thermal conductivity of Si/Ge superlattice nanowires. *J. Appl. Phys.*, *95(2)*, 682-693.

[14] Khitun, A., Liu, J. & Wang, K. L. (2004). On the modeling of lattice thermal conductivity in semiconductor quantum dot superlattices. *Appl. Phys. Lett.*, *84(10)*, 1762-1764.

[15] Morelli, D. T. & Meisner, G. P. (1995). Low temperature properties of the filled skutterudite $CeFe_4Sb_{12}$. *J. Appl. Phys.*, *77(8)*, 3777-3781.

[16] Caillat, T., Borshchevsky, A. & Fleurial, J. P.(1996) Proceedings of the 15th International Conference on Thermoelectrics (Piscataway,NJ: IEEE) p. 100.

[17] Yang, J., Meisner, G. P., Morelli, D. T. & Uher, C. (2000). Transport and magnetic properties of Iron valence in skutterudites: $Co_{1-x}Fe_xSb_3$. *Phys. Rev. B*, *63(1)*, 014410(1-11).

[18] Nolas, G. S., Slack, G. A. & Schujman, S. B. (2001). Semiconductor clathrates: a phonon glass electron crystal material with potential for thermoelectric applications. *Semicond Semimet*, *69*, 255-300.

[19] Im, J. T., Hartwig, T. K. & Sharp, J. (2004). Microstructural refinement of cast p-type Bi_2Te_3–Sb_2Te_3 by equal channel angular extrusion. *Acta Mater.*, *52(1)*, 49-55.

[20] Puyet, M., Lenoir, B., Dauscher, A., Dehmas, M., Stiewe, C. & Müller, E.(2004). High temperature transport properties of partially filled $Ca_xCo_4Sb_{12}$ skutterudites. *J. Appl. Phys.*, *95(9)*, 4852-4855.

[21] Kleber, X., Simonet, L., Fouquet, F. & Delnondedieu, M. (2005). Thermoelectric power of a two-dimensional metal/metal composite: a numerical approach. *Modelling Simul. Mater. Sci. Eng.*, *13(3)*, 341-354.

[22] Okamoto, N. L., Tanaka, K. & Inui, H. J. (2006). Crystal structure refinement of a type-I clathrate compound Ba_8Ge_{43} with an ordered arrangement of germanium vacancies. *Acta Mater.*, *54(1)*, 173-178.

[23] Liu, W. S., Zhang, B. P., Li, J. F. & Zhao, L. D. (2007). Thermoelectric property of fine-grained $CoSb_3$ skutterudite compound fabricated by mechanical alloying and spark plasma sintering. *J. Phys. D: Appl. Phys., 40(2)*, 566-572.
[24] Francis, J. D. (1999). Thermoelectric Cooling and Power Generation. *Science., 285(5428)*, 703-706.
[25] Mayadas, F. & Shatzkes, M. (1970). Electrical-Resistivity Model for Polycrystalline Films: the Case of Arbitrary Reflection at External Surfaces. *Phys. Rev. B, 1(4)*, 1382-1389.
[26] Vancea, J., Reiss, G. & Hoffmann, H. (1987). Mean-free-path concept in polycrystalline metals. *Phys. Rev. B, 35(12)*, 6435-6437.
[27] Durkan, C. & Welland, M. E. (2000). Size effects in the electrical resistivity of polycrystalline nanowires. *Phys. Rev. B, 61(20)*, 14215-14218.
[28] Kästle, G., Boyen, H. G., Schröder, A., Plettl, A. & Ziemann, P. (2004). Size effect of the resistivity of thin epitaxial gold films. *Phys. Rev. B, 70(16)*, 165414(1-6).
[29] Sondheimer, E. H. (1952). The mean free path of electrons in metals. *Adv. Phys., 1(1)*, 1-42.
[30] Mayadas, A. F., Shatzkes, M. & Janak, J. F. (1969). Electrical resistivity model for polycrystalline films: The case of specular reflection at external surfaces. *Appl. Phys. Lett., 14(11)*, 345-347.
[31] Gurtin, M. E. & Murdoch, A. I. (1975). A continuum theory of elastic materials surfaces. *Arch. Rat. Mech. Anal., 57(4)*, 291-323.
[32] Miller, R. E. & Shenoy, V. B. (2000). Size dependent elastic properties of nanosized structural elements. *Nanotechnology, 11(3)*, 139-147.
[33] Cammarata, R. C. (1994). Surface and interface stress effects in thin films. *Prog. Surf. Sci., 46(1)*, 1-38
[34] Maranganti, R. & Sharma, P. (2008). Length scales at which classical elasticity breaks down for various materials. *Phys. Rev. Lett., 98(19)*, 195504(1-4).
[35] Tritt, T. M. (2000) Recent Trends in Thermoelectric Materials Research I–III, Semiconductors and Semimetals, vol 69-71(San Diego: Academic).
[36] Caillat, T., Chung, S., Fleurial, J. P., Snyder, G. J. & Borshchevsky, A. (1998). Proceedings of the 17th International Conference on Thermoelectrics (unpublished), 298
[37] Mahan, G. D. (1997). Seebeck coefficient for the Anderson model. *Phys. Rev. B, 56(18)*, 11833-11838.
[38] Mao, W. & Bedell, K. S. (1999). Thermoelectric figure of merit of strongly correlated superlattice semiconductors. *Phys. Rev. B, 59(24)*, R15590-R15593.
[39] Terasaki, I., Sasago, Y. & Uchinokura, K. (1997). Large thermoelectric power in $NaCo_2O_4$ single crystals. *Phys. Rev. B, 56(20)*, R12685- R12687.
[40] Hicks, L. D. & Dresselhaus, M. S. (1993). Effect of quantum-well structures on the thermoelectric figure of merit. *Phys. Rev. B, 47(19)*, 12727-12731.
[41] Shi, X., Chen, L., Yang, J. & Meisner, G. P. (2004). Enhanced thermoelectric figure of merit of $CoSb_3$ via large-defect scattering. *Appl. Phys. Lett., 84(13)*, 2301-2303.
[42] Zhou, Z., Uher, C., Jewell, A. & Caillat, T. (2005). Influence of point-defect scattering on the lattice thermal conductivity of solid solution $Co(Sb_{1-x}As_x)_3$. *Phys. Rev. B, 71(23)*, 235209(1-6).

[43] Nakagawa, H., Tanaka, H., Kasama, A., Anno, H. & Matsubara, K. (1997). Proceedings of the 16th International Conference on Thermoelectrics(Piscataway,NJ: IEEE), 351.

[44] Majumdar, A. (1993). Microscale Heat Conduction in Dielectric Thin Films. *J. Heat Transfer, 115(1)*, 7-16.

[45] Chen, G. (1998). Thermal conductivity and ballistic-phonon transport in the cross-plane direction of superlattices. *Phys. Rev. B, 57(23)*, 14958-14973.

[46] Swartz, E. T. & Pohl, R. O. (1989). Thermal boundary resistance. *Rev. Mod. Phys., 61(3)*, 605-668.

[47] Chen, G. (1997). Size and Interface Effects on Thermal Conductivity of Superlattices and Periodic Thin-Film Structures. *J. Heat Transfer, 119(2)*, 220-229.

[48] Vancea, J., Reiss, G. & Hoffmann, H. (1987). Mean-free-path concept in polycrystalline metals. *Phys. Rev. B, 35(12)*, 6435-6437.

[49] Landauer, R. (1957). Spatial Variation of currents and fields due to localized scatterers in metallic conduction. *IBM J. Res. Dev., 1(3)*, 223-231.

[50] Wu, W., Brongersma, S. H., Van Hove, M. & Maex, K. (2004). Influence of surface and grain-boundary scattering on the resistivity of copper in reduced dimensions. *Appl. Phys. Lett., 84(15)*, 2838-2840.

[51] Reiss,G., Vancea, J. and Hoffmann, H. (1986). Grain-Boundary Resistance in Polycrystalline Metals. *Phys. Rev. Lett., 56(19)*, 2100-2103.

[52] Steinhögl, W., Schindler, G., Steinlesberger, G. & Engelhardt, M. (2002). Size-dependent resistivity of metallic wires in the mesoscopic range. *Phys. Rev. B, 66(7)*, 075414(1-4) (2002).

[53] de Vries, J. W. C. (1987). Resistivity of thin Au films as a function of grain diameter and temperature. *J. Phys. F: Met. Phys., 17(9)*, 1945- 1952

[54] Yu, B. L., Tang, X. T., Zhang, Q. J., Liu, T. X., Luo, P. F. & Wang, J. (2003). *Proceedings of the 22th International Conference on Thermoelectrics*, 17- 21 August 2003 (IEEE, Piscataway), 101.

[55] Kassner, M. E., Nemat-Nasser, S., Suo, Z., Bao, G., Barbour, J. C., Brinson, L. C., Espinosa, H., Gao, H., Granick, S., Gumbsch, P., Kim, K. S., Knauss, W., Kubin, L., Langer, J., Larson, B. C., Mahadevan, L., Majumdar, A., Torquato, S. & van Swol, F. (2005). New Directions in Mechanics. *Mechanics of Materials, 37(2-3)*, 231-259.

[56] Cammarata, R. C. & Sieradzki, K. (1989). Effects of surface stress on the elastic moduli of thin films and superlattices. *Phys. Rev. Lett., 62(17)*, 2005-2008.

[57] Cammarata, R. C. & Sieradzki, K. (1994). Surface and interface stresses. *Annu. Rev. Mater. Sci., 24*, 215-234.

[58] Nix, W. D. & Gao, H. (1998). Indentation size effects in crystalline materials: a law for strain gradient plasticity. *J. Mech. Phys. Solids, 46(3)*, 411-425.

[59] Streitz, F. H., Cammarata, R. C. & Sieradzki, K. (1994). Surface-stress effects on elastic properties I: thin metal films. *Phys. Rev. B, 49(15)*, 10699-10706.

[60] Sun, C. T. & Zhang, H. T. (2003). Size-dependent elastic moduli of platelike nanomaterials. *J. Appl. Phys., 93(2)*, 1212-1218

[61] Diao, J., Gall, K. & Dunn, M. L. (2004). Atomistic simulation of the structure and elastic properties of gold nanowires. *J. Mech. Phys. Solids, 52(9)*, 1935-2186.

[62] Zhou, L. G. & Huang, H. (2004). Are surfaces elastically softer or stiffer? *Appl. Phys. Lett., 84(11)*, 1940-1942.

[63] Yang, F. (2004). Size dependent effective modulus of elastic composite materials: spherical nanocavities at dilute concentrations. *J. Appl. Phys.*, *95(7)*, 3516-3520.
[64] Cuenot, S., Fretigny, C., Demoustier-Champagne, S. & Nysten, B. (2004). Surface tension effect on the mechanical properties of nanomaterials measured by atomic force microscopy. *Phys. Rev. B*, *69(16)*, 165410(1-5).
[65] Dingreville, R., Qu, J. & Cherkaoui, M. (2005). Surface free energy and its effect on the elastic behavior of nano-sized particles, wires and films. *J. Mech. Phys. Solids*, *53(8)*, 1827-1854.
[66] Guo, J. G. & Zhao, Y. P. (2005). The size-dependent elastic properties of nanofilms with surface effects. *J. Appl. Phy.*, *98(7)*, 074306(1-11).
[67] Jing, G. Y., Duan, H. L., Sun, X. M., Zhang, Z. S., Xu, J., Li, Y. D., Wang, J. X. & Yu, D. P. (2006). Surface effects on elastic properties of silver nanowires: Contact atomic-force microscopy. *Phys. Rev. B*, *73(23)*, 235409(1-6).
[68] Gurtin, M. E., Weissmuller, J. & Larche, F. (1998). A general theory of curved deformable interfaces in solids at equilibrium. *Philos. Mag. A*, *78(5)*, 1093-1109.
[69] Sharma, P. & Ganti, S. (2002). Interfacial elasticity corrections to size-dependent strain-state of embedded quantum dots. *Physique Status Solida B*, *234(3)*, R10-R12.
[70] Sharma, P., Ganti, S. & Bhate, N. (2003). Effect of surfaces on the size-dependent elastic stated of nano-inhomogeneities. *Appl. Phys. Lett.*, *82(4)*, 535-537.
[71] Sharma, P. & Ganti, S. (2004). Size-dependent Eshelby's tensor for embedded nano-inclusions incorporating surface/interface energies. *J. Appl. Mech.*, *71(5)*, 663-671.
[72] Lim, C. W., Li, Z. R. & He, L. H. (2006). Size dependent, non-uniform elastic field inside a nano-scale spherical inclusion due to interface stress. *Int. J. Solids and Struct.*, *43(17)*, 5055-5065.
[73] Duan, H. L., Wang, J., Huang, Z. P. & Karihaloo, B. L. (2005). Size-dependent effective elastic constants of solids containing nano-inhomogeneities with interface stress. *J. Mech. Phys. Solids*, *53(7)*, 1574-1596.
[74] Chen, Tungyang, Dvorak, G. J. & Yu, C. C. (2007). Solids containing spherical nano-inclusions with interface stresses: Effective properties and therma-mechanical connections. *Int. J. Solids and Struct.*, *44(3-4)*, 941-955.
[75] Wolf, D. & Kluge, M. (1990). Relationship between shear resistance and local atomic structure at grain boundaries in FCC metals. *Scr. Metall.*, *24(5)*, 907-914.
[76] Wolf, D. & Lutsko, J. F. (1989). Structurally-induced elastic anomalies in a supperlattice of (0 0 1) twist grain boundaries. *J. Mater. Res.*, *4(6)*, 1427-1443.
[77] Adams, J. B., Wolfer, W. G. & Foiles, S. M. (1989). Elastic properties of grain boundaries in copper and their relationship to bulk elastic constants. *Phys. Rev. B*, *40(14)*, 9479-9484.
[78] Bassani, J. L., Vitek, V. & Alber, I. (1992). Atomic-level elastic properties of interfaces and their relation to continua. *Acta Metall. Mater.*, *40*, S307-S320.
[79] Schioz, J., Di Tolla, F. D. & Jacobsen, K. W. (1998). Softening of nanocrystalline metals at very small grain sizes. *Nature*, *391(6667)*, 561-563.
[80] Nan, C. W., Li, X. P., Cai, K. F. & Tong, J. Z. (1998). Grain size-dependent elastic moduli of nanocrystals. *J. Mater. Sci. Lett.*, *17(22)*, 1917-1919.
[81] Sharma, P. & Ganti, S. (2003). On the grain-size-dependent elastic modulus of nanocrystalline materials with and without grain-boundary sliding. *J. Mater. Res.*, *18(8)*, 1823-1826.

[82] Wang, G. F., Feng, X. Q., Yu, S. W. & Nan, C. W. (2003). Interface effects on effective elastic moduli of nanocrystalline materials. *Mater. Sci. Eng. A, 363(1-2)*, 1-8.

[83] Qiu, X., Huang, Y., Wei, Y., Gao, H. & Hwang, K. C. (2003). The flow theory of mechanism-based strain gradient plasticity. *Mechanics of Materials*, *35(3)*, 245-258.

[84] Zhang, X. & Sharma, P. (2005). Inclusions and inhomogeneities in strain gradient elasticity with couple stress and related problems. *Int. J. Solids Struct.*, *42(13)*, 3833-3851.

[85] Smydhlyaev, V. P. & Fleck, N. A. (1996). The role of strain gradients in the grain size effect for polycrystals. *J. Mech. Phys. Solids*, *44(4)*, 465-495.

[86] Gleiter, H., Weissmüller, J., Wollersheim, O. & Würschum, R. (2001). Nanocrystalline materials: a way to solids with tunable electronic structures and properties?. *Acta Mater.*, *49(4)*, 737-745.

In: Bulk Materials: Research, Technology and Applications ISBN: 978-1-60692-963-6
Editors: T. Frias, V. Maestas, pp. 307-334

Chapter 7

ELECTRONIC PROPERTIES AND BONDING MECHANISM OF THE MERCURY VACANCY AND ARSENIC IMPURITIES IN $HG_{1-X}CD_XTE$

Xiaoshuang Chen, Lizhong Sun, He Duan, Yan Huang and Wei Lu
National Laboratory for Infrared Physics, Shanghai Institute of Technical Physics, Chinese Academy of Sciences, Yutian Road 500, Shanghai 200083, China.

ABSTRACT

As a most important material for the infrared photodetection, mercury cadmium tellurium (HgCdTe) has advantageous application in the military engineering and the space technology. With the development of the infrared detection technology, the knowledge about the properties of HgCdTe is not just confined to the empirical parameters, but extends to the physical mechanism of the crystal growth and even device performance based on the quantum theories. Some important issues for the HgCdTe, such as the evaluation of mercury vacancy (V_{Hg}) and *p*-type doping technical *et al.*, have been investigated from the first-principles calculations.

In the mercury vacancy of $Hg_{1-x}Cd_xTe$, the effects of the mercury vacancy on the electronic structure have been shown by the structural relaxation, local charge density, and densities of states. It is indicated that the nearest neighbor Te around the Hg vacancy defect contract toward to the defect site. It is attributed to the concentration of the electrons around the defect accompanying the dangling bond reconstruction, which induces extra Coulomb interactions between the defect site and the nearest neighbor Te. The Jahn-Teller distortion has been also found in some mole fraction and some disordered configurations of HgCdTe with mercury vacancy. The double acceptor levels introduced by the vacancy are determined in agreement well with the experimental results.

There are the amphoteric behaviors of arsenic *in situ* impurities in $Hg_{1-x}Cd_xTe$. It is found that the Jahn-Teller distortion gives rise to the change of the degenerate situation of the impurity levels. The amphoteric behaviors of arsenic *in situ* impurities in HgCdTe have been proved based on basic theoretical methods. At the same time, we have found that the impurity has the covalent bonding characteristics with host atoms. Based on the calculated formation energy, the chemical trends of the *in situ* arsenic impurity between As_{Te} and As_{Hg} in $Hg_{1-x}Cd_xTe$ have been studied systematically.

The understanding of the microstructures of the arsenic complex in $Hg_{1-x}Cd_xTe$ is important to explain the high electrical compensation of molecular beam epitaxy (MBE) samples and the conversion to *p*-type behavior. According to the defect formation energy and the binding energy, we predicted the most probable configurations for the absorption of As_4 and As_2 cluster defects on the grown surface. By the comparison with the experimental results, the arsenic dimer combined with the mercury vacancy is believed to account for the highly electrically compensated *n*-type characteristics in arsenic-doped $Hg_{1-x}Cd_xTe$ by MBE.

1. INTRODUCTION

Mercury cadmium telluride ($Hg_{1-x}Cd_xTe$, MCT) is currently one of the most widely used semiconductor materials for infrared detector arrays. It is well known that structural defects are introduced inevitably in MCT materials either in the process of materials growth or in the process of device fabrication. Point defects are important in controlling carrier concentrations, minority-carrier lifetimes, and noise in $Hg_{1-x}Cd_xTe$ [1,2]. In the simplest terms, the type and concentration of the free carriers can be controlled by the defects that introduce shallow levels in the band gap of $Hg_{1-x}Cd_xTe$, while the minority-carrier lifetimes and noise are controlled by the defects associated with deep levels in the band gap. The importance of native defects in $Hg_{1-x}Cd_xTe$ is undisputed. In many circumstances, the physical properties of $Hg_{1-x}Cd_xTe$ and its dopability are dominated by native point defects [3]. Vydyanath [4] has revealed that most important native defects in $Hg_{1-x}Cd_xTe$ are the double acceptor mercury vacancy. Gold *et al.* [5] have further pointed out that the mercury vacancy defect in MCT introduces not only a shallow acceptor level but also a deep level at 40% of the band gap. It is known that vacancies play role in many electrical properties in semiconductors, for example, acting as largely uncontrolled dopants and affecting the electron and hole mobilities [6,7]. The structural properties are also sensitive to the existence of vacancies [8], such as, dislocation formation and propagation, as well as self-diffusion [9] and diffusion of impurities [10,11],depending on the vacancy concentrations. Moreover the vacancy defects are not only formed in the process of materials growth, but also introduced when exposing of the device to a radiation such as electrons [12], laser [13], or some light ion [14,15] radiations. Such complicated working circumstance dominates the performance of the device. Unfortunately, the experimental evidences on the electronic properties associated with mercury vacancy defects are mainly indirect and depend on their ionization states. The neutral defects and compensating defects are even more difficult to observe.

Although there are some theoretical studies on the mercury vacancy in MCT materials or in II-VI semiconductors (CdTe, HgTe) [16–19], the characteristic of bonding mechanism around the vacancy, namely, the dangling bond rehybridization, has not been reported as yet, to our knowledge. Defects certainly induce local relaxation on the atomic structure. The relaxation may change the symmetry of the local lattice, disturb the charge distribution in the vicinity of the defect, and modify the positions of the defect energy levels and the corresponding localized states. Recently, Errico *et al.* [20,21] have revealed the influence of the charge states of the systems on the relaxation procedure and the electronic states. Up to now, there have not been reports on the effect of the mercury vacancy on the induced relaxation procedure because of different charge states and local atomic configurations, and on the dependence of atomic configuration on the mole fraction x in MCT. Since the electronic and structural properties of HgCdTe are considered to originate from vacancies, the understanding of the electronic properties of mercury vacancy in MCT is important to reveal their alloy's behavior.

One of the most important limitations of the applied prospect for the semiconductor materials is the difficulty of *n*- or *p*-type doping via incorporation of suitable impurities. For example, ZnSe materials can be fabricated as blue emitting laser diodes, which is mainly profit from the achievement of sufficient *p*-type doping [22,23]. Another typical instance is ZnO. The difficulty of efficient *p*-type doping [24] is the limitation for its practical applications until now. In general, the doping controllability is the major challenge in most semiconductors. The difficulties, generally speaking, concern two aspects: dopant stability and doping efficiency. The dopant stability, consequently low diffusivity, mainly determines the performance of homo- or heterojunction devices which rely on the sharpness of the doping profile [25]. The problem of the doping efficiency might arise from a variety of causes. Firstly, an element can be a desirable dopant in certain semiconductor, depending on whether it can produce a shallow impurity level. The shallow impurity levels can provide high thermal excitation of carriers into the valence or conduction bands, namely, it can easily be electronically activated. Secondly, the solubility of the dopant in a host lattice can be a source of doping restrict. Lastly, the compensation effect is another most important obstacle to be overcome in the doping process. The compensation can either result from the intrinsic defects with low formation energy or from the dopant-induced *AX* (*DX*) centers [3, 26–28]. As for the amphoteric dopant, the self-compensation is a crucial limitation of the doping efficiency, such as the doping behavior of arsenic in II-VI semiconductors and alloys. Due to its low diffusivity in $Hg_{1-x}Cd_xTe$ [29] compared with the native acceptor and the group I elements, attentions have focused on the group V elements, especially arsenic, as the *p*-type dopant. The experimental results have indicated that arsenic can provide a shallow acceptor level in $Hg_{1-x}Cd_xTe$ grown by bulk method [30], liquid phase epitaxy (LPE) [31], molecular beam epitaxy [32–34], and metal organic chemical vapor deposition [35]. At the same time, the amphoteric behavior of arsenic in HgCdTe has been found. Arsenic behaves as a *p*-type dopant under Hg-rich growth conditions and behaves as an *n*-type or unactive dopant under Te-rich growth conditions [30,31,34,35]. Although MBE is done at much lower temperatures than LPE, and thus offers some distinct advantages, the group V impurities are still observed to incorporate as donors [32], self-compensating the *p*-type doping. Reference [36] has shown that there is a significant fraction of the arsenic residing on the cation sublattice even under cation-saturated conditions. According to the arsenic dopant incorporation model [37] and the quasichemical predictions [29, 38], some of the doped arsenics in MCT grown by MBE are incorporated on the cation sublattice as isolated defects, As_{Hg}, unbound to vacancies, as donors. So, an annealing process is required for its electric activation [39]. However, as the doping levels reach 5×10^{18} cm^{-3}, the active efficiency of the *p*-type arsenic doping drastically drops due to the strong self-compensation effect [10,40]. So, the amphoteric behavior of the arsenic *in situ* impurities in MCT is a typical prototype to systematically study the selfcompensating effect.

The group V elements are preferred over the mercury vacancies and the group I elements for extrinsic acceptor doping due to their low diffusivity in HgCdTe. Arsenic is a group V element that has received great attention in recent years [36,40]. Although success in achieving *p*-type doping of HgCdTe using arsenic has been reported by several groups [41–

45], experimental evidence is not yet available as to the chemical nature of the arsenic impurity. The *in situ* incorporation of arsenic [46] is routine by MBE, but the observed highly compensated *n*-type behavior in the as-grown samples has never been satisfactorily explained, except for the model developed by Berding *et al.* with the suggestion that the (As_{Hg}–V_{Hg}) pair is the culprit [38]. Also, the complete arsenic activation is confined to the arsenic concentration above 10^{16} cm^{-3}, as adopted by many approaches. Otherwise, the arsenic activation appears to be masked by some unidentified *n*-type defects [47]. Unfortunately, the understanding of such defects, which is vital to the control of the arsenic concentration lower than 10^{16} cm^{-3} for high-operating-temperature infrared photodetectors, has not been established. Moreover, an examination of the activated arsenic structures is necessary to confirm the impurity states determined by Hall measurements, especially for the controversial deep acceptor levels [31].

For these purposes, we have performed the total energy calculations of full potential linear augmented plane wave (FP-LAPW) to systematically investigate the effect of the mercury vacancy defect on the electronic structure, the bonding mechanism accompanying with the dangling bond rehybridization, the relaxation procedure for different charge states in $Hg_{1-x}Cd_xTe$. Here, we choose $Hg_{0.5}Cd_{0.5}Te$ as the major prototype system. The reason for choosing such a mole fraction is that the computational results can be compared with that of the thoroughly studied II-VI semiconductors (CdTe and HgTe) to some extent. On the other hand, the results can be easily extrapolated to other mole fraction compositions. Furthermore, the results and conclusions of the calculations on $Hg_{0.5}Cd_{0.5}Te$ are confirmed, and also extended to the other mole fractions by using plane-wave pseudopotential calculations with larger supercells.

Although a simple model of the arsenic amphoteric behavior has been suggested, there is no comprehensive consensus on the microscopic mechanism of group V impurity incorporation in MCT. Moreover, the alloying effect of MCT interferes with the behavior of the amphoteric dopant, which makes the study more complicated. It is affirmable that the understanding of the amphoteric behavior of As in MCT is essentially important for qualifying the material growth and device design. In order to gain full understanding of the electronic behaviors and the compensation effect of the *in situ* As impurities in MCT, we have performed the total energy calculations of full-potential linear augmented plane wave to systematically investigate the effect of the arsenic *in situ* impurity on the electronic structure, bonding mechanism between the impurity and the host atoms, the relaxation, and the chemical trends of the As *in situ* impurity in $Hg_{1-x}Cd_xTe$. We choose $Hg_{0.5}Cd_{0.5}Te$ as the major prototype system. Furthermore, the results and conclusions of the calculations on $Hg_{0.5}Cd_{0.5}Te$ are confirmed and also extended to the other mole fractions by using planewave pseudopotential calculations with larger supercells. Our general understanding of the amphoteric impurity behaviors and its self-compensating effect in MCT from the present study is expected to be applicable also to other semiconductors and alloys. With the goal of better understanding the microscopic behavior of arsenic incorporation during MBE growth, we analyzed the properties of the complex defects in the form of arsenic tetramer, dimer, and singlet in the present work. We calculated the formation energies of possible configurations for As_n (n = 1, 2, and 4), from which the stable configurations were obtained.

2. Relaxations and Bonding Mechanism in $Hg_{1-x}Cd_xTe$ with Mercury Vacancy Defect

2.1. Computational Procedure

The density functional theory [48] calculations are performed using the FP-LAPW method implemented in the WIEN2K package [49]. We adopt Perdew-Burke-Ernzerhof functional [50] to describe the exchange-correlation interaction. We consider a 2×2×2 supercell (SC) with totally 64 atoms, consisting of eight quasi-zinc-blende crystal structure of unit cells $Hg_{0.5}Cd_{0.5}Te$ (each unit cell contains eight atoms including four Te atoms, two Hg atoms, and two Cd atoms). In the SC model, one of the 16 mercury atoms on the lattice site (0.5, 0.5, 0.5) of SC is replaced by a vacancy. A 2×2×1 SC with 32 atoms of no vacancy has been chosen as reference system for comparison purpose, and the computational parameters are set to be the same as the vacancy case. In order to ensure well convergence, the muffin-tin (MT) radius and the number of *k*-points for generating the final results have been carefully chosen after optimization. The states treated as valence are Hg($5d^{10}6s^2$), Cd($4d^{10}5s^2$), and Te($5s^25p^4$). The muffin-tin radii are chosen to be 2.60 atomic units (a.u.) for all the three kinds atoms. A satisfactory self-consistent convergence has been achieved by considering a number of FP-LAPW basis functions up to $R_{MT}K_{max}$=6.0 for the 64- atom SC and $R_{MT}K_{max}$ =7.0 for 32-atom SC. The K_{max} is the maximum value of the plane-wave vector that determines the so-called energy cutoff for the plane-wave expansion. The self-consistent iteration is considered to be converged when both the total energy and the total charge in the atomic sphere are stable within 10^{-4} eV per unit cell and 10^{-4} electron charges per atom, respectively. The localized levels are deduced from the calculations of the single-electron energy levels at Γ point and are aligned using the energy levels of core electrons. The relaxation procedures are conducted following the damped Newton dynamic schemes [51-53]. The criterion of the forces convergence for all atoms is set below 0.05 eV/Å. The relativistic effect of spin-orbit (SO) coupling is also included. To confirm the results on the supercell of 64 atoms using FP-LAPW methods and extend the conclusions to the other mole fractions of MCT, the plane-wave pseudopotential code Vienna *ab initio* simulation package [54,55] (VASP) is also used. In the VASP calculations, the core-electron interaction is modeled by the ultrasoft pseudopotentials [56], which can yield reasonably precise results with relatively low cutoff energy. Thus, it is possible to deal with the larger supercell. The supercells with 256 and 512 atoms have been chosen. The plane-wave cutoff energy is chosen as 300 eV, the standard ultrasoft pseudopotentials from the VASP package are employed for all the cases, and the Brillouin zone is sampled by using 5×5×5 and 3×3 ×3 Monkhorst-Pack grid for 256-atom and 512-atom supercells, respectively. The energy convergent criterion is 1×10^{-4} eV per unit cell, and forces on all relaxed atoms are less than 0.01 eV/Å It is well known that the local density approximation calculations typically underestimate the band gap of a semiconductor and is not able to give the absolute positions of the acceptors accompanying with the Hg vacancy defect very accurately. The emphasis here has been placed on the analysis of the defect formation energy and the transition energy levels. At temperature of zero Kelvin, the defect formation energy in a given charge state q is given by [3,57]

$$\Delta H_f(\mathrm{V_{Hg}},q) = E(\mathrm{V_{Hg}},q) + E(\mathrm{Hg}) - E(\mathrm{MCT}) + qE_{\mathrm{VBM}} + qE_{\mathrm{F}} + \mu_{\mathrm{Hg}} \qquad (2.1)$$

where $\Delta H_f(\mathrm{V_{Hg}},q)$ is the formation energy for a supercell containing the vacancy defect in charge state q. $E(\mathrm{V_{Hg}},q)$ and $E(\mathrm{MCT})$ are the total energies for the supercell containing the mercury vacancy defect in charge state q and the same size supercell without defect, respectively. It is well known that MCT is a pseudobinary alloy, which means the two types of cations occupy randomly the cation sites. Different local concentrations of cations are accompanied with the disordered effect in MCT. Within a simple model, the disordered system can be treated approximately as the mixture of MCT with different mole fractions. Just as Shih *et al.* have pointed out [58], the valence band maximum (VBM) becomes lower as the mole fraction x becomes larger. Thus, the reference of electronic Fermi energy E_{F} in the MCT alloy is chosen to the VBM of CdTe in the present work. $E(\mathrm{Hg})$ and μ_{Hg} are the total energy of Hg referencing to element and the chemical potential of Hg, respectively. Equation (2.1) indicates that the formation energy with Hg vacancy defect, $\Delta H_f(\mathrm{V_{Hg}},q)$, is a function of the E_{F} and the chemical potential of Hg, μ_{Hg}. The defect transition energy level $\varepsilon_\alpha(q/q')$ is the Fermi energy E_{F} in Eq. (2.1), at which the formation energy $\Delta H_f(\mathrm{V_{Hg}},q)$ of vacancy defect and charge q is equal to that of another charge q'. The formation energy and transition energy levels of MCT with cation vacancy defect are calculated by projector augmented wave (PAW) method for the supercell of 64 atoms. For the systems with charged defect, a uniform background charge is added to keep the global charge neutrality of the periodic supercell. The total energy of the systems with charged defect has been corrected for the interaction of the charged defect with the compensating background and its periodic images [59]. In the 64-atom supercell due to the finite cell size, basis set, and k-point sampling used, the error bars are about 0.2 and 0.1 eV for the calculated formation energy and transition energy, respectively.

2.2. Structural Relaxation and Formation Energy

The appearance of cation vacancy defects induces the structural relaxation of the host and modifies the electronic structure of the system. Before relaxation of atomic structures, from the calculated total energy versus five different volumes around the lattice constants given by Vegard's law, the equilibrium volumes have been obtained for MCT systems with and without vacancy defect, respectively. All the calculated values are fitted to the Murnaghan equation of state [60]. In comparison with the experimental results, the computed bond lengths of R(Te-Cd) and R(Te-Hg) are about 1% larger than that of experimental results of CdTe and HgTe. Geometry optimizations reveal that the cation vacancy defects lead to inward relaxation of the nearest-neighbor (NN) anions, in agreement with the previous studies [8]. Meanwhile, the bond lengths between the NN Te atoms and the next-nearest-neighbor (NNN) cations are also reduced. The result indicates that the inward relaxation of the NNN cations is larger than that of NN anions. In order to find out the reason of the NN Te inward relaxation in MCT caused by cation vacancy defects, the valence charge density and the bonding charge density have been calculated. The bonding charge density is defined as the difference between the total charge density in the solid and the superpositions of neutral atomic charge densities placed at atomic sites; i.e.,

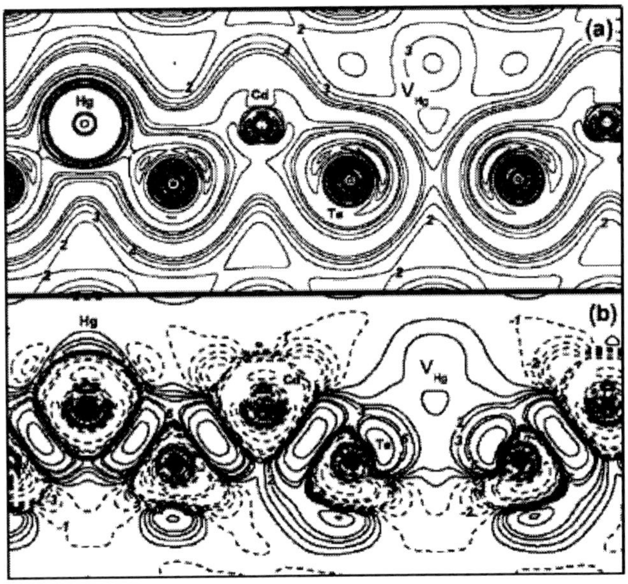

Figure. 2.1. The valence charge density (a) and the bonding charge density (b) of $Hg_{0.5}Cd_{0.5}Te$ with Hg vacancy in the plane (110), including the NNN Cd. The contour step size is 0.8×10^{-3} e/(a.u.)^3 for the valence charge density (a) and 0.2×10^{-3} e/(a.u.)^3 for the bonding-charge density (b). V_{Hg} indicates the mercury vacancy lattice site.

$$\Delta\rho(r) = \rho_{solid}(r) - \sum_{\alpha} \rho_{\alpha}(r - r_{\alpha}) \tag{2.2}$$

Therefore, the bonding charge density represents the net charge redistribution as atoms are brought together to form the crystal. The dangling bond rehybridization [61] induced by the vacancy defect also can be revealed clearly by the bonding charge density. The calculated valence charge density and the bonding charge density of $Hg_{0.5}Cd_{0.5}Te$ in the (110) plane, containing the mercury vacancy site and the NNN Cd sites are shown in the Figure 2.1. The charge accumulation on the vacancy site, and between the NN Te site and the vacancy site can be observed. Namely, the mercury vacancy becomes a negative charge center accompanying with the dangling bond rehybridization of the NN Te atoms. The Coulomb attractive potential between the negatively charged vacancy site and the ions makes the NN anions and the NNN cations move inward. The charge statistics shows that all the four NN Te and twelve NNN cations around the mercury vacancy have their charges to transfer out of the MT sphere by comparing with the total charge of the same MT sphere in MCT without the vacancy defect. There are about 0.1 electrons departing from the NN tellurium MT sphere, implying that there are more than 0.1 electrons accumulating at the vacancy site.

The density of states (DOS) calculated for the neutral vacancy defect V_{Hg}^{0} is shown in the Figure 2.2. One can see that the Fermi level does not lie within the band gap region, while it locates within the valence band. From the DOS shown in Figure 2.2(b), the integration from the Fermi level to the conduction band bottom accommodates two electrons. Accordingly, these empty states are identified as the double acceptor levels, and the details will be discussed below. In the present work, we perform the calculations for three systems with different charged states, corresponding with three different situations: (i) V_{Hg}^{0}, neutral defect state, an extremely pure crystal at low temperatures; (ii) V_{Hg}^{-1}, one charged acceptor state, providing one electron to the double acceptor state of the system; and (iii) V_{Hg}^{-2}, double charged acceptor state. In case (i), we use the 64-atom supercell described in Sec. 2.3. However, in the studies of cases (ii) and (iii), with the same 64-atom supercell, we perform self-consistent calculations with one and two additional electrons to the supercell, respectively. Finally, the self-consistent FP-LAPW calculations are performed as follows:

$$\text{(i) } V_{Hg'}^{0} \Rightarrow Hg_{15}Cd_{16}Te_{32}$$
$$\text{(ii) } V_{Hg'}^{-1} \Rightarrow Hg_{15}Cd_{16}Te_{32} + 1e^{-1}$$
$$\text{(iii) } V_{Hg'}^{-2} \Rightarrow Hg_{15}Cd_{16}Te_{32} + 2e^{-1}$$

The computational results with different charge states are summarized in the Table I. We find that the inward relaxations of the NN Te for cases (ii) and (iii) are larger than that in case (i). As mentioned above, the rehybridization of the dangling bond around the vacancy site induces the negative charge center to cause the inward relaxation for case (i). In cases (ii) and (iii), the charged states enhance the negative charge center and the Coulomb attractive interaction, thus lead to larger inward relaxation for cases (ii) and (iii) than that of the case (i). Meanwhile, the bond length between the NN Te and NNN cations almost remain the same as that of case (i). The result means that the attractive potential also makes the cations more inward relaxation than that of the case (i). The relaxation procedure corresponding to different charge states as described above also exists in the other mole fractions

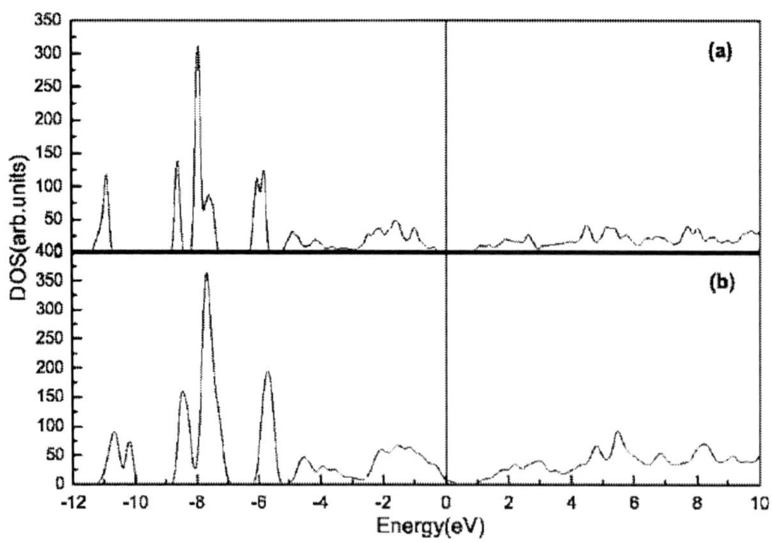

Figure 2.2. The total density of states in $Hg_{0.5}Cd_{0.5}Te$ without (a) and with (b) Hg vacancy, respectively. Fermi level is set to zero. The following Fermi level in the DOS figures is in the same way.

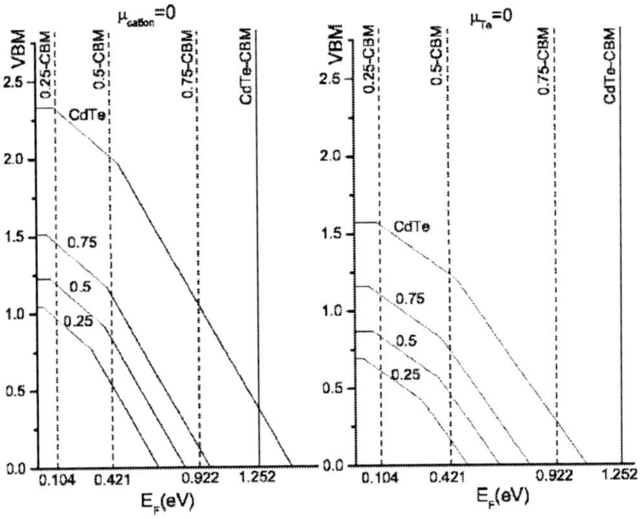

Figure 2.3. The calculated formation energies of the Hg vacancy defect for MCT with x=0.25, 0.5, 0.75, and Cd vacancy defect for CdTe as a function of the Fermi levels. The VBM of all the four materials is set to zero.

The formation energies of the mercury vacancy defect in MCT for four different mole fractions (x=0, 0.25, 0.5, 0.75), and the cadmium vacancy defect in CdTe for neutral charge state and chemical potential μ_{cation}=0 are calculated. The formation energy of Cd vacancy defect in CdTe is in good agreement with the previous results [3]. The formation energy of the Hg vacancy defect in MCT has a nearly linear relationship with the mole fraction x. The formation energy becomes smaller as the mole fractions of MCT are decreased. The result indicates that the Hg vacancy defects occur more easily as the mole fraction is decreased. Taking the atom as the reference, the formation energy of $Hg_{0.75}Cd_{0.25}Te$ is 1.72 eV, in good agreement with the experimental result [4]. Figure 2.3 shows the calculated formation energies of Hg vacancy defects for MCT with x =0.25, 0.5, and 0.75, and Cd vacancy defect for CdTe as a function of the Fermi energy. In the figure, the slope of the line gives the charge state of the defect at the Fermi energy. The transition energy level is the Fermi energy, at which the slope of the lines changes. Under the Hg- or Cd-rich condition, the chemical potential of μ_{Hg} or μ_{Cd} equals to zero. Under the Te-rich condition, the cation chemical potential becomes complicated due to the alloying effect. Here, we simply choose the formation energy of HgTe and CdTe (The calculated formation energies of HgTe and CdTe are −0.35 eV and −0.76 eV, respectively. The results are in good agreement with the experimental values [62–64].) as the cation chemical potential because our main motivation is the relation between the transition energy level and the mole fraction of MCT. The transition energy levels of V_{Cd} in CdTe are $\varepsilon(0/-1)$=0.10 eV and $\varepsilon(-1/-2)$=0.43 eV, respectively, in agreement with the previous results under the calculated error. The calculated results show that the transition energy level shifts to the deeper position within the band gap as the mole fraction becomes larger. The double acceptors caused by the cation vacancy defect are two empty states of t_{2v}, coming mainly from the p-like orbitals. The transition energy level moves more deeply into the band gap as the Cd mole fraction becomes larger which agree well with the experimental results [31,65]. As shown in Figure 2.3, either under the Hg-rich or Te-rich condition, for x=0.25 the transition energy level of $\varepsilon(0/-1)$ is 37 meV and the transition energy level of ε(−1/−2) is 0.31 eV, which is larger than the calculated band gap (0.1 eV). By considering the calculated error bar in the transition energy levels, the ε(0/−1) may locate in the band gap of the real MCT with x=0.25 and behave as a shallow acceptor. However the ε(−1/−2) is always larger than the band gap of MCT with x=0.25 (the real band gap about 0.15 eV). The results imply that the second ionization of the double acceptors merges in the direct band-to-band electronic transition process. Namely, the second order ionization of the double acceptors cannot occur in MCT with x=0.25. The experimental results indicate that the concentration of the acceptors is nearly equal to that of Hg vacancy defect around x=0.2~0.3[66]. The calculated transition energy levels and the results can be used to explain the experimental observations. Here, we give a further discussion about the results. From our calculations, the transition energy levels for ε(0/−1) of the mercury vacancy defect of MCT are all below 0.1 eV, meaning the Hg vacancy defect causes a shallow acceptor. The theoretical predications, especially the trend of the transition energy levels related to the mole fraction of MCT, are in good agreement with the experimental results[2,31,65]. The theoretical results also indicate that the p-type MCT caused by native Hg vacancy defect becomes more efficient with the Cd mole fraction becoming smaller because the acceptor becomes shallower. Moreover, the transition energy levels for ε(−1/−2) are all around 0.3 eV. By considering the calculated errors, the energy level can be seen as a trap center. The results are agreement with the experimental observation, in which a trap center around 0.15 eV (see Ref. 66) above the VBM occurs because of the Hg vacancy defects.

2.3. Bonding Mechanism

To indicate the strength of orbital hybridization in the MCT materials, one can examine the density of states and its partial density of states (p-DOS), in which it is shown how the "atomic" levels evolve into the band states. The total and partial density of states is calculated using the modified tetrahedron integration method [68] and is broadened using a Gaussian-like function with full width at half-maximum being equal to 0.1 eV. The total density of states of $Hg_{0.5}Cd_{0.5}Te$ with and without a mercury vacancy is shown in Figure 2.2. The main difference is that the valence bottom in the Figure 2.2(b) has two DOS peaks around −10.2 eV. It is known that the valence band bottom of II-VI semiconductors comes from the anion s state. The double DOS peaks demonstrate that the energy level of some Te 5s states is split in the MCT with a mercury vacancy. By comparing the anion's partial DOS of different sites, it is found that the upshift of the 5s energy states comes mainly from the four NN Te atoms.

Based on the same analysis as mentioned above, we can find that the dangling bond rehybridization induced by vacancy defect also has effect on the other electronic states of the material. We focus on the situation of the band edge states, namely, the valence band maximum (VBM) and the conduction band minimum (CBM) because the band gap of the materials is determined by them. It is known that the bands in the vicinity of the VBM and the CBM have strong Te p and cation-s characteristics for the II-VI semiconductors, respectively. The partial DOS of Te 5p and cation-s states for different neighbor shell sites of $Hg_{0.5}Cd_{0.5}Te$ are calculated as shown in Figure 2.4, in which only the Hg 6s states are shown. The figure indicates that the energy states of NN Te p in VBM around the defect move upwards as the Te 5s states, but the Hg s states in CBM nearly keep the same position at different shell sites.

Further, it is demonstrated that the energy upshift of the electronic states mainly is caused by the NN Te dangling bond rehybridization. The defect and the dangling bond rehybridization mainly have effect on the nearest-neighbor shell. The effect becomes weaker as the neighbor shell is farther from the vacancy site. From the calculations, it is also found that the band gap becomes smaller due to the structural change. As mentioned above, the vacancy is created by removing two valence electrons from cation. The removing causes the dangling bond orbitals of two electrons around the vacancy so that the vacancy is a double acceptor with two levels at different ionization energies, just above the VBM. For the tetrahedral symmetry group, the defect states consist of a low-lying singlet a_1^v state and a high-lying threefold-degenerate t_2^v state. For the case of charged neutral V_{Hg}^0, the defect center has total six electrons. Two of them occupy the a_1^v state, while the remaining four electrons occupy the t_2^v states. In this way, there is one t_2^v unoccupied state just above the the VBM, acting as double acceptor. The acceptor levels determined by single-particle energy state are 106 and 135 meV above the VBM for $Hg_{0.5}Cd_{0.5}Te$, respectively. The first acceptor level calculated here agrees well with the experimental results [31] of the ionization energy of the mercury vacancy, and the second acceptor level locates in the position of the 42% band gap from the VBM. The calculated band gap of the $Hg_{0.5}Cd_{0.5}Te$ with mercury vacancy defect by FP-LAPW method is 320 meV, and the band gap is smaller than that of PAW method due to the SO coupling effects [69]. The total DOS (shown in Figure 2.2) includes the two un-occupied single electron energy states above the VBM. Actually, the band gap of $Hg_{0.5}Cd_{0.5}Te$ with Hg vacancy defect is 90 meV larger than that of the pure materials from the calculation. As Wei *et al.* [70] have pointed out that the *p-d* coupling plays an important role in the II-VI semiconductor's VBM, behaving as the *p-d* repulsion. The magnitude of the repulsive coupling can be estimated quantitatively by considering the *p-d* repulsive energy perturbatively, as

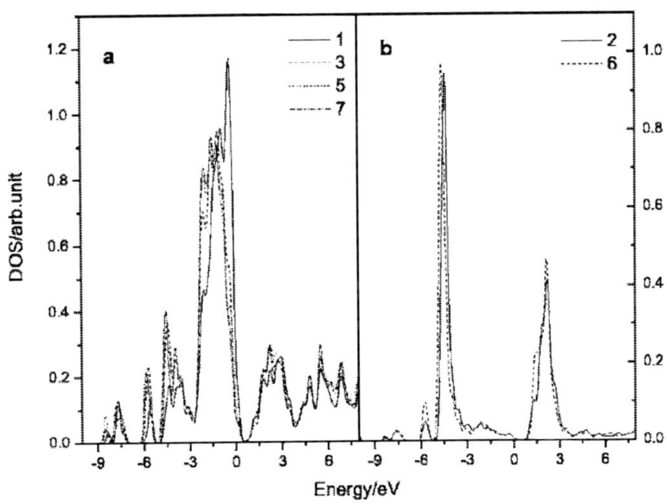

Figure 2.4. The comparison of the Te-*p* states in different shell sites (a) and the comparison between the next-nearest-neighbor (NNN) Hg *s* p-DOS and the sixth-neighbor Hg *s* p-DOS in the farthest from the vacancy (b). The number in the (a) and (b) means the neighbor shell.

$$\Delta E_{pd} \sim V_{pd}^2 /(\varepsilon_p^a - \varepsilon_d^c), \tag{2.3}$$

where the V_{pd}^2 is the coupling matrix element and ($\varepsilon_p^a - \varepsilon_d^c$) is the atomic energy differences between the anions and cations. Because one cation is lost in the system, mercury vacancy defect reduces the *p-d* repulsive energy ΔE_{pd}. The effect induces the downshift of the energy of the NN Te *p* states. Therefore, the empty states cause the response wavelength of MCT with Hg vacancy to become longer although the Hg vacancy widens the band gap.

3. Electronic Properties and Chemical Trends of the Arsenic *In Situ* Impurities in $Hg_{1-x}Cd_xTe$

3.1. Computational Procedure

The density functional theory [48] calculations were performed using the FP-LAPW method implemented in the WIEN2K package [49]. We adopt Perdew-Burke-Ernzerhof functional [50] to describe the exchange-correlation interaction. We consider a 2×2×2 supercell (SC) with a total of 64 atoms, consisting of eight quasi-zinc-blende crystal structure of the unit cells of $Hg_{1-x}Cd_xTe$ (each unit cell contains eight atoms including four Te atoms and four cations of Hg and Cd atoms corresponding to the fraction *x*). The doped system was modeled by putting one As atom at the center of a periodic supercell. A 2×2×1 supercell with 32 atoms without any impurity was chosen as a reference system for comparison purpose, and the computational parameters are set to be the same as the As doping cases. To ensure convergence well, the muffin-tin (MT) radii and the number of *k* points for generating the final results have been carefully chosen after optimization. 8 and 32 *k* points in the irreducible Brillouin zone have been chosen for doped and pure MCT, respectively. A satisfactory self-consistent convergence has been achieved by considering a number of FP-LAPW basis functions up to $R_{MT}K_{max}$=6.0 for the 64-atom supercell and $R_{MT}K_{max}$=7.0 for 32-atom supercell. K_{max} is the maximum value of the plane-wave vector which determines the energy cutoff for the plane-wave expansion. Self-consistent iteration was considered to be converged when both the total energy and the total charge in the atomic sphere are stable within 10^{-4} eV per unit cell and 10^{-4} electron charges per atom, respectively. The relaxation procedures were conducted following the damped Newton dynamics schemes [52,53]. The criterion of the force convergence for all atoms was 0.05 eV/Å. The relativistic effect of spin-orbit (SO) coupling was also included. To confirm the results from the 64-atoms supercell using FP-LAPW calculations and to extend the conclusions to the other mole fractions of MCT, plane-wave pseudopotential calculations were performed using the Vienna *ab initio* simulation package (VASP) [54,55]. In the VASP calculations, the coreelectron interaction was modeled by the ultrasoft pseudopotentials [56], which can yield reasonably precise results with relatively low cutoff energy. Thus, it is possible to deal with the larger supercell. In this work, supercells with 256 and 512 atoms have been chosen. The plane-wave cutoff energy was chosen as 300 eV, and the Brillouin zone was sampled by using 5×5×5 and 3×3×3 Gamma centered Monkhorst-Pack grids for 256- and 512-atom supercells, respectively. The energy convergent criterion was 1×10^{-4} eV per unit cell, and forces on all relaxed atoms were less than 0.01 eV/Å. It is well known that local density approximation calculations typically underestimate the band gap of a semiconductor and are unable to give the absolute positions of the impurity level accurately. Here, we emphasize on the analysis of the defect formation energy and the transition energy levels [3,57]. At temperature of 0 K, the formation energy of arsenic impurity in a given charge state q in MCT is given by

$$\Delta H_f(\mathrm{As}_\alpha, q) = E(\mathrm{As}_\alpha, q) + E(\alpha) - E(\mathrm{MCT}) - E(\mathrm{As}) + qE_{\mathrm{VBM}} + qE_{\mathrm{F}} + \mu_\alpha - \mu_{As} \quad (3.1)$$

$\Delta H_f(\mathrm{As}_\alpha,q)$ is the formation energy for a supercell containing the arsenic dopant in charge state q. $E(\mathrm{As}_\alpha,q)$ and $E(\mathrm{MCT})$ are the total energies for the supercell containing the arsenic dopant in charge state q and for the supercell of the same size without impurity, respectively. $E(\alpha)$ and $E(\mathrm{As})$ are the total energies of host atoms (Te, Cd, and Hg) and the impurity arsenic atom, respectively. The atomic chemical reservoir is Te_2 molecule, Cd and Hg gas for host atoms, and As_4 molecule for arsenic impurity, respectively, which coincide with the MBE growth conditions. μ_α and μ_{As} are the external chemical potentials of the host atoms and the arsenic impurity, respectively. Equation (3.1) indicates that the formation energy with arsenic dopant, $\Delta H_f(\mathrm{As}_\alpha,q)$, is a function of the E_{F} and the chemical potential of host and arsenic impurity, μ_α and μ_{As}. The defect transition energy level $\varepsilon_\alpha(q/q')$ is the Fermi energy E_{F} in Eq. (3.1), at which the formation energy $\Delta H_f(\mathrm{As}_\alpha,q)$ of arsenic impurity with charge q is equal to that of another charge q'. The formation energy and transition energy levels of MCT with arsenic dopant were calculated by the projector-augmented wave (PAW) method for the 64-atom supercell. For the charged systems with impurity, a uniform background charge is added to keep the global charge of the periodic supercell neutral. The total energy of the systems with charged defect has been corrected for the interaction of the charged defect with the compensating background and its periodic images [59]. For the 64-atom SC, the error bar is about 0.2 and 0.1 eV for the calculated formation energy and transition energy level, respectively, due to the finite cell size and basis set.

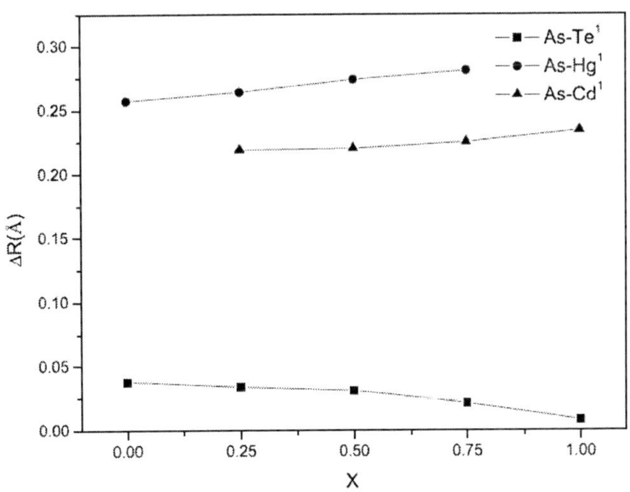

Figure 3.1. Relationship between the inward radial relaxations of the NN atoms and the mole fraction of MCT.

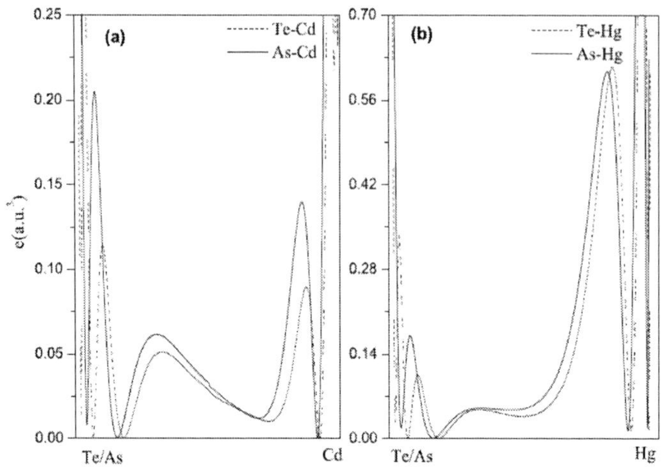

Figure 3.2. Valence charge density along the bond line of As-Hg and As-Cd for the AsTe case and the total charge density along the bond line of Te-Hg and Te-Cd for the pure system.

3.2. Structural Relaxation

The existence of the *in situ* arsenic impurity in MCT induces the structural relaxation of the host atoms and modifies the electronic structure of the system. All the internal structural parameters of the supercell in this work were optimized by minimizing the total energy and quantum mechanical forces. Geometry optimizations reveal that the As impurity, either substituting Te or Hg (Cd atom for CdTe), leads to inward relaxations of the nearest-neighbor (NN) host atoms around the impurity. The radial relaxations of the NN atoms around the impurity calculated by using the PAW method for 64-atom supercell are shown in Figure 3.1. On the one hand, the inward radial relaxations derive from the difference of the covalent radii between the As and the host atoms substituted (the covalent radii are 1.20, 1.36, 1.49, and 1.48 for As, Te, Hg, and Cd, respectively). On the other hand, the inward relaxations of the NN atoms around impurity are induced by the strong bonding mechanism between the As and the host atoms. The detail will be discussed below. The results also show that the As_{Te} impurity induces larger inward radial relaxations of the NN atoms than As_{Hg}. The relaxations of the NN atoms around the impurity calculated by using the FP-LAPW method give the same inward relaxed trend. However, the inward radial relaxations of the NN atoms derived from the PAW method are nearly three times larger than those from the FP-LAPW method. For example, as x=0.5 for the As_{Hg} case, the inward radial relaxations of NN Te is 0.01Å (Ref. 71) and 0.03Å for the FP-LAPW and the PAW method, respectively. Larger radial relaxation of the PAW calculations mainly comes from relatively small force criterion by comparison with that of FP-LAPW calculations.

3.3. Bonding Mechanism

To find out the reason of the NN atom inward relaxation in MCT caused by the As impurity, the valence and the bonding charge density have been calculated. The bonding charge density is defined as the difference between the total charge density in the solid and the superpositions of neutral atomic charge densities placed at atomic sites, such as Eq. (2.2). Therefore, the bonding charge density represents the net charge redistribution as atoms are brought together to form the crystal. The bond rehybridization induced by the As impurity can also be revealed clearly by the bonding charge density.

The bonding characteristic between the impurity and the host atoms is similar to the situation in the case of As_{Hg} [71]. The valence charge density along the line of As-Hg/Cd bond in the case of As_{Te} is shown in Figure 3.2 (refer to Ref. 71 for the results in the case of As_{Hg}). The results indicate that As-Hg/Cd bonds for the As_{Te} case are covalent with stronger ionic characteristics than that of Te-Hg/Cd bond in the pure system. Together with the bonding mechanism analysis in the case of As_{Hg} [71], we can see that, in addition to the smaller arsenic radius, the strong bonding also results in the NN host atom inward relaxation. Moreover, the strong bonding characteristics account for the stably doping behavior of As in MCT, which produces low diffusivity of the As doping. The total density of states (DOS) of As_{Te}-doped $Hg_{0.5}Cd_{0.5}Te$ is shown in Figure 3.3. Substitution of a tellurium by an arsenic atom produces the resulting system metallic. Evidently, the Fermi level of the As_{Te}-doped system (see Ref. 71 for the discussions of the As_{Hg}) does not lie within the band gap which extends to the valence band, as shown in Figure 3.3. The DOS integrated from the Fermi level to the midgap accommodates one electron. The results show that the impurity As_{Te} in MCT would behave as a single acceptor. The theoretical predications are in good agreement with the experimental results [31,34]. For the case of charged neutral As_{Te}, the defect center has a total of seven electrons. Two of them occupy the a_1^v states, five of them occupy the t_2^v states. One of unoccupied t_2^v state just above the valence band maximum (VBM) behaves as a single acceptor.

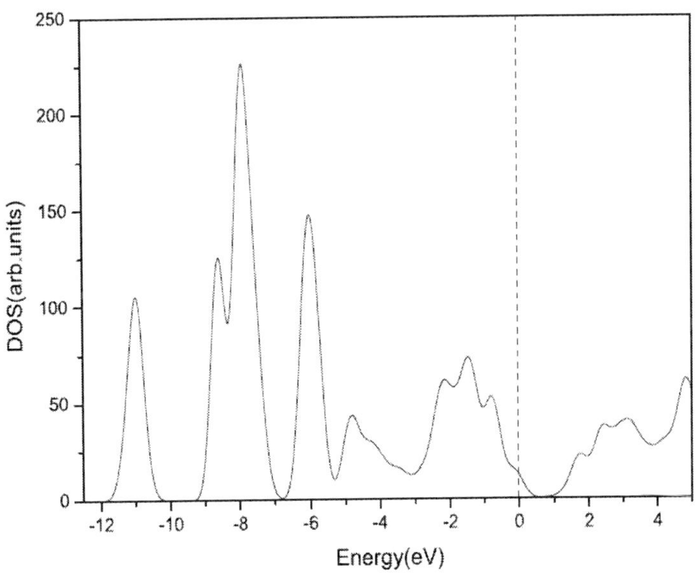

Figure 3.3. Total DOS of As_{Te}-doped MCT-0.5. The Fermi level is set to zero.

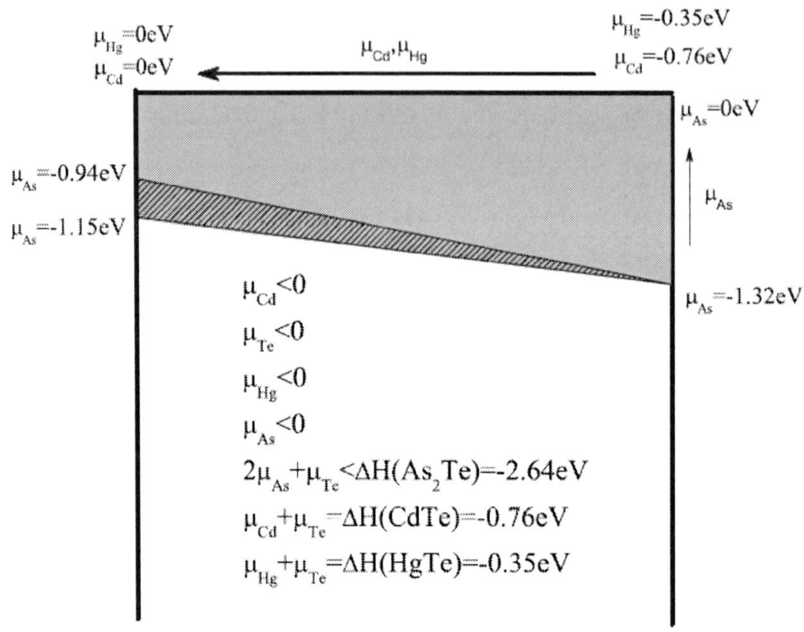

Figure 3.4. Calculated available equilibrium chemical potential region for MCT: As in the two dimensional (μ_{Cd}, μ_{Hg}, and μ_{As}) plane. The shaded and diagonal area is forbidden under equilibrium growth condition.

3.4. Formation Energy

During the defect formation, there are particles exchanged between the host and the chemical reservoir. The defect formation energy depends sensitively on the atomic chemical potential, as well as on the Fermi energy, as indicated in Eq. (3.1). Moreover, there are some thermodynamic limits on the achievable values of the host and impurity atoms' chemical potentials. In order to maintain a stable MCT compound, to avoid precipitation of host and impurity element, and to avoid the formation of secondary phases between the dopant arsenic and host elements, the chemical potentials of Hg, Cd, Te, and As are limited as shown in Figure 3.4. The analysis of bonding mechanism indicates that the As impurity forms strong bonds with the host atoms, which would easily produce secondary phases between arsenic and host elements. After thorough analysis, it shows that As forms stable compound As_2Te with Te. Comparing the MCT-0.5 with Te vacancy defect and the MCT-0.5 with double As interstitial impurities, the formation energy of As_2Te in MCT-0.5 crystal potential field was obtained to be about −2.64 eV. Here, we simply chose the formation energy of As_2Te in MCT-0.5 as the criterion for MCT. This treatment only affects the absolute values and do not influence the trend and the comparisons between the formation energies in the same mole fraction. The formation energy of As_2Te limits the chemical potential of As under the cation- or tellurium-rich conditions, as shown in Figure 3.4. The shaded and shaded plus diagonal areas in the figure are forbidden under equilibrium growth condition for arsenic in CdTe and HgTe, which can be taken as the extreme situation of MCT. Above these chemical potential limits, secondary phase As_2Te will form. Under the cation-rich condition, the chemical potential of Te in MCT becomes complicated due to the alloying effects. We simply adopt the virtual crystal approximation (VCA) to predict the chemical potential of Te based on the formation energy of HgTe and CdTe (the formation energy of HgTe and CdTe is −0.35 and −0.76 eV, respectively). As pointed out in the previous work [70], the VCA will introduce errors into the theoretical results of the formation energy. However, these errors would cancel with each other in the results by comparing between the formation energies, especially for the results of the dynamic levels.

It is known that whether an impurity can easily be incorporated in a material depends sensitively on the growth condition, such as the atomic chemical potential, the growth temperature, as well as the Fermi energy related to the doping levels. The dependence is especially important for the As dopant in MCT because it exhibits amphoteric behavior. In order to analyze the chemical trend of the As impurity in MCT, the formation energy as function of atomic chemical potential and Fermi energy has been calculated. Figure 3.5 shows the calculated formation energies of As-doped MCT for x=0.25, 0.5, 0.75, and 1 as a function of the Fermi energy under cation- and Te-rich growth conditions. The slope of the line gives the charge state of the doping at the Fermi energy. The transition energy level is the Fermi energy, at which the slope changes value. Under μ_{Te}=0 condition, as shown in Figure 3.5, the formation energy of As_{Hg} for x=0.25, 0.5, and 0.75 is smaller than that of As_{Te} as the Fermi energy shifts inside the band gap. These results indicate that the arsenic behaves as dominant n-type dopant for MCT as x=0.25, 0.5, and 0.75 under Te-rich condition. The experiments of MBE growth of HgCdTe (Refs. 10, 40, and 47) (for which the typical growth condition is Te saturated) and LPE growth of HgCdTe (Ref. 31) incorporate arsenic as dopant under Te-saturated condition, both resulting in n-type doping behavior. The results shown in Figure 3.5 indicate that the formation energies of As_{Hg} are about 1.0–1.5 eV smaller than that of As_{Te} as the Fermi level at VBM. It indicates that As easily incorporates into the cation site under the Te-saturated growth condition, behaving as n-type dopant. However, under the Te-rich limit, the formation energy of As_{Cd} for CdTe is smaller than that of As_{Te} as E_F below 0.3 eV. As the Fermi energy shifts up toward the midgap, the formation energy of As_{Te} is smaller than that of As_{Cd}. Namely, the As_{Te} and As_{Cd} compensate each other at the Te-rich limit and the Fermi energy is pinned at a level closer to the VBM; consequently, the arsenic impurity in CdTe, even at Te-rich limit, tends to be slightly p-type dopant. Although there are few experimental reports under this extreme growth condition, arsenic doped CdTe layers grown by organometallic vapor phase epitaxy produce p-type doping behavior even when the partial pressures of Cd and Te are equal [37]. Under μ_{Cd}=0 condition, as shown in Figure 3.5, the formation energy of As_{Te} is nearly 1 eV smaller than that of As_{Cd} for CdTe. The arsenic dominantly incorporates into the Te site, behaving as p-type dopant under the Cd-saturated condition. The result is in good agreement with the MBE growth CdTe doped by arsenic under Cd-saturated condition as efficient p type [72]. The difference of the formation energy of As_{Hg} and As_{Te} for x=0.25, 0.5, and 0.75 is around 0.5 eV as the Fermi energy at VBM under Hg-rich limit. As for x=0.25, the formation energy of As_{Hg} is smaller than that of As_{Te} as the Fermi energy shifts inside the band gap. These results indicate that the arsenic in MCT-0.25 dominantly behaves as n-type dopant, even under Hg-saturated growth condition. The compensation effects between As_{Hg} and As_{Te} occur for x=0.5 and 0.75 and the Fermi energy is pinned at about 0.1 eV above VBM. Namely, arsenic doping in MCT-0.5 and MCT-0.75 tends to be slight p-type dopant under Hg-rich limit.

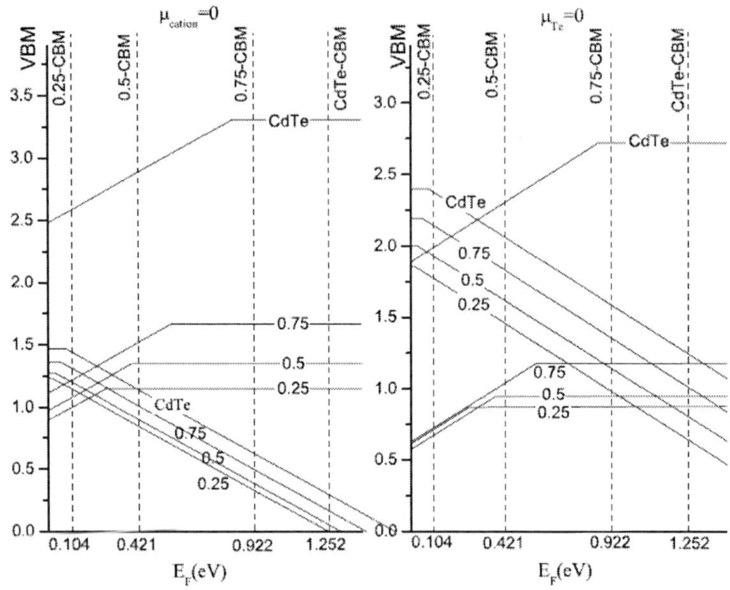

Figure 3.5. The Calculated formation energy of As_{Te} and $As_{Hg/Cd}$ according to the mole fraction of MCT as a function of the Fermi levels. The VBM of all the four mole fraction is set to zero.

As shown in Figure 3.5, the difference of the formation energy of As_{Hg} and As_{Te} is much smaller under Hg-rich limit than that under Te-rich limit. As mentioned above, with mole fraction around 0.25, the arsenic easily incorporates into the cation site, behaving as donor. In order to switch arsenic doping from *n* to *p* type, postgrowth annealing must be done. In the model developed by Berding *et al.* [38], they suggested two steps in the annealing processes. In the first step, low Hg partial pressure is maintained to enhance the amount of metal vacancies (V_{Hg}) which aid transfers of atom. In the second step, saturated Hg partial pressure is maintained to fill the Hg vacancies. According to the calculated results shown in Figure 3.5, we argued that the second annealing step must be achieve under Hg-saturated condition mainly because the Hg-rich limit minimizes the difference of the formation energy between As_{Hg} and As_{Te}, which is beneficial for the arsenic transfer from the cation to the anion site achieving the *p*-type arsenic dopant activation.

The transition energy levels of As_{Hg} and As_{Te} for x =0.25, 0.5, and 0.75 and As_{Cd} and As_{Te} for CdTe have been calculated and shown in Figure 3.6. The theoretical results show that the As_{Te} has shallow transition energy level (0/−1) for all mole fractions of MCT. The transition energy levels become deeper as the mole fraction becomes larger. The trend of the transition energy level with regard to the variation of the mole fraction x agrees well with experiments [73]. The transition energy level (0/−1) of CdTe is 78 meV, in good agreement with the experiments' ionization energy of arsenic-doped *p*-type CdTe [around 58–92 meV (Refs. 72 and 74)]. The transition energy level calculated for MCT-0.25 is 19 meV, while the experimental data have shown that the ionization energy is about 6 meV (Ref. 10) for As_{Te} acceptor level for the mole fraction around 0.3. The calculated results reasonably agree with the experiment. The transition energy levels of As_{Hg} in MCT also move deeper as the mole fraction becomes larger. The results indicate that the impurity levels of As_{cation} tend to behave as deep center for CdTe and MCT with large mole fraction.

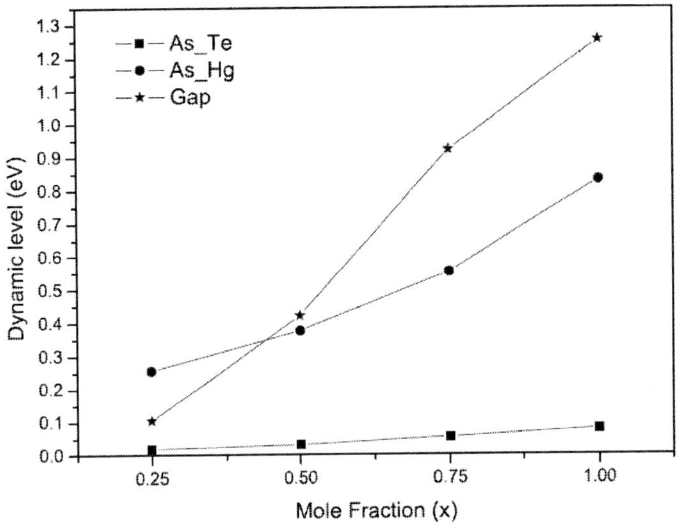

Figure 3.6. Transition energy levels of As-doped MCT. The band gap is calculated here.

The calculated results of the formation energy under different atomic chemical potential limits indicate that the compensation between As_{Te} and As_{Cd} only occurs for CdTe under Te-rich limit, and occurs for x=0.5 and 0.75 under Hg-rich limit. The difference of the formation energies between the As_{cation} and As_{anion} is minimized under cation-rich limit, which is beneficial for the *p*-type activation of arsenic-doped MCT. Moreover, the calculated transition energy levels indicate that the levels become deeper as the mole fraction becomes larger for both As_{cation} and As_{anion} cases.

4. THE ARSENIC IMPURITY CLUSTERS IN MOLECULAR BEAM EPITAXY (MBE) GROWN HGCDTE

4.1. Computational Methods

Here, we chose HgTe as the host system. The random-distributions of cation atoms in HgCdTe system is not in consideration. The choice of HgTe is feasible because of the equivalent properties between HgTe and technologically important HgCdTe with low Cd fraction. The total energy calculations were performed within the generalized-gradient approximation (GGA) implemented by the plane-wave total energy VASP code [54] with the Vanderbilt ultrasoft pseudopotentials [56]. The cutoff energy for the wave functions was 280 eV. The calculated ground lattice constant for bulk HgTe was 6.66Å , in agreement with the experimental value 6.46Å[75].We used a 64-atom supercell and a 9×9×9 k-points Monkhorst-pack mesh in the Brillouin zone of the supercell. All internal structural parameters were fully relaxed until the Hellmann–Feynman (HF) forces were converged to within 0.1 meV/Å. The defect system was modeled by putting a complex defect at the center of a periodic supercell of HgTe, and all the defect structures considered here are electronically neutral.

4.2. Chemical Potential Dependence of the Formation Energy

The formation energies of the defect system were calculated from the total energies based on the standard formalism by Zhang and Northrup [57]. For neutral defect $\alpha(q=0)$ in HgTe, the formation energy $\Delta H_f(\alpha, q=0)$ depends on the chemical potential μ_i (i refers to Hg, Te, and As).

$$\Delta H_f(\alpha, q=0) = \Delta E(\alpha, q=0) + n_{Hg}\mu_{Hg} + n_{Te}\mu_{Te} + n_{As}\mu_{As} \tag{4.1}$$

where

$$\Delta E(\alpha, q=0) = E_{tot}(\alpha, q=0) - E_{tot}(HgTe) + n_{Hg}\mu^0_{Hg} + n_{Te}\mu^0_{Te} + n_{As}\mu^0_{As} \tag{4.2}$$

Here, $E_{tot}(\alpha, q=0)$ is the total energy for the defective supercell, $E_{tot}(HgTe)$ is the total energy for the same supercell in the absence of α, the n_i is the number of Hg, Te, and As atoms transferred from the supercell to the reservoirs in forming the defect, respectively, μ_i is the chemical potential referenced to elemental liquid/solid with energy μ_i^0. However, the achievable values of μ_i are limited thermodynamically under equilibrium growth conditions. First, to avoid precipitation of the constituent i, μ_i is bound by

$$\mu_{Hg} < 0,\ \mu_{Te} < 0,\ \mu_{As} < 0. \tag{4.3}$$

Second, to maintain a stable HgTe compound, μ_i is bound by

$$\mu_{Hg} + \mu_{Te} = \Delta H_f(HgTe) \tag{4.4}$$

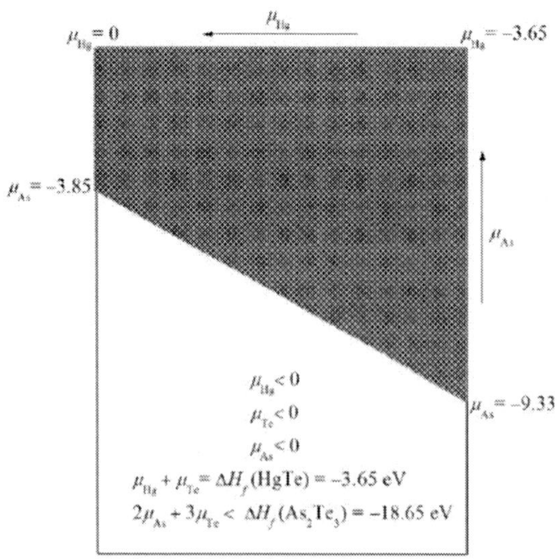

Figure 4.1. Calculated accessible equilibrium chemical potential region for HgTe doped with arsenic in the two-dimensional (μ_{Hg}, μ_{As}) plane. The shaded area is forbidden under equilibrium growth conditions.

The calculated ΔH_f (HgTe) is −3.65 eV, in agreement with the experimental value −3.3 eV [76]. Finally, to avoid the formation of secondary phases As_2Te_3, μ_i is bound by

$$2\mu_{As} + 3\mu_{Te} < \Delta H_f(As_2Te_3) \tag{4.5}$$

Figure 4.1 plots the accessible chemical potential region for HgTe:As, as defined by Eqs. (4.3)–(4.5). It is shown that because arsenic and tellurium form a very stable compound As_2Te_3 with a rather low formation energy −8.65 eV, the highest possible μ_{As}, at the Hg-rich condition ($\mu_{Hg} = 0$, $\mu_{Te} = -3.65$ eV), is −3.85 eV. Under the Te-rich condition ($\mu_{Hg} = -3.65$ eV, $\mu_{Te} = 0$), μ_{As} is further reduced to less than −9.33 eV. Above these chemical potential limits, secondary As_2Te_3 compound will be formed, thus stopping the arsenic doping process. Such low accessible μ_{As} is one of the limiting factors for arsenic doping in HgCdTe.

4.3. Possible Configurations for the Complex Defects As_n (n = 1, 2, and 4)

4.3.1. Arsenic Tetramers

Under Te-saturated condition, the density of V_{Hg} is much higher than that of tellurium vacancy (V_{Te}), so that the As_4 defects with two of the four arsenic atoms occupying mercury sites are expected to be the most reasonable configuration. In our calculations, we typified such As_4 defects into arsenic tetramers (As_t) and isolated arsenic atoms (4As), respectively (see Figure 4.2).

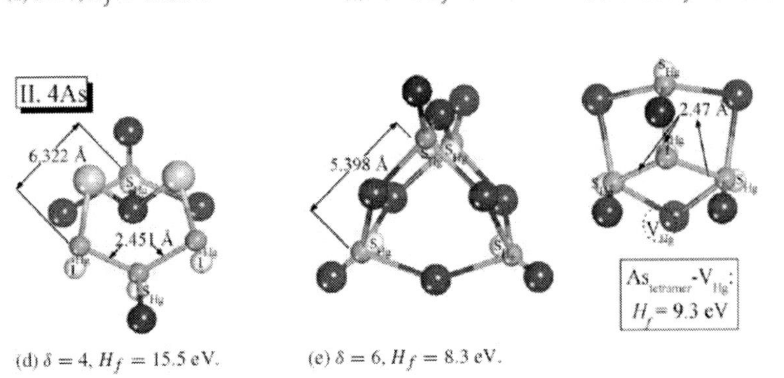

Figure 4.2. Calculated ground geometries for the As_4 defects in HgTe, with two of the four arsenic atoms occupying the Hg sites. In this figure, the Hg atoms are of the largest size (in light grey), the Te atoms of intermediate size (in black), and the As atoms are the smallest (in cyan). To analyze the atomic displacements, the As atoms in the pre-relaxed structures are also shown by the smallest atoms (in white). For the labels on the As atoms, "s" and "i" denotes the substitutional site and the interstitial site, respectively. (For interpretation of the references to colour in this figure legend).

As_t can be regarded as the distorted pyramidal structure of As_4 molecules. During typical low-temperature MBE growth, there are difficulties in cracking As_4 into As_2 or As_1 on the growing surface, so that As_4 molecules initially absorbed on the surface are subjected to the stress exerted by the HgTe lattice owing to the different properties of the dopant and host atoms. The stress can be released at the cost of the break of the partial As–As bonds. In this way the pyramidal structure distorts into the tetramer configuration. For As_t shown in Figure 4.2 the optimized As–As split changes from 2.45Å for configuration (c) to 2.555Å for configuration (a), approximately conserving the experimental value 2.435Å for the pyramidal As_4 molecular [77].

One can also see that different occupation patterns for As_t produces the dissimilar numbers of the broken As–As bonds (designated as δ), and the values of δ for all the possible As_t defects do not exceed 3 ($\delta \leqq 3$). This can be understood from the bond-length of 2.435Å for the pyramidal structure, slightly less than the Hg–Te distance of 2.798Å but far less than the spacing of the nearest neighboring Hg atoms in HgTe (4.57Å). As a result, As_{Hg}–As_{Hg} bond is to be broken up while the As_{Hg}–As_{Te} bond is left to adjust itself to accommodate the host lattice. In the same way it is conjecturable that the more As_{Hg} units in the As_4 defects leads to the larger value of δ. Figure 4.2 also gives the calculated formation energies of the As_4 defects at the Te-rich limit. In the case of As_t the configuration (a) and (b) are metastable states from the formation energies, and the configuration (c) is the most probable one. Therefore, under the equilibrium stage the former two configurations are to be converted into the latter one by means of extra V_{Hg} to release the lattice stress. The geometry of the configuration (c), optimized in a manner of $[(As_{Hg}–As_i)_d\ ]_3$ copolymer (dimer denoted as "d"), is not only more favorable in energy by 2 eV lower than the strongest configuration (b), but also stronger with 25% gain in the binding energy with respect to the quasipyramidal configuration (a).

When the As–As bonds for As_4 molecular are broken completely (δ=6) or partially (3<δ<6), the perfect pyramidal dissociates into isolated atoms and/or dimers, as demonstrated by the configurations of type II in Figure 4.2. Such a situation occurs only under high-temperature growth conditions, which is not applied for the state-of-the-art MBE. Focusing on the configuration (c), one can see that the atomic relaxation makes the three As_{Hg} atoms to move towards the common As_i atom by an amplitude of 0.348Å, which leads to the final optimized geometry in $[(As_{Hg}–As_i)_d]_3$ copolymer configuration. However, such a configuration fails to explain the experimental observation of the slight *n*-type conduction in as-grown materials because of the three As_{Hg} atoms with strong donor characteristics. The previous study [29] has given a reasonable explanation of the As_{Hg}–V_{Hg} pairs, suggesting that the behavior of the compensation conduction in as-grown samples is possibly associated with V_{Hg} which is bound to the donor-like arsenic defects. To augment our understanding of such a suggestion, we computed the complex defect of As_t–V_{Hg} (see the last section in Figure 4.2), which was modeled by placing an additional V_{Hg} to the given As_t in the configuration (c). The calculated ground geometry of As_t–V_{Hg} displays that the configuration of $[(As_{Hg}–As_i)_d]_3$ is stable, without experiencing the relaxation to other configurations with the perturbation of V_{Hg}. However, the complex defect of $[(As_{Hg}–As_i)_d]_3$–V_{Hg} is found to substantially lower the formation energy $[(As_{Hg}–As_i)_d]_3$ by about 3 eV. This reduction of the formation energy is attributed to the removing of Hg atoms from the defect system, thus effectively compensating the energy lost by the introduction of As atoms into the defect system. Also, the configuration information of the complex defect of $[(As_{Hg}–As_i)_d]_3–V_{Hg}$ is deemed to be in analogy to that of clusters of $As_{Hg}–V_{Hg}$ pairs with one common V_{Hg}, so that both of them are of compensating donor nature, as supported by our calculated transition energies (unpublished).

4.3.2. Arsenic Dimer (As_d) and Singlet (As_s)

The calculations of As_d and As_s are based on the fact that these defects in a small quality are present in a noncracking arsenic source but predominant in a cracking arsenic source. In Figure 4.3(a), we give the probable configuration of As_d with the lower formation energy. The As–As spacing is 2.228Å, in agreement with the experimental value 2.10Å for As_2 molecule [19]. We note that the As_{Hg} atom deviates significantly from the ideal lattice sites toward the interstitial site by the amplitude of 0.783Å, which suggests that the interstitial sites are favorable in energy to the absorption of As_2 molecules. However, our calculations indicate that two arsenic atoms, both residing in the interstitial sites, can never be configured in As_d unless an amount of energy barriers are overcome when an extra V_{Hg} is available; otherwise, they are in the configuration of two isolated arsenic atoms (2As), as shown in Figs. 4.3(b) and (c). So the configuration with two arsenic atoms of split interstitial is expected to be the most probable for As_d, but this assumption needs to be examined in our further work. At the same time, we have modeled the complex defect $(As_{Hg}–As_i)_d–V_{Hg}$, as demonstrated by the configuration (f) in Figure 4.3. An obvious shift of As_i toward V_{Hg} can be observed, and the amplitude of the As_i relaxation is more significant relative to the $(As_{Hg}–As_i)_d$ defect.

The displacements of As_s are negligible during the atomic relaxations. The lattice relaxation energies are of the order of meV. During the short-displacement and successive-steps of the relaxations, the work done by the HF force can be approximated to that by the first (1) and second (2) nearest neighbors of impurity while the contribution from the more distant neighbors is not taken into account.

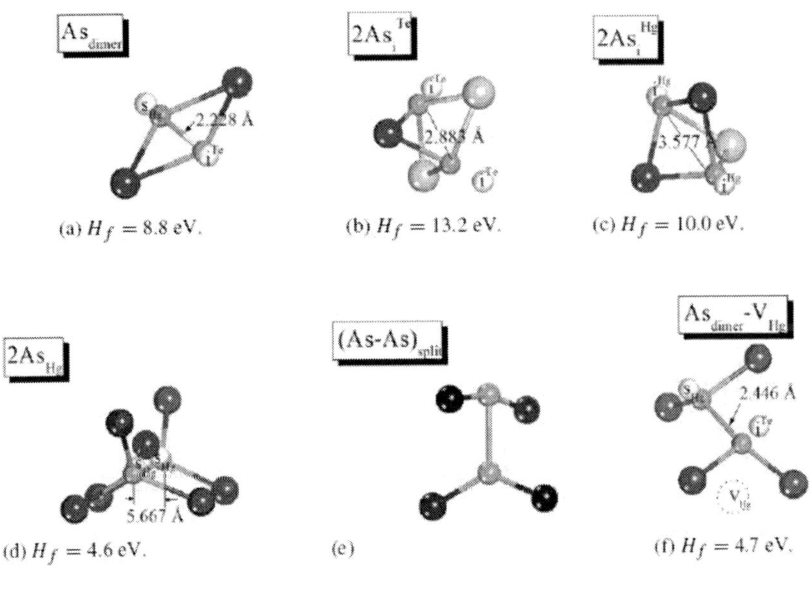

Figure 4.3. Calculated ground geometries for As_2 defects in HgTe. The configuration of (As–As) split is assumed to be more plausible than other ones for As_d. The atomic notations are the same as denoted in Figure 4.2.

Table 1. Work of the Hellmann–Feynman forces (HF) (in meV) for the first (1) and second (2) nearest neighbors as well the lattice relaxation energies ΔE_{relax} (in meV)

$(\alpha, q = 0)$	W_1^{HF}	W_2^{HF}	ΔE_{relax}
As_{Hg}	1.68	15.84	19.58
As_{Te}	143.96	530.24	709.67
As_i^{Hg}	27.76	92.13	339.97
As_i^{Te}	3.54	90.76	366.81

$$W^{HF} = W_1^{Hf} + W_2^{HF} = (\sum F_{io}^{HF} \cdot \Delta d_{io} + \sum F_{ij}^{HF} \cdot \Delta d_{ij})/2 \quad (4.6)$$

Table 1 displays the evaluation of the work done by the HF force together with the relaxation energies ΔE_{relax}. One can see that the sum of W_1^{HF} and W_2^{HF} is comparable to the value of ΔE_{relax}, and the main contributions of the work done by the HF forces come from the second-nearest neighbors [78].

4.3. Comparison of the Formation Energies for As_n (n = 1, 2, and 4)

In Figure 4.4 we compare the formation energies per arsenic atom for various types of the arsenic defects As_n (n=1, 2, and 4). It can be seen that As_t is more favorable in energy than As_d and As_s, indicating that As_t is more abundant than As_d and As_s in as-grown materials. The order of the formation energies is relevant to the atomic site configurations in arsenic defects, for example, the energy levels of the arsenic defects containing the As_{Te} atoms are much higher than those containing the As_{Hg} atoms. This is a direct effect of the chemical potentials on the formation energies. Under Te-rich condition (μ_{Te}=0 and μ_{Hg}<0), more Hg atoms involved in the given defect results in the lower energy levels. From the formation energies of As_n the values of the binding energies are also obtained for As_t and As_d. We find that the more stable configuration corresponds to the stronger binding of the arsenic atoms in As_t and As_d [78]. Since the distorted pyramidal structure of As_4 molecular relieves the lattice stress during MBE growth, the heavier degree of the distortion yields more amount of the lattice stress to be released and thus more steady configurations as well. Moreover, it can be concluded that under thermal equilibrium isolated arsenic atoms prefer to be bound into two or four nearest neighboring atoms, because the binding energies of 2As and 4As defects are at least 0.61 eV/As atom and 0.84 eV/As atom higher than the substitutional singlet (As_{Hg}), respectively.

5. CONCLUSION

The structural and electronic properties of a mercury vacancy defect in the $Hg_{1-x}Cd_xTe$ have been studied. The defect leads to inward relaxations and the reasons for the relaxations of NN tellurium have been revealed clearly by the analysis of the valence charge density and the bonding charge density [79]. The inward relaxation is mainly determined by the negative charge accumulation. We found that the charged state of the systems enhances the relaxation.

The formation energy and transition energy levels of the cation vacancy defects have been shown for different mole fractions. In combination with the analysis of the calculated acceptor levels induced by mercury vacancy, the predicated acceptor levels, especially the trend of the acceptor levels according to the mole fraction of MCT, are in good agreement with existing experimental results. Meanwhile, we also show that the rehybridization of the dangling bond of the NN Te atom around the mercury vacancy leads to upshift of the Te 5*s* states [79]. In addition, the mercury vacancy defect increases the band gap in comparison with that of the pure materials due to the weaker *p-d* coupling, but the acceptor levels in the band gap make the response wavelength become longer for the MCT material with a mercury vacancy defect.

In addition, the structural and bonding mechanisms of *in situ* arsenic impurity in $Hg_{1-x}Cd_xTe$ have been studied. The impurity maintains a relative strong bonding with the host atom both in As_{anion} and As_{cation} cases. The strong bonding mechanism between arsenic and host atom in $Hg_{1-x}Cd_xTe$ can explain its stability of doping behavior [80]. The computed formation energy supports the broadly adopted experimental annealing process, low temperature Hg-saturated annealing process, which minimizes the difference of the formation energy of As_{anion} and As_{cation}. The minimization is beneficial for the atomic transfer during the annealing process. Meanwhile, the nearly linear relationship between the formation energy, both for As_{anion} and As_{cation}, and the mole fraction is shown here. The transition energy levels for both As_{anion} and As_{cation} agree well with the experimental results [80]. In addition, the present work shows that the compensation effect occurs for CdTe under Te-saturated condition and for 0.5 and 0.75 under Hg saturated condition.

Furthermore, the theoretical calculations for arsenic tetramer (As_t), arsenic dimer (As_d), and arsenic singlet (As_s) in HgCdTe system. We calculated the defect formation energies from the total-energy calculations, thereby the most probable configurations in the form of $[(As_{Hg}–As_i)_d]_3$ for As_t , $(As_{Hg}–As_i)_d$ for As_d, and As_{Hg} for As_s have been established under Te-rich conditions [81]. In the case of $[(As_{Hg}–As_i)_d]_3$ and As_{Hg}, they are favorable in energy to combine a nearest neighboring V_{Hg}, and the corresponding $[(As_{Hg}–As_i)_d]_3–V_{Hg}$ and $As_{Hg}–As_{Hg}$ defects are thought to explain the selfcompensated *n*-type characteristics in as-grown materials. The absorption of As_2 molecules is assumed to be relevant to the complex defect of (As–As) split instead of $(As_{Hg}–As_i)_d$ because the interstitial sites seem to be preferable [81]. Further studies of the arsenic defects in charge state are necessary to deeply understand the microscopic behavior of arsenic incorporation in HgCdTe.

ACKNOWLEDGMENT

This work was supported in part by the State Key Program for Basic Research of China(2007CB613206), National Natural Science Foundation of China (10725418, 10734090, 10990104, 60976092 and 60811120169), Key Fund of Shanghai Science and Technology Foundation(08JC1421100, 09DJ1400203), Science and Technology Commission of Shanghai Municipality (09ZR1436100), Knowledge Innovation Program of the Chinese Academy of Sciences and the computational support from Shanghai Super-computer Center.

REFERENCES

[1] Mainzer, N., Lakin, E. & Zolotoyabko, E. (2002). *Appl. Phys. Lett.*, *81*, 763.
[2] Jones, C. E., James, K., Merz, J., Braunstein, R., Burd, M., Eetemadi, M., Hutton, S. & Drumheller, J. (1985). *J. Vac. Sci. Technol. A*, *3*, 131.
[3] Wei, S. H., & Zhang, S. B. (2002). *Phys. Rev. B*, *66*, 155211.
[4] Vydyanath, H. R. (1981). *J. Electrochem. Soc.*, *128*, 2609.
[5] Gold, M. C. & Nelson, D. A. (1986). *J. Vac. Sci. Technol. A*, *4*, 2040.
[6] Selamet, Y., Grein, C. H., Lee, T. S. & Sivananthan, S. (2001). *J. Vac. Sci. Technol. B*, *19*, 1488.
[7] Gopal, V. & Gupta, S. (2004). *Appl. Phys. Lett.*, *95*, 2467.
[8] Berding, M. A., van Schilfgaarde, M. & Sher, A. (1991). *MRS Symposia Proceedings No. 216*, Materials Research Society, Pittsburgh, 3.
[9] Tang, M. S., & Stevenson, D. A. (1989). *J. Vac. Sci. Technol. A*, *7*, 544.
[10] Zandian, M., Chen, A. C., Ewall, D. D., Pasko, J. G. & Arias, J. M. (1997). *Appl. Phys. Lett.*, *71*, 2815.
[11] Berding, M. A. et al. (1997). *J. Electron. Mater.*, *27*, 605.
[12] Stapp, R. S. et al. (1988). *Appl. Phys. Lett.*, *52*, 1614.
[13] Garg, A., Kapoor, A., Tripathi, K. N. & Bansal, S. K. (2004). *Proc. SPIE*, *5273*, 303.
[14] Aguirre, M. H., Cnepa, H. R. & Walse de Reca, N. E. (2002). *J. Appl. Phys.*, *92*, 5745.
[15] Cotton, V. A. & Wilson, J. A. (1986). *J. Vac. Sci. Technol. A*, *4*, 2177.
[16] Schick, J. T. & Morgan-Pond, C. G. (1990). *Semicond. Sci. Technol.*, *5*, S81.
[17] Berding, M. A., van Schilfgaarde, M. & Sher, A. (1994). *Phys. Rev. B*, *50*, 1519.
[18] Berding, M. A. (1999). *Phys. Rev. B*, *60*, 8943.
[19] Berding, M. A., van Schilfgaarde, M. & Sher, A. (1992). *J. Vac. Sci. Technol. B*, *B10*, 1417.
[20] Errico, L. A., Fabricius, G., Renteria, M., dela Piesa, P. & Forker, M. (2002). *Phys. Rev. Lett.*, *89*, 055503.
[21] Errico, L. A., Fabricius, G. & Renteria, M. (2003). *Phys. Rev. B*, *67*, 144104.
[22] Park, R. M., Troffer, M. B., Rouleau, C. M., DePuydt, J. M. & Haase, M. A. (1990). *Appl. Phys. Lett.*, *57*, 2127.
[23] Haase, M. A., Qiu, J., DePuydt, J. M. & Cheng, H. (1991). *Appl. Phys. Lett.*, *59*, 1272.
[24] Park, C. H., Zhang, S. B. & Wei, S. H. (2002). *Phys. Rev. B*, *66*, 073202.
[25] Sivananthan, S., Wijewarnasuriya, P. S. & Faurie, J. P. (1995). *Proc. SPIE*, *2554*, 55.
[26] Chadi, D. J. & Chang, K. J. (1988). *Phys. Rev. Lett.*, *61*, 873.
[27] Chadi, D. J. (1999). *Phys. Rev. B*, *59*, 15181.
[28] Limpijumnong, S., Zhang, S. B., Wei, S. H. & Park, C. H. (2004). *Phys. Rev. Lett.*, *92*, 155504.
[29] Berding, M. A. & Sher, A. (1999). *Appl. Phys. Lett.*, *74*, 685.
[30] Vydyanath, H. R., Abbott, R. C. & Nelson, D. A. (1983). *J. Appl. Phys.*, *54*, 1323.
[31] Vydyanath, H. R., Ellsworth, J. A. & Devaney, C. M. (1987). *J. Electron. Mater.*, *16*, 13.
[32] Shi, X. H., Rujirawat, S., Ashokan, R., Grein, C. H. & Sivananthan, S. (1998). *Appl. Phys. Lett.*, *73*, 638.

[33] Grein, C. H., Garland, J. W., Sivananthan, S., Wijewarnasuriya, P. S., Aqariden, F. & Fuchs, M. (1999). *J. Electron. Mater.*, *28*, 789.
[34] Sivananthan, S., Wijewarnauriya, P. S., Awariden, F., Vydyanath, H. R., Zandian, M., Edwall, D. D. & Arias, J. M. (1997). *J. Electron. Mater.*, *26*, 621.
[35] Edwall, B. D., Chen, J. S. & Bubulac, L. O. (1990). *J. Vac. Sci. Technol. B*, *9*, 1691.
[36] Harman, T. C. (1989). *J. Electron. Mater.*, *8*, 191.
[37] Vydyanath, H. R. (1990). *Semicond. Sci. Technol.*, *5*, S213.
[38] Berding, M. A., Sher, A., Van Schilfgaarde, M., Chen, A. C. & Arias, J. (1998). *J. Electron. Mater.*, *27*, 605.
[39] Garland, J. W., Grein, C. H., Yang, B., Wijewarnasuriya, P. S., Aqariden, F. & Sivananthan, S. (1999). *Appl. Phys. Lett.*, *74*, 1975.
[40] Wijewarnasuriya, P. S. & Sivananthan, S. (1998). *Appl. Phys. Lett.*, *72*, 1694.
[41] Bevan, M. J., Chen, M. C. & Shih, H. D. (1995). *Appl. Phys. Lett.*, *67*, 3450.
[42] Aqariden, F., Wijewarnasuriya, P. S. & Sivananthan, S. (1998). *J. Vac. Sci. Technol. B*, *16*, 1309.
[43] Bubulac, L. O., Edwall, D. D. & Wiswanathan, C. R. (1991). *J. Vac. Sci. Technol. B*, *9*, 1695.
[44] Aqariden, F., Shih, H. D., Kinch, M. A. & Schaake, H. F. (2001). *Appl. Phys. Lett.*, *78*, 3481.
[45] Selamet, Y., Badano, G., Grein, C. H., Boieriu, P., Nathan, V. & Sivananthan, S. (2001). *Proceedings of SPIE*, *4454*, 71.
[46] Lee, T. S., Garland, J., Grein, C. H., Sumstine, M., Jandeska, A., Selamet, Y. & Sivananthan, S. (2000). *J. Electron. Mater.*, *29*, 869.
[47] Boieriu, P., Grein, C. H., Jung, H. S. & Garland, J. (2005). *Appl. Phys. Lett.*, *86*, 212106.
[48] Hohenberg, P. & Kohn, W. (1964). *Phys. Rev.*, *136*, B864; Kohn, W. & Sham, L. J. (1965). *Phys. Rev.*, *140*, A1133.
[49] Blaha, P., Schwarz, K., Madsen, G., Kvasnicka, D. & Luitz, J. (2001). *WIEN2k, An Augmented Plane Wave + Local Orbitals Program for Calculating Crystal Properties*, Karlheinz Schwarz, Techn. University Wien, Austria.
[50] Perdew, J. P., Burke, K. & Ernzerhof, M. (1996). *Phys. Rev. Lett.*, *77*, 3865.
[51] Bernd Kohler, Steffen Wilke & Matthias Scheffler (1996). *Comput. Phys. Commun.*, *94*, 31.
[52] Yu, R., Singh, D. & Krakauer, H. (1991). *Phys. Rev. B*, *43*, 6411.
[53] Kohler, B., Wilke, S., Scheffler, M., Kouba, R. & Ambrosch-Draxl, C. (1996). *Comput. Phys. Commun.*, *94*, 31.
[54] Kresse, G. & Furthmuller, J. (1996). *Phys. Rev. B*, *54*, 11169.
[55] Kresse, G. & Furthmuller, J. (1996). *Comput. Mater. Sci.*, *6*, 15.
[56] Vanderbilt, D. (1985). *Phys. Rev. B*, *32*, 8412.
[57] Zhang, S. B. & Northrup, J. E. (1991). *Phys. Rev. Lett.*, *67*, 2339.
[58] Shih, C. K. & Spicer, W. E. (1987). *Phys. Rev. Lett.*, *58*, 2594.
[59] Lento, J., Mozos, J. L. & Nieminen, R. M. (2002). *J. Phys.: Condens. Matter*, *14*, 2637.
[60] Murnaghan, F. D. (1994). *Proc. Natl. Acad. Sci.*, U.S.A., *30*, 244.
[61] Berding, M. A., Sher, A. & Chen, A. B. (1987). *J. Vac. Sci. Technol. A*, *5*, 3009.
[62] Kittel, C. (1986). *Introduction to Solid State Physics*, 6th ed. Wiley & Sons, New York.

[63] Harrison, W. A. (1980). *Electronic Structure and the Properties of Solids*, Freeman, San Francisco.
[64] *CRC Handbook of Chemistry and Physics*, 81st ed., edited by D. R. Lide, CRC Press, New York, (2000).
[65] Hunter, A. T. & McGill, T. C. (1981). *J. Appl. Phys.*, *52*, 5779; Hunter, A. T. & McGill, T. C. (1982). *J. Vac. Sci. Technol.*, *21*, 205.
[66] Copper, D. E. & Harrison, W. A. (1990). *J. Vac. Sci. Technol. A*, *8*, 1112.
[67] Polla, D. L. & Jones, C. E. (1980). *J. Appl. Phys.*, *51*, 6233.
[68] Blöchl, P. E., Jepsen, O. & Andersen, O. K. (1994). *Phys. Rev. B*, *49*, 16223.
[69] Wang, C. S. & Klein, B. M. (1981). *Phys. Rev. B*, *24*, 3393.
[70] Wei, S. H., & Zunger, A. (1988). *Phys. Rev. B*, *37*, 8958.
[71] Sun, L. Z., Chen, X. S., Sun, Y. L., Zhou, X. H., Quan, Zh. J., He Duan & Wei Lu (2005). *Phys. Rev. B*, *71*, 193203.
[72] Arias, L. M., Shin, S. H., Cooper, D. E., Zandian, M., Pasko, J. G., Gertner, E. R., DeWames, R. E. & Singh, J. (1990). *J. Vac. Sci. Technol. A*, *8*, 1025.
[73] Yang, F. Aqariden, Grein, C. H., Jandaska, A., Lee, T. S., Nemani, A., Rujirawat, S., Shi, X. H., Sumstine, M., Velicu, S. & Sivananthan, S. (1999). *J. Vac. Sci. Technol. B*, *17*, 1205.
[74] Harper, R. L. Jr., Hwang, S., Giles, N. C., Schetzina, J. F., Dreifus, D. L. & Myers, T. H. (1989). *Appl. Phys. Lett.*, *54*, 170.
[75] Chen, X. J., Hua, X. L., Hu, J. S., Langlois, J. M. & Goddard III, W. A. (1996). *Phys. Rev. B*, *53*, 1377.
[76] Berding, M. A., Vanschilfgaarde, M. & Sher, A. (1993). *J. Electron. Mater.*, *22*, 1005.
[77] Hedberg, K. (1966). Trans. Am. Crystallogr. Assoc., 2, 79.
[78] Duan, H., Chen, X. S., Huang, Y. & Lu, W. (2007). *J. Electron. Mater.*, *36*, 890.
[79] Sun, L. Z., Chen, X. S., Sun, Y. L., Zhou, X. H., Quan, Zh. J., Duan, H. & Lu, W. (2006). *Phys. Rev. B*, *73*, 195206.
[80] Sun, L. Z., Chen, X. S., Zhao, J. J., Wang, J. B., Zhou, Y. C. & Lu, W. (2007). *Phys. Rev. B*, *76*, 045219.
[81] Duan, H., Chen, X. S., Huang, Y. & Lu, W. (2007). *Solid State Commun.*, *143*, 471.

In: Bulk Materials: Research, Technology and Applications ISBN: 978-1-60692-963-6
Editors: T. Frias, V. Maestas, pp. 335-348

Chapter 8

SEARCH FOR INCREASED PERFORMANCE IN $BiFeO_3$: EFFECT OF A-SITE SUBSTITUTION

V. A. Khomchenko and A. L. Kholkin

Department of Ceramics and Glass Engineering & CICECO, University of Aveiro, 3810-193 Aveiro, Portugal

ABSTRACT

Investigation of crystal structure, dielectric, magnetic and local ferroelectric properties of the diamagnetically- substituted $Bi_{1-x}A_xFeO_{3-x/2}$ (A= Ca, Sr, Pb, Ba; x= 0.2, 0.3) polycrystalline samples has been carried out. It has been shown that the heterovalent A^{2+} substitution result in the formation of oxygen vacancies in the host lattice. The solid solutions have been found to possess a rhombohedrally distorted perovskite structure described by the space group *R3c*. Piezoresponse force microscopy have revealed signs of existence of the ferroelectric polarization in the samples at room temperature. Magnetization measurements have shown that the magnetic state of these compounds is determined by the ionic radius of the substituting elements. A- site substitution with the biggest ionic radius ions has been found to suppress the spiral spin structure of $BiFeO_3$ giving rise to the appearance of room-temperature weak ferromagnetism. The result indicates possible way for improving multiferroic properties of $BiFeO_3$- based magnetic ferroelectrics.

1. INTRODUCTION

Multiferroics have become an object of growing interest due to the coexistence of magnetic order and ferroelectric polarization combined in a single-phase material [1-5]. The interest is motivated not only by the rich physics that is expected in this class of solids, but also by their great technological potential. Indeed, multiferroic materials with significant magnetoelectric coupling might open promising opportunities for designing novel microelectronic and spintronics devices [6, 7]. However, most magnetic ferroelectrics tend to have low magnetic ordering temperatures (for instance, in well-known hexagonal $RMnO_3$

manganites (R=Sc, Y, In, Ho-Lu) or in monoclinic $BiMnO_3$, the ferroelectric transition occurs at T_C around 600-1000 K, while the magnetic ordering takes place only below 70-130 K [8-12]). The perovskite-like $BiFeO_3$ compound, in which the stereochemical activity of the Bi lone electron pair gives rise to ferroelectric polarization, while the partially filled *3d* orbitals of the Fe^{3+} ions cause G- type antiferromagnetic order, seems to be one of the most suitable objects for multiferroic research in view of its high magnetic and ferroelectric ordering temperatures ($T_N\sim$ 640 K, $T_C\sim$ 1100 K) [13, 14]. On the other hand, spiral spin modulation, superimposed to the G-type antiferromagnetic spin ordering [14], cancels any possible net magnetization and inhibits the observation of a linear ME effect [15]. Spontaneous magnetization in $BiFeO_3$ can be induced via substitution of Fe^{3+} by other kind of transition metal. For instance, recent theoretical investigation performed for Bi_2FeCrO_6 perovskite has shown that the compound will have a polarization of ~ 80 $\mu C/cm^2$, a piezoelectric coefficient of 283 $\mu C/cm^2$, and a magnetization of 2 μ_B per formula unit (due to ferrimagnetic ordering of Fe^{3+} and Cr^{3+} magnetic moments), far exceeding the properties of any known multiferroics [16]. Unfortunately, such B-site ions substitution decreases the magnetic ordering temperature drastically, thus hampering application of the materials at room temperature [16-18]. Moreover, the high-pressure synthesis conditions are necessary to prepare the single-phase $Bi_2M^{3+}M'^{3+}O_6$ or $Bi_2M^{2+}M'^{4+}O_6$ (M, M′-transition metals) compounds [17-19].

It was recently found that, if the spiral spin structure in $BiFeO_3$ is suppressed, the system shows weak ferromagnetism [20, 21], originating from the Dzyaloshinskii-Moryia interaction [22, 23]. An enhancement of the magnetization has been reported in epitaxial $BiFeO_3$ films [24, 25], in bulk $BiFeO_3$ under application of a high magnetic field [15, 20, 26] or upon doping [27-37], indicating three possible ways to suppress the spin modulation. Spontaneous magnetization was revealed both for the $Bi_{1-x}RE_xFeO_3$ (RE= Nd, Sm, Tb) solid solutions doped by magnetically- active rare-earth ions [28-30] and for the diamagnetically- substituted $Bi_{1-x}A_xFeO_3$ (A= La, Ba, Pb, Sr) compounds [27, 31, 35-37]. It was found that value of net magnetization depends strongly on the kind of the diamagnetic substituting element. For instance, the spontaneous magnetization of the Ba- doped samples is at least an order of magnitude greater than that observed for the La- doped solid solutions at the same doping concentration [27, 31]. While the dependency between a kind of the substituting element and the magnetization value might be natural for the rare-earth substitution, which can induce additional magnetic contribution at sufficiently high concentrations (at low temperature), the reason of a correlation between the kind of A-site diamagnetic dopant and magnetic properties of samples is not quite clear.

In this work, we examine the doping route for increased ME performance and improved multiferroic properties in $BiFeO_3$ via A-site substitution. A detailed study of the defects formation mechanism as well as the crystal structure, dielectric, local ferroelectric, and magnetic properties of heterovalently- substituted $Bi_{1-x}A_xFeO_3$ (A=Ca, Sr, Pb, Ba; x=0.2, 0.3) ceramic samples will be reported.

2. Experiment

The polycrystalline samples with the nominal chemical compositions $Bi_{1-x}A_xFeO_3$ (A= Ca, Sr, Pb, Ba; x= 0.2, 0.3) were prepared by a rapid two-stage solid-state reaction method using high-purity Bi_2O_3, Fe_2O_3, $CaCO_3$, $SrCO_3$, $BaCO_3$ and PbO reagents. The compacted mixture of reagents taken in desired cation ratios was annealed at 800°C (Pb) or 850°C (Ca, Sr, Ba) for 30 min in air to obtain a single-phase perovskite structure. Final heat treatment was performed in the temperature range from 840 °C (Pb) to 990 °C (Sr) for 10 min to produce relatively dense ceramics. The maximal synthesis temperatures were close to the melting points of the solid solutions. All the compounds were quenched by removing the sample from the furnace immediately after sintering. This rapid synthesis method allowed us to prevent intensive evaporation of Bi_2O_3 and PbO and to obtain a single-phase solid solution, as was confirmed by a careful X-ray diffraction (XRD) analysis (Fig. 1). XRD patterns were collected at room temperature using a Rigaku D/MAX-B diffractometer with Cu K_α radiation. The data were analyzed by the Rietveld method using the FullProf program [38]. Mössbauer measurements were performed using a conventional constant acceleration spectrometer with a ^{57}Co-in-Rh source. The spectra were taken in transmission geometry at room temperature and were fitted using the NORMOS program [39]. The dielectric characteristics were measured in the temperature range from 80 to 400 K using a HP4194A impedance analyzer. Ferroelectric hysteresis loops were obtained with the conventional Sawyer-Tower method using a LT322 oscilloscope (LeCroy). Local ferroelectric properties of the samples were investigated with piezoresponse force microscopy (PFM). PFM measurements were performed with a commercial setup Multimode NanoScope IIIA (Veeco) equipped with a lock-in amplifier (SR- 830A, Stanford Research) and a function generator (FG-120, Yokagawa). A commercial tip-cantilever system NSG11 (NT-MDT) with a spring constant of 10 N/m and TiN-coated tip with apex radius less then 10 nm was used. Domain visualization was performed under an applied *ac* voltage with the amplitude V_{ac}= 2.5 V and frequency f= 50 kHz. Local poling was done by applying a *dc* bias of ±10 V between the tip and counter electrode followed by a consequent PFM imaging. The local piezoelectric hysteresis loops were measured inside individual grains by applying the consecutive voltage pulses and measuring the piezoelectric response as a function of the voltage. The dielectric and ferroelectric measurements were performed using cylindrical samples with a diameter (d) of 10 mm and a height (h) of 0.8 mm. Measurements of the field dependencies of the magnetization were performed with a SQUID magnetometer (MPMS-5, Quantum Design) using cylindrical sample with d= 1 mm and h= 5 mm. Ferromagnetic resonance measurements were performed with a Bruker ESP 300E spectrometer equipped with an Oxford Instruments continuous-flow helium cryostat using sample in the form of thin plate 3 mm×3 mm×0.1 mm.

3. Results and Discussion

$BiFeO_3$ is known to have a rhombohedrally distorted perovskite structure described by the space group *R3c* [40]. The rhombohedral cell is very close to the cubic one (angle α in the rhombohedral cell is about 89.4°). With respect to the cubic $Pm\bar{3}m$ structure, the rhombohedral structure is obtained by an antiphase tilt of the adjacent FeO_6 octahedra and a displacement of the ions from their centrosymmetric position along the pseudocubic $[111]_C$ axis (or corresponding $[001]_H$ axis in the hexagonal basis). The ionic displacements lead to the appearance of spontaneous polarization oriented along $[001]_H$. In accordance with the structural data obtained for the parent compound, Rietveld refinement of the X-ray powder diffraction patterns of the A-site substituted samples was performed using the same space group, and the structural model allowed us to reproduce adequately all the observed reflections. Typical results, presenting a good agreement between the observed and calculated XRD patterns are shown in Fig. 2. Unlike a number of publications on lanthanide-doped samples [27-29] (reporting substitution-induced structural phase transitions to orthorhombic or even triclinic phase), no splitting or extra reflections indicative of the necessity to reduce the lattice symmetry of synthesized ceramics were detected (Figs. 1, 2). Within the $Bi_{0.8}A_{0.2}FeO_3$ or $Bi_{0.7}A_{0.3}FeO_3$ series, unit cell volume of the samples increases in accordance with the increasing ionic radius of the substituting elements [41] (Fig. 3). Samples doped with 20 % or 30 % of the same elements have close unit cell parameters, as might be expected taking into account the influence of two opposite tendencies: A-site substitution with ions possessing larger ionic radius (with the exception of Ca^{2+}, whose ionic radius is less than that for Bi^{3+}) enlarges cell parameters; on the other hand, the same heterovalent A^{2+} substitution, which has to induce the appearance of oxygen vacancies or/and Fe^{4+} ions in the lattice, leads to the unit cell reduction.

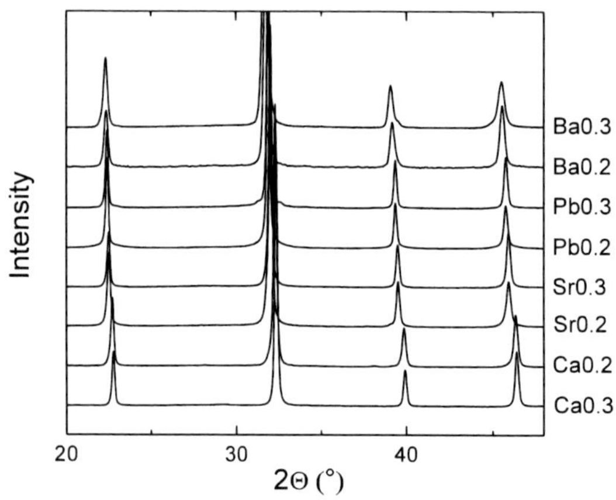

Figure 1. Experimental XRD spectra obtained for the $Bi_{1-x}A_xFeO_3$ samples at room temperature.

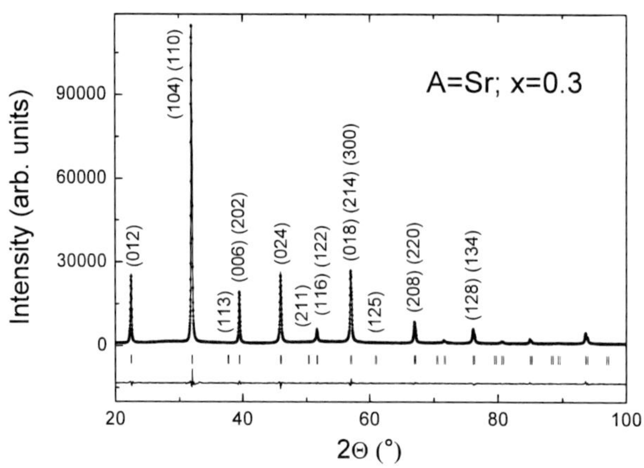

Figure 2. Observed (solid circles) and calculated (solid line) XRD patterns of the $Bi_{0.7}Sr_{0.3}FeO_3$ sample obtained at room temperature. The difference between these spectra is plotted at the bottom. Bragg reflections are indicated by ticks.

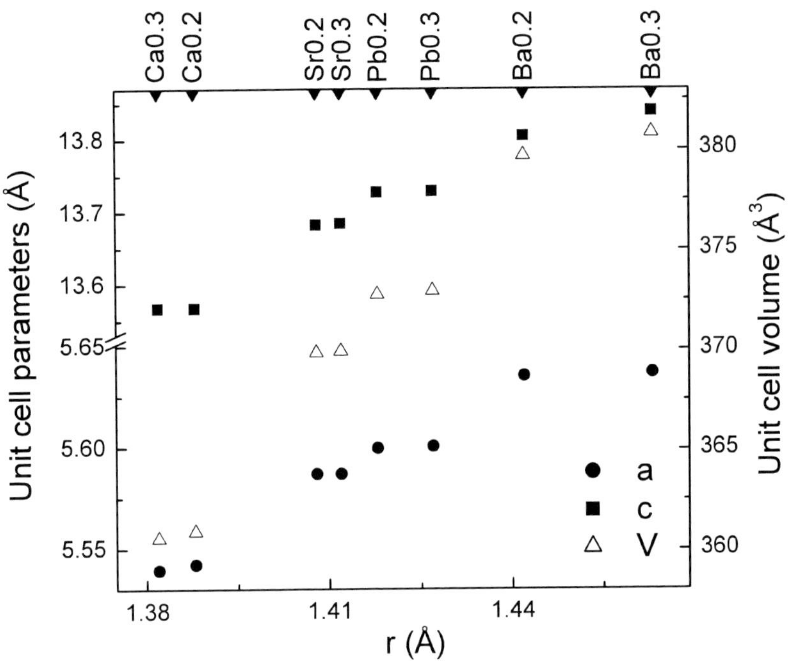

Figure 3. Room-temperature unit cell parameters of the $Bi_{1-x}A_xFeO_3$ compounds obtained by Rietveld refinement of the XRD patterns as a function of the average ionic radius of the A-site ions.

The corresponding models of the defect chemistry:

(1) $1/2(1-x)Bi_2O_3 + xAO + 1/2Fe_2O_3 \rightarrow Bi^{3+}{}_{1-x}A^{2+}{}_xFe^{3+}O^{2-}{}_{3-x/2}$

(2) $1/2(1-x)Bi_2O_3 + xAO + 1/2Fe_2O_3 \rightarrow Bi^{3+}{}_{1-x}A^{2+}{}_xFe^{3+}{}_{1-x}Fe^{4+}{}_xO^{2-}{}_3$

have been tested with Mössbauer spectroscopy measurements, performed for a number of $Bi_{1-x}A_xFeO_3$ compounds. The Mössbauer spectrum obtained for the $Bi_{0.8}Pb_{0.2}FeO_3$ sample is presented in Fig. 4. The spectrum consists of two six-line magnetic hyperfine patterns, indicating that the sample is in a magnetically ordered state. The Zeeman sextets were fitted with the hyperfine fields of H_1 = 50 T and H_2 = 46.1 T and isomer shifts of δ_1 = 0.394 mm/s and δ_2 = 0.484 mm/s (the spectral fractions of the sextets were about SF_1= 88 % and SF_2= 12 %, respectively). The values are typical of the high-spin Fe^{3+} [42]. No indication on the presence of Fe^{4+} in the lattice was detected (expected values of the Mössbauer spectrum parameters for the Fe^{4+} ions are ~ 40 T and -0.15–0.05 mm/s for hyperfine field and isomer shift, respectively [42]). Thus, the Mössbauer spectroscopy results indicate that the Pb^{2+} substitution in the $Bi_{0.8}Pb_{0.2}FeO_3$ compound is realized through the formation of oxygen vacancies (we believe that the minor spectral contribution corresponds to Fe^{3+} ions in non-octahedral (mainly, in tetrahedral) coordination, because the Mössbauer parameters for the last model are expected to be very close [42], as observed in case of our experiment). The very similar results (existence of two spectral contributions with a ratio of ~ 85: 15 and hyperfine parameters characteristic of the Fe^{3+}) were obtained for $Bi_{0.7}Ca_{0.3}FeO_3$ and $Bi_{0.7}Ba_{0.3}FeO_3$ compounds. Taking into account that the same conclusions were made on the basis of the Mössbauer spectroscopy experiments performed for $Bi_{1-x}Ca_xFeO_3$ ($x \leq 0.1$) [43] and $Bi_{1-x}Sr_xFeO_3$ ($x \leq 0.67$) [44] solid solutions, one can argue that the formation of oxygen vacancies is a general mechanism of the charge compensation in the case of the A^{2+} substitution in $Bi_{1-x}A_xFeO_3$. Accordingly, we refer below to the chemical compositions corresponding to the oxygen vacancies- related model of the defects formation. It is worth noting that neither spectral asymmetry no line broadenings peculiar to the spiral magnetic structure of pure $BiFeO_3$ [45, 46] has been found in the Mössbauer spectroscopy measurement performed for the $Bi_{0.8}Pb_{0.2}FeO_{2.9}$ sample (Fig. 4). This clearly indicates that the magnetic spiral is suppressed in the Pb- substituted compound.

The presence of the oxygen vacancies in the lattice of the $Bi_{1-x}A_xFeO_{3-x/2}$ perovskites manifests itself in the frequency/temperature dependencies of the dielectric permittivity and loss factor (Fig. 5) where both of these quantities sharply increase with decreasing frequency or increasing temperature. The behavior can be qualitatively explained in the following way [29, 47]: the oxygen vacancies- related dipoles are able to follow the alternating field at low frequencies, providing high values of ε', but begin to lag behind the field with increasing frequency, giving $\varepsilon'_\infty \sim 100$ (at room temperature), which is comparable with that observed in pure $BiFeO_3$ and isovalently substituted $Bi_{1-x}RE_xFcO_3$ ceramics [28, 29]. The increase in the apparent dielectric permittivity and loss factor with increasing temperature is then related to the thermally induced enhancement of the conductivity contribution. The values of the dc conductivity, σ_{dc}, obtained from the slope of ε'' $(1/v)$ dependencies in accordance with the relation $\sigma' = 2\pi v \varepsilon_0 \varepsilon''$ (here, σ' is real part of conductivity, ε'' is an imaginary part of dielectric permittivity, and v is a frequency) follow Arrhenius law given by $\sigma_{dc} = \sigma_0\ exp[-E_\sigma/(k_B T)]$ with activation energy $E_\sigma \sim 0.65$ eV, which is quite typical for oxygen vacancies- related conductivity [48]. Example of the measurements performed for $Bi_{0.8}Pb_{0.2}FeO_{2.9}$ sample is shown in Fig. 6.

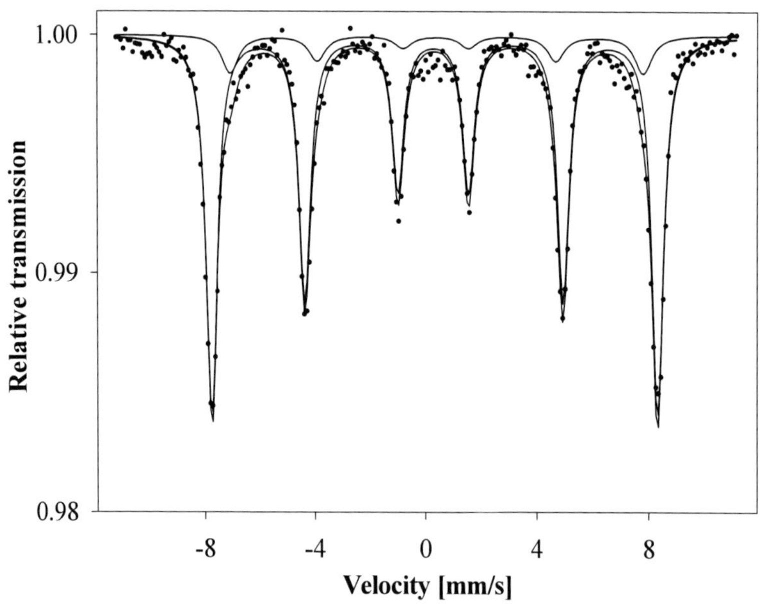

Figure 4. Mössbauer spectrum for the $Bi_{0.8}Pb_{0.2}FeO_3$ sample at room temperature.

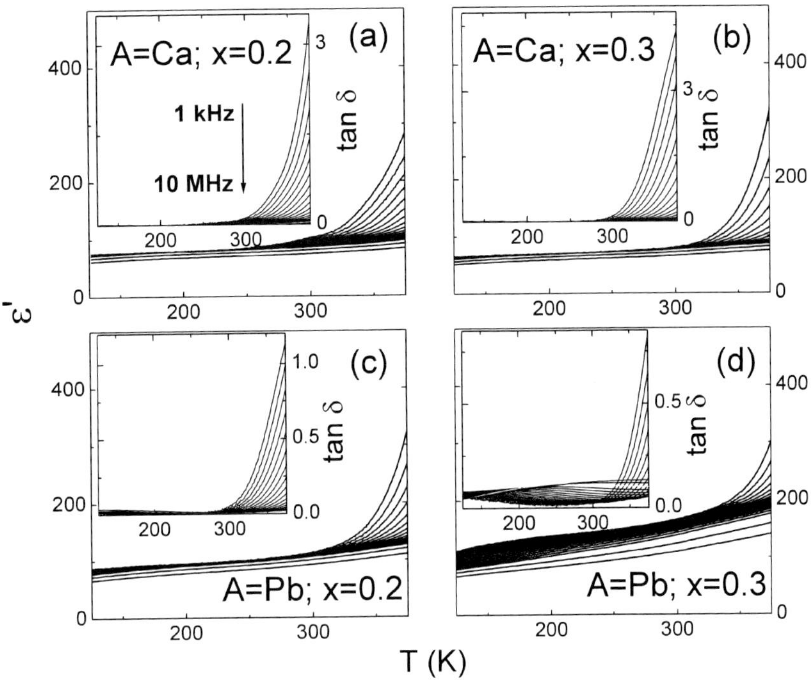

Figure 5. Temperature dependencies of the dielectric constant and dielectric loss (in insets) for a number of $Bi_{1-x}A_xFeO_{3-x/2}$ samples at different frequencies.

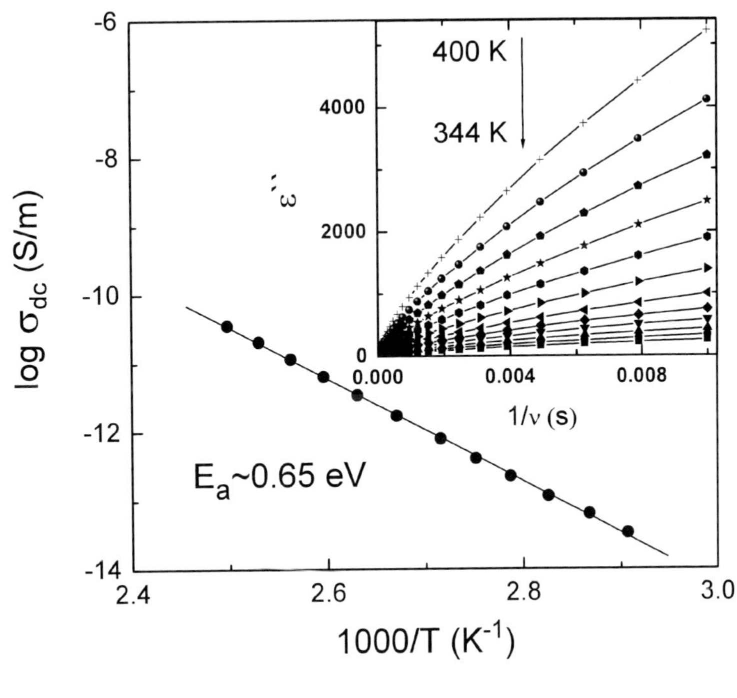

Figure 6. The Arrhenius plot of the dc conductivity for the $Bi_{0.8}Pb_{0.2}FeO_{2.9}$ sample. Symbols are the experimental points and solid line is the least-square straight-line fit. Inset shows ε'' (1/ν) dependencies at different temperatures.

The presence of the oxygen vacancies causes decrease of the dielectric breakdown field to about 100 kV/cm, thus hampering possibility to pole the samples (which are known to possess very high coercive fields [49]) and to get the ferroelectric polarization loops, so we have used the piezoresponse force microscopy technique (see, e. g. [50] and references therein) for the visualization of ferroelectric domain structure in the $Bi_{1-x}A_xFeO_{3-x/2}$ ceramics to confirm the existence of spontaneous polarization in the compounds at room temperature. The PFM method is based on the detection of thickness oscillations of ferroelectric materials under *ac* voltage applied between the conductive scanning probe microscope tip and the bottom electrode. The amplitude of the vibration signal A provides information on the magnitude of the effective piezoelectric coefficient, while the phase signal φ determines the polarization direction. Domains with oppositely oriented polarization are distinguished by different contrast (proportional to $A\ cos\varphi$). In our experiments, the bright and dark contrasts correspond to domains with polarization vector directed to the free surface of the ceramics and to the bulk, respectively. The example of these measurements performed for a number of the $Bi_{1-x}A_xFeO_{3-x/2}$ samples is shown in Fig. 7. It is seen that the areas with intermediate contrast corresponding to regions with defect structure suppressing any PFM signal coexist with areas exhibiting clear piezoresponse corresponding to antiparallel domains (Fig. 7 (a, b)). The piezoelectric contrast appears also after scanning with a *dc* voltage applied to the tip (Fig. 7 (c, d)), giving clear proof that the spontaneous polarization can be switched upon the application of electric field. In spite of the appreciable roughness of the surfaces, masking the contribution of the piezoelectric deformation to the PFM contrast, only a weak correlation has been found between the topography and piezoresponse (Fig. 8), pointing to the weak contribution of the electrostatic signal [50]. The existence of spontaneous ferroelectric polarization in $Bi_{0.8}A_{0.2}FeO_{2.9}$ ceramics was also confirmed by local piezoresponse hysteresis loop measurements (Fig. 9) in which the application of consecutive voltage pulses causes clear polarization reversal as judged from the measured local piezoelectric signal (for $Bi_{0.7}A_{0.3}FeO_{2.85}$ samples, the experiment was not carried out so far). It is worth noting that, in comparing with the $Bi_{0.8}A_{0.2}FeO_{2.9}$ compounds, notable reduction of the amplitude of the measured vibrations as well as decrease of the areas exhibiting clear piezoresponse are observed for the $Bi_{0.7}A_{0.3}FeO_{2.85}$ samples. This means that the regular polar displacements caused by stereochemical activity of the $6s^2$ lone pair of Bi^{3+} ions and giving rise to ferroelectricity in the $BiFeO_3$- based solids are suppressed with increasing the A-site dopants concentration (or/and with increasing oxygen vacancies content).

Room- temperature measurements of the magnetization of the $Bi_{1-x}A_xFeO_{3-x/2}$ compounds as a function of the applied magnetic field are shown in Fig. 10. As in the case of pure $BiFeO_3$ [51, 52], the M (H) dependencies of Ca^{2+}- substituted compounds and $Bi_{0.8}Sr_{0.2}FeO_{2.9}$ solid solution has a linear character without any spontaneous magnetization, thus confirming antiferromagnetic nature of the samples. Weak spontaneous magnetization of 0.08 emu/g appears in $Bi_{0.7}Sr_{0.3}FeO_{2.85}$ and $Bi_{0.8}Pb_{0.2}FeO_{2.9}$ compounds. Further increase of magnetization value is observed for $Bi_{0.7}Pb_{0.3}FeO_{2.85}$ and $Bi_{0.8}Ba_{0.2}FeO_{2.9}$ solid solutions. Maximal spontaneous magnetization of 1.2 emu/g (~0.06 μ_B per formula unit) has been obtained for the sample doped with 30 % of Ba^{2+} ions. Thus, investigation of magnetic properties of $Bi_{1-x}A_xFeO_{3-x/2}$ compounds reveals clear correlation between the ionic radius of substituting element and the value of spontaneous magnetization of corresponding solid solution (inset in Fig. 10).

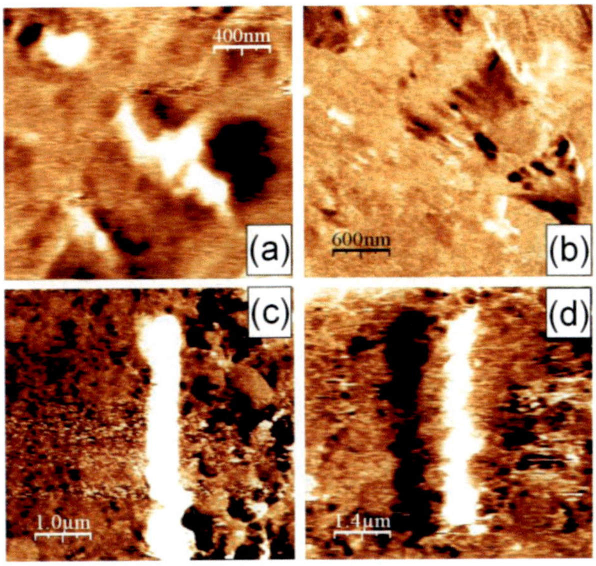

Figure 7. Piezoresponse force microscopy measurements of the $Bi_{1-x}A_xFeO_{3-x/2}$ compounds. Piezoelectric contrast for the $Bi_{0.8}Sr_{0.2}FeO_{2.9}$ (a) and $Bi_{0.7}Ca_{0.3}FeO_{2.85}$ (b) samples; electric-field-induced PFM contrast for the $Bi_{0.8}Pb_{0.2}FeO_{2.9}$ (c) $Bi_{0.7}Ca_{0.3}FeO_{2.85}$ (d) samples.

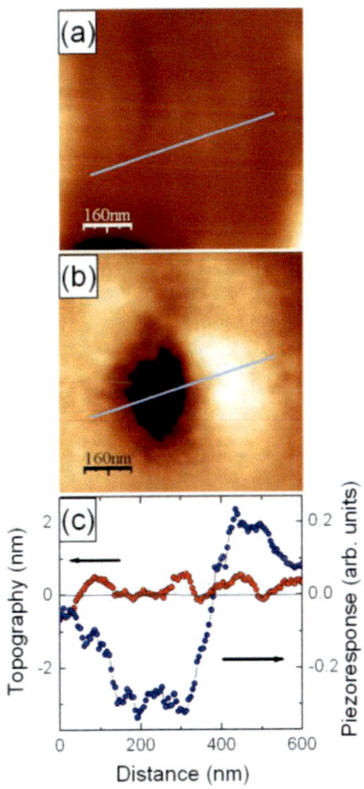

Figure 8. Scanning probe microscopy measurements of $Bi_{0.7}Ca_{0.3}FeO_{2.85}$ ceramics: (a) topography; (b) PFM image demonstrating ferroelectric domains; (c) cross-section of the PFM image showing absence of correlation between the topography and piezoresponse.

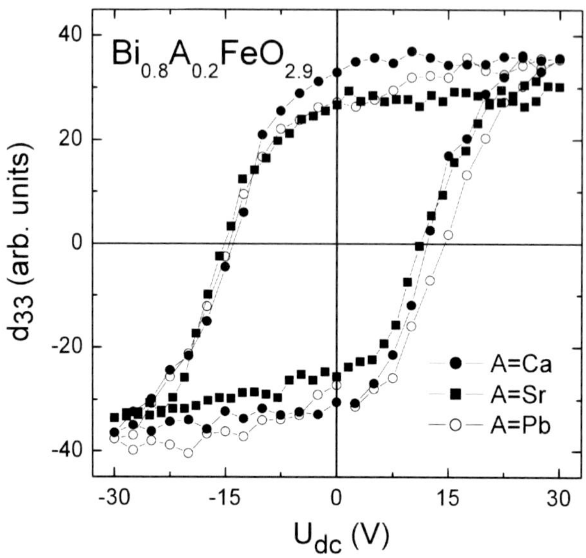

Figure 9. Local piezoresponse hysteresis loops for a number of $Bi_{0.8}A_{0.2}FeO_{2.9}$ samples.

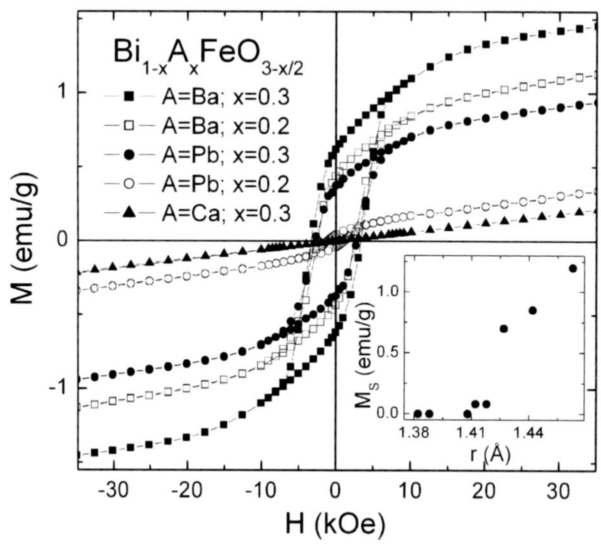

Figure 10. Field dependences of the magnetization for a number of $Bi_{1-x}A_xFeO_{3-x/2}$ samples at room temperature. Inset shows spontaneous magnetization of $Bi_{1-x}A_xFeO_{3-x/2}$ compounds as a function of the average ionic radius of the A-site ions.

The magnetically active iron sublattice entirely determines magnetic properties of diamagnetically substituted $Bi_{1-x}A_xFeO_{3-x/2}$ compounds. Since superexchange interaction Fe^{3+}-O-Fe^{3+} is antiferromagnetic [53], the only intrinsic reason of the appearance of spontaneous magnetization in the $Bi_{1-x}A_xFeO_{3-x/2}$ is a canting of the antiferromagnetic sublattices (weak ferromagnetism) [22, 23], permitted by the *R3c* symmetry if the magnetic moments of Fe^{3+} ions are oriented perpendicular to the $[001]_H$ axis [21]. The net magnetization originating from the canting can naturally exist only if the spin modulation peculiar to $BiFeO_3$ is suppressed [21]. The commonalities of the spontaneous magnetization appearance suggest that the transition from the spiral spin structure to the homogeneous magnetic state in $BiFeO_3$- based compounds can be induced by the application of high magnetic fields [15, 20, 26], epitaxial constraint [24, 25], or doping [27-37]. For all the cases, the theoretical description of the transition has been proposed based on Landau-Ginzburg formalism for the free energy [15, 20, 25, 26, 34]. However, the presented thermodynamic model seems to be unable to describe all the existing experimental data. For instance, this model, which takes into account a magnetoelectric coupling, suggests that doping- induced destruction of the spiral spin structure results in a significant enhancement of the spontaneous polarization of the sample [34]. The opposite is experimentally observed: dilution of Bi sublattice in $Bi_{1-x}Ln_xFeO_3$ solids decreases naturally both the ferroelectric Curie temperature and the spontaneous polarization value, the magnetization being increased [28, 49]. A dramatic increase in the electric-field-induced polarization of $BiFeO_3$-$PbTiO_3$-based solid solutions (P_r~ 30 μC/cm^2) in comparison with pure $BiFeO_3$ crystal (P_r is an order of magnitude lower) was probably the experimental basis of the model [34]. In reality, the spontaneous polarization of single crystal $BiFeO_3$ is close to the theoretically predicted value of 90 μC/cm^2 [54] and seems to be mainly determined by the quality of the material [45, 55]. Similar serious contradictions exist in the theoretical description adapted to the case of epitaxial films of $BiFeO_3$ [25], whose polarization and magnetization are not very sensitive to strain [55-58], while the theoretical model suggests their strong dependence. Thus, the proposed theoretical approach should be reconsidered.

The absence of net magnetization in $Bi_{1-x}Ca_xFeO_{3-x/2}$ (x= 0.2, 0.3) and $Bi_{0.8}Sr_{0.2}FeO_{2.9}$ multiferroics indicates that the existence of oxygen vacancies in the lattice can not be also considered as the main factor which favors to the suppression of the spin modulation. This agrees well with the theoretical calculations which showed that the canting of the magnetic moments was not significantly affected by the presence of the oxygen vacancies, so that no increase in the macroscopic magnetization due to oxygen vacancies could be found [56, 59]. The introduction of A-site impurities has to result in the variation of the anisotropy constant and thus provokes the transition from a spatially modulated spin structure to a homogeneously canted one. The observed correlation between the ionic radius of substituting element and magnetic properties of the $Bi_{1-x}A_xFeO_{3-x/2}$ (A= Ca, Sr, Pb, Ba; x= 0.2, 0.3) solid solutions, when spontaneous magnetization appears in compounds doped by the biggest ions (inset in Fig. 10), can reflect the capability of the substituting elements to have effective influence on the parameters controlling the magnetic anisotropy in the spin magnets (which is known to be determined by the joint action of the anisotropic crystal field and spin-orbit interaction [60]) via appropriate structural distortions, thus inducing the transition from a spiral antiferromagnetic spin structure to a weak ferromagnetic state. The variation of the anisotropy is actually comfirmed by the ferromagnetic resonance experiment performed on $Bi_{0.8}Pb_{0.2}FeO_{2.9}$ ceramics (Fig. 11). The *g*-factor, calculated in accordance with the relation $g = h\nu/(H_{eff}\mu_B)$, where h is Planck's constant, ν the resonance frequency, H_{eff} the effective magnetic field, and μ_B the Bohr magneton, is far from the value expected for spin magnets (g = 2.0023) and equal to 2.49. The result conforms to the increase of the *g*- factor above the spin-only value peculiar to $BiFeO_3$ and the simultaneous appearance of spontaneous magnetization in La-doped $BiFeO_3$ [61].

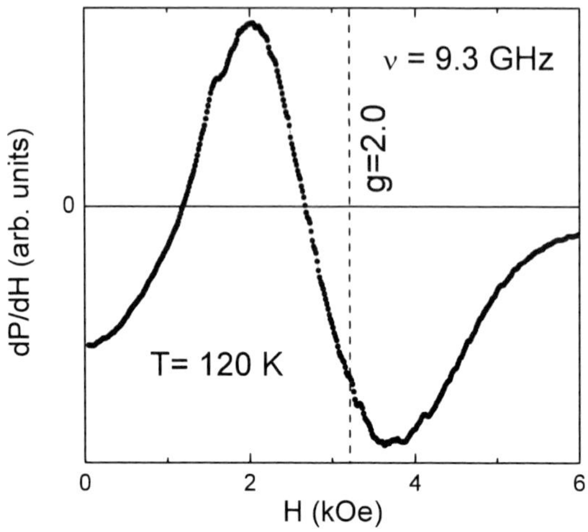

Figure 11. FMR spectrum of the $Bi_{0.8}Pb_{0.2}FeO_{2.9}$ compound at T = 120 K. The vertical line corresponds to the calculated position for $g = 2$.

4. Conclusion

The effect of A- site diamagnetic substitution on the crystal structure, dielectric, magnetic and local ferroelectric properties of the $BiFeO_3$ multiferroic perovskite has been studied. It has been found that the heterovalent A^{2+} substitution is realized through the formation of the oxygen vacancies. The $Bi_{1-x}A_xFeO_{3-x/2}$ (A= Ca, Sr, Pb, Ba; x= 0.2, 0.3) compounds have been shown to be single-phase perovskites and their crystal structure can be described within the space group *R3c*. Signs of existence of the ferroelectric polarization in the samples at room temperature has been revealed with the piezoresponse force microscopy. A correlation between the ionic radius of the substituting element and the value of the spontaneous magnetization of a corresponding solid solution has been found. Experimental results suggest that the diamagnetic A-site substitution with the biggest ionic radius ions leads to the effective suppression of the spiral spin configuration, thus resulting in the appearance of a net magnetization.

References

[1] M. Fiebig, J. Phys. D: Appl. Phys. 38 R123 (2005).
[2] W. Eerenstein, N. D. Mathur, and J. F. Scott, Nature, 442, 759 (2006).
[3] S.-W. Cheong and M. Mostovoy, Nature Mater. 6, 13 (2007).
[4] R. Ramesh and N. A. Spaldin, Nature Mater. 6, 21 (2007).
[5] Y.-H. Chu, L. W. Martin, M. B. Holcomb, and R. Ramesh, Mater. Today 10, 16 (2007).
[6] H. Bea, M. Bibes, S. Cherifi, F. Nolting, B. Warot-Fonrose, S. Fusil, G. Herranz, C. Deranlot, E. Jacquet, K. Bouzehouane, and A. Barthélémy, Appl. Phys. Lett. 89, 242114 (2006).
[7] L. W. Martin, Y.-H. Chu, Q. Zhan, R. Ramesh, S.-J. Han, S. X. Wang, M. Warusawithana, and D. G. Schlom, Appl. Phys. Lett. 91, 172513 (2007).

[8] M. Fiebig, D. Fröhlich, K. Kohn, St. Leute, Th. Lottermoser, V. V. Pavlov, and R. V. Pisarev, Phys. Rev. Lett. 84, 5620 (2000).
[9] A. V. Kimel, R. V. Pisarev, F. Bentivegna, and Th. Rasing, Phys. Rev. B 64, 201103 (2001).
[10] T. Katsufuji, M. Masaki, A. Machida, M. Moritomo, K. Kato, E. Nishibori, M. Takata, M. Sakata, K. Ohoyama, K. Kitazawa, and H. Takagi, Phys. Rev. B 66, 134434 (2002).
[11] T. Kimura, S. Kawamoto, I. Yamada, M. Azuma, M. Takano, and Y. Tokura, Phys. Rev. B 67, 180401 (2003).
[12] C.-H. Yang, J. Koo, C. Song, T. Y. Koo, K.-B. Lee, and Y. H. Jeong, Phys. Rev. B 73, 224112 (2006).
[13] J. M. Moreau, C. Michel, R. Gerson, and W. J. James, J. Phys. Chem. Solids 32, 1315 (1971).
[14] I. Sosnowska, T. Peterlin-Neumaier, and E. Steichele, J. Phys. C 15, 4835 (1982).
[15] A. M. Kadomtseva, Y. F. Popov, G. P. Vorobev, and A. K. Zvezdin, Physica B 211, 327 (1995).
[16] P. Baettig and N.A. Spaldin, Appl. Phys. Lett. 86, 012505 (2005).
[17] R. Nechache, C. Harnagea, A. Pignolet, F. Normandin, T. Veres, L.-P. Carignan, and D. Ménard, Appl. Phys. Lett. 89, 102902 (2006).
[18] M. Sakai, A. Masuno, D. Kan, M. Hashisaka, K. Takata, M. Azuma, M. Takano, and Y. Shimakawa, Appl. Phys. Lett., 90, 072903 (2007).
[19] M. Azuma, K. Takata, T. Saito, S. Ishiwata, Y. Shimakawa, and M. Takano, J. Am. Chem. Soc. 127, 8889 (2005).
[20] A. M. Kadomtseva, A. K. Zvezdin, Yu. F. Popov, A. P. Pyatakov, and G. P. Vorob'ev, JETP Lett. 79, 0571 (2004).
[21] C. Ederer and N. A. Spaldin, Phys. Rev. B 71, 060401(R) (2005).
[22] I. Dzyaloshinsky, J. Phys. Chem. Solids 4, 241 (1958).
[23] T. Moriya, Phys. Rev. 120, 91 (1960).
[24] J. Wang, J. B. Neaton, H. Zheng, V. Nagarajan, S. B. Ogale, B. Liu, D. Viehland, V. Vaithyanathan, D. G. Schlom, U. V. Waghmare, N. A. Spaldin, K. M. Rabe, M. Wuttig, R. Ramesh, Science 299, 1719 (2003).
[25] F. Bai, J. Wang, M. Wutting, J. F. Li, N. Wang, A. P. Pyatakov, A. K. Zvezdin, L. E. Cross, and D. Viehland, Appl. Phys. Lett. 86, 032511 (2005).
[26] B. Ruette, S. Zvyagin, A. P. Pyatakov, A. Bush, J.F. Li, V.I. Belotelov, A. K. Zvezdin, and D. Viehland, Phys. Rev. B 69, 064114 (2004).
[27] S.-T. Zhang, Y. Zhang, M.-H. Lu, C.-L. Du, Y.-F. Chen, Z.-G. Liu, Y.-Y. Zhu, N.-B. Ming, and X. Q. Pan, Appl. Phys. Lett. 88, 162901 (2006).
[28] G. L. Yuan, S. W. Or, J. M. Liu, and Z. G. Liu, Appl. Phys. Lett. 89, 052905 (2006).
[29] G. L. Yuan and S. W. Or, J. Appl. Phys. 100, 024109 (2006).
[30] V. R. Palkar, D. C. Kundaliya, S. K. Malik, and S. Bhattacharya, Phys. Rev. B 69, 212102 (2004).
[31] D. H. Wang, W. C. Goh, M. Ning, and C. K. Ong, Appl. Phys. Lett. 88, 212907 (2006).
[32] J. S. Kim, C. I. Cheon, Y. N. Choi, P. W. Jang, J. Appl. Phys. 93, 9263 (2003).
[33] J. S. Kim, C. I. Cheon, C. H. Lee, P. W. Jang, J. Appl. Phys. 96, 468 (2004).
[34] N. Wang, J. Cheng, A. Pyatakov, A. K. Zvezdin, J. F. Li, L. E. Cross, and D. Viehland, Phys. Rev. B 72, 104434 (2005).

[35] V. A. Khomchenko, D. A. Kiselev, J. M. Vieira, A. L. Kholkin, M. A. Sá, and Y. G. Pogorelov, Appl. Phys. Lett. 90, 242901 (2007).
[36] V. A. Khomchenko, D. A. Kiselev, J. M. Vieira, Li Jian, A. L. Kholkin, A. M. L. Lopes, Y. G. Pogorelov, J. P. Araujo, and M. Maglione, J. Appl. Phys. 103, 024105 (2008).
[37] B. Kundys, A. Maignan, C. Martin, N. Nguyen, and C. Simon, Appl. Phys. Lett. 92, 112905 (2008).
[38] J. Rodríguez-Carvajal, Physica B 192, 55 (1993).
[39] R. A. Brand, J. Lauer, and D. M. Herlach, J. Phys. F: Met. Phys. 13, 675 (1983).
[40] C. Michel, J.-M. Moreau, G. D. Achenbach, R. Gerson, and W. J. James, Solid State Commun. 7, 701 (1969).
[41] R. D. Shannon, Acta Cryst. A 32, 751 (1976).
[42] N. N. Greenwood and T. C. Gibb, Mössbauer Spectroscopy, Chapman and Hall, London, 1971.
[43] D. Kothari, V. R. Reddy, A. Gupta, V. Sathe, A. Banerjee, S. M. Gupta, and A. M. Awasthi, Appl. Phys. Lett. 91, 202505 (2007).
[44] J. Li, Y. Duan, H. He, D. Song, J. Alloys Comp. 315, 259 (2001).
[45] D. Lebeugle, D. Colson, A. Forget, M. Viret, P. Bonville, J. F. Marucco, and S. Fusil, Phys. Rev. B 76, 024116 (2007).
[46] A. Palewicz, T. Szumiata, R. Przeniosło, I. Sosnowska, I. Margiolaki, Solid State Commun. 140, 359 (2006).
[47] A. Dutta and T. P. Sinha, J. Phys. Chem. Solids 67, 1484 (2006).
[48] B. Sundarakannan, K. Kakimoto, and H. Ohsato, J. Appl. Phys. 94, 5182 (2003).
[49] H. Uchida, R. Ueno, H. Funakubo, and S. Koda, J. Appl. Phys. 100, 014106 (2006).
[50] A. L. Kholkin, S. V. Kalinin, A. Roelofs, A. Gruverman, "Review of ferroelectric domain imaging by Piezoresponse Force Microscopy", in "Scanning Probe Microscopy: Electrical and Electromechanical Phenomena at the Nanoscale", Eds. S. Kalinin, A. Gruverman, V. 1, pp. 173-214, Springer, 2006.
[51] Q.-H. Jiang, C.-W. Nan, and Z.-J. Shen, J. Am. Ceram. Soc., 89, 2123 (2006).
[52] F. Chen, Q. F. Zhang, J. H. Li, Y. J. Qi, C. J. Lu, X. B. Chen, X. M. Ren, and Y. Zhao, Appl. Phys. Lett. 89, 092910 (2006).
[53] J. B. Goodenough, A. Wold, R. J. Arnott, and N. Menyuk, Phys. Rev. 124, 373 (1961).
[54] J. B. Neaton, C. Ederer, U. V. Waghmare, N. A. Spaldin, and K. M. Rabe, Phys. Rev. B 71, 014113 (2005).
[55] V. V. Shvartsman, W. Kleemann, R. Haumont, and J. Kreisel, Appl. Phys. Lett. 90, 172115 (2007).
[56] C. Ederer and N. A. Spaldin, Phys. Rev. B 71, 224103 (2005).
[57] W. Eerenstein, F. D. Morrison, J. Dho, M. G. Blamire, J. F. Scott, and N. D. Mathur, Science 307, 1203a (2005).
[58] H. Béa, M. Bibes, S. Fusil, K. Bouzehouane, E. Jacquet, K. Rode, P. Bencok, and A. Barthélémy, Phys. Rev. B 74, 020101R (2006).
[59] C. Ederer and N. A. Spaldin, Curr. Opin. Solid St. M. 9, 128 (2005).
[60] S. V. Vonsovski, Magnetism, Wiley, New York, 1974.
[61] S. R. Das, R. N. P. Choudhary, P. Bhattacharya, R. S. Katiyar, P. Dutta, A. Manivannan, and M. S. Seehra, J. Appl. Phys. 101, 034104 (2007).

In: Bulk Materials: Research, Technology and Applications ISBN: 978-1-60692-963-6
Editors: T. Frias, V. Maestas, pp. 349-392

Chapter 9

UNUSUAL PROPERTIES OF LASER-PROCESSED STRONTIUM RUTHENATES

***Armen M. Gulian*[a] *and Vahan R. Nikoghosyan*[b]**
[a]Physics Art Frontiers, Ashton, MD 20861, USA
[b]Physics Research Institute, National Academy of Sciences, Ashtarak 0203, Armenia

ABSTRACT

In 2005, with a group of colleagues, we reported original results obtained in experimenting with laser-processed high-quality Sr_2RuO_4 crystals. These results have been treated as evidence of high-temperature superconductivity at temperatures as high as 200-250K (*cond-mat/0509313*). Both four-probe resistive transition and diamagnetic response have been reported. At the next stage (*arXiv:0705.0641*) resistive transition was documented simultaneously via four- and two-probe techniques; in the same publication it was reported that the composition of samples is more complex than it originally had been conceived: in addition to *Sr*, *Ru* and *O* it contained *Fe*, *Ni*, and *Cr*. This fact, at the time of its finding (2006) was very disappointing since the ferromagnetic compositional elements were regarded not in favor of superconductivity. The discovery of iron-based superconducting materials and the raising of their critical temperatures up to *55K* within a year or so have been quite encouraging. Moreover, in one of the iron-based compositions (*arXiv:0904.0772*) three of the four components are *Sr*, *Ru* and *Fe*! In consideration of the cross-analysis between these results it is suggestive to pursue the ways of reaching the room-temperature limit of T_c. To facilitate it, we summarize here our findings.

PREFACE[1]

Because the background of this work is critical for understanding important aspects of its content, I shall begin with its history. The story began in early 2000 when DARPA announced its famous QuIST program. At that time I was working as a consultant to the Space Science Division at the Naval Research Laboratory developing cryogenic detectors for

[1] By A.M. Gulian

future NASA missions, and our program was thirsty for funding. Any incoming announcements attracted a lot of attention. So with my NRL colleagues I decided to attend the QuIST's organizational meeting. It was a unique gathering where everybody was excitedly speaking about their lovely qubits, sharing opinions, and so on. I felt guilty that I had arrived at this meeting without any concept in hands, but this feeling did not last long. The atmosphere at the meeting was so creative that an idea came to me while standing at the long registration line. "Triplet superconductivity! Doesn't it come close to QuIST's requirements? It could be perfect!"

Readers these days are probably very familiar with the ideas of Quantum Computing (QC). Generally, QC requires basic construction blocks similar to the famous Schroedinger's cat, which can be in a superposition of at least two distinct states (*i.e.,* dead and alive, in the case of the cat). In simpler scientific terms, a quantum superposition between two states: $|0>$ and $|1>$ is required. Suppose that you have a vector, which may be in two distinct states: up and down. If this vector is a physical object which allows quantum superposition, then it can be a good embodiment for "Schroedinger's cat". Superconductors in general are macroscopic quantum objects, and *triplet* superconductors have a vectorial order parameter, which can have exactly two opposite directions.

With these distinctive features, triplet superconductors are close relatives of superfluid ^3He – a remarkable object which I explored many years ago [1,2] during a doctorate research in superconductivity at Prof. V.L. Ginzburg's group in Moscow. This superfluid requires a milli-Kelvin range of temperature to occur. Solid-state superconductors are much more preferable for QC if they can deliver similar features. As shown in Fig.1, the triplet

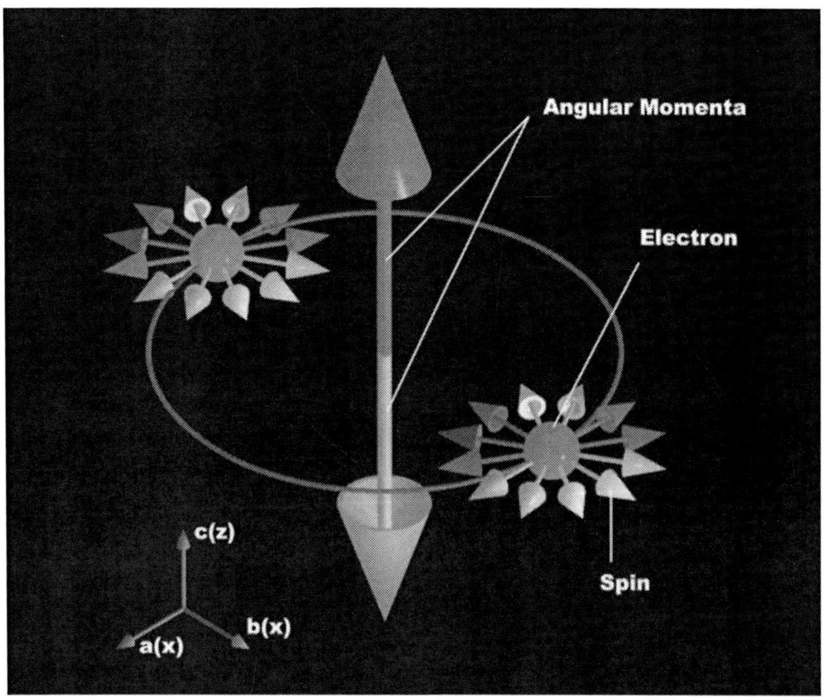

Figure 1. Superposition of Cooper-pair states with opposite angular momenta in triplet superconductors can yield "Schroedinger's cat" physics.

superconductors can do exactly that [3]. The "only" problem is that they have never been successfully produced in thin film form. This technical problem is a direct corollary of the existence of the orbital motion of electrons in the Cooper pairs: by definition, triplet superconductors posses Cooper-pairs with a wave-function corresponding to the spin $S=1$ (according to the quantum mechanical rule the number of projections $(2S+1)$ is then equal to *3*, originating the word "triplet", though in the case of planar anisotropy only *2* projections, *-1* and *+1*, are allowed, as in the case of the *A*-phase of liquid ^3He and the crystalline Sr_2RuO_4, which we will discuss below). The total (spin-orbital) wave function of Cooper pairs should be anti-symmetric at the permutation of electrons. Since $S=1$ corresponds obviously to a symmetric spin state, the orbital function should be in an anti-symmetric state, *e.g., L=1* in the simplest case, which corresponds to the macroscopic orbital motion of electrons. Very well, but then such a state is vulnerable to the presence of lattice imperfections, such as impurities, dislocations, and so on. These imperfections can be characterized in the normal state by the residual resistivity ratio (RRR). Despite great efforts (see, *e.g.,* [4-7]) the best films have not yet reached the superconductivity RRR-threshold, which has only been achieved in perfect single crystals. Thus they did not yet superconduct, and are useless for the *QC*-tasks.

But why are the films required at all? The matter is that in triplet superconductors the homogeneous states are not favorable from the thermodynamic point of view. As in ferromagnets, the domains are favorable. Small, micron-size objects are most likely required for "clean", single-domain explorations. Thin film technology is the most straightforward way of solving this task. To overcome the hurdle of its absence we have elaborated a trick [3]: to create "micron-size" mesas from good-quality single-crystals and peel them off (as explained in Fig. 2).

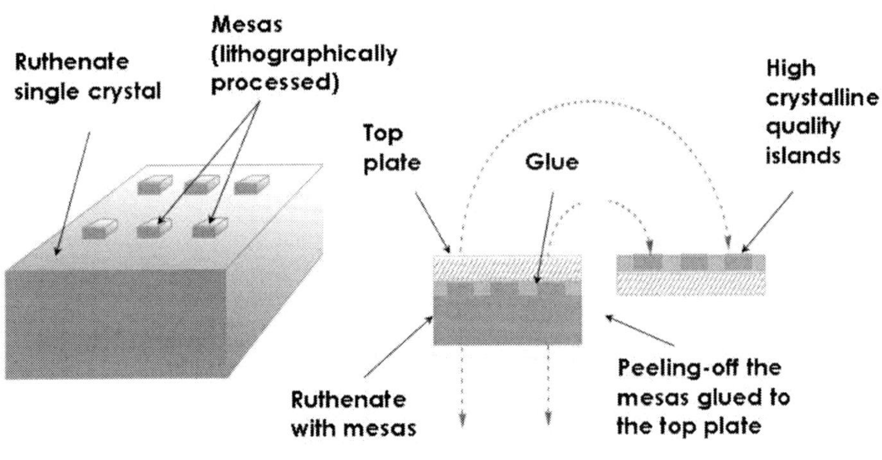

Figure 2. The trick to deriving thin film-type structures starting from bulk single crystals.

An argument in favor of this method is that the best known triplet superconductor is Sr_2RuO_4, which is a strongly anisotropic, layered material. The anisotropy makes our "peeling" technology quite appealing.

So we contacted Dr. Yoshi Maeno (who with his students discovered [8] triplet superconductivity in Sr_2RuO_4), introduced our approach, and asked for help. He was very supportive and sent us well-characterized high-quality crystals. We got some funding for this research from ONR and the work started, though success was not immediate. The strength of any of the experimented glues was not enough: the glued surfaces were disjoining instead of mesas coming off. Eventually we succeeded with the peeling off approach (Fig. 3), though we have not yet published the results.

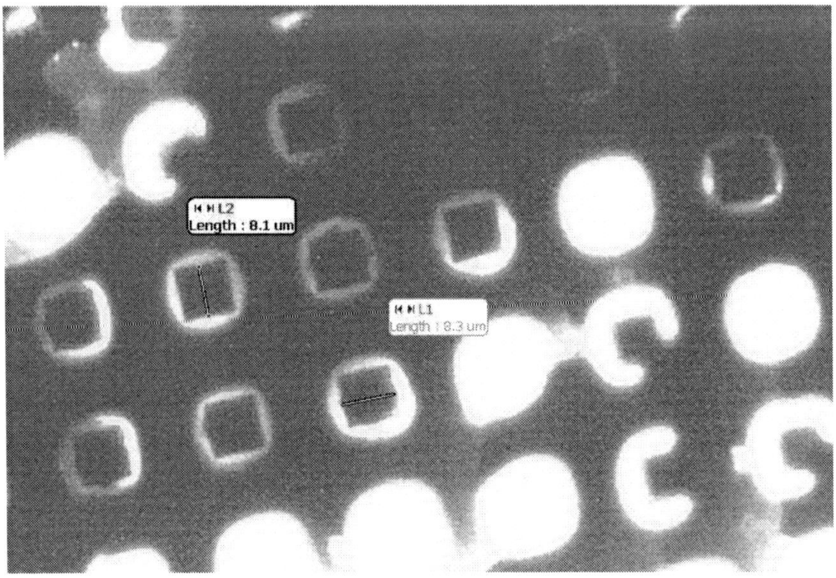

Figure. 3. Successful results of experimentation with Sr_2RuO_4 mesas. Each island is less than 1 micron thick.

But before that success, to back up our project, I started thinking of using laser micromachining and lithography to manufacture a small object testable for superconductivity by resistive transition. Happily, within the auspices of NATO's Science for Peace Program, our Armenian colleagues, supervised by Dr. J. Horwitz (DoE), developed and successfully tested this kind of machine. So I drafted a processing scheme, which we called "X-geometry" (Fig. 4), and sent it to Vahan Nikoghosyan, a friend of mine, who routinely used that machine. He agreed to help us out. The plans were to use the laser micromachining and make samples as small (and thin) as possible, to check that superconductivity is still alive in the processed objects, and then to experiment for a single domain "Schroedinger cat" test. In case the smaller sizes would be required, we were planning to use the nanotechnology instruments, like ion knives, which can go beyond the sizes available in optics. These plans have changed drastically after the first experiments.

Here is what has happened. I received the first sample in 2003 and started measurements in a ^3He-cryostat with a goal to cool it down to $0.3K$ or so. When I finished putting the liquid nitrogen into the two chambers of the cryostat and checked the integrity of the 4-probe circuits (sometimes, cooling can take the leads off the sample), I was very discouraged: the resistance of the sample looked like zero. Needless to say, a superconducting transition as a possible reason was the last in the list. I told myself that, although unlikely, perhaps some wires inside the cryostat were touching each other and I should check them out. So I dumped the liquid nitrogen out with intent of opening the cryostat after it warmed up. To obtain more

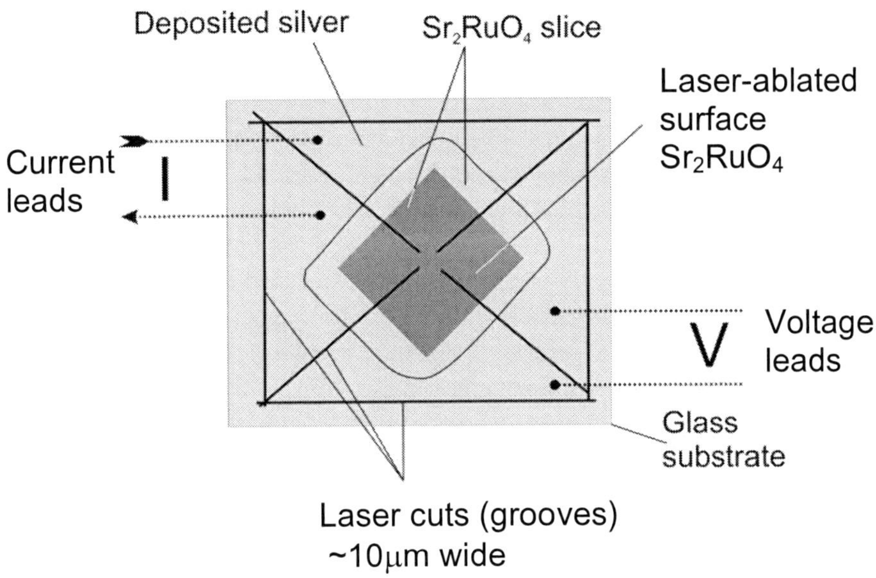

Figure 4. Sample design in the X-geometry (not to scale).

information, I switched on the resistivity measurement circuit, with the data recording under the LabVIEW, and let it warm throughout the weekend. When I had returned, I was shocked: the computer had recorded a transition-type curve to the initial normal resistance (Fig. 5).

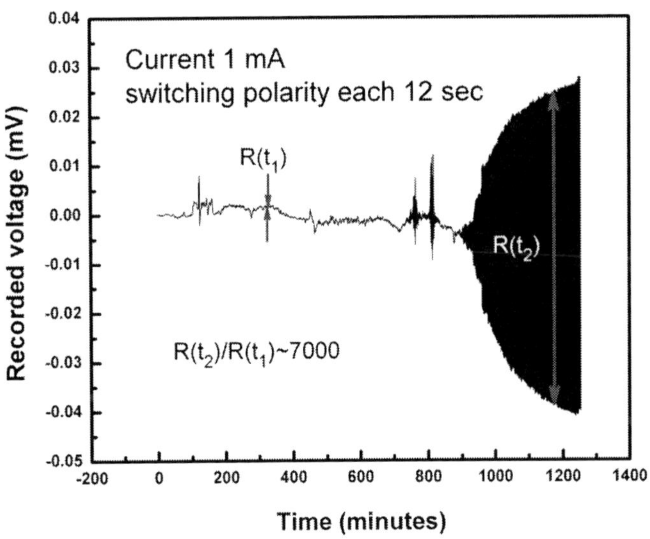

Figure 5. At the *warming* from *T=77K* (*time t=0min*) to *T=300K* (*t>1200min*), everywhere below *t~850min* the line reveals no dependence on the current flips, while above it V(t) flips with I(t).

Since these cryostats are specialized for extreme-low temperature measurements, the sensitive thermometry is routinely used only below *10K*: based on germanium diodes, this thermometry is not applicable at higher temperatures. From room temperature to *10K* the temperature measurements are pretty rough: they are typically measured via a charcoal pump (resistive) thermometer, which is located rather far from the cold finger on which the sample is mounted. So the recorded curve was pretty far from typical publication standards, but its beauty was in its scientific value: a resistive transition took place at about *200K*! At this point I decided not to open the cryostat but to cool it down again, this time watching the recorded data closely, and as soon the transition would start, to apply some more current and see what would happen. If the transition is related to the sample, then a higher current should remove zero resistance. And that is exactly what happened! I switched to higher current amplitudes manually (using a Keithley-220 current source) and during the cooling time was able to enhance the current twice (from *1mA* to *10mA*, then to *30mA*) thus restoring twice the non-zero resistivity in both attempts. I did not apply higher currents for fear that the higher current would damage the sample. Subsequent warming repeated the transition temperature and the approximate shape of the previous *R(T)* curves. I swapped current and voltage circuits, still getting this apparent zero resistance. Regretfully, I did not permute the leads, since triax cables were used and such permutation would have required additional efforts for which I did not have justification: I still did not believe that this effect could be of a superconducting origin.

Nevertheless, I reported the results to the ONR sponsor and got feedback to transfer the samples to Dr. John Claassen (NRL) for additional measurements involving the magnetic field. Unfortunately, the magnetic part of John's apparatus was not functional at that time, and he made only resistive measurements, at a current 10 times higher than in my case, and got the same type of transition, with bigger hysteresis. I believe[2] that some permutation of leads occurred at this time since I completely dismantled the sample and John connected the contact leads at his discretion. His cryostat was using *He*-filled chamber (in my case the sample was in vacuum). The sample lost mechanical integrity when leaving his cryostat after the measurement, and it was not possible to obtain further resistive measurements on it, although later we made some other measurements on this sample using other methods. Also, some other samples provided further data, close to this original one. The reader will be able to trace all these measurements throughout this chapter. We have tried to make the description as detailed as possible, and at the same time we regret that we cannot yet provide a strict, conclusive answer as to whether or not we have superconductivity.

Let me try to explain this better. The classical recipe for confirmation of superconductivity has three major components: resistive transition to the level $R=0$ (ideal conductivity), the Meissner effect (ideal diamagnetism), and reproducibility. The real world is not ideal. We were able to demonstrate a transition to conductivity orders of magnitude better than the best metals. Moreover, following the recommendation by Dr. T. Geballe (Stanford U.), who met our results with encouragement [9], and recommended to perform *2*-probe measurements of resistance, we produced and tested (thanks to the immediate financial support by Dr. J. Montgomery, the Director of Research, NRL) more samples and demonstrated the transition simultaneously in *2*- and *4*-probe geometries. We were able to see the Meissner-type behavior, but the presence of magnetic component in our samples made the analysis more complex and still demands more research. The reproducibility assumes that other researchers have been able to repeat our results. To the best of our knowledge that is not yet the case. Indeed, after publishing our initial results on the web in 2005 [10] and 2006 [11]

[2] This belief can be substantiated by the comparative analysis of John's and my own data presented in Section 3.

we accumulated more results and sent a manuscript to "Physica Status Solidi". It was rejected within 12 hours after submission with the reasoning that "the essential claim was published nearly two years ago… and has not been confirmed independently since". We were quite satisfied by that explanation: it seems to mean that web-publication really plays a good role in the widest dissemination of scientific results, as no journal publications are required to gain acknowledgment for them.

However, when Mr. Frank Columbus, the Publisher of "Nova Science", suggested writing a chapter for his book, I accepted his offer gladly: a detailed publication will make it easier for interested researchers to follow our approach. It is definitely not an easy technology for scientific research but is still a very promising one. We ourselves are also eager to reproduce our results on a higher level of confidence using the available facilities of Physics Art Frontiers and the National Academy of Sciences, Armenia[3]. This time, the research is being organized much better, since we have passed through its "spontaneous" discovery phase.

Many people have contributed to these findings. Though the major and decisive contribution came from V. Nikoghosyan, who is co-authoring this chapter, many other researchers contributed a lot of their skills, time and energy. The reason why they are not formal co-authors here is mainly related to my intent to *remove any responsibility* for the statements that will be made below from these persons. They could be considered as informal co-authors. I should mention in particular Dr. D. Van Vechten, who worked on this project under the ONR ROPO program. Dr. K. Wood (NRL) created and tried to maintain the laboratory environment at NRL for this work[4], and spent a lot of effort deciphering puzzles of experimental measurements, which sometimes seemed really inexplicable. Dr. G. Badalyan (Physics Research Institute, Armenia) together with his colleagues (Dr. A. S. Kuzanyan and V.S. Kuzanyan) contributed a lot toward determining the oxygen content of laser-treated samples. Special thanks go to Ms. S. Petrosyan (Physics Research Institute), who by accident deposited nichrome adhesive layers instead of planned adhesive layers of titanium prior to the requested silver film depositions, and to Dr. S.-F. Cheng (NRL) who first detected existence of magnetic particulates in our samples. Dr. Cheng was also very helpful in deciphering X-ray diffractograms, joining efforts with Dr. S. Qadri (NRL) who made a big volume of X-ray data acquisition and analysis with our samples. The contribution by Dr. Claassen has been acknowledged already. A large volume of magnetic measurements and analysis was relentlessly carried out by Dr. M. Osofski (NRL), Dr. R. Soulen, Jr. (NRL), Prof. A. Goldman and his students (U. Minnesota), and Dr. J. O'Brien (Quantum Design corporation). We had very useful discussions with Dr. A. Harutyunyan (Honda USA Research Institute, Inc.) on our magnetic measurements; with Dr. S. Terzyan (Oklahoma Medical Research Foundation) on crystallographic aspects of our studies; with Dr. G. Movsesyan (Yerevan Physics Institute) on artificial diamond growth at laser ablation; and with Dr. I. Mazin (NRL) on the limits of analogy between cuprates and ruthenates, as well as on general aspects of searching for higher-temperature superconductivity. Dr. G. Luepke and Dr. A. Lucarelli (College of William and Mary, Williamsburg) contributed essentially in adding excitement to our research by their magnetic visualization method: it is truly a great method! The contribution

[3] Physics Art Frontiers supports a web-site www.superconducting.org for news on results and collaborations.

[4] As a matter of curiosity, NRL submitted a patent application for our results and cut funding for the research itself. Because of that we are sorry for not being able to supply research samples to our colleagues per their requests.

by Dr. Y. Maeno was invaluable: he provided us high-quality single crystals of Sr_2RuO_4, and some other ruthenates, and with his colleague, Dr. H. Yaguchi (U. Kyoto), performed some magnetic measurements and helped us with the discussion of very delicate aspects of processed samples measurements.

I am very sorry I cannot express my gratitude directly to the late Dr. Herb Gursky, a renowned X-ray astrophysicist: he was the Superintendent of the Space Science Division (NRL) where my part of the work was done. He was very supportive of it and shortly before passing away encouraged me in his e-mail: ***"As far as your own work, it's the most exciting thing that's happened at NRL in a long time, however it turns out."*** For many of my colleagues and for myself this is the highest level of appreciation. We are determined to continue this struggle with Nature, using all available to us resources, on the thorny path to room-temperature superconductivity.

1. INTRODUCTION

Laser processing of materials opens new avenues for researchers. From the point of view of material science, the application of short laser pulses with high power creates conditions for synthesizing novel substances, which are inaccessible via the traditional equilibrium phase diagrams in laboratory conditions. A well-known example that can elucidate our approach is the creation of artificial diamonds using graphite [12-15]. Both diamond and graphite minimize Gibbs's free energy of carbon atomic system, though the graphite state corresponds to lower energy at ordinary conditions (room temperature and atmospheric pressure). At very high pressures and temperatures the diamond state has lower energy, and as soon as that state is reached, rapid cooling can yield a "frozen" diamond configuration metastable at ordinary conditions. This actually happens during laser processing of graphite; that is, the laser action creates both high pressure and high temperature, as well as abrupt cooling due to short pulse duration.

An important issue is that even though laser action takes place on the surface, its action modifies the properties of the materials in essential depth of about some microns, so that the properties to be considered are really bulk properties of the material (though of course, we must keep in mind that in this modified *3-D* crust one may have nanostructures yet to be explored, – again, there is a direct analogy with the graphite turning onto carbon nanotubes at laser processing).

In this chapter we report on unusual properties displayed by Sr_2RuO_4 perovskite ruthenates after laser processing. Sr_2RuO_4 is a remarkable superconducting material [8] since its pairing corresponds to the triplet state with T_c up *to 1.5K* (details can be found in [16]). Because of the orbital motion of Cooper pairs in spin-triplet state, in materials with low T_c the pairing is very sensitive to lattice imperfections. Thus, despite many efforts (see, *e.g.*, [4-7]), thin films of this material have never demonstrated superconductivity. Meanwhile, some physical tasks, such as studying quantum superposition of chiral states in triplet superconductors [3], require mesoscopic samples. To achieve in practice this goal we undertook laser micro machining of thin crystalline slabs with a goal to make them thinner and smaller in successive steps of laser ablation. The outcome, which we will describe in this chapter, was quite different.

A key procedure in our approach involves laser ablation of the top surface of crystalline slabs ripped off from thicker crystals. Laser processing at energy densities consistent with substantial ablation is known to be capable of altering the physical properties of the residual samples. In addition to removing atoms from the surface, the left behind re-crystallized surface layer may have altered composition and properties. Moreover, because of the extreme non-equilibrium phase conditions during the pulse (ours was *10ns* long) and the extremely rapid re-solidification thereafter, the re-crystallized material can contain compositions inaccessible via equilibrium phase formation.

Our design for sample preparation was chosen to allow four-probe measurements of resistivity, which was a mandatory first step towards further exploration of properties of micron-scale samples of Sr_2RuO_4. To our surprise, the first sample demonstrated zero resistance (or more strictly, zeroing of voltage at a constant applied current) slightly below *200K*. Although the measurement was repeatedly reproducible on this sample, it was not until the second sample demonstrated the same type of transition at almost the same temperature when we started regarding superconductivity as an alternative to other possible explanations of the transition. Magnetic measurements provided a correlation with resistive transitions, the EDX microanalysis revealed oxygen enhancement. The accumulated knowledge provided grounds for taking the occurrence of superconductivity seriously [10]. Since then, a lot more facts have been discovered. Taken together, they provide strong grounds to believe if not in superconductivity above *200K* (we leave the final answer to future studies: our current opinion is that it should not be considered as already proven) then in strong changes of the properties of the material. We hope that this disclosure of the accumulated results will accelerate the progress in this area.

2. Sample Preparation

2.1 X-Geometry

The regular method of sample preparation started with polishing a Sr_2RuO_4 plate on opposite *ab*-plane faces into a freestanding plate approximately *25* microns thick, and gluing it onto a supporting substrate for further processing. Then a thin adhesive layer and subsequently a good conducting layer (*e.g.,* silver) are deposited on top of the crystal and substrate. At the next step laser micromachining was used to make diagonal grooves to define four conductive pads extending from the outer region to within a small area in the center of the plate. Another, rectangular cut separated the internal part of the sample from the edges. Finally, the laser ablation removed the deposited layers from the central part of the crystal (together with the adjacent surface layer of the parent crystal). This left a small region at the center thinned and available for four-probe *R(T)*-measurements – the probes are made of the same material as the center spot. This excluded to the maximal extent any possible proximity effect influencing the central part. It also made the task of attaching external leads easy and mechanically non-invasive. The processed sample (in the X-geometry design, Fig. 4) is shown in Fig. 6. We will next bring in more details on its preparation.

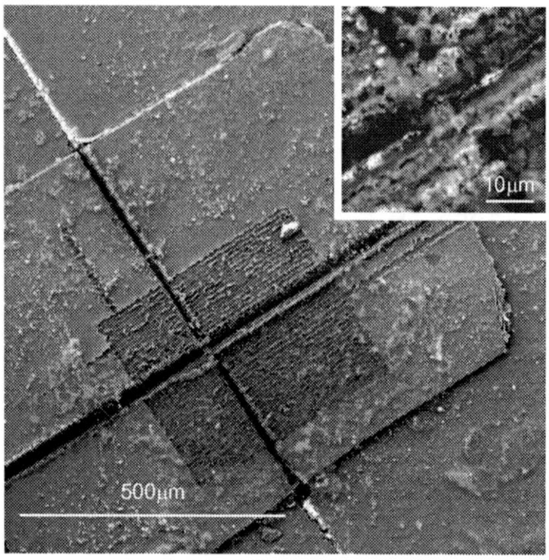

Figure 6. SEM of the first processed sample (L1-C148-Ga1) and of its central part (*inset*).

2.2 Parent Materials

The crystals used in our experiments were grown in Kyoto by floating-zone technique [17-19]. The parameters of the crystals are summarized in Table 1.

Table 1. Parameters of explored parent crystals

Number	T_c(K)	RRR	Composition	Abbreviation
C148	1.13	343	Sr_2RuO_4	214
C622b			$Sr_3Ru_2O_7$	327
Cfvϕ7			Sr_2RuO_4 - $Sr_3Ru_2O_7$	Eutectic
C120	1.30	576	Sr_2RuO_4	214
C128	1.44	1623	Sr_2RuO_4	214
C149 (X)	1.27	527	Sr_2RuO_4	214
C156	<0.3	128	Sr_2RuO_4	214
C171	1.06	291	Sr_2RuO_4	214
C167	1.43	1506	Sr_2RuO_4	214

(In this Table, RRR stands for residual resistivity ratio, defined as the room temperature resistivity related to the resistivity before superconducting transition below *1.5K*.)

2.3 Cutting and Polishing

Crystals were cleaved, preferentially along the *ab*-planes using the strong anisotropy of their layered structure, into relatively thin slices (about *100μm* thick). Some but not all of these slices were then polished using standard technologies (by "Mindrum Precision", Rancho Cucamonga, CA). This polishing was done on one or both sides and produced samples as thin as *25μm*. The importance of this polishing is discussed later in the text.

2.4 Substrates, Glues and Waxes

The mechanically prepared slices were glued onto substrates. Different substrates were used including glass, sapphire, $SrTiO_3$, and fused silica. Cyanoacrylic family glues (like "Future glue" and "Super Jet") were used for attaching samples to the substrates. In some cases acetone, isopropyl alcohol and debonding agent "Jet DE-SOLV" were used to clean the surface of glued samples. For mechanical protection of samples in some of the magnetic measurements, they were covered by wax at the final stage. All waxes used underwent special tests to exclude the possibility of magnetic contamination of the samples.

2.5 Metal Deposition and Composition Details

A thin, metallic layer of silver (typically *200-300nm*) was deposited on top of these substrate-mounted crystals. In order to increase the adhesion, an ultra-thin sub-layer of nichrome (typically *40-70nm* thick) was deposited prior to the silver[5]. The nominal composition of the nichrome target was *Ni~10%(w)*, *Fe~68%*, *Cr~17%*, and *5%*-other impurities. All metallic layers were deposited by the vacuum evaporation technique at a pressure of $5x10^{-5} Torr$.

2.6 Parameters of Laser Processing

After deposition of metallic layers, the samples were laser micro-machined using a special laser-patterning system. This is based on a solid state *UV*-laser ($YAG\text{-}Nd^{3+}$ generator, *355nm* wavelength) with *10ns* pulse duration and *0.1mJ* pulse energy. The beam was focused by a short focal length lens into a spot as small as 1 µm in diameter (typically, *~3µm*). This produces power density up to $10^{12} W/cm^2$ in the focal spot (the more typical values were $\sim 10^{11} W/cm^2$). The sample was located on a computer-controlled moving table. The laser pulsed at a *30Hz* repetition rate with the sample moving linearly at *20µm/sec* speed.

3. Resistive Transitions

3.1 Measurement Technique

For resistive measurements we used a very simple and straightforward method: a DC current with constant amplitude and alternating polarity. Alternation of polarity is required for the exclusion of thermoelectric pick-up, which is essential at small measurement currents. It was applied to two of the four conductive pads and voltage acquisition was made between the two other pads as shown in Fig. 4. The voltage difference between outputs, corresponding to the consecutive cycles of positive and negative current amplitudes, being free from thermoelectric pickup, characterizes the resistance of the bridge (*i.e.*, of the central part of the

[5] The choice of this material by sample manufacturers as an adhesion layer was accidental. Its presence was not recognized at the time of the first web-publication [10].

sample). A Keithley-220 programmable current source was used for the current generation, and Keithley DMM-2000, DMM-196 and Nanovoltmeter-2182 were used for voltage circuit measurements under the LabVIEW data acquisition.

3.2 Transitions in the First Sample

A microphotograph of the first sample (#L1-C148-Ga1) is shown in Fig. 6 above.

Resistivity vs. temperature of this sample was measured at NRL in two cryostats (by two different groups).

3.2.1 Cryostat 1 Data

Multiple resistive transitions of the first sample in the cryostat 1 are shown in Fig. 7.

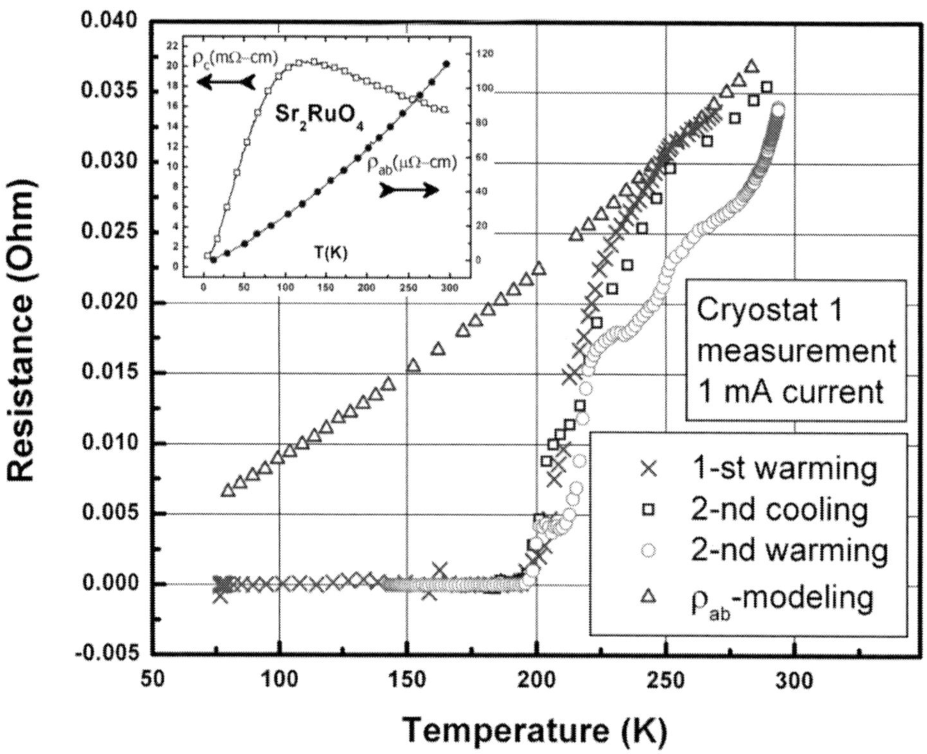

Figure. 7. Transitions recorded with the first sample (# L1-C148-Ga1). Resistance dropped about factor of *7,000* down to the noise level (see text). Some 'noise' noticeable on the first warming *R(T)* curve (at *77K*, *130K* and *160K*) is related to the noise in the registering system but not to a real deviation of *R* from zero. At modeling (triangles) it is assumed that $R(T)=A\rho_{ab}(T)$, where $\rho_{ab}(T)$ is shown in the inset, and the modeling constant *A* was chosen so as to make the *R*-curve match the room temperature value.

Measurements were performed in a He^3 cryostat ("Infrared Labs") with the sample in a vacuum. The sample was mounted on the cold stage while the temperature was registered by the (resistive) charcoal pump thermometer. Because of spatial separation between the sample and the roughness of the charcoal pump thermometer at these high temperatures, one can expect a relatively high error in temperature values, but this error is definitely less than *10K* (cooling and warming especially were slow). The first cooling also resulted in zero resistance, but it was not recorded since this happened quite unexpectedly, as was described in the Preface). All these measurements were performed with a *1mA* DC current having alternating polarity (a set of raw data obtained at the first warming are shown in Fig. 5).

3.2.2 Resistivity Above and Below the Transition

The data in Fig. 7 reveal that the resistance *R* of the bridge in the normal state was about *35mOhm*. The inset to this figure plots the ρ_{ab} and ρ_c of the parent material. A special plot in Fig. 7 (triangles) demonstrates that the high-temperature data are consistent with the current flowing solely in the *ab*-plane. So one can estimate the value $\rho \sim \rho_{ab} \sim 100 \mu Ohm\text{-}cm$ for the bridge resistance. While the shape of the bridge is not regular, one can roughly approximate it as of length $L \sim 10 \mu m$, width $w \sim 10 \mu m$, and height $d \sim 25 \mu m$. Then $R = \rho L/wd \sim 4 \cdot 10^{-2} Ohm$. This estimate is in surprisingly good agreement with the measured value. The temperature at which the *R(T)* curve separates from the R_{ab}-curve could be called the *onset temperature* of the transition. It is about *250K* in Fig. 7.

The next question is how much the value of resistivity ρ has changed after the full transition. Based on the recorded resistivity data for these measurements in cryostat 1, one can recognize that after the transition the voltage falls to the noise floor of the voltmeter, which was about $10^{-8}V$. Before the transition the signal amplitude was about $7x10^{-5}V$. Thus the reduction of the voltage amplitude between the voltage leads across the bridge was a factor of about *7,000*. If we assume that this reduction took place because of the modified properties of the surface layer which was no thicker than $5\ \mu m$ (and thus the effective cross-section of the bridge is $50 \mu m^2$) then the resistivity of the current-carrying material is about *3nOhm-cm, i.e.*, more than a factor of a hundred smaller than the resistivity of the best-conducting metals at *200K* which points to superconductivity. Possible alternative interpretations will also be considered below.

3.2.3 Cryostat 2 Data

For measurements in cryostat 2 (performed in helium atmosphere) the sample was completely rewired since this cryostat had a smaller sample holder. No attempt was made to keep the same configuration of leads for the current and voltage circuits. It is quite unlikely that by occasion they turned out to be identical to those in cryostat 1. It looks like a permutation of current and voltage leads took place. Indeed, the shape of *R(T)* above the transition is different: descending (Fig. 8) rather than ascending (Fig. 7) at the highest temperatures.

It may be caused by a partial flow of the current along the crystallographic *c*-axis in this case. That could have taken place if the crystallographic *c*-axis was not perpendicular to the surface of the sample. Since the resistance ρ_c is much larger than ρ_{ab}, a small tilt is enough for the domination of the *c*-axis contribution. This difference in the slopes, as well as in the *R*-values of the bridge resistance are indicative of the fact that the current paths in these two cryostat measurements were truly not identical.

3.2.4 Cracking or Transitions?

We can now address possible alternative interpretations of these four-probe resistive measurements. They all are developed by misbeliefs in the occurrence of transitions.

One obvious misbelief is that the current leads were broken: that possibility is excluded in our measurements since the current circuits were monitored in both cases.

The next possible misbelief is to assume that a crack had developed within the bridge which acted as an internal switch and thus separated the current circuit from the voltage leads (Fig. 9)[6]. This type of phenomenon was found to cause artifacts in the past with the first claims of superconductivity in organic materials (see, *e.g.*, [20]). However, as explained in Fig. 9, cracking cannot cause transitions in both cryostats given the different circuits measured, as noticed in Section 3.2.3. This dismisses the argument.

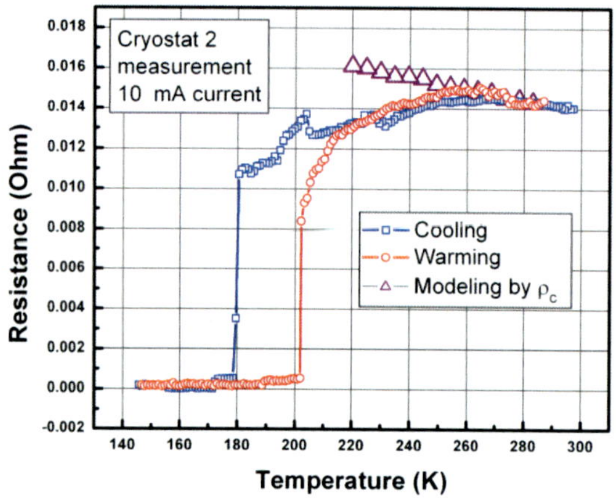

Figure 8. Transitions recorded in the first sample (#L1-C148-Ga1) in cryostat 2. Different lead configuration could be expected from the *R(T)* slope comparison with Fig. 7. The modeling curve (triangles) corresponds to $R(T)=B\rho_c(T)$, where $B=const$ and $\rho_c(T)$ is given in Fig. 7 (*inset*).

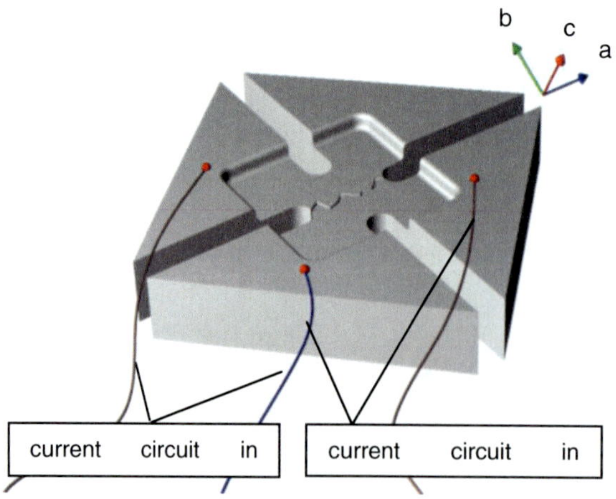

Figure 9. Schematics of the measurements in the two cryostats. A possible crack is shown for illustration. Arrows show crystallographic axes. If this crack is causing a fictitious "transition" in one of the cryostats (with the current in the *ab*-plane), in the case of the other cryostat (with current partially along the *c*-axis) the crack should have precluded the current flow. This contradicts observations.

[6] We are grateful to Dr. D. Gubser (NRL) who pointed out this argument to us.

3.2.5 Self-Heating and Transition Hysteresis

Measurements in Fig. 8 were performed at a *10mA* current. For these measurements of cooling vs. warming there is an obvious thermal hysteresis of ΔT_c of about *20K*. It is interesting to compare this shift with special tests that we performed in cryostat 1 during the second cooling run (Fig. 7). During this run and just after the transition at *1mA* we switched to higher currents (successively to *10mA* and *30mA*) as the sample continued to cool down. The higher-measuring currents temporarily restored the normal state two times. The results of this test are shown in Fig. 10.

The first of these shifts (from *1mA* to *10mA*) is in accordance with the hysteresis in Fig. 8. Indeed, as in Fig. 10, by switching from a *1mA* to a *10mA* current, the T_c is shifted by about *40K*, *i.e.*, twice more than in the case shown in Fig. 8. But the bridge resistances are also different by a factor of two. Since the self-heating is proportional to the value of the resistance ($\sim I^2R$), and the value of the current is the same in both cases, then the difference seems quite reasonable.

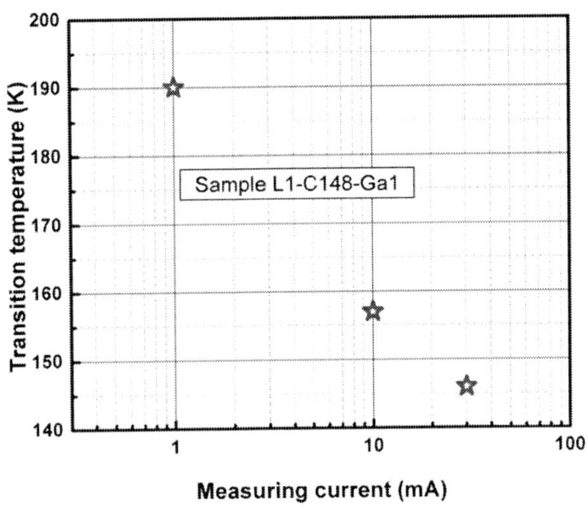

Figure 10. Suppression of the transition temperature in the first sample caused by two intentional increases of measuring current as the sample cooled down in a common thermal cycle.

Another peculiarity in Fig. 8 is the sharper transition curves compared to those in Fig. 7. At the first glance it contradicts superconductivity: higher currents typically make the superconducting transitions broader. But that takes place when the order parameter is suppressed by the magnetic field. This is not the case here because the suppression has a thermal origin: the discussed thermal hysteresis indicates such a mechanism: the magnetic suppression should be free of hysteresis. In our case the current is big enough so that that the hysteresis is plainly seen, and the higher currents generate more self-heating in the normal state. At a broad transition, as soon as superconductivity sets up, the current gradually transforms into the non-dissipative flow mode, which reduces the Joule heat and facilitates faster cooling. Thus a sharper transition occurs. A similar argument can obviously explain the sharpness of the transition at higher currents during the warm up: as soon as fluctuations create the seeds of the normal state, the heat will be injected into the system (proportional to the square of the current amplitude), which will create a spreading normal area, so that *R(T)* will jump up faster than in the case of smaller currents.

3.3 Transition in the Second Sample

The second resistive transition was recorded on the second sample (#L2-C148-Gb1) that was prepared very similarly to the first sample (Fig. 11).

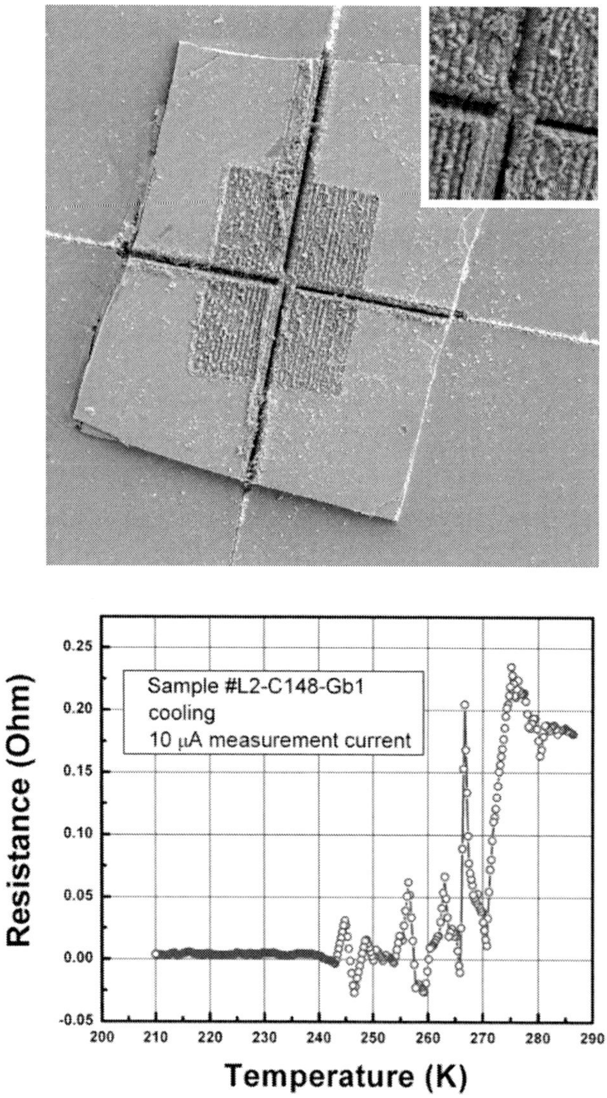

Figure 11. The second sample (#L2-C148-Gb1): (*top*) - processed slice and its central part (*inset*); resistance curve *R(T)* (*bottom*). This 4-probe technique record corresponds to the cooling; measuring current is only *10μA* (factor of hundred lower than in case of the previous sample).

The major difference between them was that the surface was ablated twelve-fold in the second sample (the laser beam went back and forth 6 times each way) vs. two-fold in the first sample. The second sample was measured in cryostat 1. The room-temperature resistance of the bridge was about an order of magnitude higher than that of the previous sample. One can hypothesize that it was higher partly because the *12* ablation passages made the bridge thinner, and partly because the current trajectory was not completely in the *ab*-plane, as the comparison of Fig. 11 (*bottom panel*) with Fig. 7 indicates. The higher values of resistance allowed the application of the $10 \mu A$ current for its characterization, though the signal-to-noise level was essentially lower, which caused a more visible noise pattern in Fig. 11 compared to the transitions recorded in the first sample. The transition temperature was about the same (technically, the temperature uncertainty in this case was higher than in the previous case). At *200K* the transition was down to the noise level of the voltmeter circuit. This sample was measured only once since it cracked apart during a deeper cooling (below *200K*).

3.4 Transitions in the Third Sample

The disadvantage of the four-probe technique is that the current and voltage circuits are spatially separated within the sample, and, moreover, temperature variation can cause further changes in this separation, particularly in inhomogeneous samples. One example of this effect, potential cracking, as discussed in Section 3.2.4, should always be taken into account when considering the results. In contrast, two-probe measurements exploit the same trajectory for current and voltage circuits, and thus are free from the misleading effects caused by the discussed reasons. The price to pay for this benefit is the inclusion of contact resistance into the measuring circuit, which itself can create artifacts and confusion. To address this issue we undertook *simultaneous* two- and four-probe measurements on a specially prepared third sample (Fig. 12)[7].

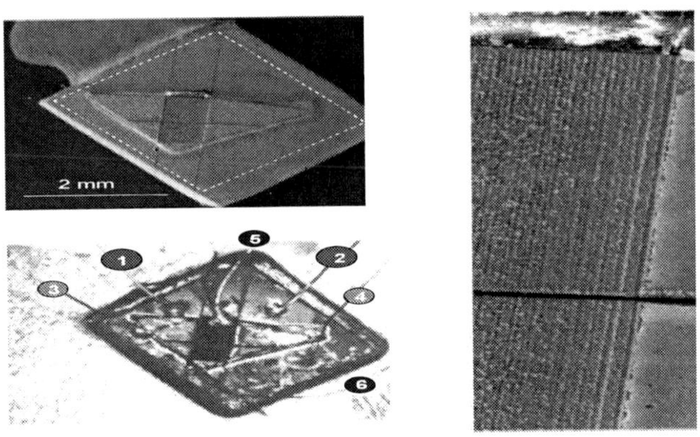

Figure 12. The third sample (#L13-C128-G1): microphotographs and lead arrangement for the simultaneous four-probe and two-probe measurement: current leads (*1* and *2*) and longitudinal voltage leads (*3* and *4*) are connected correspondingly via high conductance silver film yielding two-probe measurement, while transverse leads (*5* and *6*) are connected through the sample only.

[7] We are grateful to Dr. T. Geballe and the participants of his seminar at Stanford for bringing in the idea of two-probe measurements.

This sample (#L13-C128-G1) was prepared in a bar geometry using a thin non-polished (cleaved) plate of Sr_2RuO_4 about $100\mu m$ thick. The metals (nichrome and silver) were deposited exactly in the same order and to the same thickness as were previous samples. The laser-processed part has furrows in the vertical (longer) direction (Fig. 12, *right*). We also laser-processed a single groove (placed horizontally in Fig. 12), which does not go all the way through the sample thickness[8]. The additional laser cuts formed four silver contacts connected via the sample: the left and right triangle pads with contacts (1,3) and (2,4) respectively, and the top (5) and bottom (6) rectangular pads in the middle. The area outside the white dotted line (Fig. 12) was mechanically scratched off to secure connection between all the pads solely through the sample. Then the wiring leads were arranged as shown in Fig. 12, *bottom*.

3.4.1 Two- and Four-Probe Transitions

As with the first sample, for measurements we used a *1mA* DC-current, reversing its polarity every *12* seconds, so that possible thermoelectric effects could be excluded by the subsequent subtraction of negative from positive branches of voltage output. Voltage was measured in a longitudinal geometry (which constituted two-probe measurement between pads (1,3) and (2,4) in Fig. 12, *bottom*) and transverse geometry (between pads 5 and 6, which constituted a four probe measurement, see also Fig. 15 below). The *simultaneously recorded* results are shown in Fig. 13, and some details are explained in Figs. 14 and 15.

3.4.2 Data Consistency of Transition

Being a *simultaneous record* of the voltage pattern describing the *resistive transition* obtained *across two different conductivity channels*, Figs. 13-15 contain a lot of useful information deserving quantitative consideration. First of all, let us take into account that the voltage between two pads corresponding to the 4-probe measurements (*bottom panel*, Fig. 13), visually does not include any contact contribution, and after the transition falls to zero (as we will see below, in case of this sample the lead arrangement is not ideal, and there is a very small contact contribution to this "zero"). Its value before the transition is about *1mV*. That directly indicates that at least *1mV* of total *3.5mV* voltage between the pads of the 2-probe circuit (*top panel*, Fig. 13) is caused by the passage of the current through the sample and is not caused by the contact resistance. In view of the *1mA* measuring current this corresponds to $R=0.5Ohm$ of resistance (because of sign alternation, the registered voltage equals the double of its amplitude). Could our sample hold such a resistance[9]? In the measuring geometry of a bar with width~length, so that $R\sim\rho/d$, where ρ is the resistivity, and $d\sim 100\mu m$ is the thickness, one gets $R=0.5Ohm$ at $\rho\sim 5mOhm\text{-}cm$. This value of ρ is within the limits $\rho_{ab}<\rho_{eff}<\rho_c$, as it should be. Being above ρ_{ab}, this ρ_{eff} tells us that the current does not flow exclusively along the *ab*-plane, but that an essential part of its trajectory goes along the *c*-axis. The *inset* in Fig. 13 confirms this conclusion: it proves that the current path is neither completely within the *ab*-plane nor completely within the orthogonal *c*-direction.

[8] The reason for placing a groove there was just to create conditions possibly closer to the X-geometry of the first two samples. Conditions for the laser plume action inside the groove are different from the surface: the laser plume is maximally influential on the parent sample inside the groove. In principle, the topology of the grooves allowes the current to flow via the walls, causing the transitions in the case of the considered samples.

[9] Indeed, if the sample cannot hold such a resistance that would mean that the overall picture of the electric potential distribution is caused by contacts. We are grateful to Dr. D. Gubser (NRL) who raised this question, as dismissed by the coming arguments.

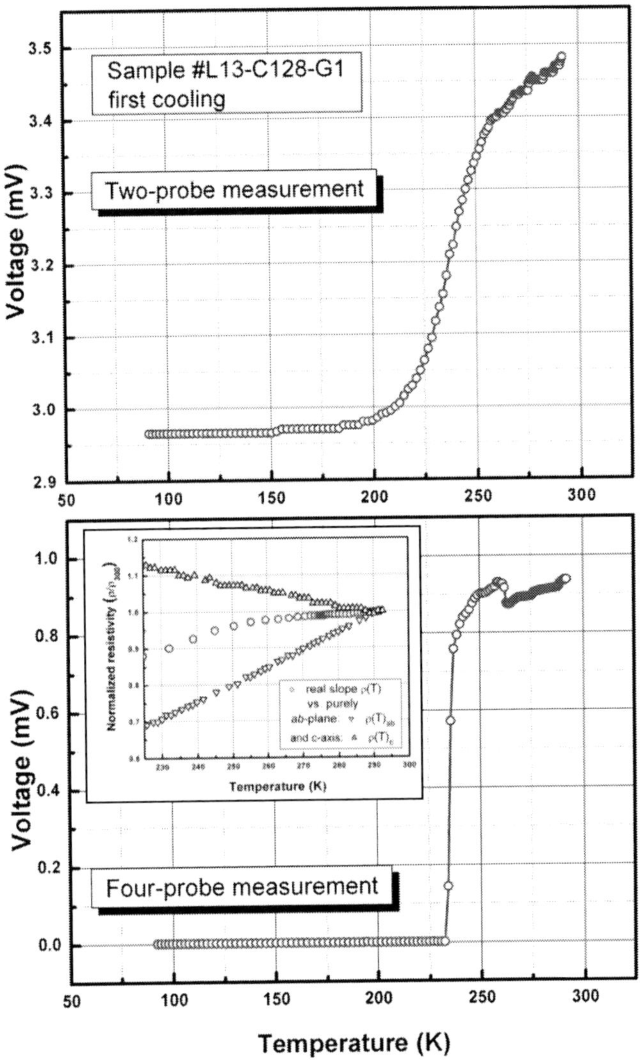

Figure 13. Transitions in the third sample (#L13-C128-G1) in the two-probe (*top panel*) and four-probe (*bottom panel*) configurations at the first cooling. The transitions are recorded simultaneously. The *inset* compares the slope of $R(T)$ of this record with that of the $\rho_{ab}(T)$ and $\rho_c(T)$ of Sr_2RuO_4 (Fig.7, *inset*) confirming that the current path corresponds to a complex trajectory inside the sample.

Also it is very likely, that the substitution $d=100\mu m$ is an exaggeration; the effective thickness may be smaller because of the unpolished surface profile and the potential crack inclusions caused by cleaving while fabricating the sample: smaller effective values of d would yield values of ρ_{eff} closer to $\rho_{ab.}$ In any case, the estimates for ρ_{eff} following from Fig. 13 are non-contradictive.

Let us consider more features of these data plots. As we see, the voltage between the *4*-probe voltage pads drops by a larger amplitude (ΔV_4~$1mV$) than between the *2*-probe pads (ΔV_2~$0.5mV$). There is no logical contradiction in $\Delta V_2<\Delta V_4$, while it would have been a contradiction in case of $\Delta V_2>\Delta V_4$. Indeed, there is not much current between the *4*-probe voltage pads, so the transition, if caused by superconductivity, is fully accomplished at some temperature, while the presence of a much bigger current between the *2*-probe pads keeps some areas normal and prohibits full transition. We also pay attention (see Fig. 12) to that the *4*-probe voltage pads are along the laser-processed furrows on the surface of the sample, while the *2*-probe current flows in the perpendicular direction, which enhances the likelihood of normal areas along its path.

Analysis given by Fig. 14 convinces us that this sample's geometry is not ideal for the *4*-probe measurement. Indeed, suppose that the laser-processed part becomes a superconductor; then only the edges of the *4*-probe voltage contact pads would touch this superconducting area (Fig. 14, *bottom panel*): the material under the pads, as well as the pads themselves, have non-zero resistance; thus, the pads should sense a small voltage drop along themselves ($=V_{56}$) even after the transition. This is indeed the case, as the *top right panel* in Fig. 14 demonstrates: the voltage drops sharply at the transition (compare the two *top panels*), but not to the noise floor: some structure caused by the alternation of the current is still seeable despite being reduced by a factor of *300*. This distinguishes it from the X-geometry case (Figs. 5, 7 and 11), where the voltage dropped to the noise floor.

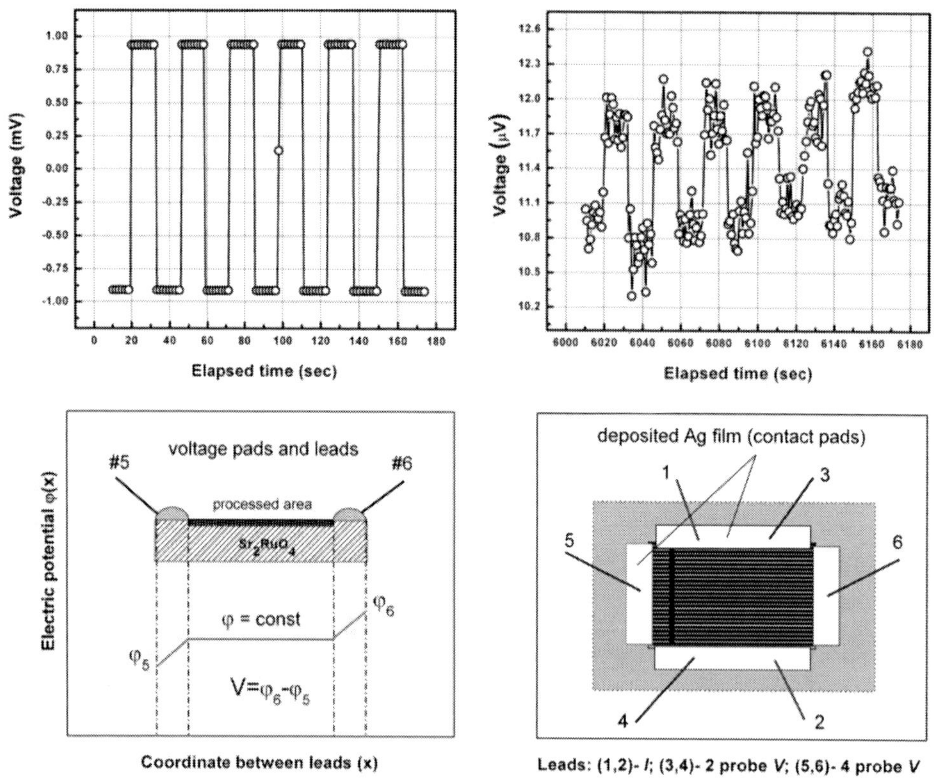

Figure 14. Alternating current through leads *1* and *2* creates alternating voltage between contact pads *5* and *6*. The top left panel demonstrates this voltage vs. time before the transition, and the top right panel – after the transition. Because of much smaller amplitude, in Fig. 13 the latter is invisible. Bottom panels explain the reasons (see also the text).

3.4.3 Cracking

Cracking in general complicates the exploration of these samples. It seems that there is a large difference between the coefficients of thermal expansion (CTE) of the Sr_2RuO_4 and the laser-processed material. (One should notice that this is indirect evidence in favor of modifying the physical properties of the parent material by the laser processing.) That there was significant CTE-caused stress was indicated by the total lack of reproducibility between the first cooling and warming for the third sample (Fig. 15, *top right panel*). To understand this result, one should recall the unpolished and heavily terraced, on the microscopic level, surface of this sample with the laser-processed "crust" on its top. A scenario that can possibly explain such behavior is given in Fig. 15 (*left*). This is very different from the polished surfaces of the first two samples, one of which survived multiple thermal cycles.

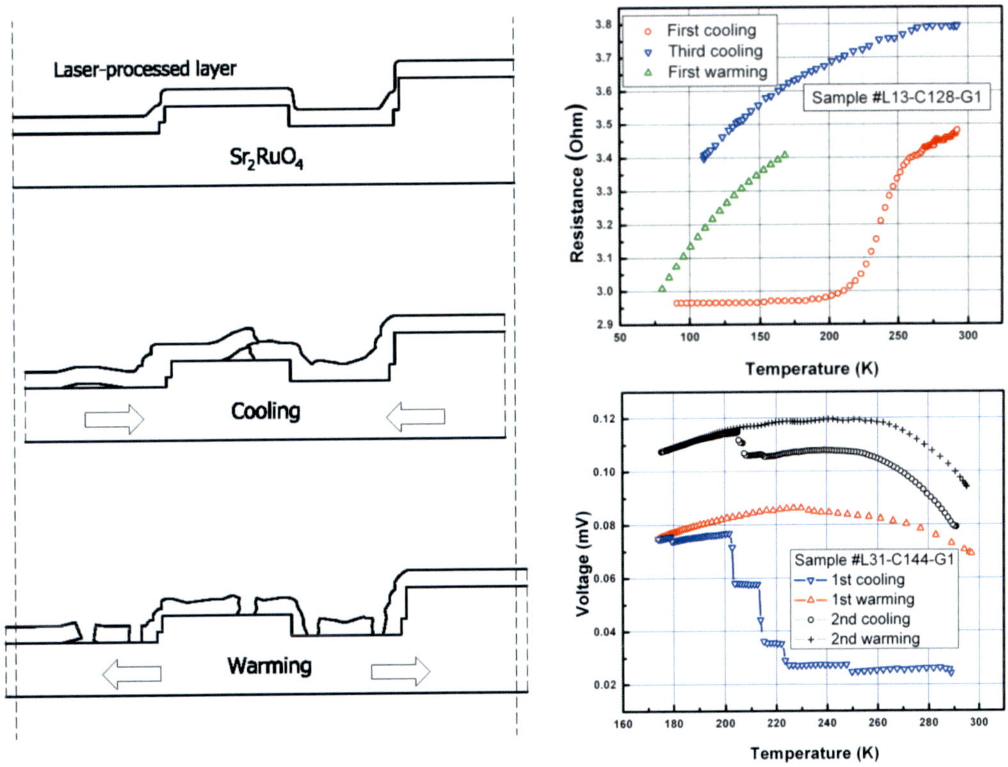

Figure 15. (*Left panel*) Possible cracking scheme which may explain the anomalous behavior in Fig. 14: *top* – initial state of sample with the terraced surface and laser-produced crust; *middle* – at cooling the contraction is stronger for Sr_2RuO_4, and the crust is squeezed, but the via-crust current trajectory is not yet broken; *bottom* – at warming the integrity is lost and the initial current trajectories are broken. Effects caused by cracking are exposed for the sample #L13-C128-G1 (*top right panels*) and for the sample #L31-C144-G1 (*bottom right*) at two runs.

Typically, cracking is easy to recognize by an odd pattern in the resistive behavior, as shown in Fig. 15 (*right bottom*). However, sometimes the situation is more complex, and it is not obvious whether the cracking could even be the cause of the observed features at all. A lot depends on how many cracks are being developed and where are they located relative to the current and voltage leads. It is not only the cracking of the parent sample, which becomes precracked at cleaving. The laser ablation stimulates a mechanical stress, and cracks, according to our observations, often begin to develop as soon as the cooling starts. Of more than a dozen explored samples only the three described here were free of obvious cracking during the cool-down process; others failed to behave consistently. Obviously, the smaller the active area the smaller the probability of cracking. Thus, it is not just an accident that the original X-geometry sample with $10x10\mu m^2$ of active area survived many thermal cycles. Reproducing resistive transitions looks hard but not hopeless!

4. Magnetic Transitions

4.1 Magneto-Optical Imaging

To better understand the laser-processed samples, magneto-optical imaging (MOI) facilities developed at the College of William and Mary were employed.

4.1.1 Method Description

The experimental setup (Fig. 16) consisted of a custom-assembled polarizing microscope built around a continuous-flow cryostat. On top of the specimen an epitaxially grown bismuth-substituted ferrite-garnet film with in-plane magnetization was mounted. Its surface was illuminated with polarized monochromatic light. The photons reflected off the bottom of the iron garnet experience a pronounced Faraday rotation of their polarization, the magnitude of which is proportional to the local magnetic field. Therefore the spatial distribution of the polarization of the light acts to image magnetic flux lines.

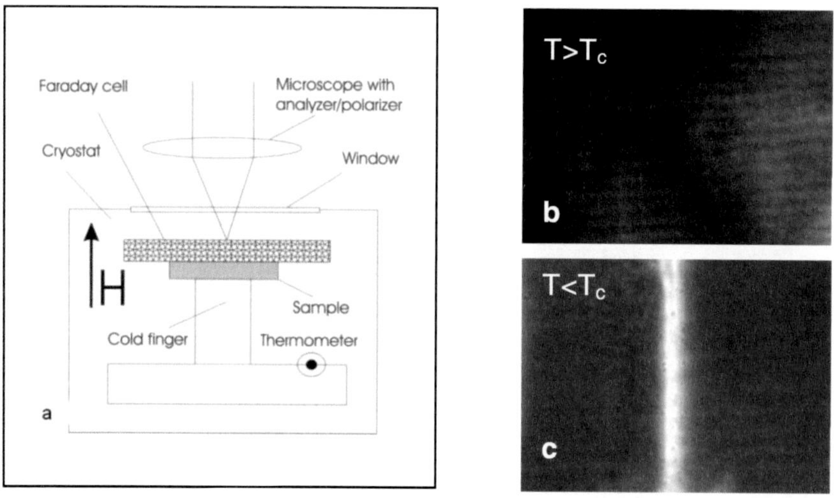

Figure 16. *a*: Experimental setup for visualization of magnetic flux distribution via Faraday effect. The structure labeled as "Faraday cell" includes the *Bi*-substitute ferrite garnet film described in the text. (The thermometer arrangement and the window geometry are such that the sample is at a somewhat elevated temperature compared with the thermometer.) *b* and *c*: Visualization of a scratch in a *YBCO* film.

4.1.2 Results

MO images of the first sample were taken at various temperatures from *300K* down to *60K*, with an applied external field of about *100Oe* parallel to the *c*-axis (Fig. 17).

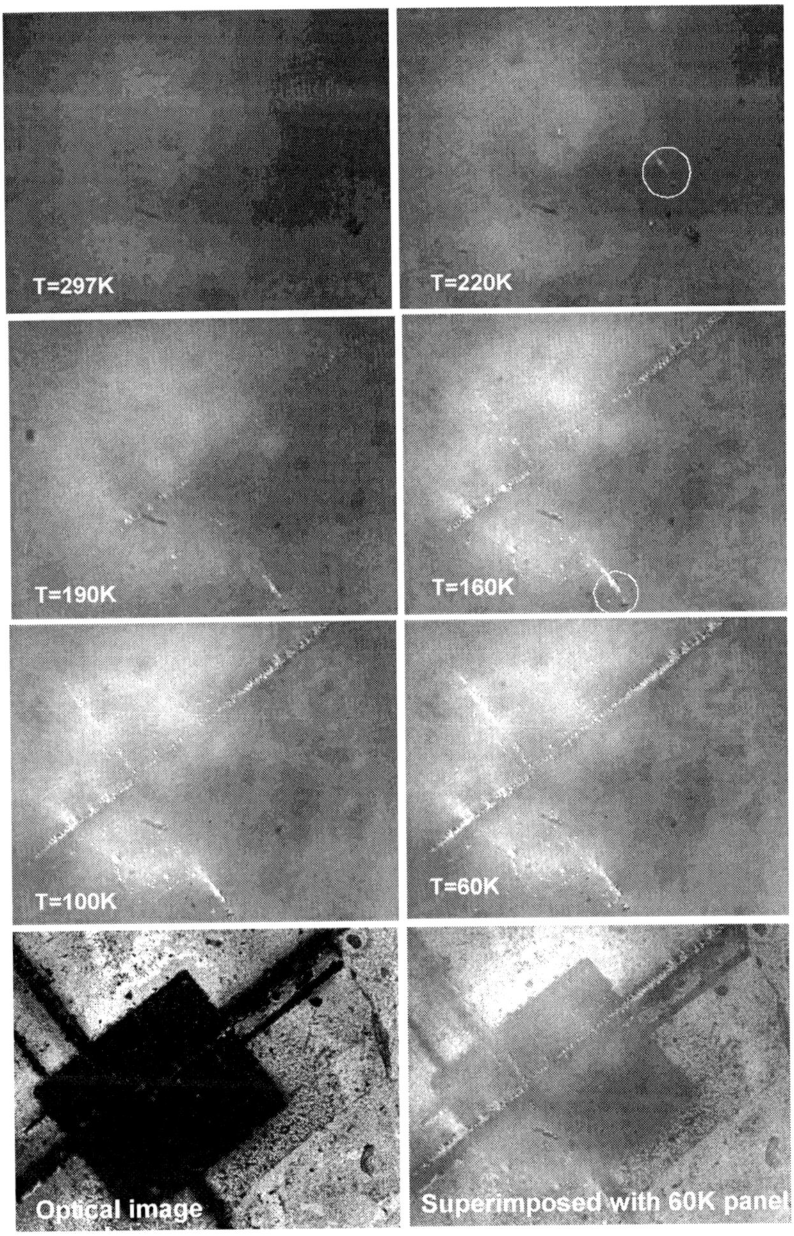

Figure 17. Magneto-optic images of the sample #L1-C148-Ga1 (*top 6 panels*) at different temperatures. At preparing the measurements, the right quadrant of the sample was lost. Before this happened, the actual picture of the sample (*bottom left*) was taken. This picture superimposed with the *60K* image is also shown (*bottom right*). The absence of the right quadrant causes the dark unstructured area in the MO panels. Also, the *top two panels* have images shifted (to the right and up, as the white circles indicate) compared to others.

Clear signatures of magnetic field lines start to appear at *220K* as bright small spots in contrast with the homogeneous background. At cooling, these features grow in brightness and in number indicating that magnetic field lines are accumulating in an inhomogeneous way in specific spots of the sample surface. A comparison between different panels in Fig. 17 reveals that the path formed by magnetic field lines mainly follows the crossed trenches created in the sample by laser ablation. Some of the field lines align along the perimeter of the processed area. This effect is clearer at the lowest temperature (*60K*), where the border of the lower left quarter of the doped rectangle is marked by continuous dotted lines. At this temperature magnetic field lines are well formed and their alignment along the trenches appears to follow the local roughness of the walls. In making these measurements the thermally induced stress caused the lower right quadrant of the sample to explosively delaminate (the optical picture in the lowest row in Fig. 17 was taken before that event, while MO images were taken afterwards thus causing unstructured images in the right bottom panel). The accumulation of magnetic field lines identified as bright spots by the MO imaging can arise from either diamagnetism or ferromagnetism. The alternative hypothesis that the image was caused by the surface roughness and mechanical stress was excluded by the absence of MO response to intentionally roughened samples.

4.1.3 Correlation with Resistive Transition

To move from the visual impression of these patterns to more informative digital analysis the normalized intensity as a function of temperature $I_N(T)$ was plotted (Fig. 18)[10]. This plot shows the increase of the intensity related to the features which start to appear simultaneously with the resistive transition, between *250K* and *200K*.

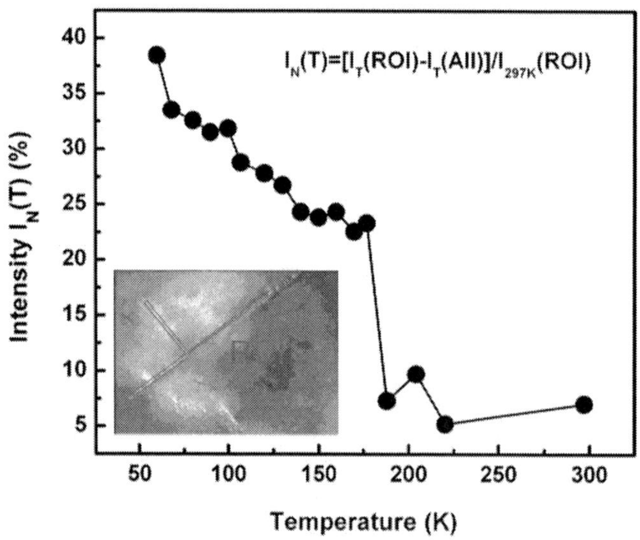

Figure 18. Digitally processed normalized image intensity (I_N) vs. temperature. I_N is calculated by first averaging intensity at various temperatures (I_T) in the range of interest (ROI), showed as a polygonal in the *inset*. Then the average intensity $I_T(All)$ in the complete image was calculated for the same temperature values. The two sets of data are subtracted and divided by the intensity of the ROI at *297K*.

[10] We are grateful to Dr. A. Lucarelli (College of William and Mary) for plotting these data.

4.2 Magnetic Susceptibility

We undertook the SQUID-magnetometer measurements of our samples at five different groups: at NRL, at Kyoto University, at Honda US Research Institute, Inc., at the University of Minnesota, and at the "Quantum Design" corporation. All the measurements were done using various Quantum Design MPMS models. The magnetic moments of samples were investigated vs. temperature in the field cooled (f.c.) and the zero-field cooled (z.f.c.) configurations, typically for $H=10Oe$ and $H=100Oe$, with the $\boldsymbol{H}$-field parallel to the sample surface.

4.2.1 Magnetization M(T)

Let us now consider the *M(T)*-response of the first sample (Fig. 19) measured at NRL.

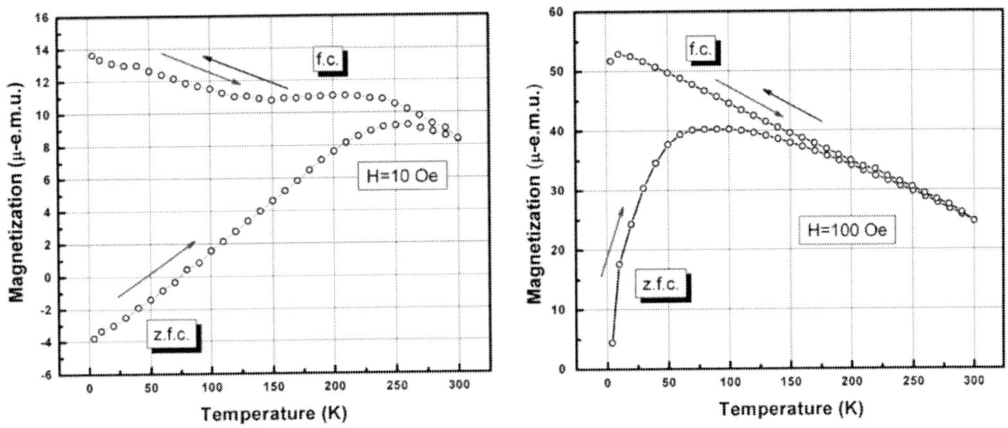

Figure 19. Magnetization *M(T)* of the first sample (#L1-C128-Ga1) vs. temperature in the field cooled (f.c.) and zero field cooled (z.f.c.) cases for two values of the magnetic field. Arrows indicate cooling and warming correspondingly.

The z.f.c. curve at the smaller field ($H=10Oe$) starts in the diamagnetic region, as it should for superconductors, and changes its curvature to merge with the f.c. curve at about *250K*. The transition is not abrupt, as it would be for homogeneous single-phase superconductors. Rather, it resembles that of oxide cuprates with inhomogeneous oxygen content and distributed values of critical temperatures from the maximal value down to very low temperatures. This is consistent with the presumed inhomogeneity of the sample. At the higher value of the field ($H=100Oe$) the merging takes place at lower temperatures, which, if the diamagnetism is due to superconductivity, can be explained by the temperature dependence of H_{c1}. On both curves ferromagnetism is apparently superimposed on the diamagnetic signal. (Ferromagnetism may be due to the nichrome remnants and/or other causes – we will discuss them later.) In addition, because of nonlinear behavior of the ferromagnetic component, its relative contribution increases as H increases, and diamagnetism, though essential, cannot completely compensate for the ferromagnetic signal in the higher field $H=100Oe$ (vs. $10Oe$), so the total z.f.c. signal is positive in the entire temperature region.

4.2.2 M(H): the "Butterfly"

Another footprint of superconductivity is its characteristic *M(H)* dependence, sometimes called a "butterfly" [21]. An example of a classical butterfly is shown in Fig. 20 (*left*) for a MgB_2 sample. The measured *M(H)*-curves of the first sample (#L1-C128-Ga1) are shown in the *right* panel. As Fig. 19 suggests, the largest diamagnetic signal could be expected at the lowest temperature. Keeping in mind that in some cases Sr_2RuO_4 reveals superconductivity at temperatures up to *3K* [22], we measured *M(H)* at *T=4K* (Fig. 20) in a field *H=10 Oe* parallel to the processed surface.

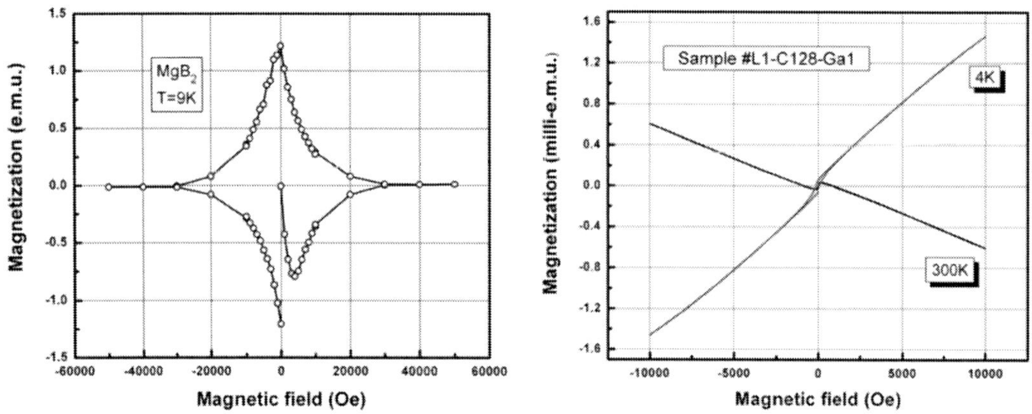

Figure 20. Classic magnetic behavior ("butterfly") of MgB_2 films (*left*) and the experimental record for the sample #L1-C128-Ga1 (*right*). These should be compared with Fig. 24 below.

The difference between two curves seems drastic, but we will see next that in reality they are almost similar qualitatively. The matter is that in addition to the signal from the laser-processed material, there are contributions from the unprocessed parent material, the glass substrate, the nichrome film remnants, and, possibly, inherent and ablation-induced ferromagnetism. To understand things better we explored a *control sample* (Figs. 21 and 22).

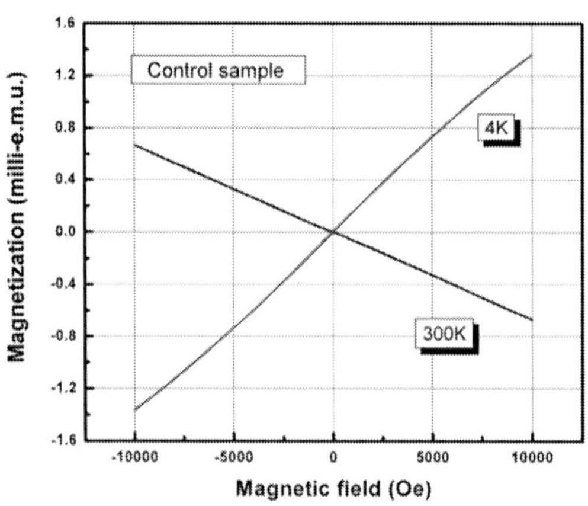

Figure 21. Magnetization of the control sample with no laser processing.

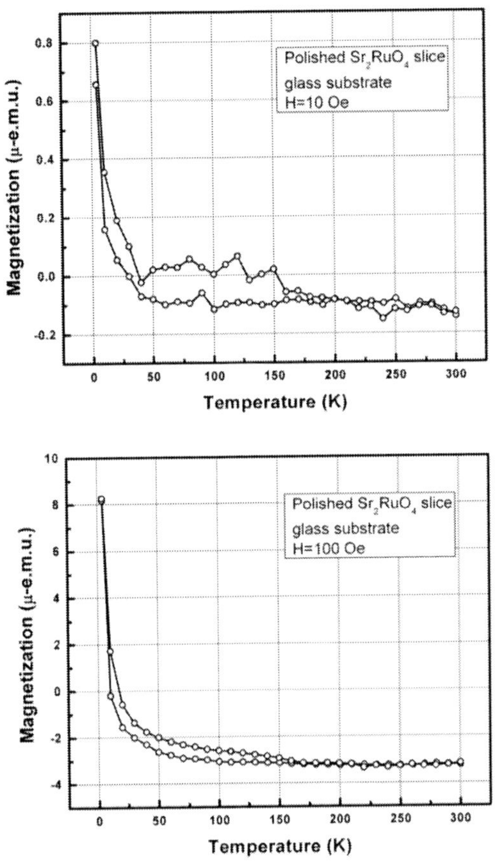

Figure 22. Magnetization vs. temperature of the control sample. Diamagnetism (caused by the substrate) at high temperatures is clearly distinguishable. At low temperatures paramagnetism overpowered it, which caused the rotation of the curves in Fig. 21.

It consisted of a polished parent crystal glued on the same type of substrate with the same glue and covered by the same wax as the first sample at magnetic measurements. *M(T)* results from *T=4K* to *300K* are shown in Fig. 22 for two values of the magnetic field. It is clear from Fig. 22 that the response at high temperatures is mainly diamagnetic (the signal from the substrate is dominating), and somewhere at lower temperatures it becomes paramagnetic (the parent material's paramagnetism is dominating). That yields the "rotation" of the *M(H)* curve when the temperature reduces from *300K* to *4K* and paramagnetism overrides the diamagnetism. Both processed (Fig. 20, *right*) and control samples (Fig. 21) reveal this feature. Let us consider them in greater detail.

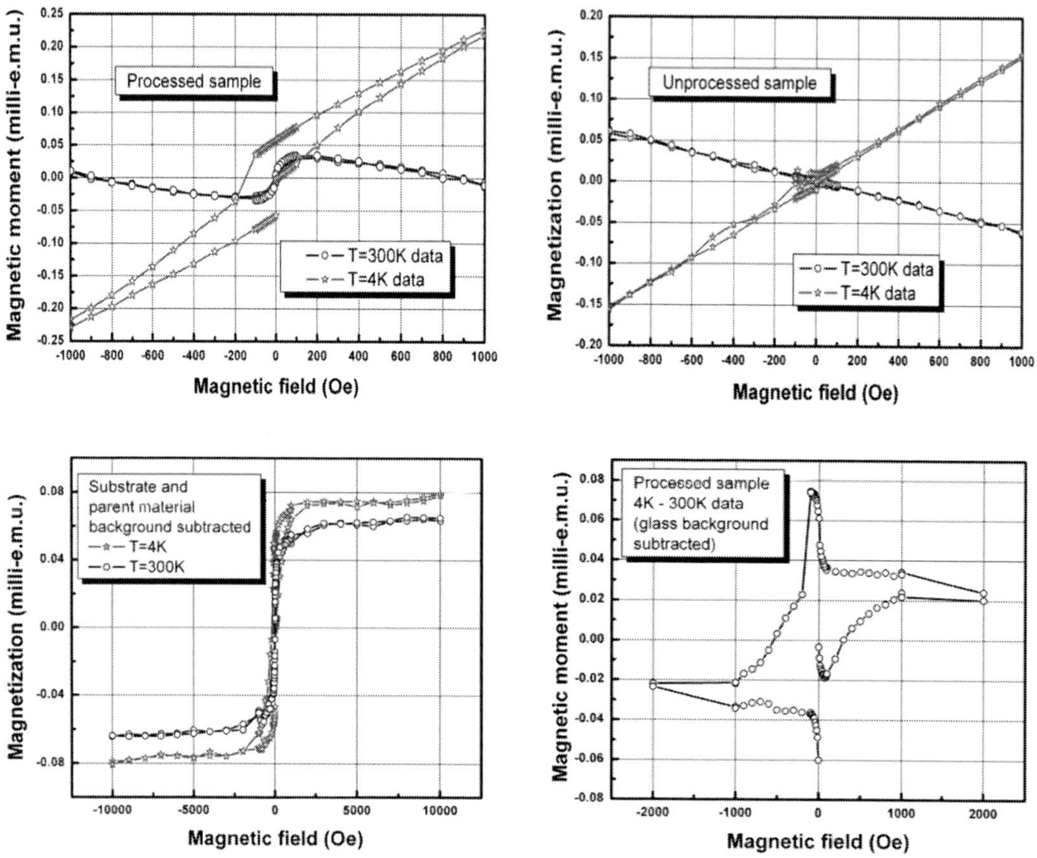

Figure 23. (*Top left*) -The first sample (#L1-C148-Ga1) data. (*Top right*) The substrate and parent material background. (*Bottom left*) Subtracted curves resulting in two sets of data. One of them (*at 300K*) is just an *S*-shape ferromagnetic *M(H)*-loop. The second one (*at 4K*) also contains a butterfly, which appears (*bottom right*) as soon as the *300K* curve is digitally subtracted from *4K* curve.

A simple analysis of the processed (Fig. 23, *top left*) and unprocessed (Fig. 23, *top right*) sample data allows us to exclude the contribution of the substrate and the parent material by digitally subtracting the right panel data from the left panel data[11]. The resultant curves are plotted in Fig. 23 (*bottom left*). The final result (Fig. 23, *bottom right*) stunningly resembles the "classic" superconductor butterfly (*cf.* Fig. 20, *left*), though a bit distorted. The distortion may be caused by our tacit assumption that the ferromagnetic contribution went unchanged from *300K* to *4K*. In reality, the ferromagnetism gains at cooling. Indeed, as is clear from the *left bottom panel* of Fig. 23, the horizontal (saturated) value of magnetization at *300K* is smaller than that at *4K* by about *20%*. Multiplying *300K* data by a factor of *1.2* (so that the horizontal "wings" of *M(H)* for both cases will coincide), and only afterwards subtracting the data, one can obtain a result shown in Fig. 24.

[11] Since magnetization is proportional to the volume factor, we tried to make the control sample possibly identical to the first sample. Nevertheless, it was required to multiply the curves in Fig. 21 by a numeric factor slightly different from 1 to have an exact match with the processed sample data.

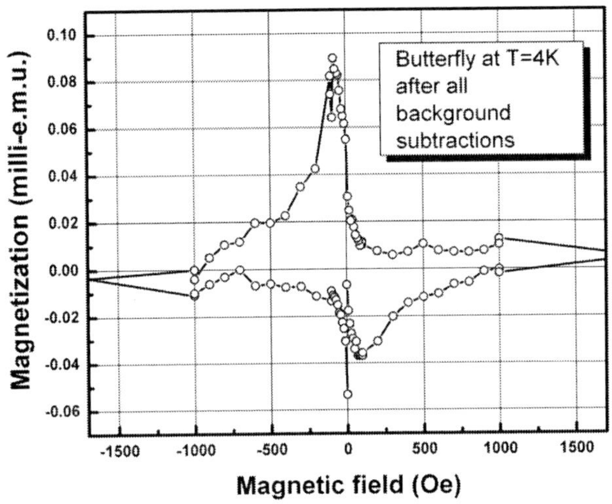

Figure 24. The magnetic behavior M(H) of the first sample after extraction of all relevant background contributions.

This figure demonstrates that after some simple logical steps of processing, the Meissner "butterfly" arrives for the first sample (L1-C128-Ga1). It is less perfect than classical examples, such as the one shown in Fig. 20 (*left*), but taking into account how many factors have been eliminated: the unprocessed parent's (super-) paramagnetic contribution, the diamagnetism of the glass substrate, the nichrome film remnants, and, possibly, inherent and ablation-induced ferromagnetism, and also how small the amplitude of the signal is, the overall result is not so bad.

On subsequent samples we replaced the glass by magnetically neutral substrates and were able to get more butterflies, such as shown in Fig. 25, at much higher temperatures just by removing the high-temperature signal background. Similar results were obtained with the sample #L19-C128-G1 (measured at the University of Minnesota). In this case the sample was mounted on a $SrTiO_3$ substrate.

Since all the samples were manufactured with a nichrome sublayer, one cannot remove completely the remnants of the ferromagnetic materials. Therefore, the subtraction becomes mandatory to reveal the features of the potential superconductor. Performing the subtraction is technically straightforward, but physically unsound if we do not take into account all the critical factors, as we did in the case of the first sample. Since the curves are obtained at different temperatures the temperature variation of the ferromagnetic response is the most uncertain factor. If treated improperly, it can create an artifact upon subtraction.

How unique is the butterfly to superconductivity? Can one achieve the butterfly pattern by applying the same subtraction procedures to, say, nichrome films? We will address these topics next.

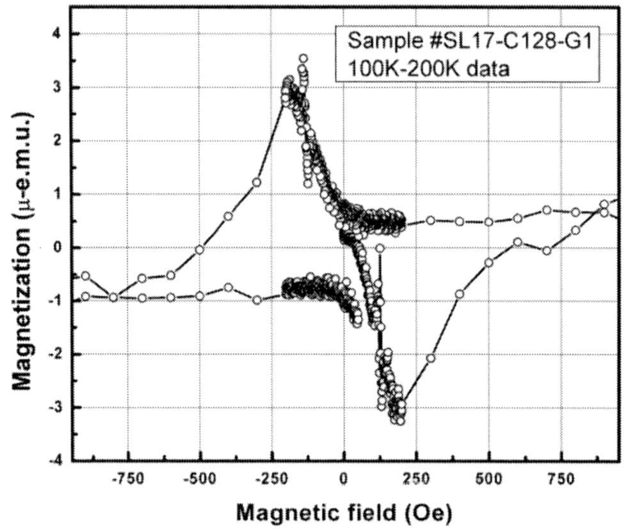

Figure 25. Butterfly of sample #L17-C128-G1 at *100K* (measured at NRL). Sample is mounted on a sapphire substrate. Single subtraction of *200K* data is used with no further corrections.

4.2.3 Nichrome Generated "Butterflies"

To find an answer to this question we made depositions of *70nm* thick nichrome films on glass substrates and measured their magnetic susceptibility in a manner in which it was done for ruthenates. Because the nichrome film has a Curie temperature between *200K* and *300K*, one should subtract the *200K*-curve from the lower temperature data. The outcome is shown in Fig. 26.

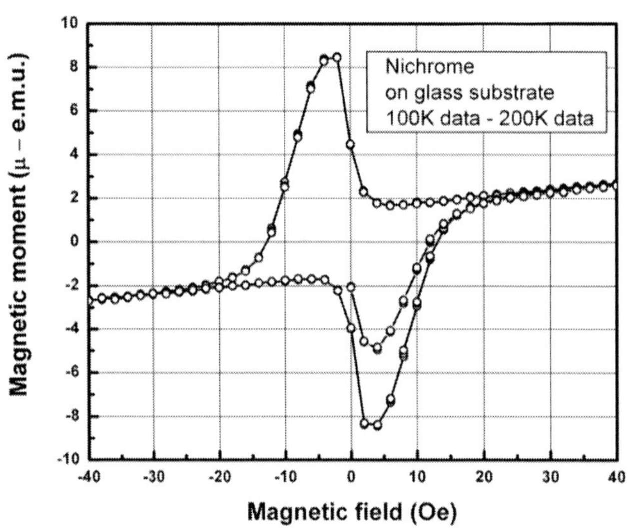

Figure 26. A spurious "butterfly" from the nichrome films on a glass substrate, with silver film on top of the nichrome. (Measurements with similar results at the Honda US Research Institute, Inc. and at the Quantum Design corporation.).

This is a very important result. It says that in principle the difference between two ferromagnetic *S*-loops at different temperatures could generate a butterfly. The role of the quantitative difference between the butterflies in Figs. 24, 25 and 26, especially the difference in the scales of the characteristic magnetic field, is not yet clear. An additional question, "Is it possible that the nichrome film itself at a nanoscale level became a superconductor?" seems not absurd in view of recent findings of iron-based superconductors. We will come back to this topic in the Discussion and Conclusions section.

4.2.4 Diamagnetism vs. (Ferro/Superpara/Meta) Magnetism

Since the existence of an *M(H)* butterfly turned out to be a non-unique criterion of superconductivity, it is instructive to re-consider the *M(T)* data and the related hysteresis below the transition temperatures (like that shown in Fig. 19). Again, our studies of nichrome film responses made it clear that one needs to distinguish not only between diamagnetism and ferromagnetism, but also consider superparamagnetism and metamagnetism. This type of magnetism is inherent in some conditions for representatives of the Ruddlesden-Popper series of ruthenates $Sr_{n+1}Ru_nO_{3n+1}$ with $n>1$ [22,23]. Indeed, despite the fact that our starting material is *2-1-4* ($n=1$), the samples may have some inclusions of *3-2-7* ($n=2$) and higher n representatives either originally or induced by the laser processing. What is important here is that both superparamagnets and metamagnets may provide curves topologically similar to Fig. 19. The only difference is that they cannot enter the diamagnetic region as our samples did.

Could there be any reasons other than superconductivity causing a behavior like that in Fig. 19? In principle, frozen magnetic moments at z.f.c. measurements may mimic negative susceptibility (magnetic momentum) or create an overall shift of both z.f.c. and f.c. curves. The likelihood of such an occasion seems relatively small, especially if we take into account the correlation between magnetic and resistive transition temperatures in this same sample. But we cannot exclude it completely, especially for such complex entities as these samples.

4.2.5 The New Material Confirmation

To get more understanding, we prepared for magnetic measurements a laser-processed sample without any nichrome layer involved. The processed area was small ($0.2x0.2mm^2$) so that the signal-to-noise ratio was also small, but it still provided a detectable magnetic response. The result of the measurement is shown in Fig. 27. The optimistic comment on this result is that the z.f.c. and f.c. curves join each other at $T\sim 250K$, and hysteresis starts at lower temperatures. Pessimists would say that the hysteresis starts immediately at *300K*. We leave it to the readers to judge for themselves.

As was stated, the observed diamagnetic response cannot be caused by superparamagnetism or metamagnetism because the z.f.c. enters the negative region. In any case, to explain the curves in Fig. 27 on a basis other than superconductivity, one needs to assume that there was a systematic shift in the vertical axis of the measuring device. That, in principle, can be the case, though the measurements were done in Quantum Design's own laboratory by an MPMS device with a residual field compensation module. If, indeed, this systematic mistake has occurred, then we have a superparamagnet with a ***hysteresis at 100K higher than ever reported in strontium ruthenates***. It is evident, therefore, that a new material has been created during the laser processing, whether it is a superconductor or a superparamagnet. Since this is established now, it is very important to obtain the structural parameters of this new material.

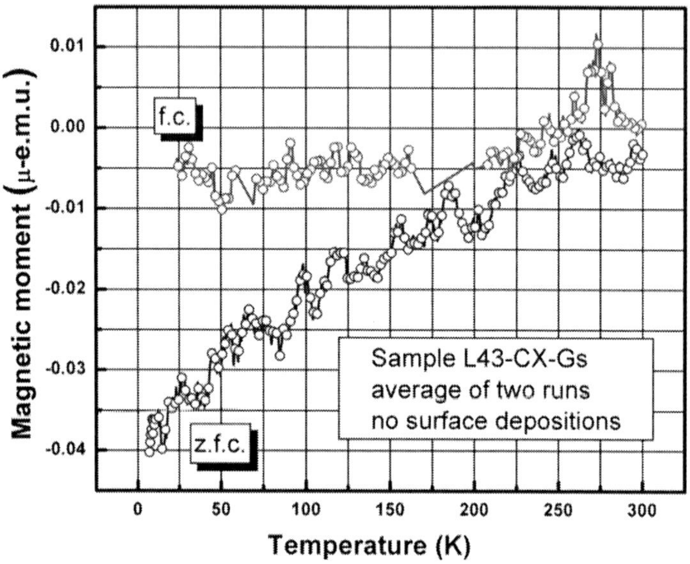

Figure 27. Magnetic hysteresis on Sr_2RuO_4 sample #L43-CX-Gs, laser-processed without the deposition of any metallic layer (the measurements were performed at Quantum Design corporation MPMS facilities with a module compensating the residual magnetic field).

5. Structural Explorations

X-ray diffraction data are very important for understanding the modifications caused by the laser-treatment in the ruthenate samples. A typical diffraction scan taken on the processed sample (sample #L22-Cfvϕ7, which represents the eutectic of *2-1-4*, *3-2-7* and *3-4-10* phases) is shown in Fig. 28, with the data in Table 2.

The *00l*-reflections corresponding to the phases *2-1-4*, *3-2-7* and *4-3-10* before and after the processing are shifted a little bit, which can be attributed to a systematic measurement drift. In general, these reflections are in good agreement with the data reported in the literature. For the sake of comparison we recall that for $Sr_3Ru_2O_7$: $a = b = 3.8875Å$, $c = 20.717Å$ [24]; for Sr_2RuO_4: $a = b = 3.8703Å$, $c = 12.7397Å$ [25]; and for $Sr_4Ru_3O_{10}$: $a = b = 3.9001Å$, $c = 28.573Å$ [26]. The background level after the laser processing became two orders of magnitude noisier. This noise may hide any new reflections caused by the laser treatment. Meanwhile, one new reflection is clearly noticeable in Fig. 28 (marked as an '*X*'). At the current moment we are unable to identify the *X*-peak, which appears after laser processing.

Table 2. Major peaks taken from sample #L22-Cfvϕ7 before and after laser processing [a)]

#	Before			After		
#	Miller Index	2θ	c (Å)	Miller Index	2θ	c (Å)
	Corresponding to Sr_2RuO_4					
1	0 0 2	13.92	12.72	0 0 2	13.94	12.71
2	Uncertain			0 0 4	28.04	12.73
3	0 0 6	41.74	12.98	0 0 6	41.78	12.97
4	0 0 8	59.84	12.36	0 0 8	59.84	12.36
	Corresponding to $Sr_3Ru_2O_7$					
5	0 0 2	8.58	20.61	0 0 2	8.62	20.52
6	0 0 4	17.14	20.69	0 0 4	17.18	20.64
7	0 0 6	25.8	20.72	Uncertain		
8	0 0 8	34.6	20.74	0 0 8	34.64	20.72
9	0 0 10	43.64	20.74	0 0 10	43.68	20.72
10	0 0 12	52.98	20.74	0 0 12	53.02	20.72
11	0 0 14	62.7	20.74	0 0 14	62.74	20.73
12	0 0 16	72.98	20.74	0 0 16	72.98	20.74
13	0 0 18	83.96	20.75	0 0 18	83.98	20.74
14				0 0 20	96.34	20.69
15				0 0 22	109.68	20.74
	Corresponding to $Sr_4Ru_3O_{10}$					
13	0 0 12	37.8	28.56	0 0 12	37.8	28.56
14	0 0 16	50.62	28.85	0 0 16	50.66	28.83
15	0 0 18	57.84	28.69	0 0 18	57.84	28.69
16	0 0 24	80.38	28.67	0 0 24	80.40	28.66
	Unknown phase					
X					42.60	

[a)] Peaks at $2\theta = 7.74$, 15.48, 31.2, 39.26, 47.56, 56.1, and 82.94 in Fig. 28 correspond to the K_β reflections.

In general, we are not yet able to interpret all the specific changes that the processing introduces to the crystalline lattice. But we can outline some peculiarities. Figure 28 represents a composition initially considered as the eutectic of *2-1-4, 3-2-7* phases (see Table 1). As it turned out the sample also contains a *3-4-10* phase. This multiphase content is typical for samples we have investigated (see also Figs. 29 and 30).

In processing the data we used the well-known relations:

$$(1/d)^2 = (h/a)^2 + (k/b)^2 + (l/c)^2 \tag{1}$$

$$2d \cdot \sin\theta = \lambda \tag{2}$$

where h, k, l are the Miller indexes, a, b, c are the lattice parameters, λ is the x-ray wavelength. Assuming also that for the "unknown" peaks $h = k = 0$, we can find:

$$\sin\theta_i / \sin\theta_j = (l_i \cdot \lambda_1) / (l_j \cdot \lambda_2), \tag{3}$$

if these reflections correspond to the two different wavelengths λ_1 and λ_2, and

$$\sin\theta_i/\sin\theta_j = l_i/l_j, \tag{4}$$

if they are caused by the same wavelength λ_1 or λ_2.

As follows from Table 2, the relations of $\sin\theta_i/\sin\theta_1$ can be represented as relations of integers for all the peaks except the one with $2\theta = 98.0^o$. Using the wavelength values ($Cu\ K_\alpha=1.54178Å$ and $K_\beta = 1.39217Å$) one can see that peak *#4* ($2\theta = 98.0^o$) corresponds to the K_β reflection of peak *#5*. Hence all the peaks correspond to the same phase. The minimal even integers that can be assigned to l are given in Table 3. They yield the lattice parameter $c = 66.58Å$. It is not clear how this lattice parameter contributes to the whole picture.

It is interesting to notice the very low-energy peak on the diffraction scan which corresponds to the angle *0.6* degrees. This line was existent at almost all of our diffraction scans of the ruthenate samples, and was referred to in [10] as a superstructure existent in the processed materials. However, we now know that it also exists in non-processed samples, although, to our knowledge, it was never reported in ruthenates[12].

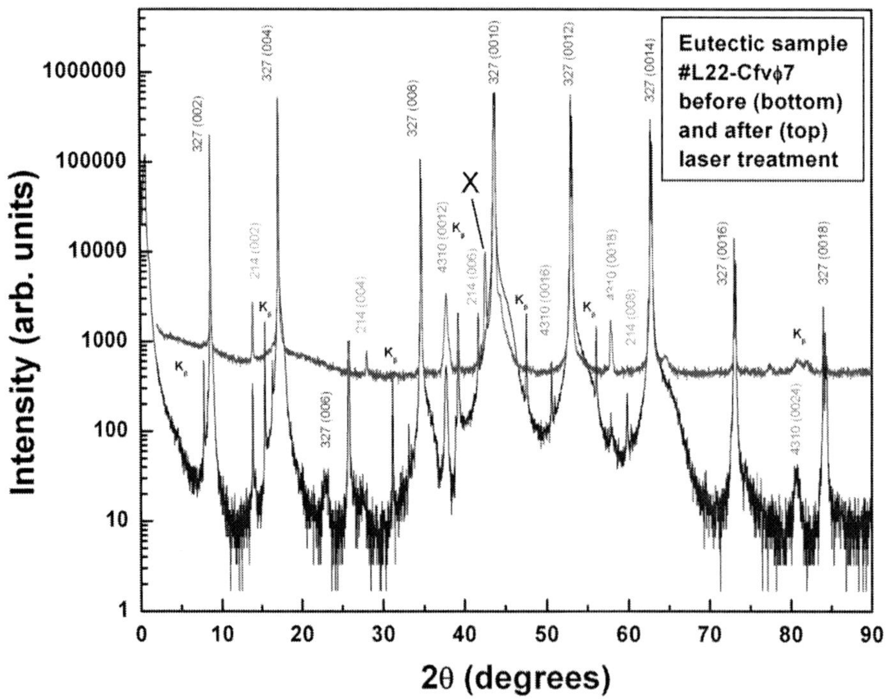

Figure 28. X-ray diffraction data of sample #L22-Cfvϕ7.

[12] This peculiarity was pointed out by Dr. S. Qadri (NRL).

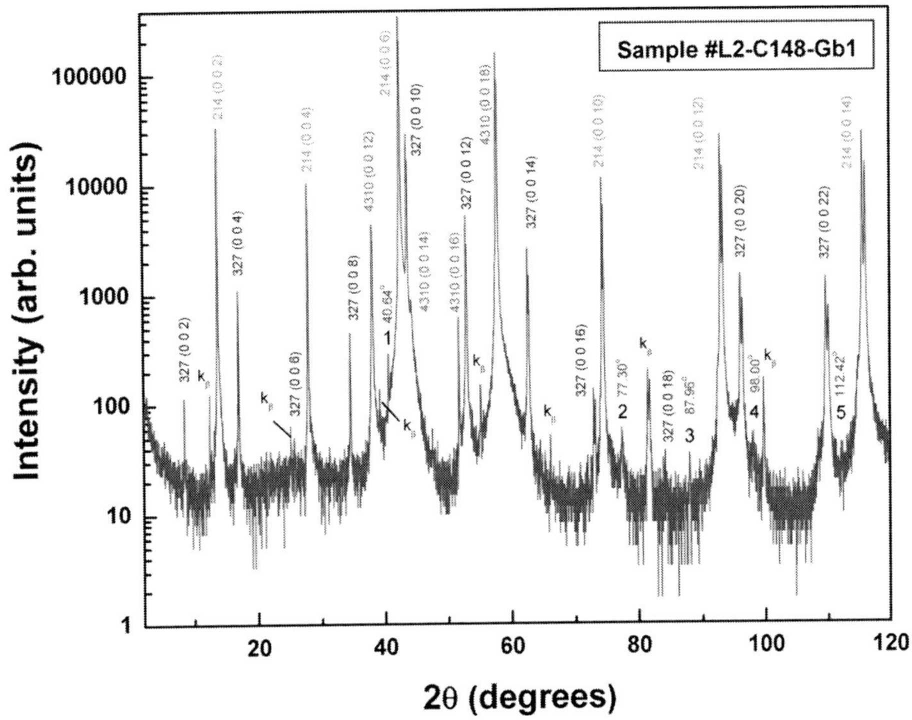

Figure 29. X-ray diffraction data of sample #L2-C148-Gb1 after laser processing.

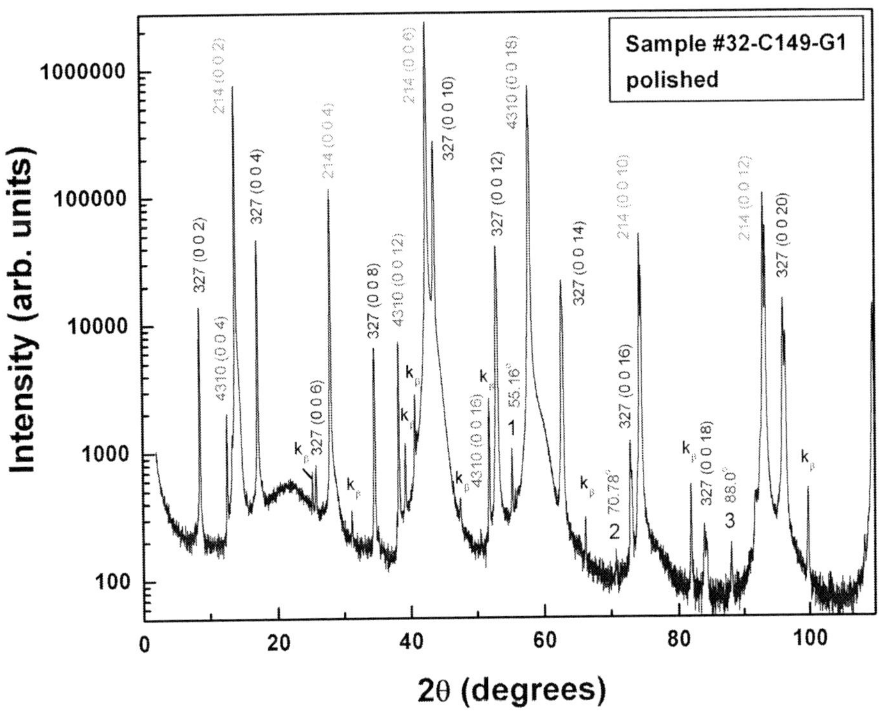

Figure 30. X-ray diffraction data of sample #32-C149-G1 before laser processing.

In these figures the peaks are all marked up and identified except for a few. Those are also indicated and marked up by numerals and detailed in Table 3.

Table 3. Unknown phase peaks in Figures 28-30

Peak #	2θ (degree)	sinθ	$\sin\theta_i/\sin\theta_1$	Ratio and accuracy	I
#L2-C148-Gb1					
1	40.64	0.3472	1	1	30
2	77.30	0.6245	1.799	9/5 ± 0.001	54
3	87.96	0.6944	2.000	2	60
4	98.00	0.7547	2.174		
5	112.42	0.8311	2.394	12/5 ± 0.006	72
#32-C149-G1					
1	55.16	0.4630	1.334	4/3 ± 0.001	40
2	70.78	0.5791	1.668	5/3 ± 0.001	50
3	88.00	0.6947	2.001	2 ± 0.001	60

6. Oxygen Content

Since the laser treatment took place in air, and extremely high temperatures exist in the laser plume, it is meaningful to expect oxygen enhancement in the modified samples. Meanwhile, precise quantitative analysis of the oxygen concentration using energy dispersive spectroscopy (EDX) is quite doubtful for these *Sr-Ru-O* samples and requires additional studies, as we have proven.

6.1 Problems with EDX Oxygen Analysis

Because oxygen is a light element, it can re-emit the radiation absorbed from the emission of other heavier elements prior to being detected by the spectrometer. This re-emitted radiation will effectively enhance the signal indicating oxygen presence. Untreated samples typically are polished and all the laser-treated samples have a very rough surface. Hence, the comparison of oxygen levels between two samples with such differing surface morphology is subject to great error.

Keeping this in mind, we looked at the EDX data from the processed and unprocessed ruthenates of different sample compositions (Table 4). Some of the difference between different treated samples could be partly attributed to the known, progressive detuning of the laser and reduction of its plume energy density. The error bars of EDX data, obviously, should be set high-enough: since *Sr* may only have a valence of *+2*, and the maximal valence of *Ru* is *+8*, the possible ratio of *O/Ru* cannot be higher than *6*. Meanwhile, Table 4 has entries of $O/Ru>7$. That means that the accuracy of determining the *O/Ru* ratio is not better than *20%*. To properly quantify the value of uncertainty in the measurements, one should address the surface-scattering effects caused by the laser processing induced roughness.

Table 4. "As measured" ratios before and after laser processing

Sample # and composition (nominally Sr_2RuO_4 +Ag+ nichrome, if otherwise not stated)	Data before treatment		Data after treatment	
	Sr/Ru	O/Ru (nominal)	Sr/Ru	O/Ru (max)
L0-C156-Ga	1.97	3.8	No data	No data
L2-C148-Gb1	No data	No data	2	6.48
L3-C148-Ga2	No data	No data	2.15	5.35
L13-C128-G1	2	3.97	2.07	6.61
L14-C128-G1	1.98	3.9	2.09	6.52
L15-C128-G1 (No deposited metals)	1.97	3.91	2.04	6.34
L16-C120-G1	No data	No data	2.06	7.18
L17-C128-G1	2.03	3.85	2.02	6.5
L18-C128-G1 (Nichrome only deposited)	2	3.88	2.07	7.08
L19-C128-G1	2	4	2.05	6.6
L21-C622-G1 ($Sr_3Ru_2O_7$)	1.47	3.34	1.49	5.18
L22-Cfvϕ7 (Eutectic: $Sr_3Ru_2O_7 + Sr_2RuO_4$)	1.5	3.29	1.55	5.5
L31-C149-G1	2	3.79	2.03	5.26
L32-C149-G1	1.99	3.95	2	5.42
L34-C149-G1	No data	No data	1.99	5.31
L35-C149-G1	2.06	4.03	2.06	5.89
L37-C149-G1	1.98	3.89	No data	No data
L38-C149-G1	1.99	3.91	No data	No data
L41-CX-Gs (No metals deposited)	1.99	4	2.05	5.38

6.2 Laser Processing vs. Surface Roughening

To clarify the roughness issue, we performed comparative investigations of oxygen content in five families of processed and unprocessed oxides, namely: sapphire (Al_2O_3), lead molibdate ($PbMoO_4$), lithium niobate ($LiNbO_3$), garnet ($Y_3Al_5O_{12}$) and strontium ruthenate (Sr_2RuO_4 and $Sr_3Ru_2O_7$). These data were received by standard EDX-analysis, using various values of magnification (electron probe energy *10* and *20 keV*). The nonconducting crystals were covered by approximately *20nm* of carbon. The data were taken first from initially polished samples (and used as standard for subsequent measurements), then from laser-processed samples, and, eventually, from mechanically scratched samples. The scratching was performed by abrasives, on many samples, so that we had surfaces corresponding to different roughness profiles.

Sapphire, since both *Al* and *O* are light elements, did not demonstrate much variation of composition in either laser-processed samples or scratched samples. Since this is in accordance with expectations, we will not dwell more on this case. Other materials, having one or more heavy elements in their compositions, demonstrated oxygen enhancement in laser-ablated crystals. This enhancement in some non-ruthenate cases was as high as *17%*. But it was possible in these same non-ruthenates to use surface roughness to get as much or even more enhancement. In contrast, in the ruthenates the laser-ablation enhanced the values by *60%* and more while the highest result obtained by scratching was *20%*. We should acknowledge that because of the mechanical properties of this material *it was not possible to scratch it deep enough* to obtain higher "density" and deeper profiles of scratches (Fig. 31).

Figure 31. Comparison between laser-treated (*first column*) and scratched by abrasive (*two next columns*) samples of oxide materials: $LiNbO_3$ (*first raw*), $PbMoO_4$ (*second raw*), $Y_3Al_5O_{12}$ (*third raw*), and Sr_2RuO_4 (*forth raw*). Laser ablation results are shown for three different intensities for Sr_2RuO_4 in the left bottom panel. Scale is the same in all panels.

6.3 Special Tests in Argon Atmosphere

To get more reliable evaluation of oxygen content in laser-processed samples, we undertook special treatments in a chemically neutral atmosphere. To treat the samples in an atmosphere other than air, we prepared a miniature bell-jar addition to the objective lens of the laser ablation device. The bell-jar was filled by nitrogen or argon under pressure to diminish the amount of oxygen available for reaction. Processing in the argon atmosphere, in addition to the possible chemical changes, introduced also a very definite change in the surface roughness (Fig. 32).

Figure 32. Laser treatment of polished S_2RuO_4 in air and in argon. In the *left panel* the left corner area is treated in argon gas, while the right half contains an area treated in air. These areas are shown in the *right* and *middle panels* respectively. The scales are *400μm* on the left panel and *100μm* on others.

As one can notice, the roughness is higher in the case of processing in argon, and the effect is bigger than any potential changes in the oxygen content. Indeed, the EDX results showed that both the treatments in air and in argon provide very high oxygen composition, with higher values in the argon case (Table 5). This is certainly nonsense.

Table 5. EDX data on average atomic ratios [a)]

	Treatment in Air	Treatment in Argon
O/Ru	6.48	7.26
Sr/Ru	2.04	2.13

[a)] Data taken from 6 areas in each of the treated regions in Fig. 32.

That confirms the conclusions of the previous section about the crucial influence of surface profile on the accuracy of the EDX analysis. Obviously, the judgment on the oxygen content in rough-surface strontium ruthenates should be considered correct only after further and more elucidating studies. The research is this area is still ongoing and the results will be published separately after completion.

7. Discussion and Conclusions

The text above already has a lot of discussion of specific experimental details. We will try to create the grander pictures now.

First of all we should acknowledge that being limited in available resources we have not yet been able to explore as many samples as required, in a way we would like to study them, for a well-accomplished scientific research. But we think that we have accumulated enough facts and results to demonstrate the promises, which this method and this particular material have opened for researchers.

Of course, there are major factors, such as the zeroing of resistance or the Meissner signatures, the presence of which are critical for claiming superconductivity. But there are also other, at the first glance, minor factors, which can reveal a lot to prepared minds. For example, let us consider the accidental presence of ferromagnetic inclusions in our samples.

For us, their initial, unexpected discovery caused shock and disbelief in our own magnetic results. Indeed, the logical purity of conclusions was seemingly violated! But shortly afterwards, the new iron-containing superconductors have been discovered [27], and the critical temperature in these bulk materials climbed above *50K* in a very short while [28]! In view of the "de facto" physics of these materials the presence of ferromagnetic dopants may be crucial[13] for the delivery of superconductivity at 250K in our case. Thus, a crucial argument against out results unexpectedly became a strong argument in favor of them – this is what Einstein called the "drama of ideas"; it occurs not only in theory, but in experiment as well, being a fundamental and inherent property of Nature herself.

Let us come back to summarizing our approach. We tried hard to squeeze out from the available samples as much information as possible. It was not possible to describe all the facts in this chapter without the risk of boring the reader. Rather, we spoke of essentials. Of the small number of samples measured, three out of eight for which the data from the first thermal cycle were available have demonstrated resistive transitions to the $R=0$ state[14]. By this we mean that the voltage measured went to zero at a constant current. We made efforts to distinguish possible misleading effects, such as a zeroing of the voltage by a hypothetical separation of current and voltage circuits in a *3D*-sample. Such a separation looks unlikely, since the transition was measured many times in different cryostats, and most likely, with permutated lead configurations. To eliminate this cause completely we also undertook a two-probe approach. In this method the measured circuit has contact resistance added, and as a result may show *reduction* of resistance both in the case of superconductivity and in the case of the reduction of contact resistance at cooling. To eliminate the contact effect we registered the transition simultaneously in the four- and two-probe techniques. Given the positive results the only alternative to a superconducting transition is cracking in the four-probe circuit and *anti-cracking* (*i.e.,* healing of the cracks) in the two-probe circuit, which must have taken place simultaneously. These worries would be considered extraneous in an ordinary case, but superconductivity above *200K* cannot be considered an ordinary claim... Even so, the occurance of resistive transition, taken alone, cannot prove superconductivity, even if reproduced limitlessly. They should be complemented by magnetic susceptibility measurements. It is very attractive to put the resistive and magnetic transition curves from our first sample on a joint panel and consider them to be caused by very-high temperature superconductivity. This optimistic conclusion can even be enhanced with the data obtained by the magneto-optical analysis of the first sample (Fig. 33).

After having some fun with this, one should exercise good caution, by recalling in mind that this sample has a noticeable ferromagnetic signal at room temperature. Despite our great efforts to eliminate the ferromagnetic contribution from the magnetic response this may not be accurate enough to really distill superconductivity. Then one can suspect that the negative values of susceptibility are caused by frozen magnetic moments in the sample, and that the correlation in Fig. 33 is "just an occasion by chance". These are the realities of "unbeaten path" research, which makes its results vulnerable to all sorts of skepticism until certified by others. Skepticism may have some specific grounds here since the magnetic materials, as we

---

[13] A very recent paper [29] reported on superconductivity in the composite material that contains 4 elements three of which (Fe, Ru and Sr) are present in our case (our material has actually a bigger palette including Cr and Ni in addition to Fe).

[14] The other five showed strong evidence of cracking during the first thermal cycle.

demonstrated, can in principle reveal the butterfly-type feature when the high-temperature data are subtracted from the low-temperature data. Of course, conversely, that does not prove that the butterflies recorded on processed ruthenates are caused by ferromagnetism. Moreover, the spurious butterflies from the nichrome samples may completely disappear as soon as one will apply the same procedures of adjustment of magnetic signals at different temperatures, as we did for the laser-processed sample described in Section 4.2.2.

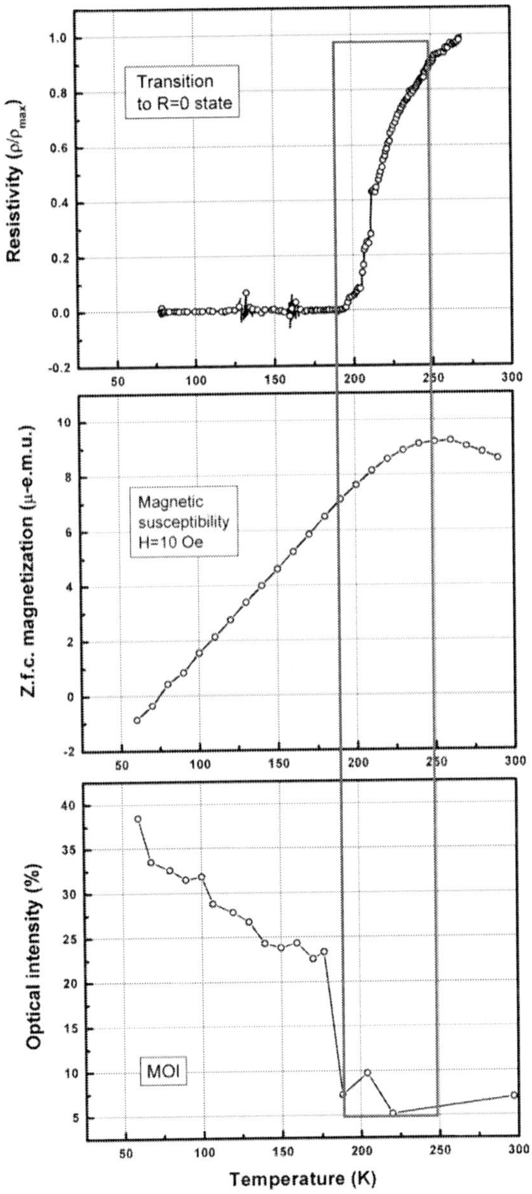

Figure 33. Correlation between resistive and magnetic transitions in the sample #L1-C148-Ga1.

Two questions are now closely related: "Do we have any reasons to expect higher T_c in these laser-processed ruthenate samples?" and "Which material is responsible for the transitions?" We discussed above the oxygen enhancement effect. Though it is not yet possible to reliably quantify this effect, one can speculate about the analogy with cuprates, in which the oxygen content (and doping in general) is crucial for superconductivity. For example, for most known $YBa_2Cu_3O_{6+\delta}$, relatively small variations of oxygen can shift the properties of this material from dielectric to superconductive (it is enough to change δ from *0* to *0.4*, see, *e.g.*, [21]). A closer example is $Sr_2CuO_{4-\delta}$, which is known as a semiconductor at $\delta=0.1$ [30]. It exhibits superconductivity at *50K* when its oxygen content is enhanced [31]. The ability of *Cu* to take on different valences plays a critical role in allowing these modified species to exist. Ruthenium is known to have valences higher than *+4* [32], so surprises are possible from general theoretical grounds.

More probable is the existence of small volumes with high-oxygen content and/or a structural unit other than *2-1-4*. Perhaps, in our case, short laser pulses and high local fluencies helped to attain non-equilibrium regimes for the oxygen doping to create new crystalline structure. We cannot be certain of this idea since the x-ray diffractometry (XRD) still keeps the question open, though the change of the oxygen content in $Sr_2RuO_{4+\delta}$ is possible up to $\delta=2$. In the previous attempts of others, it was not easy to change the oxygen content in this material [33]. Our attempt and investigations still keep the door open for it. Another opportunity may involve the lattice deformation (*e.g.*, tilting or compression). The coexistence of different crystallographic structures, noticeable on the XRD-data, makes the range of possibilities even wider. More samples, more explorations, and more time are definitely required for a final understanding of what is going on.

Since its inception, the history of superconductivity has been full of very dramatic events. The level of drama seems to grow with each closer step to room temperature. The impression is that the level of efforts required for getting to a higher critical temperature (T_c) rises exponentially with T_c.

"The most interesting yet are trials aimed at raising critical temperature to create room-temperature superconductors. There is no recipe for success here. However, the existing theoretical grounds do not contradict this possibility in any way... It is quite possible that as far as the room-temperature superconductivity goes, the skeptics would this time prove to be right. Still, it is imperative to keep searching for that kind of superconductivity, and I, for one, believe in it."[34]

REFERENCES

[1] A.M. Gulyan, Paramagnons in imbalanced ^{3}He-A, Sov. Phys. - *Lebedev Inst. Reports* **9**, 45 (1986).

[2] A.M. Gulyan and O.N. Nersesyan, Phonon-pumped nonequilibrium states in superfluid He^3, Sov. *Phys.-JETP* **68**, 756 (1989).

[3] A.M. Gulian and K.S.Wood, Triplet superconductors as the basis for solid-state quantum computing, *J. Opt. B* **5**, S619 (2003).

[4] S. Madhavan, D.G. Schlom, A. Dabkowski, H.A. Dabkowska, and Y. Liu, Growth of epitaxial a-axis and c-axis oriented Sr_2RuO_4 films, *Appl. Phys. Lett.* **68**, 559 (1996).

[5] M.A. Zurbuchen, Y.F. Jia, S. Knapp, A.H. Carim, D.G. Schlom, L.N. Zou, Y. Liu, Suppression of superconductivity by crystallographic defects in epitaxial Sr_2RuO_4 films, *Appl. Phys. Lett.* **78**, 2351 (2001).

[6] M.A. Zurbuchen, Y.F. Jia, S. Knapp, A.H. Carim, D.G. Schlom, and X.Q Pan, Defect generation by preferred nucleation in epitaxial $Sr_2RuO_4/LaAlO_3$, *Appl. Phys. Lett.* **83**, 3891 (2003).

[7] F. Lichtenberg, The story of Sr_2RuO_4, *Progr. Solid State Chem.* **30,** 103 (2002).

[8] Y. Maeno, H. Hashimoto, K. Yoshida. S. Nishizaki, T. Fujita, J.G. Bednorz, and F. Lichtenberg, *Nature* (London) **372**, 532 (1994).

[9] T. H. Geballe, The never-ending search for high-temperature superconductivity, *J. Supercond. and Novel Magnetism* **19**, 261 (2006).

[10] A.M. Gulian, K.S. Wood, D. Van Vechten, J. Claassen, R.J. Soulen, Jr., S. Qadri, M. Osofsky, A. Lucarelli, G. Luepke, G.R. Badalyan, V.S. Kuzanyan, A.S. Kuzanyan, V.R. Nikoghosyan, Evidence for High-Temperature Superconductivity in Doped Laser-*Processed Sr-Ru-O, cond-mat*/0509313 (2005).

[11] A.M. Gulian, V.R. Nikoghosyan, D. Van Vechten, and K.S. Wood, Simultaneous recording of two- and four-probe resistive transitions in doped laser-processed Sr-Ru-O, arXiv:0705.0641 (2006); published in: J. Contemp. Phys. **42**, 1068 (2007).

[12] M.C. Polo, J. Cirfe, G. Sanchez, R. Anguliar, M. Varela, and J. Estive, Pulsed laser deposition of diamond from graphite targets, *Appl. Phys. Lett.* **67**, 435 (1995).

[13] M.D.Shink, P.A.Motian. Ultra-short pulsed laser ablation of highly oriented pyrolytic graphite, *Carbon* **39**, 1183 (2001).

[14] A. Mechler, P. Heszler, Z. Kantor, T. Szorenyi, and Z. Bor, Excimer laser irradiation induced formation of diamond-like carbon layer on graphite, *Applied Surface Science* **139**, 174 (1999).

[15] M.E. Kozlov, K. Yase, N. Minami, P. Fons, H.-A. Durand, A. N. Obraztsov, K. Nozaki, and M. Tokumoto, Observation of diamond crystallites in thin films prepared by laser ablation of hard fullerene-based carbon, *J. Phys. D: Appl. Phys.* **29**, 929 (1996).

[16] A.P. Mackenzie, Y. Maeno, The superconductivity of Sr_2RuO_4 and the physics of spin-triplet pairing, *Rev. Mod. Phys.* **75**, 657 (2003).

[17] Z.Q. Mao, Y. Maeno, H. Fukuzawa, Crystal growth of Sr_2RuO_4, *Mater. Res. Bull.* **35**, 1813 (2000).

[18] R.S. Perry, Y. Maeno, Systematic approach to the growth of high-quality single crystals of $Sr_3Ru_2O_7$, *J. Crystal. Growth* **271**, 134 (2004).

[19] R. Fittipaldi, A. Vecchione, S. Fusanobori, K. Takizawa, H. Yaguchi, J. Hooper, R.S. Perry, Y. Maeno, Crystal growth of the new Sr_2RuO_4-$Sr_3Ru_2O_7$ eutectic system by a floating-zone method, *J. Crystal Growth* **282**, 152 (2005).

[20] J. Matricon and G. Waysand, *The cold wars: a history of superconductivity,* (Rutgers University Press, New Brunswick, 2003), p. 196.

[21] K. Fossheim and A. Sudbø, *Superconductivity: physics and applications*, (Wiley, 2005), p. 41.

[22] H. Yaguchi, R.S. Perry, and Y. Maeno, Ferromagnetism induced by uniaxial pressure in the itinerant metamagnet $Sr_3Ru_2O_7$ (LT-24).

[23] R.S. Perry, L.M. Galvin, S.A. Grigera, L. Capogna, A.J. Schofield, A.P. Mackenzie, M. Chiao, S.R. Julian, S.I. Ikeda, S. Nakatsuji, Y. Maeno, and C. Pfleiderer, Metamagnetism and critical fluctuations in high quality single crystals of the bilayer ruthenate $Sr_3Ru_2O_7$, *Phys. Rev. Lett.* **86**, 2661 (2001).

[24] H. Shaked, J.D. Jorgensen, S. Short, O. Chmaissem, S.I. Ikeda, Y. Maeno, Temperature and pressure effects on the crystal structure of $Sr_3Ru_2O_7$: Evidence for electronically driven structural responses, *Phys. Rev. B* **62**, 8725 (2000).

[25] O. Chmaissem, J. D. Jorgensen, H. Shaked, S. Ikeda, and Y. Maeno, Thermal expansion and compressibility of Sr_2RuO_4, *Phys. Rev. B* **57**, 5067 (1998).

[26] M. K. Crawford, R. L. Harlow, W. Marshall, Z. Li, G. Cao, R. L. Lindstrom, Q. Huang, and J. W. Lynn, Structure and magnetism of single crystal $Sr_4Ru_3O_{10}$: a ferromagnetic triple-layer ruthenate, *Phys. Rev. B* **65**, 214412 (2002).

[27] Y. Kamihara, T. Watanabe, M. Hirano, and H. Hosono, Iron-based layered superconductor $La[O_{1-x}F_x]FeAs(x=0.05\text{-}0.12)$, *J. Am. Chem. Soc.* **130**, 3296 (2008).

[28] Z.A. Ren, J. Yang, W. Lu, W. Yi, X.L. Shen, Z.C. Li, G.C. Che, X.L. Dong, L.L. Sun, F. Zhou, and Z.X. Zhao, Superconductivity at 55K in iron-based F-doped layered quarternary compound $Sm[O_{1-x}F_x]FeAs$, *Cin. Phys. Lett*, **25**, 2215 (2008).

[29] Y.Qi, L. Wang, Z. Gao, D. Wang, X. Zhang, and Y. Ma, Superconductivity induced by doping Ru in $SrFe_{2-x}Ru_xAs_2$, arXiv: 0904.0772 (2009).

[30] R. C. Lobo, F. J. Berry, and C. Greaves, The synthesis and structural characterization of Sr_2CuO_{4-x}, x~0.1, *J. Solid State Chem.* **88**, 513 (1990).

[31] S. Karimoto, H. Yamamoto, H. Sato, A. Tsukada, and M. Naito, T_c versus lattice constants in MBE-grown M_2CuO_4 (M=La, Sr, Ba*), J. Low Temp. Phys*. **131**, 619 (2003).

[32] http://www.webelements.com/webelements/compounds/text/Ru/O4Ru1-20427569.html.

[33] S. Nishizaki, Y. Maeno, and T. Fujita, Effect of annealing on the superconductivity of Sr_2RuO_4, *J. Phys. Soc. Jpn.* **65**, 1876 (1996).

[34] V.L. Ginzburg (Foreword) in: A.M. Gulian and G.F. Zharkov, *Nonequilibrium electrons and phonons in superconductors* (Kluwer Academic/Plenum, New York, 1999), p. vi.

In: Bulk Materials: Research, Technology and Applications ISBN: 978-1-60692-963-6
Editors: T. Frias, V. Maestas, pp. 393-422

Chapter 10

EFFECTIVE-MASS THEORY OF NARROW-GAP SEMICONDUCTORS

Y. H. Zhu, X. W. Zhang and J. B. Xia
State Key Laboratory for Superlattices and Microstructures, Institute of Semiconductors, Chinese Academy of Sciences, Beijing 100083, China.

ABSTRACT

The **k·p** method is particularly appropriate to study many physical properties of semiconductors which depend on a small region of **k** space. Since it was proposed, the **k·p** effective-mass envelope function theory has been extensively used in all kinds of semiconductor structures, such as bulk materials, heterojunctions, superlattices and even nanostructures. In this section the energy levels in the conduction band and valence bands of narrow-gap semiconductor is studied in the framework of the eight-band effective-mass method. Especially the eight-band effective mass Hamiltonian is written in the spherical and cylindrical coordinates in order to study the electronic structures of quantum dots and wires. The interaction between conduction and valence bands has been treated exactly, and a finite spin-orbital coupling effect is taken into account. This method is particularly adapted to such problems as arise in connection with the electric and magnetic properties in the external fields. The eight-band Hamiltonian of an electron or hole in a diluted magnetic semiconductor, in the presence of an external homogeneous magnetic field, is given using a mean-field approximation. We applied the effective-mass theory to the electronic structures of semiconductor quantum dots and quantum wires, and discussed some physical properties in these nanocrystal structures. Exciton transition behaviors, polarization properties of emission, Landé electron g factors, and giant Zeeman splitting effect were discussed, and our theoretical results are compared with the available experimental data.

I. INTRODUCTION

The **k·p** method is primarily a method to explore the band structure in the vicinity of a given point, $\mathbf{k}_0$, in great detail. As such it is particularly appropriate for semiconductor studies because in semiconductors the free carriers (electrons or holes) are also confined to a small region of **k** space and the detailed properties of this small region have been investigated very thoroughly experimentally by transport studies, cyclotron resonance, optical absorption and reflection, etc.

Kittel and Mitchell [1], Luttinger and Kohn [2] developed a new method of effective-mass equation of electron (or hole) movement in a perturbed periodic structures using the **k·p** method. They used the second-order perturbation theory to obtain the effective-mass equations for the conduction band and the valence bands separately. This method is especially suitable for those problems associated with impurity states and cyclotron resonance, etc. in Si and Ge. Kane [3] used the **k·p** method to calculate the energy band structure of InSb. Because the energy gap between the conduction band and the valence bands is very small, he dealt with the interaction between the conduction band and the valence bands accurately, and used the perturbation theory to consider the influence of higher bands. Hence, the effective-mass equation is 8×8 dimensional, including the conduction band and the valence bands, and the spin doublet states. Using this equation, Kane found that the conduction band of InSb is highly non-parabolic. The calculated cyclotron resonance mass, absolute amplitude of optical absorption, and their dependence on the impurity concentration, are in good agreement with the experimental results. This verified the accuracy of the Kane's 8 bands model.

In the long development period of the semiconductor physics from 60th of last century to now the effective-mass theory based on the **k·p** method plays an important role in understanding the electronic structures of various semiconductor structures under different external fields. For example, the shallow donor and acceptor states, the cyclotron resonance , the magnetic optical spectra, transport properties, the g factors, and the Curie temperatures, etc. The materials and structures are from bulk material to quantum wells, superlattices, quantum dots, quantum wires, and dilute magnetic semiconductors, etc.

Quantum dots and wires of semiconductors which are 1 to 10 nm in diameter exhibit quantum confinement effects, for example: 1. Absorption or luminescence peak energy as function of nanocrystal size (diameter); 2. New optical, electric, magnetic, and transport properties; 3. Large non-linear optical response and luminescence property of Si, etc. They are the zero- and one-dimensional analog of quantum wells, they have been fabricated mainly by: 1. Molecular beam epitaxy (MBE) and metal organic chemical-vapor deposition (MOCVD); 2. Chemical decomposition of organometallic precursors in an appropriate solvent; 3. Etching of electron beam, etc.

In this chapter we first introduce the eight-band effective mass Hamiltonian. The magnetic energy levels of the conduction band and valence bands of narrow-gap semiconductors and diluted magnetic semiconductors are studied in the framework of the eight-band effective-mass method. The eight-band Hamiltonian of diluted magnetic semiconductors in the presence of an external magnetic field is given using a mean-field approximation. Then the eight-band model is written in the spherical and cylindrical coordinates in order to study the electronic structures of quantum dots and wires grown by the chemical method. This model is extended to such problems as arise in connection with the electric and magnetic properties in the external fields. We applied the eight-band model to study the electronic structures of narrow gap semiconductor quantum dots and quantum wires, and discussed some their physical properties. The rest of the chapter is organized as follows: In Sec. II. we present the eight-band Hamiltonian. The magnetic energy levels of narrow gap semiconductors are given in Sec. III. The electronic structures of quantum dots are given in Sec. IV. The giant Zeeman splitting in Mn-doped quantum dots and wires are given in Sec. V. Finally, we draw a brief summary in Sec. VI.

II. Effective-Mass Eight-Band Model

The eight-band effective-mass Hamiltonian for the zinc-blende semiconductors with the **k·p** method at Γ point of the Brillouin zone was derived for the case of neglecting the spin-orbit coupling.[4] It is represented in the following form for the Bloch basic functions of $|S\rangle$, $|X\rangle$, $|Y\rangle$, and $|Z\rangle$,

$$H=\frac{1}{2m_0}\begin{pmatrix} \varepsilon_g+\alpha p^2 & ip_0p_x & ip_0p_y & ip_0p_z \\ -ip_0p_x & -Lp_x^2-M(p_y^2+p_z^2) & -Np_xp_y & -Np_xp_z \\ -ip_0p_y & -Np_xp_y & -Lp_y^2-M(p_x^2+p_z^2) & -Np_yp_z \\ -ip_0p_z & -Np_xp_z & -Np_yp_z & -Lp_z^2-M(p_x^2+p_y^2) \end{pmatrix},$$

(1)

where the energy origin is at the valence-band top of bulk material, and $\varepsilon_g/2m_0$ is the energy of the conduction-band bottom, *i.e.*, the energy gap. m_0 is the free electron mass. The effective-mass parameters L, M, N are resulted from the second-order interactions due to Löwdin renormalization involving remote states.[5] $p_0=\sqrt{2m_0E_p}$, E_p is the matrix element in Kane's theory.

The spin-orbit coupling Hamiltonian H_{so} for the basic functions $|S\uparrow\rangle$, $|X\uparrow\rangle$, $|Y\uparrow\rangle$, $|Z\uparrow\rangle$, $|S\downarrow\rangle$, $|X\downarrow\rangle$, $|Y\downarrow\rangle$, and $|Z\downarrow\rangle$, has the form of

$$H_{so}=\begin{vmatrix} 0&0&0&0&0&0&0&0\\ 0&0&-i\lambda&0&0&0&0&\lambda\\ 0&i\lambda&0&0&0&0&0&-i\lambda\\ 0&0&0&0&0&-\lambda&i\lambda&0\\ 0&0&0&0&0&0&0&0\\ 0&0&0&-\lambda&0&0&i\lambda&0\\ 0&0&0&-i\lambda&0&-i\lambda&0&0\\ 0&\lambda&i\lambda&0&0&0&0&0 \end{vmatrix}, \tag{2}$$

where

$$\lambda=-i\frac{\hbar}{4m_0^2c^2}\langle X|(\nabla V\times\mathbf{p})_y|Z\rangle=\frac{\Delta_{so}}{3}, \tag{3}$$

Δ_{so} is the spin-orbital splitting energy of the valence band.

III. Magnetic Energy Levels of Narrow-Gap Semiconductors

3.1. Magnetic Energy Levels of InAs

The theory of the magnetic energy levels in the conduction and valence bands is carried out using a modification of the effective-mass method by Luttinger.[5] Assuming the magnetic field is in the (001) direction, and choosing the symmetric gauge, the vector potential of the applied magnetic field $\boldsymbol{A} = \frac{1}{2}B(-y, x, 0)$, the momentum operator in Eq.(1) changes into $\boldsymbol{P} = \boldsymbol{p} + e\boldsymbol{A}/c$, where e is the proton charge. Because the different components of $\boldsymbol{P}$ do not commute, the $p_\alpha p_\beta$ (α, $\beta = x, y, z$) terms in Eq.(1) are not symmetric. There will be extra antisymmetric terms which enter the energy level formulas. Then the Hamiltonian matrix in Eq.(1) can be written in the following form for the Bloch basic functions of $|S\rangle$, $|11\rangle = (X + iY)/\sqrt{2}$, $|10\rangle = Z$, $|1-1\rangle = (X - iY)/\sqrt{2}$ near Γ point of the Brillouin zone:

$$H_{\text{int}} = \frac{1}{2m_0}\begin{pmatrix} s\alpha(a^+a + \frac{1}{2}) + P_z^2 + \varepsilon_g & i\sqrt{\frac{s}{2}}p_0 a^+ & ip_0 P_z & i\sqrt{\frac{s}{2}}p_0 a^- \\ -i\sqrt{\frac{s}{2}}p_0 a^- & -s[(\gamma_1+\gamma_2)(a^+a+\frac{1}{2}) + (3\kappa+1)] - (\gamma_1 - 2\gamma_2)P_z^2 & -\sqrt{\frac{s}{2}}Na^- P_z & -\frac{3}{2}(\gamma_2+\gamma_3)a^2 \\ -ip_0 P_z & -\sqrt{\frac{s}{2}}Na^+ P_z & -s(\gamma_1 - 2\gamma_2)(a^+a+\frac{1}{2}) - (\gamma_1 + 4\gamma_2)P_z^2 & -\sqrt{\frac{s}{2}}Na^- P_z \\ -i\sqrt{\frac{s}{2}}p_0 a^+ & -\frac{3}{2}(\gamma_2+\gamma_3)a^{+2} & -\sqrt{\frac{s}{2}}Na^+ P_z & -s[(\gamma_1+\gamma_2)(a^+a+\frac{1}{2}) - (3\kappa+1)] - (\gamma_1 - 2\gamma_2)P_z^2 \end{pmatrix} \tag{4}$$

The operators a^+ and a are the creation and destruction operators, respectively, defined as: $a^+ = \frac{1}{\sqrt{2s}}(P_x + iP_y)$, $a = \frac{1}{\sqrt{2s}}(P_x - iP_y)$, where $s = \frac{eB}{c}$. These operators have the commutation rule $(a, a^+) = 1$. And they have the following properties with harmonic oscillator eigenfunctions:

$$a^+u_n = \sqrt{n+1}u_{n+1}, \qquad au_n = \sqrt{n}u_{n-1}, \qquad a^+au_n = nu_n. \tag{5}$$

The valence band parameters γ_1, γ_2, and γ_3 are not those of Luttinger effective-mass parameters ($\gamma_1^L, \gamma_2^L, \gamma_3^L$), which related to L, M, N in Eq.(1) as

$$\gamma_1^L = \frac{1}{3}(L + 2M), \qquad \gamma_2^L = \frac{1}{6}(L - M), \qquad \gamma_3^L = \frac{1}{6}N. \tag{6}$$

κ is the antisymmetric constant which introduces antisymmetric splitting in the presence of magnetic field. Since the contributions of the conduction band, which is treated exactly and not as a higher band in this model, the valence band effective-mass parameters should be modified as:

$$\begin{aligned} \gamma_1 &= \gamma_1^L - E_p / 3E_g, \\ \gamma_2 &= \gamma_2^L - E_p / 6E_g, \\ \gamma_3 &= \gamma_3^L - E_p / 6E_g, \\ \kappa &= \kappa^L - E_p / 6E_g. \end{aligned} \tag{7}$$

Taking the spin-orbit coupling into account, the matrix elements of the ***k·p*** Hamiltonian in presence of the external magnetic field form an 8×8 matrix. We can write the total energy in Bloch function bases $|S\uparrow\rangle, |1\ 1\uparrow\rangle, |1\ 0\uparrow\rangle, |1\ -1\uparrow\rangle, |S\downarrow\rangle, |1\ 1\downarrow\rangle, |1\ 0\downarrow\rangle, |1\ -1\downarrow\rangle$ as:

$$H = \begin{pmatrix} H_{\text{int}} & \\ & H_{\text{int}} \end{pmatrix} - H_{so} + \frac{1}{2} g\mu_B \sigma \cdot B \tag{8}$$

The last term in Eq.(8) is new, and represents the interaction of the electron's spin magnetic moment with the external magnetic field. By inspection, the eigenvectors of matrices H can be sought in the form of:

$$\Psi = \sum_n \begin{pmatrix} a_1 u_n \\ a_2 u_{n-1} \\ a_3 u_n \\ a_4 u_{n+1} \\ a_5 u_n \\ a_6 u_{n-1} \\ a_7 u_n \\ a_8 u_{n+1} \end{pmatrix} e^{ik_z z}. \tag{9}$$

The first four basic functions correspond to spin-up while the last four basic functions correspond to spin-down. k_z is wave vector along applied external magnetic field. We then obtain the eigenenergies by solving the Schrödinger equation. These equations hold for Landau quantum number $n \geq 1$. For $n=-1$ we set all coefficients a_i in Eq.(9) equal to zero except for a_4 and a_8, while for $n=0$ we set $a_2 = a_6 = 0$.

Figure 1(a) shows the electron and hole magnetic energy levels of the InAs as a function of external magnetic field B. We make the approximation that the momentum along the magnetic field (k_z) is zero. The symbols of energy levels represent the main components of its wave function. We can see that the electron energy levels split in the present of external field, and the ground electron state is spin up state with harmonic oscillator quantum number $n=0$.

For comparison, we plotted the electron energy levels when the magnetic field is applied along the (111) crystallographic direction in Figure 1(b). We need to make the coordinate transform to express p in appropriate coordinate system in the calculation. The theoretical method has been exactly introduced in previous, so we do not repeat here for the sake of conciseness. The order of higher energy levels is different from the case that magnetic field is applied along (001) direction, which is related to the crystal structure anisotropy, $\gamma_2 \neq \gamma_3$.

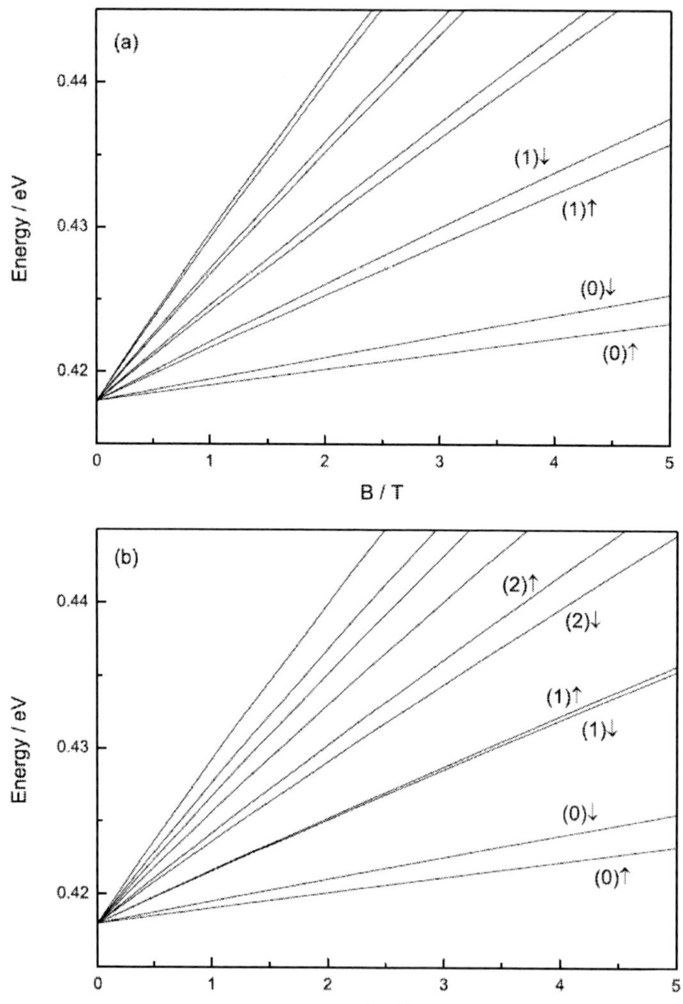

Figure 1. The energies of the minima of the conduction subbands (k_z=0) in InAs as a function of an external magnetic field along (a) (001) and (b) (111) crystallographic direction.

3.2. Magnetic Energy Levels of InMnAs

Having described the model of the band structure of a narrow-gap semiconductor in quantizing magnetic field, let us focus on the problem of a diluted magnetic semiconductor. In manganese doped semiconductors, the electrons from the half-filled 3d shell of manganese form the total spin of an ion and produce magnetic moments localized on Mn. Beside the 8 band effective-mass Hamiltonian, the new model take additionally into account the exchange interaction of carriers (electrons and holes) with magnetic ions. It is possible to represent this interaction by a simple Heisenberg-like Hamiltonian

$$H_{exch} = -\sum_j J(r - R_j)\sigma \cdot S_j \ , \tag{10}$$

where σ are spin operators of an band electron at position r, S_j is the total spin operator of the j-th Mn ion at R_j and $J(r\text{-}R_j)$ is an appropriate exchange constant rapidly varying over a unit cell. Treating $H_{\rm exch}$ in the mean-field approximation, we replace the total spin operator S_j by the thermo-dynamical average of the z-component of manganese spin $\langle S_z \rangle$. Usually, two of the basic exchange constants are encountered in diluted magnetic semiconductors. They involve s-like electrons from conduction band of Γ_6 symmetry in semiconductors with the zinc blende lattice, and p-like electron from the Γ_8 symmetry valence bands. Traditionally, they are commonly called *s-d* and *p-d* exchange interaction, respectively. In the basic functions of $|S\uparrow\rangle$, $|1\ 1\uparrow\rangle$, $|1\ 0\uparrow\rangle$, $|1\ {-1}\uparrow\rangle$, $|S\downarrow\rangle$, $|1\ 1\downarrow\rangle$, $|1\ 0\downarrow\rangle$, and $|1\ {-1}\downarrow\rangle$, the additional exchange interaction matrix which should be add to Eq.(8) before the eigenvalues are calculated has the form of:

$$H_{exch} = \begin{pmatrix} A &&&&&&& \\ & B &&&&&& \\ && B &&&&& \\ &&& B &&&& \\ &&&& -A &&& \\ &&&&& -B && \\ &&&&&& -B & \\ &&&&&&& -B \end{pmatrix} \tag{11}$$

Where

$$\begin{aligned} A &= -\frac{1}{2} N_0 \alpha \langle S_z \rangle x \quad \text{(for electrons)}, \\ B &= -\frac{1}{2} N_0 \beta \langle S_z \rangle x \quad \text{(for holes)}. \end{aligned} \tag{12}$$

N_0 stands for the volume of a unit cell, and x is the effective Mn^{2+} ions concentration. The adjustable parameters $\alpha = \langle S|J|S \rangle$ and $\beta = \langle X|J|X \rangle = \langle Y|J|Y \rangle = \langle Z|J|Z \rangle$ can be determined experimentally. The value of $\langle S_z \rangle$ is always calculated with some simple theoretical model. In the approximation of non-interacting magnetic moments we have

$$\langle S_z \rangle = SB_{5/2}\left(\frac{g\mu_B SB}{k_B T}\right). \tag{13}$$

where S=5/2 for Mn, and $B_s(y)$ is the Brillouin function

$$B_S(y) = \frac{2S+1}{2S}\coth(\frac{2S+1}{2S}y) - \frac{1}{2S}\coth(\frac{y}{2S}). \tag{14}$$

The Landé factor g of the manganese ion can be taken as equal to 2 with a very good accuracy.

The electron energy levels of InMnAs near the band gap in the present of magnetic field B=3T is shown in figure 2 at two temperatures. We can see that the spin splitting is largely enhanced due to the strong exchange interaction between the band carriers and the localized magnetic ions at low temperature. The splitting energy is the sum of magnetic exchange splitting energy and Landau energy differences originating from the effect of the external magnetic field on the orbital degrees of freedom of an electron or a hole. Since the value of magnetic exchange splitting energy is much larger, it can easily exceed the cyclotron splitting and change the order of Landau energy levels. Therefore, the spin properties of DMS may show some important differences compare to the nonmagnetic semiconductors. The variation of the band structure may be qualitatively understood via the exchange interaction terms $<S_z>$ involving the Brillouin function. The exchange-induced modification of the band structure depends on the effective magnetic ion concentration and also on the magnetic field strength and temperature. Considering a perfect paramagnet system where the magnetic moments do not interact with each other, the $<S_z>$ varies as $1/T$, so that at higher temperatures the additional terms brought by Eq.(13) may become negligible compare to the cyclotron splitting, shown in Figure 2(b). The magnetic field tends to align the magnetic moments along its direction. When all the moments are aligned, a saturation behavior which is described by a simple Brillouin function is taken place.

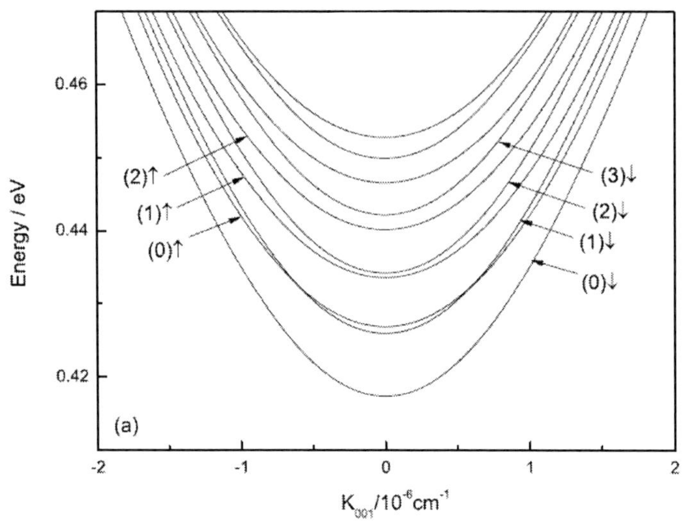

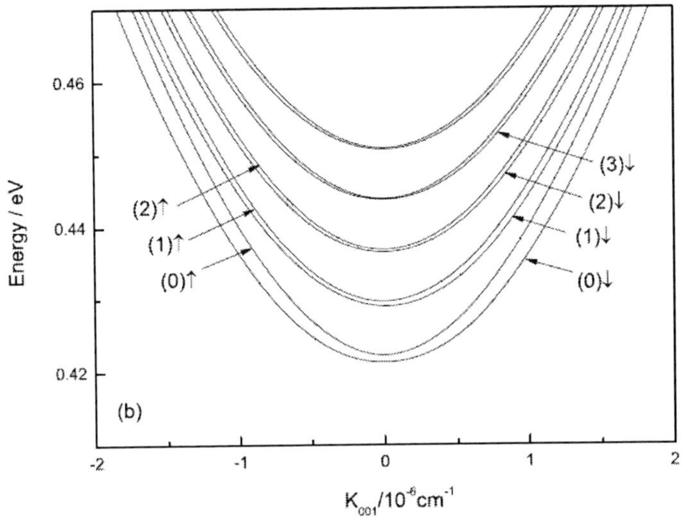

Figure 2. The Landau energy levels of $In_{1-x}Mn_xAs$, x=0.025 calculated at a constant magnetic field B=3T and at temperature (a) T=4.2K and (b) T=30K as functions of the wave vector k along the magnetic field direction. The field is applied along (001) crystallographic axis. The exchange integral constants are $N_0\alpha$=−0.25eV and $N_0\beta$=2.7eV.

IV. Electronic Structures of Narrow-Gap Semiconductor Quantum Dots

4.1. Electronic States in InAs Quantum Spheres and Ellipsoids [6]

Some semiconductors are narrow energy gap semiconductors, for example InAs、InSb of zinc-blende structure, HgTe of wurtzite structure, etc. In these semiconductors there are strong interactions between the conduction band and the valence bands, so that we should use the effective-mass theory of 8 bands model, including the conduction band and the valence bands. In order to calculate the electronic structure of quantum spheres, we write the 8-band Hamiltonian (1) for spin up and down states in the spherical coordinate. Taking the Bloch functions of the conduction band bottom and the valence band top $|S\rangle$, $|11\rangle = (X+iY)/\sqrt{2}$, $|10\rangle = Z$, $|1-1\rangle = (X-iY)/\sqrt{2}$ as basis functions, the Hamiltonian is written as

$$H_{\text{int}} = \frac{1}{2m_0}\begin{pmatrix} \varepsilon_g + \alpha p^2 & -ip_0 P_1^{(1)} & ip_0 P_0^{(1)} & ip_0 P_{-1}^{(1)} \\ -ip_0 P_{-1}^{(1)} & -P_1 & -S & -T \\ -ip_0 P_0^{(1)} & -S^* & -P_3 & -S \\ ip_0 P_1^{(1)} & -T^* & -S^* & -P_1 \end{pmatrix}, \tag{15}$$

where

$$
\begin{aligned}
P_1 &= \gamma_1 p^2 - \sqrt{\frac{2}{3}}\gamma_2 P_0^{(2)}, \\
P_3 &= \gamma_1 p^2 + 2\sqrt{\frac{2}{3}}\gamma_2 P_0^{(2)} + 2m_0 E_\Delta, \\
S &= AP_{-1}^{(1)} + \sqrt{2}\gamma_3 P_{-1}^{(2)}, \\
T &= \eta P_2^{(2)} + \delta P_2^{(2)},
\end{aligned} \tag{16}
$$

$P^{(1)}$ and $P^{(2)}$ are the first- and second-order irreducible tensors of the momentum operator, respectively. $\gamma_1, \gamma_2, \ldots$are effective-mass parameters related to $L, M, \ldots$.

$$
\begin{aligned}
&\gamma_1 = \frac{L+2M}{3}, \quad \gamma_2 = \frac{L-M}{3}, \quad \gamma_3 = \frac{N}{6}, \\
&\eta = \gamma_2 + \gamma_3, \quad \delta = \gamma_2 - \gamma_3.
\end{aligned} \tag{17}
$$

Taking the spin-orbit coupling into account, we represent the 8-band effective mass Hamiltonian in Bloch bases |S↑⟩, |11↑⟩, |10↑⟩, |1-1↑⟩, |S↓⟩, |11↓⟩, |10↓⟩, |1-1↓⟩ as

$$
H_{eb} = \begin{pmatrix} H_{\text{int}} & \\ & H_{\text{int}} \end{pmatrix} - H_{so}, \tag{18}
$$

where H_{so} is the spin-orbit coupling Hamiltonian,

$$
H_{so} = \begin{pmatrix}
0 & 0 & 0 & 0 & 0 & 0 & 0 & 0 \\
0 & 0 & 0 & 0 & 0 & 0 & 0 & 0 \\
0 & 0 & \lambda & 0 & 0 & \sqrt{2}\lambda & 0 & 0 \\
0 & 0 & 0 & 2\lambda & 0 & 0 & -\sqrt{2}\lambda & 0 \\
0 & 0 & 0 & 0 & 0 & 0 & 0 & 0 \\
0 & 0 & \sqrt{2}\lambda & 0 & 0 & 2\lambda & 0 & 0 \\
0 & 0 & 0 & -\sqrt{2}\lambda & 0 & 0 & \lambda & 0 \\
0 & 0 & 0 & 0 & 0 & 0 & 0 & 0
\end{pmatrix}. \tag{19}
$$

As we have taken into account the coupling of valence and conduction bands, the Luttinger parameters γ_i^L (i=1,2,3) should minus the contribution from the conduction band to the hole effective masses, and the contribution from the valence band to the electron effective mass. Taking the isotropic approximation of the valence bands, $\gamma_2^L = \gamma_3^L = \gamma^L$, then

$$
\gamma_1 = \gamma_1^L - E_p/3E_g, \quad \gamma = \gamma^L - E_p/6E_g. \tag{20}
$$

And the electron effective mass,

$$\alpha = \frac{1}{m_c} - \frac{1}{m_0}\left[\frac{E_p}{3}\left(\frac{2}{E_g} + \frac{1}{E_g + \Delta_{so}}\right)\right]. \tag{21}$$

They are modified further due to the nonlocal contribution which are absent in narrow gap nanocrystallites, but existed in bulk material. [7] Therefore, using these parameters requires taking special precautions. The nonlocal contributions are

$$\begin{aligned} &\Delta\gamma_1 = -5\delta_{nl}, \quad \Delta\gamma = -4\delta_{nl}, \\ &\Delta\alpha = -10\delta_{nl}, \quad \delta_{nl} = \frac{2}{15\pi\varepsilon_r E_g}\sqrt{\frac{E_B E_p}{3}}, \end{aligned} \tag{22}$$

Where E_B=27.2eV is the Bohr energy. For InAs, using the dielectric constant ε_r=14.6, we found δ_{nl}=0.0971, which finally leads to the remote band contributions γ_1=3.01, γ=0.74, and α=2.62.

We assume that the electrons and holes are confined in an infinitely high potential barrier. Because the momentum terms in the Hamiltonian (15) consists of the first- and second-order momentum tensor operators, whose matrix elements are not zero only for states with orbital angular momentum l and $l'=l\pm1$, or l and $l'=l\pm2$, respectively. Therefore, for a state of the specific z component of the total angular momentum $J_z=m+1/2$, the envelope function including the electron and hole states is expanded as

$$\Psi_{m+(1/2)} = \sum_{l,n} \begin{pmatrix} e_{l,n\uparrow} A_{l,n} j_l\left(k_n^l r\right) Y_{l,m}(\theta,\phi) \\ a_{l,n\uparrow} A_{l,n} j_l\left(k_n^l r\right) Y_{l,m-1}(\theta,\phi) \\ b_{l,n\uparrow} A_{l,n} j_l\left(k_n^l r\right) Y_{l,m}(\theta,\phi) \\ c_{l,n\uparrow} A_{l,n} j_l\left(k_n^l r\right) Y_{l,m+1}(\theta,\phi) \\ e_{l,n\downarrow} A_{l,n} j_l\left(k_n^l r\right) Y_{l,m+1}(\theta,\phi) \\ a_{l,n\downarrow} A_{l,n} j_l\left(k_n^l r\right) Y_{l,m}(\theta,\phi) \\ b_{l,n\downarrow} A_{l,n} j_l\left(k_n^l r\right) Y_{l,m+1}(\theta,\phi) \\ c_{l,n\downarrow} A_{l,n} j_l\left(k_n^l r\right) Y_{l,m+2}(\theta,\phi) \end{pmatrix}, \tag{23}$$

where $j_l(x)$ is the spherical Bessel function of order l, and Y_{lm} is the spherical harmonics. The spherical Bessel functions satisfy the boundary condition of the infinitely high potential barrier, are always zero at the boundary, and its nth zero point is $\alpha_n^l=k_n^l R$, R is the sphere radius. $A_{l,n}$ is the normalization constant,

$$A_{l,n} = \sqrt{\frac{2}{R^3}} \frac{1}{j_{l+1}\left(\alpha_n^l\right)}. \tag{24}$$

The electron energy levels of InAs quantum spheres as functions of the radius R are shown in Figure 3, where the energy unit is $\varepsilon_0=(1/2m_0)(\hbar/R)^2$. Hence the electron energy is not proportional to $1/R^2$, due to the coupling between the conduction and valence bands. The notation in Figure 1 $nQ_{e\sigma}^{(Jz)}$ and $nQ_{h\sigma}^{(Jz)}$ represent the main component of an electron or a hole state, where n is the main quantum number, Q=S, P, D,..., refers to the angular momentum state, the foot index e denotes an electron state and h=X, Z denotes a hole state with the $X\pm iY$, Z Bloch states of valence band top, and σ=↑, ↓ represents the spin up and down, respectively. From Figure 3 we see that the electron states are mixed with some valence band states, for example, the electron $1P^{(3/2)}$ state (c) includes the valence band $1S_X^{(3/2)}$ state, up to 11% when R=2.5nm, and this mixture increases with the increasing R.

The hole energy levels of M=1/2 and 3/2 states as functions of R are shown in Figure 4. Similar to the electron levels, they are not constants in units of ε_0. One kind of hole levels decreases quickly as the radius increases, which is produced by the spin-orbit splitting band due to the constant splitting energy Δ_{so}. Another kind of hole levels decreases slowly with R increasing, which is produced by the heavy and light hole bands. In the range of R=1~5nm, the hole ground state is always P state, hence the transition is the indirect transition.

Figure 5 shows the calculated exciton transition energies minus the first exciton transition energy E_1 as functions of E_1, where the circles represent the experimental results. All the experimental dots [8] can be assigned to different transitions, especially the inflexion point in the f line of exciton transition.

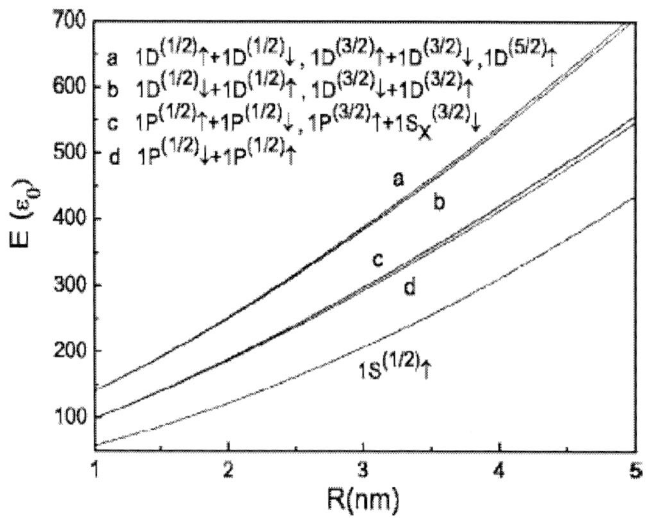

Figure 3. Electron energy levels of InAs quantum spheres as functions of the radius R.

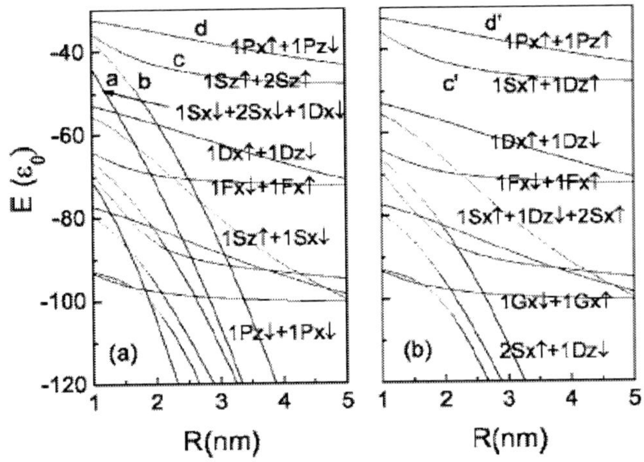

Figure 4. Hole energy levels of M=1/2 and 3/2 states of InAs quantum spheres as functions of R

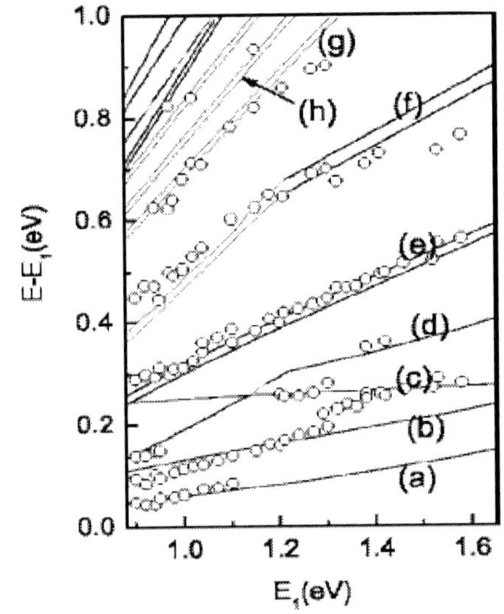

Figure 5. Calculated exciton transition energies minus the first exciton transition energy E_1 as functions of E_1, where the circles represent the experimental results. [8].

Experimentally, CdSe quantum rods with diameters ~4.2nm and length of 11, 20, and 40nm has been fabricated. [9] The size distribution was ~5% in diameter and <10% in length. Luminescence polarization measurements from individual quantum rods at room temperature show that the polarization factor $P=(I_{\|}-I_{\perp})/(I_{\|}+I_{\perp})$ changes rapidly from nearly zero to ~70% when the aspect ratio increases from 1 to 2, then remains almost constant afterward ($I_{\|}$ is the intensity parallel to the optical c axis, $I_{\perp}$ the intensity perpendicular to the optical axis). This discovery gives colloidal quantum dots a much more promising futurc because linearly polarized emission has a much wider range of applications, such as biological labeling, and optoelectronic devices etc.

We extended the effective-mass theory of quantum spheres to the quantum ellipsoids. [10] For the case of quantum ellipsoids, we introduce a coordinate transformation, which transforms the ellipsoidal boundary to the spherical one: $x'=x$, $y'=y$, $z'=z/e$, where e is the aspect ratio of the ellipsoid. In the new coordinate system x', y', z', the boundary is spherical, and the electronic Hamiltonian changes into,

$$H_e = \frac{p^2}{2m_a} - \frac{1}{2m_b}\sqrt{\frac{2}{3}}P_0^{(2)}, \tag{25}$$

where

$$\frac{1}{m_a} = \frac{1}{3m^*}\left(2 + \frac{1}{e^2}\right), \quad \frac{1}{m_b} = \frac{1}{3m^*}\left(1 - \frac{1}{e^2}\right). \tag{26}$$

Similarly we can transform the hole Hamiltonian to the new coordinate, for example the P_1 term in the hole Hamiltonian (15),

$$P_1 = \left[\frac{(\gamma_1 + \gamma_2)(1+e^2)}{3e^2} - \frac{\gamma_2}{e^2}\right]p^2 - \sqrt{\frac{2}{3}}\left[\frac{\gamma_2}{e^2} - \frac{(1-e^2)(\gamma_1 + \gamma_2)}{3e^2}\right]P_0^{(2)}, \tag{27}$$

etc.

Then we calculate the electronic and hole states for different respect ratio e as in the spherical case. [9] Take into account the Boltzmann distribution of the electronic and hole states, and sum up all contribution to the transitions of z and x, y polarizations, we obtain the strengths of optical transition for two polarizations: $I_{\parallel}$ and $I_{\perp}$, and the polarization factor P. Taking the temperature T= 300K, the rod radius R=2.1nm, and the crystal field splitting energy Δ_c=25meV, we obtained that P increases rapidly as the e increases from 1 to 3, then approaches to a saturation value 0.5, in agreement with the experimental results. [8]

The linear polarization factors P of optical transitions of InAs quantum ellipsoids at T=300K as functions of e for different radii when $\cos\theta=0$ (θ is the angle between the orientation of the wave propagation and the main axis z of the ellipsoid) are shown in Figure 6(a). Those of InAs ellipsoids with R=2 nm for different temperature are shown in Figure 6(b). We see that when e=1 (sphere), P=0 due to the cubic symmetry, in agreement with the experimental results. [11] When e>1, P increases, and approaches a saturation value. The P is larger for smaller radius and lower temperature. The saturation value of P for InAs ellipsoid with R=2.0 nm is 0.86. For comparison, the linear polarization factors of optical transitions of CdSe ellipsoids with R=2.0 nm at T=300 K as functions of e when $\cos\theta=0$ are shown in Figure 6 (c). We see that for CdSe ellipsoids, when e=1, P<0. This is due to the hexagonal crystal symmetry and the crystal field splitting energy Δ_c, which lowers the energies of the states with the Z-Bloch state of the valence-band top, resulting in P<0. With a bigger Δ_c=40 meV, the curve moves down more. The P of InAs ellipsoids with R=2.0 nm at T=300K as functions of $\cos\theta$ for different e are shown in Figure 6 (d). At $\cos\theta=1$ (light propagates along the z direction), they all equal 0.

4.2. Electronic Structure of InSb Quantum Ellipsoids in External Magnetic Field

We extended the Luttinger's effective-mass theory of semiconductors of zinc-blende structure in magnetic field [5] to the case of spherical quantum dots of wurtzite structure. [12] Assume that the magnetic field is applied in the *x-z* plane of the crystal structure, θ is the angle between the magnetic field and the *z* axis. The vector potential of the magnetic field,

$$\mathbf{A} = \left(-\frac{1}{2}B_z y, \frac{1}{2}B_z x - \frac{1}{2}B_x z, \frac{1}{2}B_x y \right). \tag{28}$$

The momentum in the Hamiltonian $\mathbf{p} \Rightarrow \mathbf{p} + e\mathbf{A}/c$, whose components do not commute. Luttinger divided the Hamiltonian into two parts. The symmetric part is the original Hamiltonian, in which the operator $p_\alpha p_\beta$ is replaced by the symmetric product,

$$\{p_\alpha p_\beta\} = \frac{1}{2}\left(p_\alpha p_\beta + p_\beta p_\alpha\right). \tag{29}$$

Each matrix element in the symmetric part becomes more complicated, for example,

$$\begin{aligned} p^2 \Rightarrow p^2 + eB_z L_z - \frac{e^2}{2} B_z^2 r_1^{(1)} r_{-1}^{(1)} - \frac{ie}{\sqrt{2}} B_x \left(P_1^{(1)} r_0^{(1)} + P_{-1}^{(1)} r_0^{(1)} - P_0^{(1)} r_1^{(1)} - P_0^{(1)} r_{-1}^{(1)}\right) \\ - \frac{e^2}{2\sqrt{2}} B_z B_x \left(r_{-1}^{(1)} r_0^{(1)} - r_1^{(1)} r_0^{(1)}\right) - \frac{e^2}{8} B_x^2 \left(r_1^{(1)2} + r_{-1}^{(1)2} - 2r_0^{(1)2} + 2r_1^{(1)} r_{-1}^{(1)}\right), \end{aligned} \tag{30}$$

where $r_i^{(1)}$ (i=−1, 0, 1) is the first-order tensor of the coordinate operator. The other matrix elements are given in the appendix of the paper [12].

The antisymmetric part is written as

$$H_{asym} = K\mu_B \mathbf{I} \cdot \mathbf{B}, \tag{31}$$

where $\mathbf{I}$ is an angle momentum matrix. If the basic functions are $|1,1\rangle$, $|1,0\rangle$, and $|1,-1\rangle$, it's components are

$$I_x = \frac{\sqrt{2}}{2}\begin{pmatrix} 0 & -1 & 0 \\ -1 & 0 & 1 \\ 0 & 1 & 0 \end{pmatrix}, \quad I_y = \frac{\sqrt{2}}{2}\begin{pmatrix} 0 & i & 0 \\ -i & 0 & -i \\ 0 & i & 0 \end{pmatrix}, \quad I_z = \begin{pmatrix} 1 & 0 & 0 \\ 0 & 0 & 0 \\ 0 & 0 & -1 \end{pmatrix}. \tag{32}$$

The total Hamiltonian including electron and hole can be written as

$$H = H_{eb} + H_{mm} + H_{asym}^{a} + H_{Zeeman}, \tag{33}$$

where H_{eb} is the Hamiltonian independent of magnetic field [Eq.(15)], H_{mm} is the remained symmetrical part, H^{a}_{asym} is the anti-symmetrical part caused by the magnetic field,

$$H^{a}_{asym} = \begin{pmatrix} 0 & & & \\ & H_{asym} & & \\ & & 0 & \\ & & & H_{asym} \end{pmatrix}, \tag{34}$$

and $H_{Zeeman}=(1/2)g\mu_B\boldsymbol{\sigma}\cdot\mathbf{B}$ is the intrinsic spin Zeeman Hamiltonian.

Due to the requirement of the interpretation of magneto-optical experiments, magneto-transport phenomena, resonant spectroscopy and spin-flip inelastic light scattering, electron g factors of some low-dimensional systems were investigated. [13,14] But, only few papers are reported with the results of the energy spectrum of highly confined colloidal InSb quantum dots in external magnetic field. We studied the g factors of narrow gap semiconductor InSb quantum dots. [15]

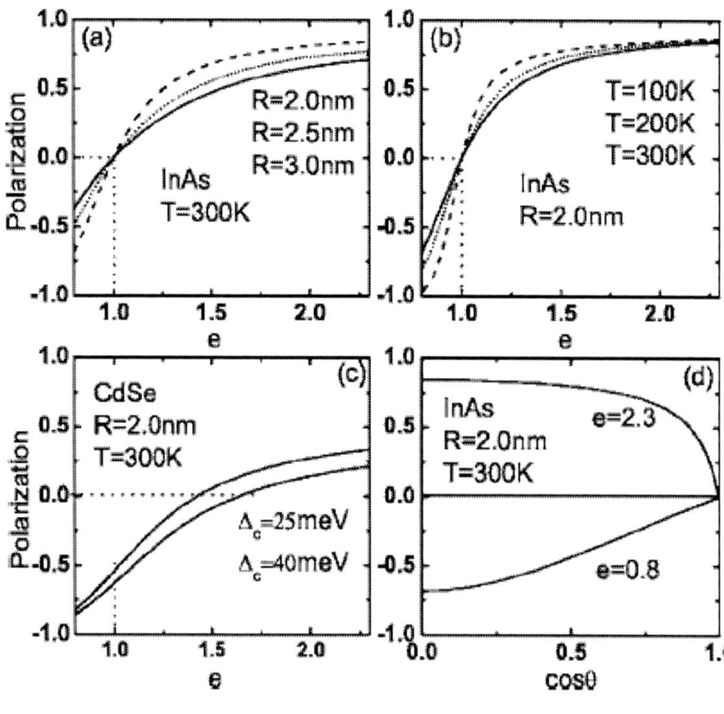

Figure 6 (a) Linear polarization factors of optical transitions of InAs quantum ellipsoids at T=300 K as functions of e for different radii when cosθ=0.

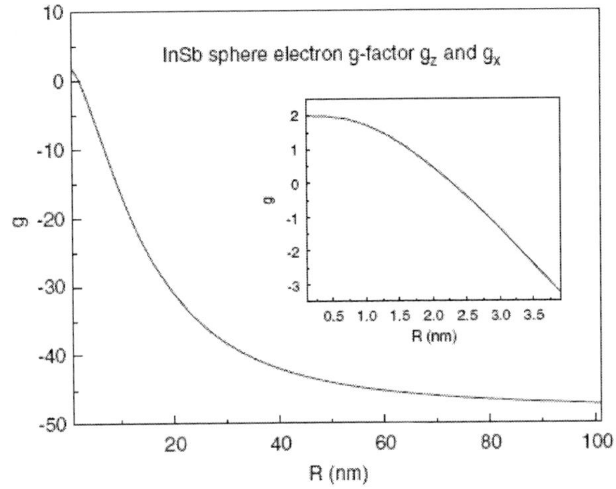

Figure 7. Electron g factors of InSb spheres as functions of R, the inset shows the enlarged electron g factors in the smallest sized spheres.

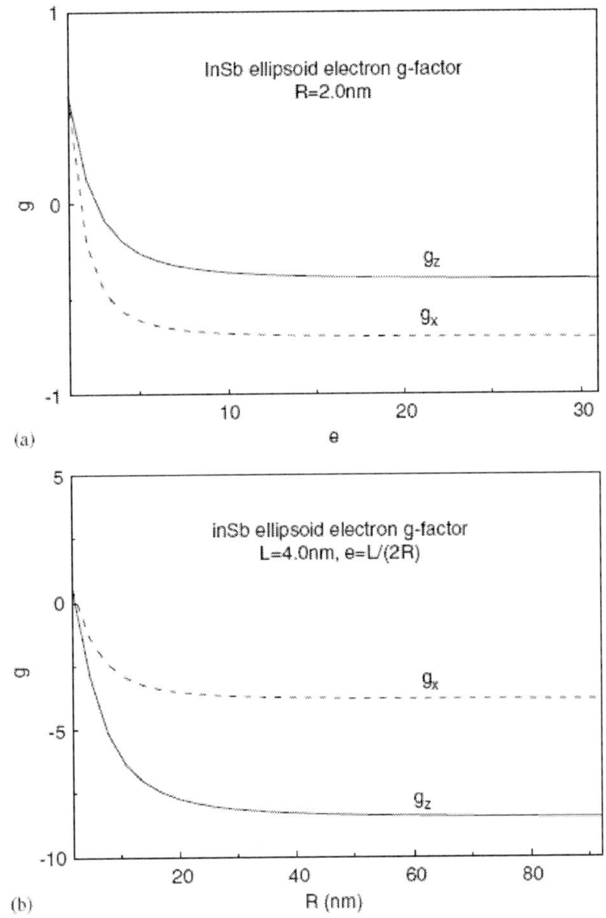

Figure 8. Electron g factors g_z and g_x of InSb ellipsoids (a) as functions of e for R=2.0 nm, (b) as functions of R for L=4.0 nm.

Figure 7 shows the electron g factors of InSb spheres as functions of R, the inset shows the enlarged electron g factors in the smallest sized spheres. The g factor increases as the decreasing R. When R is very small, the g factors are about 2. As R increases, the g factor approaches to -47.2, corresponding to the bulk-material value. The g factor of the bulk material is due to the large spin-orbit coupling in narrow-gap semiconductors. When the radius of the sphere is very small, the orbital movement is quenched, the g factor approaches to the value of free electron. The g factor inverts the sign at R=2.25 nm, it means that the g factor can be tuned to be positive, zero, and negative values at certain radius of quantum dots.

Figure 8 shows the electron g factors g_z and g_x of InSb ellipsoids (a) as functions of e for R=2.0 nm, (b) as functions of R for L=4.0 nm. The g_z and g_x are the g factor for the magnetic field along the z and x directions, respectively. For the long ellipsoids, as e increases, $g_z > g_x$, and the g_z and g_x approach to -0.399 and -0.704, respectively. For the oblate ellipsoids, as R increases, $g_z < g_x$, and the g_z and g_x approach to -8.455 and -3.836, respectively. Therefore, the asymmetric shape of quantum dots results in anisotropic g factor, the g factor for the magnetic field along one direction is affected more by the confinement in the plane perpendicular to that direction.

4.3. Electronic Structure and g Factors of Narrow-gap Zinc-Blende Nanowires and Nanorods

Narrow-gap semiconductor nanowires and nanorods have been synthesized by colloidal chemistry techniques. [11,16] Energy spectrum, electron and hole g factors, linear polarized absorption and emission, spin-orbit coupling effects, and one-dimesional systems were investigated. (see citations of Ref.17) We extended the 8-band model to the quantum wires and rods under magnetic field. [17] In the Bloch function bases $|S\uparrow\rangle$, $|11\uparrow\rangle$, $|10\uparrow\rangle$, $|1\text{-}1\uparrow\rangle$ $|S\downarrow\rangle$, $|11\downarrow\rangle$, $|10\downarrow\rangle$, and $|1\text{-}1\downarrow\rangle$, The 8-band Hamiltonian written in the cylindrical coordinate as

$$H_{eb} = \begin{pmatrix} H_{\text{int}} & \\ & H_{\text{int}} \end{pmatrix} - H_{so}, \tag{35}$$

where

$$H_{\text{int}} = \frac{1}{2m_0} \begin{pmatrix} \varepsilon_g + P_e & \left(i/\sqrt{2}\right) p_0 p_+ & i p_0 p_z & \left(i/\sqrt{2}\right) p_0 p_- \\ -\left(i/\sqrt{2}\right) p_0 p_- & -P_1 & -S & -T \\ -i p_0 p_z & -S^* & -P_3 & -S \\ -\left(i/\sqrt{2}\right) p_0 p_+ & -T^* & -S^* & -P_1 \end{pmatrix}, \tag{36}$$

$$\begin{aligned}
P_e &= \alpha p_- p_+ \alpha p_z^2, \\
P_1 &= \frac{L+M}{2} p_- p_+ + M p_z^2, \\
P_3 &= M p_- p_+ + L p_z^2, \\
S &= \frac{1}{\sqrt{2}} N p_- p_z, \\
T &= \frac{L-M-N}{4} p_+^2 + \frac{L-M+N}{4} p_-^2, \\
p_\pm &= p_x \pm i p_y.
\end{aligned} \tag{37}$$

H_{so} is the spin-orbit coupling Hamiltonian Eq.(19).

The wave function is expanded with the Bessel functions,

$$\Phi_{J,k_z} = \sum_n \begin{pmatrix} e_{n\uparrow} A_{l,n} J_l\left(k_n^l r\right) e^{il\theta} \\ b_{n\uparrow} A_{l-1,n} J_{l-1}\left(k_n^{l-1} r\right) e^{i(l-1)\theta} \\ c_{n\uparrow} A_{l,n} J_l\left(k_n^l r\right) e^{il\theta} \\ d_{n\uparrow} A_{l+1,n} J_{l+1}\left(k_n^{l+1} r\right) e^{i(l+1)\theta} \\ e_{n\downarrow} A_{l+1,n} J_{l+1}\left(k_n^{l+1} r\right) e^{i(l+1)\theta} \\ b_{n\downarrow} A_{l,n} J_l\left(k_n^l r\right) e^{il\theta} \\ c_{n\downarrow} A_{l+1,n} J_{l+1}\left(k_n^{l+1} r\right) e^{i(l+1)\theta} \\ d_{n\downarrow} A_{l+2,n} J_{l+2}\left(k_n^{l+2} r\right) e^{i(l+2)\theta} \end{pmatrix} e^{ik_z z}, \tag{38}$$

where $A_{L,n}$ is the normalization constant,

$$A_{L,n} = \frac{1}{\sqrt{\pi} R J_{L+1}\left(\alpha_n^L\right)}, \tag{39}$$

$\alpha_n{}^L = k_n{}^L R$ is the zero point of $J_L(r)$, R is the radius of the wire. $J=l+1/2$ is the total angular momentum, which is a good quantum number in the absence of transversely applied magnetic field.

When the magnetic field is along the z direction, the vector potential is taken as

$$\mathbf{A} = \left(-\frac{1}{2} B_z y, \frac{1}{2} B_z x, 0\right). \tag{40}$$

The Hamiltonian is divided into symmetric and asymmetric parts, the symmetric part is that the operator product P_+P_- or P_-P_+ are replaced by their symmetrized product,

$$\{P_+P_-\} = \frac{1}{2}\left(P_+P_- + P_-P_+\right), \tag{41}$$

where

$$\begin{aligned} P_\pm &= \left(p_x - \frac{eB}{2c}y\right) \pm i\left(p_y + \frac{eB}{2c}x\right) = p_\pm \pm i\frac{eB}{2c}r_\pm, \quad r_\pm = x \pm iy, \\ \{P_+P_-\} &= p_+p_- + \beta^2 r^2 + i\beta\left(r_+p_- - r_-p_+\right), \quad \beta = \frac{eB}{2c}. \end{aligned} \tag{42}$$

The matrix elements of the Hamiltonian under the magnetic field are given in Appendix of Ref.17. The whole Hamiltonian is just as Eq.(33), but in the cylindrical coordinate.

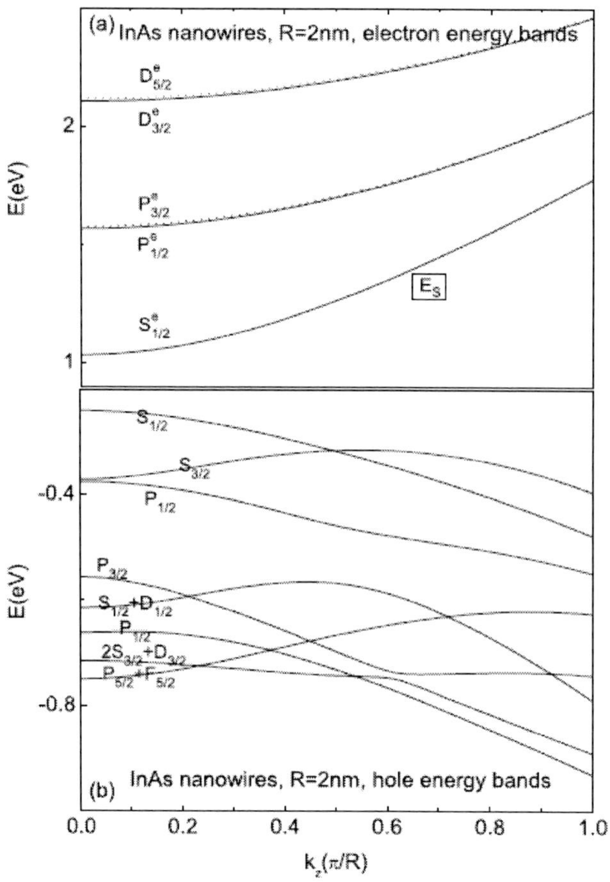

Figure 9. Energy bands of the InAs nanowires with R=2 nm as functions of k_z

Figure 9 shows the energy bands of the InAs nanowires with R=2 nm as functions of the wave vector along the wire direction k_z. The symbol of each energy band represents the main component of it's wave function. For example, $S_{1/2}$ means that the state is a hole state and consists of the J=1/2, l=0 and n=1 state of the envelope function [Eq.(36)] multiplied with the Bloch state and the spin state. From Figure 9 we see that the electron and hole ground states are $S^e_{1/2}$ and $S_{1/2}$ states with the bottom and top at the k=0 point, respectively. Therefore, the InAs nanowires are direct band-gap semiconductor. The electron energy bands are simple parabolic bands with the normal order S, P, D. The valence bands are not simple parabolic bands, and some bands rise with the increasing k_z, due to the interaction between the heavy hole states and light hole states.

Figure 10 shows the energy bands of InSb nanowires (a) R=2 nm, (b) R=4 nm as functions of k_z. From Figure 10 we see that for InSb nanowire of R=2 nm the band gap is direct, while for that of R= 4nm, the highest hole state is not at k_z=0 point, while at k_z=0.1021 nm^{-1}. We found that for InSb nanowires of R=8 and 64 nm, the highest hole states are also not at k_z=0 point. The indirect band gap is mainly due to the coupling of the electron $S_{1/2}$ and hole $P_{1/2}$ bands. When the radius is very small, the $S_{1/2}$ and $P_{1/2}$ bands are too far away, the coupling is so small that it can not bring an indirect band gap, as shown in Figure 10 (b).

Figure 11 shows the g factors of InSb rods (a) electron g factors as functions of R for length L=2 nm, (b) electron g factors as functions of L for R=2 nm, (c) hole g factors as functions of L for R=2 nm. In Figure 11 (a) the InAs rods are oblate, so $g_z < g_x$, in Figure 11 (b) the rods are long, so $g_z > g_x$, similar to the case of ellipsoids (Figure 8). It is noticed that in Figure 11 (c) the g_z factors of the hole ground state increases from 0.8 for L=10 nm to 1.05 for L=80 nm, i.e. the nanowire limit.

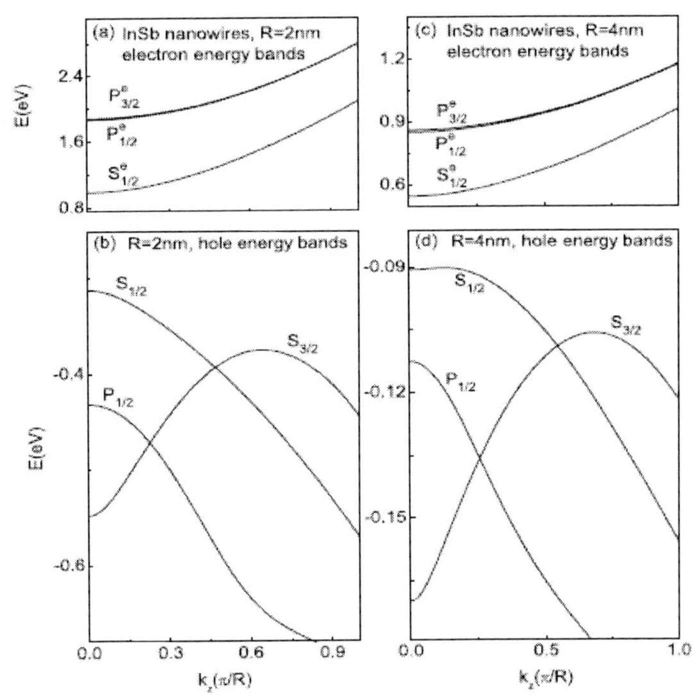

Figure 10. Energy bands of InSb nanowires (a,b) R=2 nm, (c,d) R=4 nm as functions of k_z

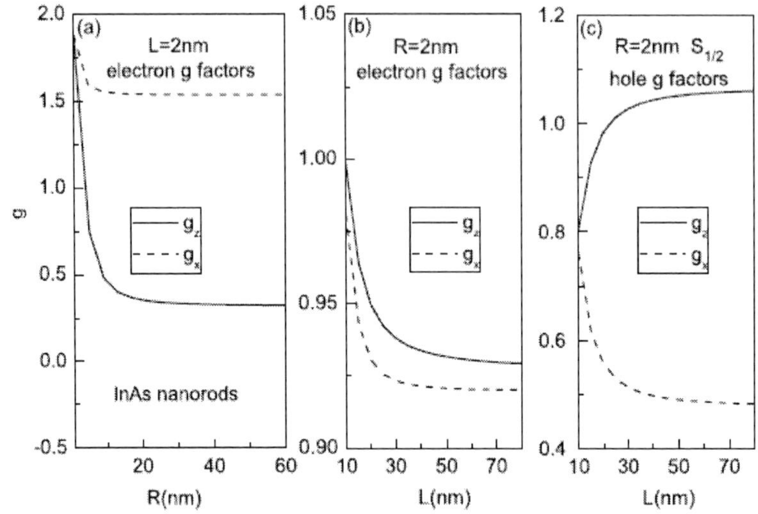

Figure 11. *g* factors of InSb rods (a) electron *g* factors as functions of *R* for *L*=2 nm, (b) electron *g* factors as functions of *L* for *R*=2 nm, (c) hole *g* factors as functions of *L* for *R*=2 nm.

V. Giant Zeeman Splitting in Mn-Doped Quantum Dots and Wires

5.1. Giant Zeeman Splitting in ZnMnSe Quantum Spheres

Norris et al. [18] demonstrated that high-quality ZnSe colloidal nanocrystals which are doped paramagnetic Mn ion impurities can be prepared. The exchange field between the Mn ions and carriers inside the nanocrystals will result in giant Zeeman splitting of the spin sublevels of the electron and hole. We studied the giant Zeeman effect in $Zn_{1-x}Mn_xSe$ quantum spheres.

For the dilute magnetic semiconductor quantum sphere under the magnetic field, the total Hamiltonian is Eq.(31) plus a term H_{exch}. H_{exch} is the *sp-d* exchange interaction between the carriers and magnetic ions Mn^{2+}, in the mean-field approximation,

$$\begin{aligned} H^e_{exch} &= -\bar{x}N_0\alpha\sigma_z\langle S_z\rangle, \\ H^h_{exch} &= -\bar{x}N_0\beta\sigma_z\langle S_z\rangle \end{aligned} \tag{43}$$

for electron and hole, respectively. $\bar{x}$ is the effective Mn^{2+} ions concentration, $N_0\alpha$ ($N_0\beta$) is the exchange integral for the conduction band (valence band). $\langle S_z\rangle$ is the thermal average of the Mn^{2+} spin, given by the Brillouin function $B_{5/2}$,

$$\langle S_z\rangle = \frac{5}{2}B_{5/2}\left(\frac{5\mu_B B}{k_B\left(T+T_0\right)}\right). \tag{44}$$

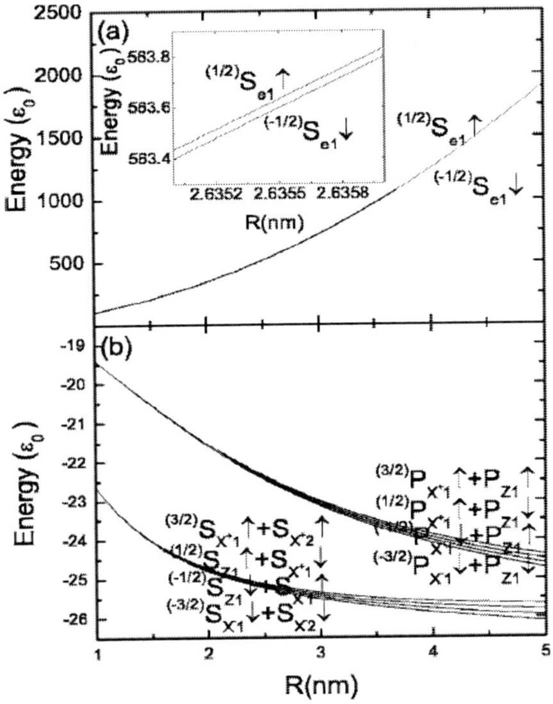

Figure 12. Energy levels of electron states (a) and hole states (b) of ZnSe quantum spheres as functions of R in B=2T.

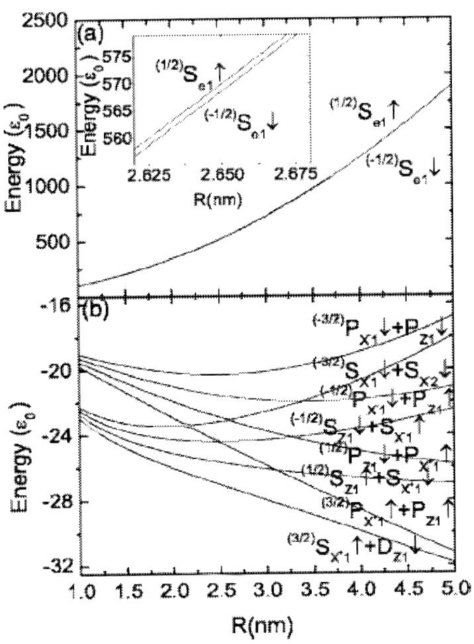

Figure 13. Energy levels of electron states (a) and hole states (b) of ZnMnSe (x=0.018) quantum spheres as functions of R in B=2T.

Figure 12 and Figure 13 show the energy levels of electron states (a) and hole states (b) of ZnSe and ZnMnSe (x=0.018) quantum spheres as functions of R in B=2T, respectively. The energy unit is $\varepsilon_0=(1/2m_0)(\hbar/R)^2$. The splitting of energy levels in the doped case is much larger than that in the undoped case under the same magnetic filed strength. From Figure 12 we see that the electron ground state is ${}^{(-1/2)}S_{e1}\downarrow$, and the hole ground state is ${}^{(-3/2)}P_{X1}\downarrow+P_{Z1}\downarrow$ split in the external magnetic field. Comparing Figs. 12 and 13, we see that the splitting of energy levels in the doped case is much larger than that in the undoped case under the same magnetic field strength. Because the energy unit is ε_0, the actual energy splitting should be divided by R^2.

The total Zeeman splitting energy is achieved by the energy difference between the transition from the hole state ${}^{(-3/2)}S_{X1}\downarrow+S_{X2}\downarrow$ to the electron ground state ${}^{(-1/2)}S_{e1}\downarrow$ and the transition from the hole state ${}^{(-1/2)}S_{Z1}\downarrow+S_{X1}\uparrow$ to the electron state ${}^{(1/2)}S_{e1}\uparrow$. This transition energy differences ΔE is the sum of the electron and hole Zeeman splitting energies. Figure 14 shows the total Zeeman splitting energies in ZnMnSe quantum spheres with different R and $\bar{x}$ as functions of the magnetic field B. The square points are the results of magnetic circular dichroism. The calculated results are qualitatively in agreement with experimental data. The splitting energies increase linearly with fields at low fields, depending on doping concentration, and saturate as modest fields, determined by the Brillouin function of $\langle S_z \rangle$, Eq.(44).

Figure 15 shows the total Zeeman splitting energy of ZnMnSe quantum spheres as functions of R in B=2 T, the curve of the undoped case is scaled by 10. The Zeeman splitting energy of the Mn doped case is much larger than that the undoped case. The total Zeeman splitting energies are basically independent on the R, but mainly depend on the effective Mn concentration $\bar{x}$. Recent experiment [18] has shown that the saturation value of the Zeeman splitting energy of the dot with R=2.56 nm is about 10 meV. We think the difference between the experimental and theoretical values may be due to the uncertainty of the effective Mn concentration $\bar{x}$.

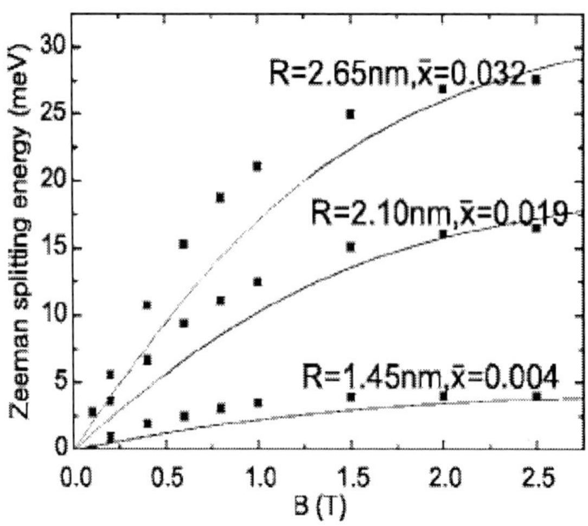

Figure 14. Total Zeeman splitting energies in ZnMnSe quantum spheres with different R and $\bar{x}$ as functions of the magnetic field B.

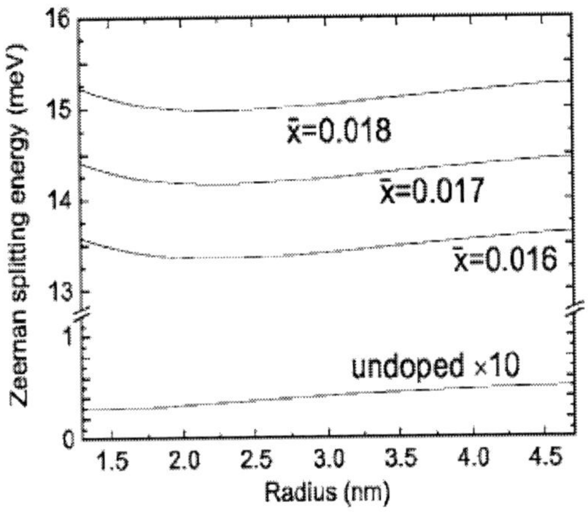

Figure 15. Total Zeeman splitting energy of ZnMnSe quantum spheres as functions of R in B=2 T.

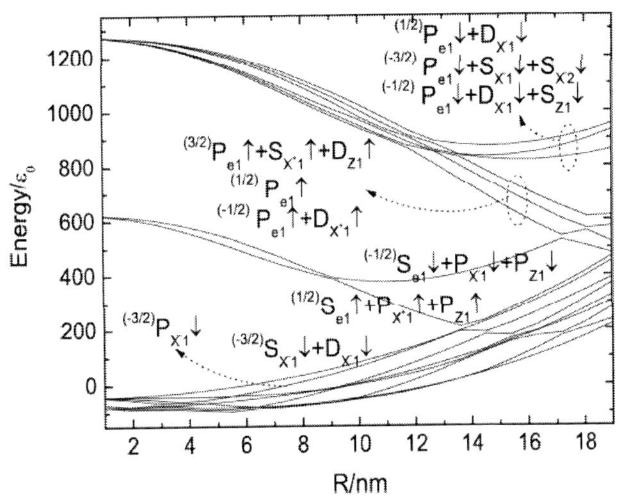

Figure 16. Energies of the lowest electron and hole states as functions of the sphere radius R for HgMnTe quantum spheres in the external magnetic field B=2T.

5.3. Controllable Electron g Factors in HgMnTe Quantum Spheres

HgTe has an inverted (zero-gap) band structure, the conduction band Γ_6 is under the valence band Γ_8. Krenn *et al.* have reported the observation of light-induced magnetization in HgMnTe and analyzed its connection with the spin-exchange and the spin-lattice-relaxation processes. [20] Bastard *et al.* declared that a drastic increase of the electron g factor is observed by interband magnetoabsorption experiments. [21] We studied the electronic structure of HgMnTe quantum dots, and found the giant and controllable g factor. [22]

In quantum dots with radius smaller than a critical value the conduction band states becomes higher than the valence band states due to the quantum confinement effect. Figure 16 shows the energies of the lowest electron and hole states as functions of the sphere radius R for HgMnTe quantum spheres in the external magnetic field B=2T. We have plotted curves in the unit of $\varepsilon_0=(1/2m_0)(\hbar/R)^2$, Mn-doped effective concentration $\bar{x}=0.06$ and temperature T=4.2K. The lowest degenerated electron states with the S character split in the presence of magnetic field. The electron states with Γ_6 symmetry merge with the hole states with Γ_8 symmetry as the radius increases. When the spherical radius is smaller than 13.6nm, the electron state $^{(1/2)}S_{e1}\uparrow+P_{X1}\uparrow+P_{Z1}\uparrow$ is separated from valence bands, indicating the transition from the semimetal to the semiconductor structure. When the spherical radius decrease to R_c=8.87 nm, electron state $^{(-1/2)}S_{e1}\downarrow+P_{X-1}\downarrow+P_{Z1}\downarrow$ is lower than the state $^{(1/2)}S_{e1}\uparrow+P_{X1}\uparrow+P_{Z1}\uparrow$, and turns out to be the electron ground state. This crossing behavior indicates that electron *g* factor could change from a negative value to a positive one, due to the quantum confinement effect.

Figure 17 shows the size dependences of the *g* factors of electron ground states with Γ_6 symmetry in the applied magnetic field B=2.0T for different $\bar{x}$. The electron *g* factor, $g=\Delta E/\mu_B B$, where the Zeeman splitting ΔE denotes the total spin splitting caused by the intrinsic Zeeman effect, quantum confinement effect and the *sp-d* exchange interaction in DMS nanocrystals, is considerably shifted from the bulk value $g=-41$ in HgTe. The *g* factor decrease with the increasing spherical radius, and can be turned to be zero at the spherical radius R_c. Due to the exchange interaction, the *g* factor is greatly enlarged and it decreases from 580 to −374 as the radius changes from 1 to 13.6nm. The absolute value of the *g* factor increases linearly with the doped effective concentration $\bar{x}$ as shown in the inset.

Figure 18 shows the size dependence of the electron *g* factors in the external field B=1, 2, and 3T for $\bar{x}$=0.06 and T=4.2 K. We see that the electron *g* factors are strongly influenced by the external field as well. In the insets of Figure 18 we see that the *g* factor decreases with increasing magnetic field when R is small (R=2 nm), and increases (negative) when R is large (R=12 nm). This is caused by the Brillouin function $B_{5/2}$ of the $\langle S_z\rangle$ [Eq.(44)]. The variation of the $\langle S_z\rangle$ with the magnetic field is linear when the magnetic field is small, and saturates when it becomes larger, resulting the absolute value of the *g* factor decreasing.

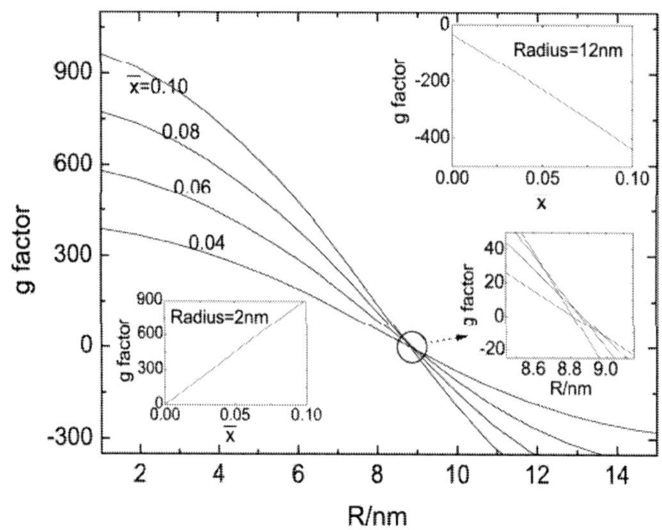

Figure 17. Size dependences of the *g* factors of electron ground states with Γ_6 symmetry in the applied magnetic field B=2.0T for different $\bar{x}$.

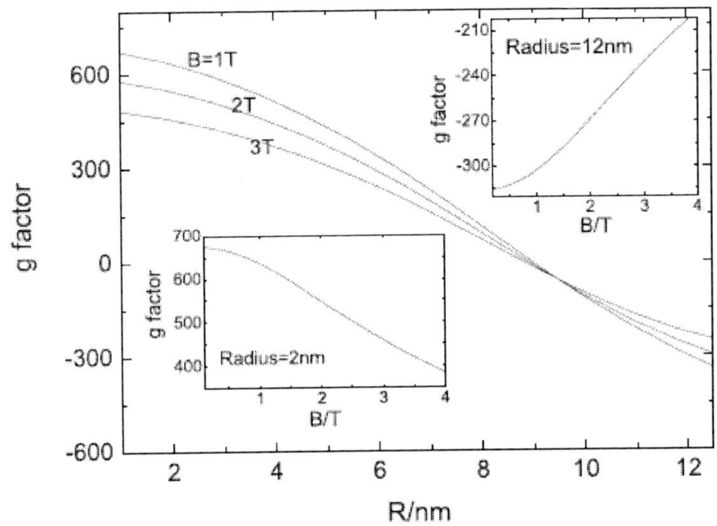

Figure 18. Size dependence of the electron g factors in the external field B=1, 2, and 3T for $\bar{x}$ =0.06 and T=4.2 K.

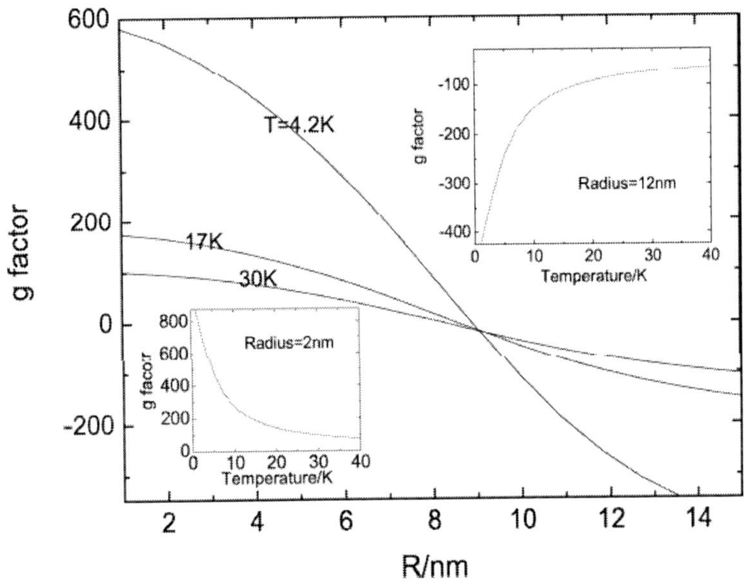

Figure 19. Size dependence of the g factors of electron states at temperature T=4.2, 17 and 30K for $\bar{x}$ =0.06 and B=2.0T.

Figure 19 shows the size dependence of the g factors of electron states at temperature T=4.2, 17 and 30K for $\bar{x}$=0.06 and B=2.0T. The g factors depend sensitively on the temperature, and they decrease more quickly with increasing temperature. The temperature dependence of the g factor is shown in the inset. The absolute value of g factor decreases linearly at low temperature, and decreases further with a smaller slope at higher temperature. The complex temperature dependence of the g factor is also resulted from the Brillouin function $B_{5/2}$ of the $\langle S_z \rangle$.

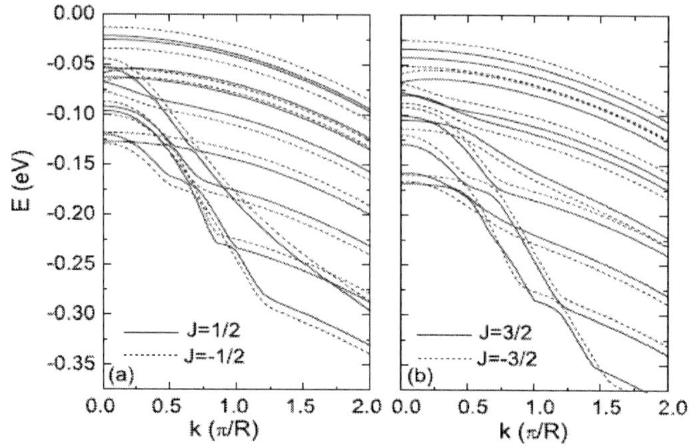

Figure 20. Hole levels of ZnMnO (x=0.0018) quantum wires with R=3 nm in B=2 T as functions of k_z: (a) J=±1/2, (b) J=±3/2.

5.3. Giant Zeeman Splitting in ZnMnO Quantum Wires

Nowadays, Mn-doped ZnO is of great interest as a new class of DMS due to the prediction of possible ferromagnetic properties at room temperature. [23] We studied the electronic structure of DMS quantum wires in magnetic field. [24]

Figure 20 shows the hole levels of ZnMnO (x=0.0018) quantum wires with R=3 nm in B=2 T as functions of k_z: (a) J=±1/2, (b) J=±3/2. The degenerated hole states split in the magnetic field due to the exchange interaction between the magnetic ions and holes, and generally the negative J states are higher than the positive J states (the solid lines for positive J and dot lines for negative J). Some of hole energy levels are in opposite order due to the quantum confinement effect and the coupling effect between hole states. The hole ground state is $(1,-1)X^+\downarrow+(1,1)X^-\downarrow$, so it is indirect transition.

Figure 21 shows the splitting energy of the hole ground states as a function of R at k_z=0 for (a) undoped, (b) x=0.0018 doped ZnO quantum wires. The splitting energy of the doped wires is much larger than that of the undoped wires, it equals 9.13 meV for R=11.4 nm and x=0.0018, 127 times that in the undoped case (0.072 meV). The splitting energies increase with increasing R, and all approach to the value of the bulk material. But the increments of splitting energies are small for both undoped and doped cases (<1 meV), it means that the splitting energy is basically independent of the R, and determined mainly by $\bar{x}$.

Figure 22 shows the energies of Zeeman components of the excitons of ZnMnO quantum wires as functions of B in the Faraday configuration: (a) $\bar{x}$=0.0018, R=3.0 nm; (b) $\bar{x}$=0.0056, R=3.0 nm; (c) $\bar{x}$=0.0018, R=1.5 nm. For larger radius (R=3.0 nm) the ground exciton transition is σ polarization (the electric field of light $F\|x$), and for smaller radius (R=1.5 nm) the ground transition is π polarization ($F\|z$). Because due to the quantum confinement effect perpendicular to the z direction, when the radius is smaller than 2.3 nm, the hole states $(1,0)Z\uparrow$ and $(1,0)Z\downarrow$ become higher than the $(1,0)X^+\downarrow+(1,0)X^-\downarrow$ and $(1,0)X^+\uparrow+(1,0)X^-\uparrow$ states. The splitting energies are larger for larger $\bar{x}$ (0.0056), and saturate at modest magnetic field.

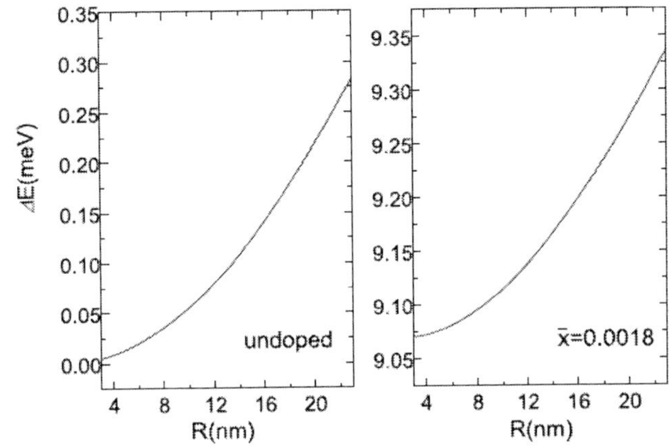

Figure 21. Splitting energy of the hole ground states as a function of R at k_z=0 for (a) undoped, (b) $\bar{x}$ =0.0018 doped ZnO quantum wires.

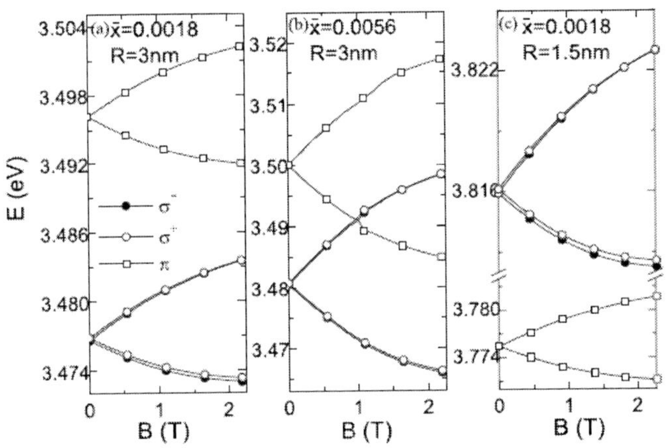

Figure 22. Energies of Zeeman components of the excitons of ZnMnO quantum wires as functions of B in the Faraday configuration: (a) $\bar{x}$ =0.0018, R=3.0 nm; (b) $\bar{x}$ =0.0056, R=3.0 nm; (c) $\bar{x}$ =0.0018, R=1.5 nm.

VI. Summary

There are strong interaction between the conduction band and the valence bands in narrow-gap semiconductors, which resulting in strong spin-orbit coupling in conduction band and many special physical properties, for example, large g factor, non-parabolic band, dependence of electron effective mass on the energy or carrier concentration, etc. In this chapter we studied the electronic structures, magnetic energy levels, g factors, polarization of emission of narrow-gap semiconductor bulk material, quantum spheres, quantum ellipsoids, and quantum wires with the $\boldsymbol{k}\cdot\boldsymbol{p}$ 8-band model proposed by Kane. In order to study various systems we introduce in detail the 8-band Hamiltonian written in the rectangular, spherical, and cylindrical coordinates. The 8-band Hamiltonian in a diluted magnetic semiconductor, in the presence of an external homogeneous magnetic field, is given using a mean-field approximation. We applied the effective-mass theory to the electronic structures of narrow-gap semiconductor bulk material, quantum dots and quantum wires, and discussed some physical properties in these nanocrystal structures. Exciton transition behaviors, polarization properties of emission, Landé electron *g* factors, and giant Zeeman splitting effect were discussed, and our theoretical results are compared with the available experimental data.

REFERENCES

[1] Kittel, C; Mitchell. *Phys. Rev.*, 1954, 96, 1488.
[2] Luttinger, JM; Kohn, W. *Phys. Rev.*, 1954, 97, 869.
[3] Kane, EO. *J. Phys. Chem. Solids*, 1957, 1, 249.
[4] Kane, EO. In *Handbook on Semiconductors*, edited by Paul W and Moss TS (North-Holland, Amsterdam, 1982), vol. 1.
[5] Luttinger, JM. *Phys. Rev.*, 1956, 102, 1030.
[6] Zhu, YH; Zhang, XW; Xia, JB. Electronic states in InAs quantum spheres and ellipsoids, *Phys. Rev. B*, 2006, 73, 165326.
[7] Gel'mont, BL; Seisyan, RP; Efros, AlL. *Sov. Phys. Semicond.*, 1982, 16, 776.
[8] Banin, U; Lee, JC; Guzelian, AA; Kadavanich, AV; et al. *J. Chem. Phys.*, 1998, 109, 2306.
[9] Alivisatos, AP. *J. Phys. Chem.*, 1996, 100, 13226.
[10] Li, XZ; Xia, JB. Electronic structure and optical properties of quantum rods with wurtzite structure, *Phys. Rev.*, 2002, B6, 115316.
[11] Kan, S; Mokari, T; Rothenberg, E; Banin, U. *Nature* (London), 2003, 2, 155.
[12] Zhang, XW; Xia, JB. Effects of magnetic field on the electronic structure of wurtzite quantum dots: Calulations using effective-mass envelope function theory, *Phys. Rev.*, 2005, B72, 075363.
[13] Snelling, MJ; Flinn, GP; Plaut, AS; Harley, RT; Tropper, AC; Eccleston, R; Phillips, CC. *Phys. Rev.*, 1991, B44, 11345.
[14] Destefani, CF; Ulloa, SE. *Phys. Rev.*, 2005, B71, 161303(R).
[15] Zhu, YH; Zhang, XW; Xia, JB. Electronic structure of InSb quantum ellipsoids in an external magnetic field, *Physica*, 2006, E35, 57.
[16] Banerjee, S; Dan, A; Chakravorty, D. *J. Mater. Sci.*, 2002, 37, 4261.
[17] Zhang, XW; Zhu, YH; Xia, JB. Electronic structure and *g* factors of narrow-gap zinc-blende nanowires and nanorods, *Euro. Phys. J.*, 2006, B52, 133.
[18] Norris, DJ; Yao, N; Charnock, FT; Kennedy, TA. *Nano Lett.*, 2001, 1, 3.
[19] Zhu, YH; Xia, JB. Giant Zeeman splitting in manganese-doped ZnSe quantum spheres: Eight-band effective-mass calculations, *Phys. Rev.*, 2006, B74, 245321.
[20] Krenn, H; Zawadzki, W; Bauer, G. *Phys. Rev. Lett.*, 1985, 55, 1510.
[21] Bastard, G; Rigaux, C; Guldner, Y; Mycielski, J; Mycielski, A. *J. Phys. (Paris)*, 1978, 39, 87.
[22] Zhu, YH; Xia, JB. Giant and controllable g factor in Manganese-doped HgTe quantum spheres, *A Letters Journal Exploring the Frontiers of Physics (EPL)*, 2008, 82, 37004.
[23] Dietl, T; Ohno, H; Matsukura, F; Cibert, J; Ferrand, D. *Science*, 2000, 287, 1019.
[24] Zhu, YH; Xia, JB. Electronic structure of Mn-doped ZnO quantum wires: A mean-field theory study, *Phys. Rev.*, 2007, B75, 205113.

INDEX

A

B

C

D

G

H

I

J

K

L

M

N

O

P

Q

R

S

T

U

V

W

X

Y

Z